de Gruyter Studies in Mathematics 3

Editors: Heinz Bauer · Peter Gabriel

Ludger Kaup · Burchard Kaup

Holomorphic Functions of Several Variables

An Introduction to the Fundamental Theory

With the Assistance of Gottfried Barthel

Translated by Michael Bridgland

W DE G
Walter de Gruyter
Berlin · New York 1983

Authors

Dr. Ludger Kaup
Professor of Mathematics
Universität Konstanz

Dr. Burchard Kaup
Professor of Mathematics
Universität Fribourg

Unseren Eltern gewidmet

Library of Congress Cataloging in Publication Data

Kaup, Ludger, 1939–
Holomorphic functions of several variables.
(De Gruyter studies in mathematics ; 3)
Bibliography: p.
Includes index.
1. Holomorphic functions. I. Kaup, Burchard, 1940– . II. Barthel, Gottfried. III. Title. IV. Series.
QA331.K374 1983 515.9 83-10120
ISBN 3-11-004150-2

CIP-Kurztitelaufnahme der Deutschen Bibliothek

Kaup, Ludger:
Holomorphic functions of several variables : an introd. to the fundamental theory / Ludger Kaup ; Burchard Kaup. With the assistance of Gottfried Barthel. Transl. by Michael Bridgland. – Berlin ; New York : de Gruyter, 1983. –
(De Gruyter studies in mathematics ; 3)
ISBN 3-11-004150-2
NE: Kaup, Burchard: ; GT

Cover design: Rudolf Hübler, Berlin. Typesetting and Printing: Tutte Druckerei GmbH, Salzweg-Passau.
Binding: Lüderitz & Bauer, Berlin.

Foreword

χαλεπὰ τὰ καλά

In the preface to his pioneering book, Die Idee der Riemannschen Fläche [Wl], Hermann Weyl wrote

"Erst Klein[1)] hat ja jene freiere Auffassung der Riemannschen Fläche recht zur Geltung gebracht, welche ihre Verbindung mit der komplexen Ebene als eine über der Ebene sich ausbreitende Überlagerungsfläche aufhebt, und hatte dadurch den Grundgedanken Riemanns erst seine volle Wirkungskraft gegeben. ... Ich teilte seine Überzeugung, daß die Riemannsche Fläche nicht bloß ein Mittel zur Veranschaulichung der Vieldeutigkeit analytischer Funktionen ist, sondern ein unentbehrlicher sachlicher Bestandteil der Theorie; nicht etwas, was nachträglich mehr oder minder künstlich aus den Funktionen herausdestilliert wird, sondern ihr prius, der Mutterboden, auf dem die Funktionen erst wachsen und gedeihen können."[2)]

Likewise in the theory of holomorphic functions of several complex variables, the investigation of holomorphic functions and their natural domains of existence is rooted in an abstract version of a Riemann surface, called a "complex space".

Two phenomena appearing only in the multidimensional theory are particularly striking:

i) For $n > 1$, a domain in $\mathbb{C}^n$ is not in general a *domain of holomorphy*; that is, it need not be the maximal domain of definition of a holomorphic function. Complements of finite point-sets in $\mathbb{C}^n$ fall into that category.

ii) In the generalization of concrete Riemann surfaces to higher dimensions, called "Riemann domains", ramification points need not possess local uniformizations, as can be seen even in such a simple example as the "origin" in the Riemann domain of $\sqrt{z_1 z_2}$. Such points are called *singularities.*

Those two phenomena profoundly influenced the development of (multidimensional) complex analysis; they also determine the content and organization of our book:

i) In the study of domains of holomorphy in $\mathbb{C}^n$ and their function theory, it soon became evident that the concept of "*analytic convexity*" is of fundamental importance. In attempting to free oneself, in the sense indicated by Weyl, of the concrete realization of domains of holomorphy as subsets of $\mathbb{C}^n$, one is led naturally to the concept of

---

1) Felix Klein (1849–1925).

2) Klein had been the first to develop the freer conception of a Riemann surface, in which the surface is no longer a covering of the complex plane; thereby he endowed Riemann's basic ideas with their full power. ... I shared his conviction that Riemann surfaces are not merely a device for visualizing the many-valuedness of analytic functions, but rather an indispensable essential component of the theory; not a supplement, more or less artificially distilled from the functions, but their native land, the only soil in which the functions grow and thrive. [Wl]$_2$

Stein spaces. Due to their rich function-theoretic structure, such spaces also are called "holomorphically complete"; they play a leading role in complex analysis. Both Chapter 1 in the first part of the book and the entire third part are devoted to the topics just outlined.

ii) Even at its singular points, a Riemann domain can be described locally as an "*analytic set*", i.e., the solution set of a system of holomorphic equations. Also the systematic study of complex manifolds depends heavily on analytic sets, for example, in many inductive arguments. Consequently, function theory on analytic sets is indispensible for an understanding of Riemann domains. A further attempt to obtain a "freer conception" à la Weyl, this time to describe analytic sets without reference to an embedding in a complex number space, leads to the construction of the complex spaces mentioned previously. That explains why they have become the central object of investigation in complex analysis – or, as some authors say in analogy to algebraic geometry, in complex analytic geometry.

A systematic treatment of complex spaces entails the analysis of *punctual*, *local*, and *global* properties. With its investigation of power series algebras and their homomorphic images, Chapter 2 can be viewed as a central portion of the *punctual* theory. The "*Weierstrass Preparation Theorem*" occupies such a key position in that investigation that Chapter 2 might well have been entitled, "Variations on the Weierstrass Preparation Theorem".

In Chapter 3, we prepare for the transition to the local and global theories of complex spaces by developing a more general formal framework in the concept of a "ringed space", which is a topological space with a distinguished "structure sheaf". We then develop a geometric intuition through the presentation of numerous examples of manifolds and reduced complex spaces.

In Chapter 4, the *local* theory is the main topic. The concept of *coherent sheaves* plays a decisive role in the step from "punctual" to "local"; the central results, aside from the actual Coherence Theorems, are the Representation Theorem for Prime Germs 46.1 and the Local Characterization Theorem for Finite Morphisms 45.4.

The *global* theory is particularly well developed for two classes of complex spaces that can be characterized in terms of topological or function-theoretic "completeness", namely, compact spaces and Stein spaces. In this book, the emphasis in the global theory is on Stein spaces, since their function theory yields fundamental results for the local investigation of complex spaces as well. For the step from "local" to "global", we apply cohomology theory; the central result is *Theorem B*: "Stein spaces have trivial analytic cohomology".

In a supplement, Chapter 7, we treat a class of complex spaces whose function theory closely resembles that of manifolds, namely, the "*normal spaces*", characterized by the validity of the Riemann Removable Singularity Theorems. In the Normalization Theorem, we show how to modify an arbitrary reduced complex space so that it becomes a normal space (that may be viewed as a first step toward "regularizing" it into a manifold). A much deeper result is Hironaka's Theorem on the resolution of singularities [Hr], whose proof lies well beyond the scope of this book; however, we do present the tools for the resolution of singularities of complex curves and surfaces: normalization and quadratic transformation.

The coherence of the structure sheaves of complex spaces leads to an interplay between algebra and geometry that we find especially charming, and we have used it as a methodical leitmotiv for Chapter 4. The algebraic objects "*analytic algebras*" (which appear punctually as stalks of structure sheaves) correspond to the local geometric objects "*germs of complex spaces*". As a result, geometric statements have algebraic proofs (which are frequently easier and more transparent), and algebraic statements can be interpreted geometrically. In order to exploit this "antiequivalence" to the fullest, it is necessary to drop the restrictive condition "reduced" from the definition of complex spaces in Chapter 3. Consequently, stalks of structure sheaves of complex spaces, which are simply analytic algebras, may contain *nilpotent elements*. We thereby have an appropriate tool for the treatment of solution sets of systems of holomorphic equations with "multiplicities". For example, consider the zero of the function z^m; its multiplicity is reflected in the analytic algebra $\mathbb{C}\{z\}/(z^m)$, rather than in the corresponding reduced algebra $\mathbb{C}\{z\}/(z)$. That more general approach does not complicate the proofs of the essential results; on the contrary, it is advantageous – for example, for the proof of Cartan's Coherence Theorem. There is a similar anti-equivalence, which we discuss to a limited extent in connection with the Character Theorem (§ 57), between Stein spaces on the geometric side and their algebras of global holomorphic functions on the algebraic side.

This introduction to complex analysis is intended both as a textbook and as a guide for independent study. In general, we neither discuss the historical development of the theory, nor mention the discoverers of propositions or proofs. However, those familiar with the subject cannot fail to notice the tremendous influence on our presentation, both direct and indirect, of the Münster school that grew up around Heinrich Behnke (1898–1979), particularly through the ideas of Grauert and Remmert.

The limited scope of an introductory textbook permits the presentation of only a small selection of topics from a rich and living branch of mathematics; the resulting omissions are all the more apparent if the discipline under consideration distinguishes itself through numerous connections to other subjects, as is the case with complex analysis and, say, commutative algebra, differential geometry, algebraic geometry, functional analysis, the theory of partial differential equations, and algebraic topology. The reader who wishes to pursue the subject in greater depth should consult the relevant monographs (such as [Fi] and the appendices in [BeTh] for an overview, [GrRe] and [Ab] for the punctual theory, $[\mathrm{GrRe}]_2$ and [BaSt] for the global theory, [Hö] for $\bar{\partial}$-theory, etc.).

Since this book is an introduction to the foundations of the theory, we generally do not treat subjects that have not already appeared in one form or another in textbooks (two major exceptions are the investigation of quotient structures and part of the discussion of normalization). We have given considerable attention to the (generally simple) exercises, which serve to test the reader's understanding of the material, to vary or complete simple proofs, and sometimes to augment the text. In general, an exercise whose conclusion is needed later in the book is provided with hints, if necessary, and an indication of the place at which it is applied for the first time.

We use small print to indicate both material from other disciplines and supplementary material that may be left out during a first reading of the text without sacrificing rigor.

Aside from sheaf theory and cohomology theory (for which we summarize the necessary concepts and results in such a way as to motivate their application), we assume very little prior knowledge of the reader: some elements of one-dimensional complex analysis, and some basic results from differential calculus, algebra, and functional analysis. Even for those topics, we frequently give explicit references.

We are indebted to many colleagues for valuable suggestions and detailed comments on the text; in particular, we should like to mention Milos Dostal, George Elenczwajg, Gerd Fischer, Wilhelm Kaup, Leopoldo Nachbin, and Reinhold Remmert. We are pleased to acknowledge the work of Karl-Heinz Fieseler and Ernst-Ulrich Kolle, who read the German manuscript with great care, and corrected many inaccuracies. We are also grateful to our students for their contributions, in the form of questions and comments, to the betterment of the text. Mrs. Gisela Schroff patiently and carefully typed the various drafts of both the German and English manuscripts, for which we are particularly grateful. Most especially, our thanks go to Michael Bridgland, who, under circumstances that were not always of the best, has rendered the German manuscript into English with so much care. Finally, we would like to thank Heinz Bauer for suggesting to us that we write this book. The publishers have earned our gratitude both for their patience and understanding during the many delays that accompanied the writing of the book, and for their friendly cooperation during the printing. We also thank everyone else, colleagues and institutions, who have contributed either directly or indirectly to the completion of the book, not least of all our families.

Konstanz/Fribourg, Summer 1983

Ludger Kaup
Burchard Kaup

Contents

Part Two: Function Theory on Analytic Sets

Interdependence of chapters

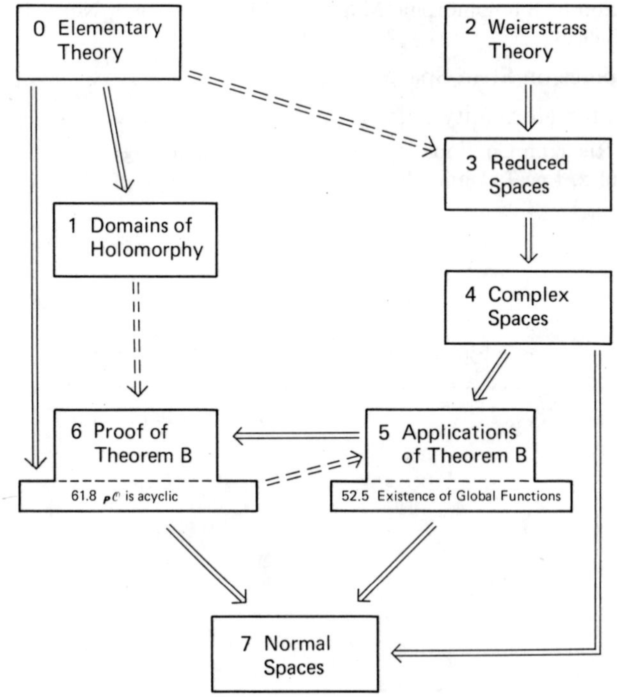

Courses and Seminars: Guide to the Essentials of Specific Topics

Here we propose ways of using the book for various courses or seminars:

1) *Function theory on domains in* $\mathbb{C}^n$: §§ 1–14.
2) *Complex manifolds*: §§ 1, 3, 4, 6.1–6.6, 8; definition of manifolds with E. 32a; § 32 without the results on singular spaces; § 32 A, B.
3) *Analytic algebras and their dimension theory*: §§ 21–24; 45.6 proved as in [GrRe II.2 Satz 2]; all of the algebraic results of § 48, up to and including 48.11_{alg}, with proofs; possibly E.48e, (32.11.1)–32.13, 45.7, and 45.8, as well; for a geometric interpretation of algebraic statements, we recommend the definition of germs of complex spaces and their morphisms as geometric realizations of analytic algebras and their homomorphisms, to be followed with 44.7–44.14 and 45.1–45.14, possibly without the proof of 45.2.
4) The *theory of complex spaces* can be appended to 3): central themes: Coherence Theorems (Oka, Cartan, Finite), Representation Theorem for Prime Germs, decomposition into irreducible components – § 31, examples from § 32, § 33, and the remainder from Chapter 4 (without the supplement).
5) *Domains of holomorphy*, either in $\mathbb{C}^n$ or in Stein manifolds X: the goal is either 63.7 or 63.2: Chapter 1; §§ 30, 41, 42; specialization to domains in $\mathbb{C}^n$ (resp., manifolds) of §§ 50–53, § 55, 56.2, and Chapter 6. (The Finite Coherence Theorem in the proof of 63.3 can be avoided for manifolds X that admit local coordinates by global functions: then, for each $K = \bar{K} = \hat{K} \subset U \Subset X$, there exists a $\varphi \in \mathrm{Hol}(X, \mathbb{C}^m)$ such that

$$\varphi(K) \subset \boldsymbol{P} := \boldsymbol{P}^m(1), \ \varphi(\partial U) \cap \boldsymbol{P} = \emptyset,$$

and such that $\varphi|_U$ is an injective mapping that induces on every stalk a surjective homomorphism; hence, $\varphi\colon W := \varphi^{-1}(\boldsymbol{P}) \cap U \to \boldsymbol{P}$ is a (closed) embedding, i.e., W is a B-space and Runge in X, thus in every analytic polyhedron in X that includes W.)
6) *Solution of Levi's problem*: Chapter 1, §§ 61, 62, 63 A (prerequisites: Approximation Theorem, sheaf theory, cohomology theory).
7) *Quotients of complex spaces*: 32 B.3 and its applications in § 32 B, § 49 A (prerequisites: ringed spaces, coherent sheaves, Finite Coherence Theorem).
8) *Normalization*: §§ 45 B, 46 B, 71, 72, possibly 74.
9) *Finiteness Theorem of Cartan-Serre*: §§ 30, 41, 42, 50, 51, 52, and 55; 56.2; §§ 61, 62.
10) *Finite Mappings*: 33.1, §§ 45, 45 B, 46.

Part One: Function Theory on Domains in $\mathbb{C}^n$

We begin our exposition of complex analysis by adapting familiar propositions and methods of proof from the one-dimensional theory to the multidimensional case in Chapter 0. In order to make the analogy between the two cases particularly transparent, we introduce holomorphic functions as continuous partially holomorphic functions; we then show that approach to be equivalent to the other usual definitions.

In Chapter 1, we work out basic differences between the two theories, the existence of which is discernible in Chapter 0 only in connection with the Kugelsatz. Thus, we show that, contrary to the impression left by Chapter 0, the multidimensional theory does not consist of mere adaptations, but rather that completely new methods must be developed.

Chapter 0: Elementary Properties of Holomorphic Functions

In analogy to the historical development of multidimensional function theory, we begin with the basic concepts and results that carry over from the one-dimensional theory relatively easily.

The set $\mathcal{O}(X)$ of holomorphic functions on an open set[1] $X \Subset \mathbb{C}^n$ is an algebra that includes the polynomial algebra $\mathbb{C}[Z_1, \ldots, Z_n]$; in the metric topology of compact convergence, it is complete, and Montel's Theorem holds. From Cauchy's Integral Formula, it follows that holomorphic functions are precisely the ones that are represented locally by their Taylor series. Multidimensional versions of the Identity Theorem and the Maximum Principle hold; in particular, every nonconstant holomorphic function is an open mapping. In addition, standard theorems from real analysis, such as the Implicit Function Theorem, the Inverse Function Theorem, and the Rank Theorem, extend to the complex case. In contrast to real differential calculus, a holomorphic injection between equidimensional domains is "biholomorphic" (8.5).

In the multidimensional version of the classical Removable Singularity Theorem, the exceptional sets are not merely isolated points, but rather arbitrary analytic subsets, in other words, solution sets of systems of holomorphic equations; that leads us to investigate the elementary properties of such sets. The First Riemann Removable Singularity Theorem has an interesting extension that is vacuous in the one-dimensional case: if a function is holomorphic outside an exceptional set of codimension at least two, then it is holomorphically extendible (Second Riemann Removable Sin-

[1] For an open subset A of B, we use the notation $A \Subset B$.

gularity Theorem). The proof is based on Hartogs's *Kontinuitätssatz*, which we analyze more carefully in Chapter 1, where we are concerned with fundamental differences between the one- and multidimensional theories.

§ 1 Definition of Holomorphic Functions

Throughout this introductory chapter, X denotes an open subset, called a *region*, of the *complex number space* $\mathbb{C}^n$. If X is also connected, then X is called a *domain*; in that case we often use the letter G (for *Gebiet*). We usually use $z = (z_1, \ldots, z_n)$ and $z_j = x_j + iy_j$ to denote the coordinates of $\mathbb{C}^n$ and the decomposition of z_j into real and imaginary parts, respectively. We frequently use the *Euclidean norm* $\|z\|$ and the *maximum norm* $|z|$ defined as follows:

$$\|z\|^2 := \Sigma z_j \bar{z}_j \quad \text{and} \quad |z| := \max_{1 \leqq j \leqq n} |z_j|.$$

We recall the one-dimensional theory (i.e., $X \subset \mathbb{C}^1$): a continuous function $f: X \to \mathbb{C}$ is called *holomorphic* if it satisfies the following equivalent conditions:

i) f is complex differentiable: $\dfrac{df}{dz}$ exists and coincides with

$$\frac{\partial f}{\partial z} := \frac{1}{2}\left(\frac{\partial f}{\partial x} + \frac{1}{i} \cdot \frac{\partial f}{\partial y}\right).$$

ii) Locally, f is representable by a convergent power series.

iii) For each contractible piecewise-smooth closed curve γ in X, $\int_\gamma f(z)dz = 0$ (*Cauchy's Theorem*).

iv) f has continuous partial derivatives with respect to x and y at each point in X, and they satisfy the Cauchy-Riemann differential equation,

$$\frac{\partial f}{\partial \bar{z}} := \frac{1}{2}\left(\frac{\partial f}{\partial x} - \frac{1}{i} \cdot \frac{\partial f}{\partial y}\right) = 0.$$

With that background, we can generalize to the n-dimensional case.

1.1 Definition. *A function $f: X \to \mathbb{C}$ is called partially holomorphic if, for each fixed $(z_1^0, \ldots, z_n^0) \in X$, and each $j = 1, \ldots, n$, the function of one variable determined by the assignment*

$$z_j \mapsto f(z_1^0, \ldots, z_{j-1}^0, z_j, z_{j+1}^0, \ldots, z_n^0)$$

is holomorphic. A continuous partially holomorphic function is called holomorphic, and the set of holomorphic functions on X is denoted by $\mathcal{O}(X)$.

In the one-dimensional case, sums, products, and complex multiples of holomorphic functions are holomorphic; $\mathcal{O}(X)$ is thus an algebra whose unit element is the constant function with value 1. The units of $\mathcal{O}(X)$ are the holomorphic functions with no zeros.

We call a complex vector space A an *algebra*[2] if there is a multiplication on A,

$$A \times A \to A, \quad (a, b) \mapsto a \cdot b,$$

that, with vector space addition, makes A a commutative ring with a unit element such that ring multiplication and scalar multiplication satisfy the condition

$$\lambda(a \cdot b) = (\lambda a) \cdot b = a \cdot (\lambda b), \quad \forall \lambda \in \mathbb{C}, \quad a, b \in A.$$

The multiplicatively invertible elements of an algebra are called *units*. It should be noted that $\varphi(1) = 1$ for every *algebra-homomorphism* $\varphi: A \to B$, by definition.

Thus, even in the general n-dimensional case, $\mathcal{O}(X)$ is an algebra that contains the coordinate functions z_j; with those, it must also include the polynomial algebra $\mathbb{C}[z_1, \ldots, z_n]$. In fact, $\mathcal{O}(X)$ contains every function that can be constructed from some polynomial by replacing each z_j with a holomorphic function f_j of one variable.

1.2 Proposition. *$\mathcal{O}(X)$ is an algebra whose set of units $\mathcal{O}^*(X)$ consists precisely of those holomorphic functions on X with no zeros.* ■

According to a nontrivial theorem of Hartogs, a partially holomorphic function is necessarily continuous. Since we do not need that theorem, we refer the reader to the literature for its proof (see [Ha], [Hö 2.2.8]). It is easy to prove the following weaker statement:

1.3 Proposition. *A function $f: X \to \mathbb{C}$ is holomorphic iff it is locally bounded and partially holomorphic.*

(A function f is called *locally bounded* if each point of its domain has a neighborhood on which f is bounded.)

Proof of 1.3. Since we are concerned with a local statement, we may assume that f is partially holomorphic on $X = \{z \in \mathbb{C}^n; |z_j| < r, \forall j\}$ and that $M := \sup\{|f(z)|;\ z \in X\}$ is finite; we wish to show that f is continuous at the point $z = 0$. Observe that

$$f(z_1, \ldots, z_n) - f(0, \ldots, 0) = \sum_{j=1}^{n} (f(0, \ldots, 0, z_j, \ldots, z_n) - f(0, \ldots, 0, z_{j+1}, \ldots, z_n)).$$

For fixed j and fixed $z_{j+1}, \ldots, z_n$, the function

$$f_j: \mathbb{C} \mapsto \mathbb{C},\ z_j \mapsto f(0, \ldots, 0, z_j, \ldots, z_n) - f(0, \ldots, 0, z_{j+1}, \ldots, z_n),$$

is holomorphic in the variable z_j, so it follows easily from the one-dimensional version of Schwarz's Lemma (see [Co VI.2.1] or E.4h) that

$$|f_j(z_j)| \leqq 2\frac{M}{r}|z_j|.$$

That establishes the inequality

$$|f(z_1, \ldots, z_n) - f(0, \ldots, 0)| \leqq 2\frac{M}{r} \sum_{j=1}^{n} |z_j|,$$

which implies that f is continuous at $z = 0$. The reverse implication is trivial. ■

---

[2] Specifically, it is a $\mathbb{C}$-algebra; as with functions, vector spaces, differentiability, manifolds, etc., we mention the field of scalars only if it is not $\mathbb{C}$!

E. 1a. Suppose that the function $f\colon X \to \mathbb{C}$ has continuous (first) partial derivatives with respect to the real variables x_j and y_j (so, in particular, f is continuous). Then the following statements hold:

i) f is holomorphic iff

$$\frac{\partial f}{\partial \bar{z}_j} = \frac{1}{2}\left(\frac{\partial}{\partial x_j} - \frac{1}{i}\frac{\partial}{\partial y_j}\right) f = 0 \quad \text{for each } j \text{ such that } 1 \leqq j \leqq n$$

(Cauchy-Riemann differential equations).

ii) $$\overline{\left(\frac{\partial f}{\partial z_j}\right)} = \frac{\partial \bar{f}}{\partial \bar{z}_j} \quad \text{and} \quad \overline{\left(\frac{\partial f}{\partial \bar{z}_j}\right)} = \frac{\partial \bar{f}}{\partial z_j} \quad \text{(for 8.2).}$$

iii) If f has continuous second partial derivatives, then

$$\frac{\partial^2 f}{\partial z_j \partial z_k} = \frac{\partial^2 f}{\partial z_k \partial z_j}, \quad \frac{\partial^2 f}{\partial z_j \partial \bar{z}_k} = \frac{\partial^2 f}{\partial \bar{z}_k \partial z_j}, \quad \text{and} \quad \frac{\partial^2 f}{\partial \bar{z}_j \partial \bar{z}_k} = \frac{\partial^2 f}{\partial \bar{z}_k \partial \bar{z}_j}.$$

iv) *Chain Rule*: For $Y \Subset \mathbb{C}^m$, if each component g_k of the mapping $g := (g_1, \ldots, g_n)\colon Y \to X$ has continuous partial derivatives, then

$$\frac{\partial}{\partial w_k}(f \circ g) = \sum_{j=1}^{n} \left(\frac{\partial f}{\partial z_j} \circ g\right) \frac{\partial g_j}{\partial w_k} + \sum_{j=1}^{n} \left(\frac{\partial f}{\partial \bar{z}_j} \circ g\right) \frac{\partial \bar{g}_j}{\partial w_k} \quad \text{and}$$

$$\frac{\partial}{\partial \bar{w}_k}(f \circ g) = \sum_{j=1}^{n} \left(\frac{\partial f}{\partial z_j} \circ g\right) \frac{\partial g_j}{\partial \bar{w}_k} + \sum_{j=1}^{n} \left(\frac{\partial f}{\partial \bar{z}_j} \circ g\right) \frac{\partial \bar{g}_j}{\partial \bar{w}_k} \quad \text{(for 4.3).}$$

v) The ($\mathbb{R}$-linear) *derivative* $d_a f\colon \mathbb{C}^n \to \mathbb{C}$ of f at a is defined to be

$$d_a f = \sum_{j=1}^{n} \frac{\partial f}{\partial x_j}(a)dx_j + \sum_{j=1}^{n} \frac{\partial f}{\partial y_j}(a)dy_j = \sum_{j=1}^{n} \frac{\partial f}{\partial z_j}(a)dz_j + \sum_{j=1}^{n} \frac{\partial f}{\partial \bar{z}_j}(a)d\bar{z}_j,$$

with $dx_j(z) = x_j$, $dy_j(z) = y_j$, $dz_j(z) = z_j$, and $d\bar{z}_j(z) = \bar{z}_j$ (for § 8).

E. 1b. For $Y \Subset \mathbb{C}^m$ and $X \Subset \mathbb{C}^n$, let $g := (g_1, \ldots, g_n)\colon Y \to X$ and $f\colon X \to \mathbb{C}$ be continuously partially differentiable functions. *If f and all of the g_j are holomorphic, then so is $f \circ g$* (for 3.2; note that the assumed differentiability actually follows from the fact that the functions are holomorphic; see 4.8).

E. 1c. Show that, for each domain $X \Subset \mathbb{C}^n$,

$$\dim_{\mathbb{C}} \mathcal{O}(X) < \infty \Leftrightarrow \dim_{\mathbb{C}} \mathcal{O}(X) = 1 \Leftrightarrow n = 0 \quad \text{(for § 5).}$$

§ 2 $\mathcal{O}(X)$ as a Topological Algebra

The algebra $\mathcal{O}(X)$ possesses a natural topology that is of fundamental importance for limit processes. We first introduce such a topology on the algebra $\mathscr{C}(T) := \mathscr{C}(T, \mathbb{C})$ of continuous complex-valued functions on a *locally compact* topological space T (see [Bou GT I § 9.7]).

For $A \subset T$, the mapping $\|\cdot\|_A\colon \mathscr{C}(T) \to \mathbb{R} \cup \{\infty\}$ defined by setting

$$\|f\|_A := \sup_{t \in A} |f(t)|$$

has the following properties:

i) $\|f\|_A \geqq 0$.

ii) $\|\lambda f\|_A = |\lambda| \|f\|_A, \quad \lambda \in \mathbb{C}$ (with $0 \cdot \infty = 0$).

iii) $\|f+g\|_A \leqq \|f\|_A + \|g\|_A$.

For a vector space V, a function $s: V \to \mathbb{R}$ that satisfies conditions i)–iii) is called a *seminorm* on V (note that s cannot assume the value ∞). If $s(v) = 0$ only at the point $v = 0$, then s is called a *norm.* Some examples of seminorms on $\mathscr{C}(T)$ are the mappings of the form $\|\cdot\|_A$ with $A \subset\subset T$.[3)]

If T is compact, then $\|\cdot\|_T$ is a norm on $\mathscr{C}(T)$; it induces a metric topology on $\mathscr{C}(T)$ such that every Cauchy sequence converges (i.e., $\mathscr{C}(T)$ is *complete*). A sequence $(f_j)_{j \in \mathbb{N}}$ in $\mathscr{C}(T)$ is called a *Cauchy sequence* if it satisfies this condition:

$$\forall \varepsilon > 0 \quad \exists n_0(\varepsilon) \quad \forall j, k \geqq n_0(\varepsilon) \quad \|f_j - f_k\|_T < \varepsilon .$$

For such sequences, the assignment $f(t) := \lim_{j \to \infty} f_j(t)$ determines a function f in $\mathscr{C}(T)$, and (f_j) converges to f.

If the topology of the (locally compact) space T is countable[4)], then T can be written as a countable union $T = \bigcup_{j=1}^{\infty} T_j$ with compacta $T_j \subset \mathring{T}_{j+1}$[5)] (see 51A.2). It follows that $\mathscr{C}(T)$, provided with the family of seminorms $\|\cdot\|_{T_j}$, is still complete, since every Cauchy sequence in $\mathscr{C}(T)$ (i.e., Cauchy sequence with respect to every $\|\cdot\|_{T_j}$) has a unique limit on each T_j. If $\|f\|_{T_j} = 0$ for every j, then $f = 0$.

A family of seminorms $(s_j)_{j \in J}$ on a vector space V determines a unique topology, for which the family of sets of the form

$$W(v, \varepsilon; s_j) = \{w \in V;\ s_j(v - w) < \varepsilon\}$$

is a subbasis. A vector space with a Hausdorff topology is called a *Fréchet space* if its topology is determined by an at most countable family of seminorms in the manner described above, and if it is complete. For (locally compact) T with a countable topology, $\mathscr{C}(T)$ is thus a Fréchet space.

E. 2a. All families $(T_j)_{j \in J}$ with $T_j \subset\subset T$ and $\bigcup_J \mathring{T}_j = T$ determine the same topology on $\mathscr{C}(T)$ (for 4.3).

That topology is called the *topology of compact convergence.* Every $\|\cdot\|_A$ is *submultiplicative*; i.e.,

$$\|fg\|_A \leqq \|f\|_A \|g\|_A .$$

E. 2b. Find polynomials $f, g \in \mathbb{C}[z]$ and a $K \subset\subset \mathbb{C}$ such that

$$\|fg\|_K < \|f\|_K \|g\|_K .$$

If a Fréchet space topology is defined on an algebra V by means of submultiplicative seminorms, then V is called a *Fréchet algebra.* In particular, $\mathscr{C}(T)$ is a Fréchet algebra if T has a countable topology.

---

3) $A \subset\subset T$ means that A is a *relatively compact* subset of T; i.e., its closure, $\bar{A}$, is compact.

4) A *countable topology* is one that satisfies the second axiom of countability (i.e., there exists a countable basis of open sets).

5) $\mathring{T}$ denotes the interior of T.

2.1 Proposition. *For $X \Subset \mathbb{C}^n$, the subalgebra $\mathcal{O}(X)$ of $\mathscr{C}(X)$ is a Fréchet algebra (in the induced topology).*

Proof. As a subalgebra of $\mathscr{C}(X)$, $\mathcal{O}(X)$ is certainly a Fréchet algebra if it is closed in the topology of compact convergence. That is known to hold in the one-dimensional theory [Co VII.2.3]; thus, it also holds in higher dimensions. ∎

E. 2c. Show that $\mathscr{C}(T)$ is a *topological vector space*; i.e., the vector space operations $\mathbb{C}\times\mathscr{C}(T)\to\mathscr{C}(T)$, $(\lambda, f)\mapsto\lambda f$, and $\mathscr{C}(T)\times\mathscr{C}(T)\to\mathscr{C}(T)$, $(f,g)\mapsto f+g$, are continuous. Since $\mathscr{C}(T)\times\mathscr{C}(T)\to\mathscr{C}(T)$, $(f,g)\mapsto f\cdot g$, also is continuous, $\mathscr{C}(T)$ is a *topological algebra.*

In a topological vector space V, the translation $V\to V$, $f\mapsto f+h$, is a homeomorphism for each $h\in V$. Many local topological statements are translation-invariant; such statements need only be investigated at $0\in V$.

E. 2d. Show that $\mathscr{C}(T)$ is a *locally convex vector space*; i.e., every $f\in\mathscr{C}(T)$ has arbitrarily small convex neighborhoods U. A convex neighborhood U is one such that for each pair of functions $f, g\in U$, the segment joining them, $\{\alpha f+(1-\alpha)g;\ 0\leqq\alpha\leqq 1\}$, is included in U.

E. 2e. If a Hausdorff topology is given for a vector space V by means of seminorms $(s_j)_{j\in\mathbb{N}}$, then

$$d(v,w):=\sum_{j=1}^{\infty}2^{-j}\min(1, s_j(v-w))$$

is a metric that generates the topology of V (hints: i) d is translation-invariant: $d(h+v, h+w)=d(v,w)=d(v-w,0)$, $\forall h\in V$; ii) $V\to\mathbb{R}_{\geqq 0}$, $v\mapsto d(v,0)$ is continuous; iii) $2^{-j}\min(1,s_j(v))\leqq d(v,0)$, $\forall j\in\mathbb{N}$; for 5.2).

Conversely, one can show that the topology of a metrizable locally convex vector space is determined by an at most countably infinite family of seminorms [Bou EVT II, § 5 Prop. 6].

E. 2f. *A pointwise convergent sequence $(f_j)_{j\in\mathbb{N}}$ in $\mathcal{O}(\mathbb{C})$ does not necessarily converge.* Hint: Apply the following form of Runge's Theorem [Co VIII.1.19]: if $X\Subset\mathbb{C}$, and if $\mathbb{C}\backslash X$ is connected, then $\mathbb{C}[z]$ is dense in $\mathcal{O}(X)$. It follows that, for each $j\in\mathbb{N}$, there exists a $P_j\in\mathbb{C}[z]$ such that

$$\|P_j\|_{A(j)}<1/j \quad\text{and}\quad \|1-P_j\|_{B(j)}\leqq 1/j,$$

where $\boldsymbol{A}(j):=\{z\in\mathbb{C};\ |z|<j,\ \operatorname{im} z>1/j\}$ and $\boldsymbol{B}(j):=\{z\in\mathbb{C};\ |z|<j,\ \operatorname{im} z<1/2j\}$.

E. 2g. $\mathcal{O}(X)$ *as a projective limit.* Let $\mathfrak{U}=(U_k)_{k\in I}$ be an open cover of X. Prove the following statements:

i) A sequence (f_j) in $\mathcal{O}(X)$ converges iff the sequence $(f_j|_{U_k})$ converges in $\mathcal{O}(U_k)$ for each k (for 4.8).

ii) The mapping $\mathcal{O}(X)\to\prod_k\mathcal{O}(U_k)$, $f\mapsto(f|_{U_k})_{k\in I}$, induces an isomorphism of topological vector spaces onto a closed linear subspace of $\prod_k\mathcal{O}(U_k)$.

iii) Consider the partial order on the index set I defined by

$$j\leqq k :\Leftrightarrow U_j\subset U_k,$$

and suppose that this order *directs I downward*; i.e., for every $k, l\in I$, there exists a $j\in I$ such that $j\leqq k$ and $j\leqq l$. With the restriction-homomorphisms $i^0_{jk}:\mathcal{O}(U_k)\to\mathcal{O}(U_j)$ for $j\leqq k$, the family $(\mathcal{O}(U_k), i^0_{jk})_{k\in I}$ is an *inverse* (or *projective*) *system*; i.e., $i^0_{kk}=\mathrm{id}_{\mathcal{O}(U_k)}$, and $i^0_{jk}i^0_{kl}=i^0_{jl}$ for $j\leqq k\leqq l$. The *inverse* (or *projective*) *limit* of the system, $\varprojlim_{k\in I}\mathcal{O}(U_k)$, is that subspace of $\prod_k\mathcal{O}(U_k)$

which consists of all "sequences" $(f_k)_{k \in I}$ satisfying the following condition:

If $j \leqq k$ and $j \leqq l$, then $i^0_{jk} f_k = i^0_{jl} f_l$.

Prove that $\mathcal{O}(X)$ *is isomorphic to* $\varprojlim_{k \in I} \mathcal{O}(U_k)$ *as a topological vector space.*

§ 3 Holomorphic Mappings

Let $Y \subset\subset \mathbb{C}^m$ be a nonempty region.

3.1 Definition. *A mapping* $f = (f_1, \ldots, f_m)\colon X \to Y$ *is called* <u>*holomorphic*</u> *if every component* f_k *of* f *is a holomorphic function. If, in addition,* f *is bijective and* $f^{-1}\colon Y \to X$ *is holomorphic, then* f *is called biholomorphic, and* X *and* Y, *biholomorphically equivalent. The set of holomorphic mappings from* X *to* Y *is denoted by* $\mathrm{Hol}(X, Y)$.

In particular, we have that $\mathrm{Hol}(X, \mathbb{C}) = \mathcal{O}(X)$.

3.2 Proposition. *The mapping* $f\colon X \to Y$ *is holomorphic iff* $g \circ f \in \mathcal{O}(X)$ *for every* $g \in \mathcal{O}(Y)$.

Proof. If. For $1 \leqq k \leqq m$, the projections $\mathrm{pr}_k\colon w \mapsto w_k$[6] are holomorphic on Y; if every $\mathrm{pr}_k \circ f$ is holomorphic, then $f = (\mathrm{pr}_1 \circ f, \ldots, \mathrm{pr}_m \circ f)$ is holomorphic, by definition. *Only if.* Let f be holomorphic, and choose a $g \in \mathcal{O}(Y)$. Then, by E.1b, we know that $g \circ f \in \mathcal{O}(X)$. ■

3.3 Corollary. *If* $f\colon X \to Y$ *and* $g\colon Y \to Z$ *are holomorphic mappings, then* $g \circ f\colon X \to Z$ *is a holomorphic mapping.* ■

By 3.2, a holomorphic mapping $f\colon X \to Y$ induces a mapping

$$f^0\colon \mathcal{O}(Y) \to \mathcal{O}(X), \quad g \mapsto g \circ f.$$

If $X \subset\subset Y$ and if f is the inclusion mapping, then $f^0(g)$ is the restriction $g|_X$.

3.4 Proposition. *If* $f\colon X \to Y$ *is holomorphic, then* f^0 *is a continuous algebra-homomorphism. If* f *is biholomorphic, then* f^0 *is an isomorphism of topological algebras.*

Proof. The consistency of f^0 with addition and multiplication can easily be checked by substitution. Furthermore, $f^0(1_Y) = 1_X$, so f^0 is an algebra-homomorphism. Let $(g_j)_{j \in \mathbb{N}}$ be a convergent sequence in $\mathcal{O}(Y)$. For each compact set $K \subset X$, $f(K)$ is compact, and $\|f^0(g_j)\|_K = \|g_j\|_{f(K)}$; it follows that $f^0(g_j)$ converges in $\mathcal{O}(X)$. If

[6] For $A \subset \prod_{j \in J} X_j$, $\mathrm{pr}_i\colon A \to X_i$ denotes the i-th projection.

f is biholomorphic, then f^{-1} and $(f^{-1})^0$ exist; it is easy to see that $(f^{-1})^0 = (f^0)^{-1}$, and the second assertion follows immediately. ■

E. 3a. Suppose that the set-mapping $f: X \to Y$ of one-dimensional domains induces an isomorphism $f^0: \mathcal{O}(Y) \to \mathcal{O}(X)$ of topological algebras. Show that f is bijective, and thus biholomorphic.

3.5 Remarks. i) According to 57.7, one need not demand the existence of f in E. 3a: each isomorphism $\varphi: \mathcal{O}(Y) \to \mathcal{O}(X)$ of the topological algebras in E. 3a is induced by a unique holomorphic mapping $f: X \to Y$. Thus, for biholomorphically nonequivalent regions $X, Y \subset\subset \mathbb{C}$, $\mathcal{O}(X)$ and $\mathcal{O}(Y)$ are nonisomorphic as topological *algebras*; however, they can be isomorphic as topological *vector spaces*: if $\mathbb{C} \setminus X$ has only finitely many connected components, then $\mathcal{O}(X)$ is, as a topological vector space, $\mathcal{O}(\mathbb{C})$, $\mathcal{O}(\boldsymbol{P}^1(1))$, or $\mathcal{O}(\mathbb{C}) \times \mathcal{O}(\boldsymbol{P}^1(1))$ [Ro Th 2.1].

ii) E. 3a does not hold for higher dimensions, as is shown, for example, by 7.8.

The one-dimensional Riemann Mapping Theorem [Co VII. 4.2] provides this description of simply connected domains $G \subset\subset \mathbb{C}$: if $G \neq \mathbb{C}$, then there exists a biholomorphic mapping from G onto the open unit disk in $\mathbb{C}$. To clear the way for a generalization, we introduce the following higher-dimensional analogue of the open unit disk:

3.6 Definition. *Let* $\varrho = (\varrho_1, \ldots, \varrho_n) \in \mathbb{R}^n_{>0}$ *and* $a \in \mathbb{C}^n$ *be given. Then*

$$\boldsymbol{P}(a; \varrho) := \boldsymbol{P}^n(a; \varrho) := \{z \in \mathbb{C}^n;\ |z_j - a_j| < \varrho_j,\ 1 \leqq j \leqq n\}$$

is called the polydisk (or polycylinder) about a with polyradius ϱ.

We adopt the abbreviations

$$\boldsymbol{P}(a; r) := \boldsymbol{P}(a; r, \ldots, r),\ \ \boldsymbol{P}(\varrho) := \boldsymbol{P}(0; \varrho),\ \text{and}\ \ \boldsymbol{P}(a; w) := \boldsymbol{P}(a; |w_1|, \ldots, |w_n|),$$

with $\mathbb{C}^* := \mathbb{C} \setminus \{0\}$ and $w \in \mathbb{C}^{*n}$.

3.7 Riemann Mapping Theorem. *If* $G = G_1 \times \ldots \times G_n$ *is a product of simply connected domains* $G_j \subset\subset \mathbb{C}^1$, *then there exists a* $k \geqq 0$ *such that* G *can be mapped biholomorphically onto the product of a* k*-dimensional unit polycylinder and an* $(n-k)$*-dimensional complex number space.*

The theorem follows from the one-dimensional version and the following simple observation:

3.8 Remark. *If* $f_i: X_i \to Y_i$, $i = 1, 2$, *are holomorphic mappings between regions in complex number spaces, then the mapping*

$$f_1 \times f_2: X_1 \times X_2 \to Y_1 \times Y_2, \qquad (x_1, x_2) \mapsto (f_1(x_1), f_2(x_2)),$$

is holomorphic. ■■

For $n \geqq 2$, simply connected proper domains are not necessarily biholomorphically equivalent. To see that, we consider the *ball* about a with radius $r > 0$,

$$B(a;r) := B^n(a;r) := \{z \in \mathbb{C}^n;\ \|a-z\| < r\}.$$

We see that $P(r)$ and $B(r) := B(0;r)$ are topologically equivalent, since $P(r)$ may be regarded as the ball of radius r with respect to the maximum norm; however, by 3.11, they are not biholomorphically equivalent for $n \geqq 2$.

Both are natural generalizations of the open unit disk in $\mathbb{C}^1$. Their nonequivalence leads to the application of fairly varied methods of proof in the local analytic theory, according to the generalization chosen.

It is also useful to introduce for continuous mappings $f = (f_1, \ldots, f_m): X \to \mathbb{C}^m$ the real-valued functions defined by setting

$$\|f\| := \sqrt{\sum_{j=1}^{m} f_j \bar{f}_j} \quad \text{and} \quad |f| := \sup_{1 \leqq j \leqq m} |f_j|.$$

For $A \subset X$, then, $\|f\|_A := \sup_{a \in A} \|f(a)\|$ has properties i)–iii) of § 2.

For the nonequivalence of balls and polydisks, we show first that each holomorphic mapping between them induces a linear mapping between them. In applying the linear mapping $\frac{\partial f}{\partial z}(0)$ (see 8.1), we have the following facts at our disposal (see also 8.1 and 8.2):

3.9 Lemma. *For* $B := B^n(1)$ *and* $P := P^m(1)$, *the following statements hold*:

i) If $f \in \mathrm{Hol}(P, B)$ *and* $f(0) = 0$, *then* $\frac{\partial f}{\partial z}(0)(P) \subset B$.

ii) If $f \in \mathrm{Hol}(B, P)$ *and* $f(0) = 0$, *then* $\frac{\partial f}{\partial z}(0)(B) \subset P$.

Proof. In each case, we construct a $g \in \mathrm{Hol}(P^1(1), P^1(1))$ with $g(0) = 0$; it follows from the one-dimensional version of Schwarz's Lemma (see E.4h) that $|g'(0)| \leqq 1$. For i), put $g(\lambda) := \sum_j f_j(\lambda z) w_j =: f(\lambda z) \cdot w$ for fixed $z \in P$ and $w \in B$. Then $g'(0) = \sum_{i,j} \frac{\partial f_j}{\partial z_i}(0) z_i w_j$, and thus

$$\left|\frac{\partial f}{\partial z}(0)(z) \cdot w\right| = |g'(0)| \leqq 1.$$

Since $\bar{B} = \{z \in \mathbb{C}^n; |z \cdot w| \leqq 1, \forall w \in B\}$, it follows that $\frac{\partial f}{\partial z}(0)(P) \subset \bar{B}$; the linearity of $\frac{\partial f}{\partial z}(0)$ implies that $\frac{\partial f}{\partial z}(0)(P) \subset B$. For ii), we may assume that $m = 1$; choose the function $\lambda \overset{g}{\mapsto} f(\lambda z)$.

Since $g'(0) = \sum_{j=1}^{n} \frac{\partial f}{\partial z_j}(0) z_j$, we see that $\left|\frac{\partial f}{\partial z}(0)(z)\right| = |g'(0)| \leqq 1$; then $\frac{\partial f}{\partial z}(0)(B) \subset \bar{P}$, and the assertion follows as in i). ■

The biholomorphic mappings from a region X onto itself are called *automorphisms* of X. They obviously form a group, $\mathrm{Aut}(X)$. A domain X is called *homogeneous* if, for each pair of points $a, b \in X$, there exists an $f \in \mathrm{Aut}(X)$ such that $f(a) = b$. (One also says that $\mathrm{Aut}(X)$ *acts transitively* on X.)

3.10 Example. $P^n = (P^1(1))^n$ *is a homogeneous domain*: For $a \in P^1$,

$\{z \mapsto e^{i\vartheta} \cdot \dfrac{z-a}{\bar{a}z-1};\ \vartheta \in \mathbb{R}\}$ is the set of those automorphisms of $\boldsymbol{P}^1$ which send a to 0 [Co VI.2.5]. Specifically, $\mathrm{Aut}(\boldsymbol{P}^1)$ is a group with three real parameters.

3.11 Proposition. *If $f: \boldsymbol{B}^n \to \boldsymbol{P}^n$ is biholomorphic, then $n = 1$.*

Proof. Since $\boldsymbol{P}^n$ is homogeneous, we may assume that $f(0) = 0$. Moreover, by 3.9 and the Chain Rule 8.2, we may assume that f is linear, and thus defined on all of $\mathbb{C}^n$. In particular, $f(\partial \boldsymbol{B}^n) = \partial \boldsymbol{P}^n$. If $n \geqq 2$, then the boundary $\partial \boldsymbol{P}^n$ includes the real segment $\{(1, t, 0, \ldots, 0) \in \mathbb{C}^n; -1 \leqq t \leqq 1\}$; since f^{-1} is linear, the boundary $\partial \boldsymbol{B}^n = f^{-1}(\partial \boldsymbol{P}^n)$ must include a real segment as well. ↯ ■

The identity mapping on a region $X \subset \mathbb{C}^n$ is holomorphic, and the composition of holomorphic mappings is holomorphic. That can be stated concisely: the regions in complex number spaces and their holomorphic mappings form a *category*. The sets $\mathrm{Hol}(X, Y)$ are called the *sets of morphisms* of the category. Similarly, the Fréchet algebras with the sets of morphisms

$$\mathrm{Hom}_{alg}(E, F) := \{\varphi : E \to F;\ \varphi \text{ is a continuous algebra-homomorphism}\}$$

form a category.

E. 3b. For $f \in \mathrm{Hol}(X, Y)$, $g \in \mathrm{Hol}(Y, Z)$, verify that
i) $(\mathrm{id}_X)^0 = \mathrm{id}_{\mathcal{O}(X)}$, and
ii) $(g \circ f)^0 = f^0 \circ g^0$.
The correspondence $\mathcal{O}: X \mapsto \mathcal{O}(X)$, $f \mapsto \mathcal{O}(f) := f^0$, between the category of regions and the category of Fréchet algebras is called a (*contravariant*) *functor*, and i) and ii) are the functor properties.

E. 3c. Determine φ^0 for the following φ's:

i) $\mathbb{C}^n \times \mathbb{C}^n \to \mathbb{C}^n$, $(z, w) \mapsto z + w$,
ii) $\mathbb{C} \times \mathbb{C}^n \to \mathbb{C}^n$, $(\lambda, z) \mapsto \lambda z$,
iii) $\mathbb{C}^* \times \mathbb{C}^n \to \mathbb{C}^n$, $(\lambda, z) \mapsto z/\lambda$.

E. 3d. For $Y \subset \mathbb{C}^m$, the subset $\mathrm{Hol}(X, Y)$ of $\mathrm{Hol}(X, \mathbb{C}^m) = \mathcal{O}(X)^m$ has a canonical topology. With respect to that topology, show the continuity of
i) evaluation, $\varepsilon_x : X \times \mathrm{Hol}(X, Y) \to Y$, $(x, f) \mapsto f(x)$, and
ii) composition, $\mathrm{Hol}(X, Y) \times \mathrm{Hol}(Y, Z) \to \mathrm{Hol}(X, Z)$, $(f, g) \mapsto g \circ f$
(hint: identify X with $\mathrm{Hol}(\text{point}, X)$).

§ 4 Cauchy's Integral Formula

By Cauchy's Integral Formula, every continuous function f on the *topological boundary* ∂D of an open disk D in $\mathbb{C}$ determines a unique holomorphic function on $\bar{D}$. In the generalization of that result to a polydisk $\boldsymbol{P} = \boldsymbol{P}^n(a; \varrho)$ with $n \geqq 2$, it suffices to prescribe f on a proper subset of the topological boundary, namely, on

$$\boldsymbol{T}^n(a; \varrho) := \{z \in \mathbb{C}^n;\ |z_j - a_j| = \varrho_j,\ j = 1, \ldots, n\},$$

called the *distinguished boundary* of $\boldsymbol{P}$. For $S \subset \mathbb{C}^n$, a function $f: S \to \mathbb{C}$ is called *holomorphic*[7)] if f is the restriction of a holomorphic function defined near S[8)]. We also use the abbreviation $z^1 := z_1 \cdot \ldots \cdot z_n$ (observe the difference between that and $z = (z_1, \ldots, z_n)$).

4.1 Cauchy's Integral Formula. *If f is a continuous function on $\boldsymbol{T} := \boldsymbol{T}^n(a; \varrho)$, then the assignment*

$$h(z) := \left(\frac{1}{2\pi i}\right)^n \int_T \frac{f(\zeta)}{(\zeta - z)^1}\, d\zeta :=$$

$$:= \left(\frac{1}{2\pi i}\right)^n \int_{|\zeta_n - a_n| = \varrho_n} \cdots \int_{|\zeta_1 - a_1| = \varrho_1} \frac{f(\zeta_1, \ldots, \zeta_n)}{(\zeta_1 - z_1) \ldots (\zeta_n - z_n)}\, d\zeta_1 \ldots d\zeta_n$$

determines a holomorphic function h on $\boldsymbol{P} := \boldsymbol{P}(a; \varrho)$. If f can be extended holomorphically to $\bar{\boldsymbol{P}}$, then $f|_{\boldsymbol{P}} = h$.

Proof. The function h is continuous, since $f(\zeta) / (\zeta - z)^1$ is continuous on $\boldsymbol{T} \times \boldsymbol{P}$. By Fubini's Theorem [Ry 12.21], the multiple integral does not depend on the order of integration. Hence, for every contractible piecewise smooth curve γ_j that depends only on z_j, the one-dimensional versions of Cauchy's Theorem and Cauchy's Formula [Co IV. 6.3] imply that $\int_{\gamma_j} h dz_j = 0$. Thus h is partially holomorphic. Now suppose that f is holomorphic near $\bar{\boldsymbol{P}}$; we prove that $f|_{\boldsymbol{P}} = h$ by inducting on n, beginning with the one-dimensional theorem.

"$n - 1 \Rightarrow n$". For fixed $(z_1, \ldots, z_n)$ and $F(\zeta_n) := f(z_1, \ldots, z_{n-1}, \zeta_n)$,

$$F(z_n) = \frac{1}{2\pi i} \int_{|\zeta_n - a_n| = \varrho_n} \frac{F(\zeta_n)}{\zeta_n - z_n}\, d\zeta_n .$$

If ζ_n is fixed and $z_1, \ldots, z_{n-1}$ are variable, then it follows from the induction hypothesis that

$$f(z_1, \ldots, z_{n-1}, \zeta_n) =$$

$$= \left(\frac{1}{2\pi i}\right)^{n-1} \int_{|\zeta_{n-1} - a_{n-1}| = \varrho_{n-1}} \cdots \int_{|\zeta_1 - a_1| = \varrho_1} \frac{f(\zeta_1, \ldots, \zeta_{n-1}, \zeta_n)\, d\zeta_1 \ldots d\zeta_{n-1}}{(\zeta_1 - z_1) \ldots (\zeta_{n-1} - z_{n-1})} .$$

The assertion follows by substitution. ■

4.2 Remarks. i) As the nature of the proof indicates, 4.1 holds for a product $\bar{D}_1 \times \ldots \times \bar{D}_n$ in place of $\bar{\boldsymbol{P}}$ if the one-dimensional version of Cauchy's Formula holds for each $\bar{D}_j$.

---

7) For locally analytic subsets of $\mathbb{C}^n$, we shall introduce a more general notion in 32.3.

8) "*Near*" a subset A of a topological space means „in some neighborhood of A".

ii) To ensure that $f|_{\boldsymbol{P}} = h$ in 4.1, it suffices that f be continuous on $\overline{\boldsymbol{P}(a;\varrho)}$ and holomorphic on $\boldsymbol{P}(a;\varrho)$ [Os I § 45].

For a *multi-index* $\nu = (\nu_1, \ldots, \nu_n) \in \mathbb{N}^n$, we use the following notation:

$$|\nu| := \sum_{j=1}^{n} \nu_j; \quad \nu! := \nu_1! \cdot \ldots \cdot \nu_n!; \quad z^\nu := z_1^{\nu_1} \cdot \ldots \cdot z_n^{\nu_n}; \quad \nu + 1 := (\nu_1 + 1, \ldots, \nu_n + 1).$$

If $f: X \to \mathbb{C}$ is a $|\nu|$-fold partially differentiable function, then $D^\nu f$ denotes the function

$$D^\nu f = \frac{\partial^{|\nu|} f}{\partial z_1^{\nu_1} \ldots \partial z_n^{\nu_n}} : X \to \mathbb{C}.$$

4.3 Corollary. *For each multi-index* ν, D^ν *is a continuous endomorphism of* $\mathcal{O}(X)$. *For each polydisk* $\boldsymbol{P} = \boldsymbol{P}^n(a;\varrho) \subset\subset X$ *with distinguished boundary* $\boldsymbol{T}$, *and for each holomorphic function* f *on* X, *the following statements hold*:

i) $$D^\nu f(z) = \frac{\nu!}{(2\pi i)^n} \int_{\boldsymbol{T}} \frac{f(\zeta)}{(\zeta - z)^{\nu+1}} d\zeta \quad \textit{for} \quad z \in \boldsymbol{P}.$$

ii) $$|D^\nu f(a)| \leqq \frac{\nu!}{\varrho^\nu} \|f\|_{\boldsymbol{T}} \quad \textit{(Cauchy's Estimate)}.$$

iii) *Fix* $w \in \boldsymbol{P}$ *and* $m \in \mathbb{N}$; *if, for every* $z \in \boldsymbol{P}$, X *includes the "closed complex segment"* $\{(1-\lambda)w + \lambda z;\ \lambda \in \overline{\boldsymbol{P}^1(1)}\}$, *then*

$$\sup_{z \in \boldsymbol{P}} \Big| \sum_{|\nu| = m} \frac{1}{\nu!} D^\nu f(w)(z-w)^\nu \Big| \leqq \|f\|_X.$$

The following fact is an immediate consequence of the continuity of D^ν:

4.4 Weierstrass's Theorem. *If* (f_j) *is a sequence in* $\mathcal{O}(X)$ *that converges to* f, *then* $(D^\nu f_j)$ *converges in* $\mathcal{O}(X)$ *to* $D^\nu f$ *for each* $\nu \in \mathbb{N}^n$. ■

Proof of 4.3. The continuous function $f(\zeta)/(\zeta - z)^1$ has continuous first partial derivatives with respect to z and $\bar{z}$, so differentiation can be performed under the integral in 4.1. Thus, the integral in i) is partially holomorphic and continuous; for $\nu = (\nu_1, \ldots, \nu_n)$, we have that $D^\nu = D^{\nu_1} \cdots D^{\nu_n}$, so i) follows by induction on ν (with a suitably chosen order), and we conclude that D^ν is a vector space endomorphism on $\mathcal{O}(X)$. Assertion ii) follows from the fact that $\int_{\boldsymbol{T}} d\zeta = \prod_{k=1}^{n} 2\pi\varrho_k = (2\pi)^n \varrho^1$, since we know that

$$|(D^\nu f)(a)| \underset{\text{i)}}{\leqq} \frac{\nu!}{(2\pi)^n} \frac{\|f\|_{\boldsymbol{T}}}{\varrho^{\nu+1}} \Big| \int_{\boldsymbol{T}} d\zeta \Big| = \frac{\nu!}{\varrho^\nu} \|f\|_{\boldsymbol{T}}.$$

iii) Fix $w \in \boldsymbol{P}$, and define a polynomial P_m by means of the assignment

$$P_m(z) := \sum_{|\nu|=m} \frac{1}{\nu!} D^\nu f(w) z^\nu .$$

Now fix $z \in \boldsymbol{P}$; by assumption, then, $V := \{\lambda \in \mathbb{C}; (1-\lambda)w + \lambda z \in X\}$ is an open neighborhood of $\overline{\boldsymbol{P}^1(1)}$ in $\mathbb{C}$. The function $g: V \to \mathbb{C}$, $\lambda \mapsto f(\lambda z + (1-\lambda)w)$, is holomorphic; by the Chain Rule,

$$\frac{d^m g}{d\lambda^m}(0) = \sum_{|\nu|=m} \frac{m!}{\nu!} D^\nu f(w)(z-w)^\nu = m!\, P_m(z-w) .$$

The one-dimensional Cauchy Estimate for g, and the inequality

$$|P_m(z-w)| = \frac{1}{m!} \left| \frac{d^m g}{d\lambda^m}(0) \right| \leqq \sup_{|\lambda|=1} |g(\lambda)| \leqq \|f\|_X ,$$

imply the validity of assertion iii). The continuity of D^ν remains to be shown; to that end, we apply the following simple fact:

4.5 Lemma. *For a linear mapping $\varphi: E \to F$ of topological vector spaces whose topologies are given by families of seminorms p_i and q_j, respectively, if*[9] *there exist for each q_j an $r \in \mathbb{R}$ and a p_i such that $q_j \circ \varphi \leqq r p_i$, then φ is continuous.* ∎

By E.2a, it suffices to estimate $\|D^\nu f\|_{\boldsymbol{P}(a;r)}$ for each pair a, r such that $\boldsymbol{P}(a; 2r) \subset\subset X$. Then, by ii), we see that

$$|D^\nu f(b)| \leqq \frac{\nu!}{r^\nu} \|f\|_{\boldsymbol{P}(b;r)} \leqq \frac{\nu!}{r^\nu} \|f\|_{\boldsymbol{P}(a;2r)}$$

for each $b \in \boldsymbol{P}(a; r)$, and it follows that

$$\|D^\nu f\|_{\boldsymbol{P}(a;r)} \leqq \frac{\nu!}{r^\nu} \|f\|_{\boldsymbol{P}(a;2r)} .$$ ∎∎

With the help of 4.3, we want to show that a function is holomorphic iff it can be represented locally by its Taylor series (Weierstrass's definition of holomorphic functions). Before doing that, we need to clarify the concept of convergence of an infinite series of holomorphic functions, by which one could understand, say, pointwise convergence or convergence in $\mathcal{O}(X)$. The following result is an easy consequence of [Di V.3]:

4.6 Remark. *Consider a family of functions $(f_\nu)_{\nu \in \mathbb{N}^n}$ with $f_\nu \in \mathcal{O}(X)$. If, for some order of summation, $\sum_{\nu \in \mathbb{N}^n} |f_\nu|$ converges in $\mathscr{C}(X)$, then, for any order of summation,*

i) $\sum_{\nu \in \mathbb{N}^n} f_\nu(z)$ *converges absolutely for each $z \in X$,*

---

[9] The converse is also easy to show; see [Bou EVT II.5.6 Prop. 9].

ii) $\sum_{\nu \in \mathbb{N}^n} |f_\nu|$ *converges in* $\mathscr{C}(X)$, *and*

iii) $\sum_{\nu \in \mathbb{N}^n} f_\nu$ *converges in* $\mathcal{O}(X)$

to a limit that is independent of the order of summation. ■

As indicated by the next lemma, pointwise convergence and convergence in $\mathcal{O}(\boldsymbol{P})$ are equivalent for a power series on a polydisk $\boldsymbol{P}$ (however, see E.2f).

4.7 Abel's Lemma. *If* $\sum_{\nu \in \mathbb{N}^n} a_\nu z^\nu$ *is a power series and* $w \in \mathbb{C}^{*n}$, *then the following statements hold*[10]*:*

i) If the set $\{a_\nu w^\nu;\ \nu \in \mathbb{N}^n\} \subset \mathbb{C}$ *is bounded, then* $\Sigma |a_\nu z^\nu|$ *converges in* $\mathscr{C}(\boldsymbol{P}^n(w))$ *and* $\Sigma a_\nu z^\nu$ *converges in* $\mathcal{O}(\boldsymbol{P}^n(w))$.

ii) Otherwise, the series $\Sigma |a_\nu z^\nu|$ *and* $\Sigma a_\nu z^\nu$ *diverge for every* $z \in \mathbb{C}^n$ *such that* $w \in \boldsymbol{P}(z)$.

Proof. i) For $z \in \boldsymbol{P}(w)$, we have that $\frac{|z_j|}{|w_j|} < 1$ for every j; thus, by E. 4a, for each $\varepsilon > 0$, there exists a $k \in \mathbb{N}$ such that

$$\Big|\sum_{|\nu| \geqq k} a_\nu z^\nu\Big| \leqq \sum_{|\nu| \geqq k} |a_\nu z^\nu| = \sum_{|\nu| \geqq k} |a_\nu w^\nu| \cdot \Big|\frac{z^\nu}{w^\nu}\Big| \leqq C \cdot \Sigma \Big|\frac{z^\nu}{w^\nu}\Big| \leqq C \cdot \varepsilon,$$

where $C := \sup_\nu |a_\nu w^\nu|$.

ii) If the series $\Sigma a_\nu z^\nu$ were convergent, then $\{a_\nu z^\nu;\ \nu \in \mathbb{N}^n\}$ would be bounded; by i), that would imply the boundedness of $\{a_\nu w^\nu;\ \nu \in \mathbb{N}^n\}$ ↯. ■

E. 4a. *Geometric series.* The series $\sum_{\nu \in \mathbb{N}^n} |z^\nu|$ converges in $\mathscr{C}(\boldsymbol{P}^n(1))$, and $\sum_{\nu \in \mathbb{N}^n} z^\nu = \prod_{j=1}^{n} \frac{1}{1 - z_j}$ in $\mathcal{O}(\boldsymbol{P}^n(1))$.

Convergent power series obviously determine continuous partially holomorphic (i.e., holomorphic) functions[11]; in fact, the converse holds as well:

4.8 Corollary. *If* f *is holomorphic on the polydisk* $\boldsymbol{P} = \boldsymbol{P}(a; \varrho)$, *then the* <u>*Taylor series*</u> *of* f, $\sum_{\nu \in \mathbb{N}^n} \frac{1}{\nu!} D^\nu f(a)(z - a)^\nu$, *converges in* $\mathcal{O}(\boldsymbol{P})$ *to* f.

Proof. Without loss of generality, set $a = 0$. The series $\sum_{\nu \in \mathbb{N}^n} \frac{1}{\nu!} D^\nu f(0) z^\nu$ converges in $\mathcal{O}(\boldsymbol{P})$: for each $z \in \boldsymbol{P}$, the set $\{\frac{1}{\nu!} D^\nu f(0) z^\nu;\ \nu \in \mathbb{N}^n\}$ is bounded by 4.3 ii), since $\|f\|_{T(0;z)} < \infty$; the series converges in $\boldsymbol{P}(z)$, by Abel's Lemma, and in $\boldsymbol{P}$, by E.2g.

---

[10] For $w \in \mathbb{C}^n$, see E. 11a.

[11] They also have continuous partial derivatives with respect to the real variables x_j and y_j.

The Taylor series represents f: for $\boldsymbol{T} := \boldsymbol{T}(a; \varrho)$ we have that

$$f(z) \underset{4.1}{=} \frac{1}{(2\pi i)^n} \int_T \frac{f(\xi)}{(\xi - z)^1} d\xi \underset{\text{E.4}a}{=} \frac{1}{(2\pi i)^n} \int_T f(\xi) \Big(\sum_{\nu \in \mathbb{N}^n} \frac{z^\nu}{\xi^{\nu+1}} \Big) d\xi$$

$$= \sum_{\nu \in \mathbb{N}^n} \left[\frac{1}{(2\pi i)^n} \int_T \frac{f(\xi)}{\xi^{\nu+1}} d\xi \right] z^\nu \underset{4.3\, ii)}{=} \sum_{\nu \in \mathbb{N}^n} \frac{1}{\nu!} D^\nu f(0) z^\nu ,$$

since $\boldsymbol{T}$ is a compact set on which the geometric series converges absolutely and uniformly. Thus we may interchange summation and integration. ∎

E. 4b. A function $f \in \mathcal{O}(\mathbb{C}^n)$ is called an *entire* (holomorphic) *function.* Prove *Liouville's Theorem: Every bounded entire function is constant.*

E. 4c. Let f be an entire function, and suppose that there exist a multi-index ν and a constant $\gamma > 0$ such that $|f(z)| \leqq \gamma |z^\nu|$ for every $z \in \mathbb{C}^n$. Show that f is a polynomial of degree at most $|\nu|$.

E. 4d. *Identity Theorem.* Let f and g be holomorphic functions on a connected region X; assume that there is an $x_0 \in X$ such that $D^\nu f(x_0) = D^\nu g(x_0)$ for every $\nu \in \mathbb{N}^n$. Prove that $f = g$ (hint: 4.8; for 6.1).

E. 4e. Let $\boldsymbol{P} = \boldsymbol{P}(a; \varrho)$ and $f \in \mathcal{O}(\boldsymbol{P})$ be given. Show that f has a unique representation of the form $f = \sum_{k=0}^{\infty} P_k$ with *homogeneous polynomials* (see § 21) P_k of degree k in $(z - a)$ such that $\sum_{k=0}^{\infty} P_k$ converges in $\mathcal{O}(\boldsymbol{P})$. Then the series $\sum_{k=0}^{\infty} |P_k|$ converges in $\mathscr{C}(\boldsymbol{P})$. The minimal k such that $P_k \neq 0$ is called the (*total*) *order of f at a* (for 4 A.1).

E. 4f. „*Overconvergence*" (enlarging the region of convergence by skillfully juggling the parentheses). Consider $f(z_1, z_2) = \sum_{j=0}^{\infty} (z_1 + z_2)^j$.

i) Determine the maximal $G \subset\subset \mathbb{C}^2$ such that the series f converges in $\mathcal{O}(G)$.

ii) Give the power series expansion g of f about 0.

iii) Show that the region of absolute convergence of the series g is included in $\{|z_1| + |z_2| \leqq 1\}$ (it follows from 11.4 that g converges absolutely in $\{|z_1| + |z_2| < 1\}$).

E. 4g. If $f = \sum_{k=0}^{\infty} P_k$ is a formal series (see § 21) of homogeneous polynomials P_k of degree k in z, and if there exists a $w \in \mathbb{C}^n \backslash 0$ such that the series $\sum_{k=0}^{\infty} P_k(w)$ converges, then f converges on $\boldsymbol{B}^1(1) \cdot w := \{\lambda w; \lambda \in \boldsymbol{B}^1(1)\}$ and determines on $\boldsymbol{B}^1(1)$ a holomorphic function of λ.

E. 4h. *Schwarz's Lemma.* Let the function $f \in \mathcal{O}(\boldsymbol{B}(0; r + \varepsilon))$ have order k at the point $0 \in \mathbb{C}^n$. Then, for each $z \in \boldsymbol{B} := \boldsymbol{B}(0; r)$,

$$|f(z)| \leqq \|f\|_{\boldsymbol{B}} \|\frac{z}{r}\|^k$$

(hint: Fix $z \neq 0$ in $\boldsymbol{B}$ and consider the function defined by setting $g(\lambda) := \lambda^{-k} f\left(\lambda \frac{z}{\|z\|}\right)$ for nonzero $\lambda \in \boldsymbol{P}^1(r)$).

E. 4i. Compare the holomorphic functions $f \in \mathcal{O}(\mathbb{C}^*)$, $z \overset{f}{\mapsto} 1/z$, and $g \in \mathcal{O}(\boldsymbol{P}^1(1))$,

$$z \overset{g}{\mapsto} \frac{1}{2\pi i} \int_{T(0;1)} \frac{f(\zeta)}{\zeta - z} d\zeta .$$

§ 4 A Supplement: Cartan's Uniqueness Theorem and Automorphisms of Bounded Domains

According to Schwarz's Lemma [Co VI.2.1], holomorphic mappings f from the unit disk $\boldsymbol{P}^1(1)$ into itself satisfy this condition: if $f(0) = 0$ and $|f'(0)| = 1$, then f is determined by $f'(0)$. In the following generalization of that fact, $I_n \in GL(n, \mathbb{C})$ denotes the identity matrix:

4 A. 1 Cartan's Uniqueness Theorem. *Let $f: G \to G$ be a holomorphic mapping from the bounded domain G into itself. If there exists a fixed point a of f such that $\frac{\partial f}{\partial z}(a) = I_n$, then f is the identity mapping.*

Proof. Without loss of generality, assume that $a = 0$. By E. 4e, on an arbitrary fixed $\boldsymbol{P} = \boldsymbol{P}(r) \subset\subset G$, f admits a representation of the form $f = \sum_{j=0}^{\infty} P_j$ in which the components of $P_j = (P_{j1}, \dots, P_{jn})$ are homogeneous polynomials of degree j. By E. 4e, the assumption that $f(0) = 0$ and $\frac{\partial f}{\partial z}(0) = I_n$ ensures the existence of a $k \geqq 2$ such that

$$f = \mathrm{id}_{\boldsymbol{P}} + \sum_{j \geqq k} P_j .$$

For $f^m := f \circ \dots \circ f$, induction on m yields that

$$f^m = \mathrm{id}_{\boldsymbol{P}} + mP_k + \text{terms of higher order}$$

on $\boldsymbol{P}$. By applying 4.3 iii) to the components of f^m, it can be shown that

$$\|mP_k\|_{\boldsymbol{P}} = \| \sum_{|\nu| = k} \frac{1}{\nu!} D^\nu f^m(0) z^\nu \|_{\boldsymbol{P}} \leqq \|f^m\|_{\boldsymbol{P}} \leqq \|f\|_G < \infty .$$

Since $\sup_{m \in \mathbb{N}} \|mP_k\|_{\boldsymbol{P}} < \infty$, it follows that $P_k = 0$, so $f = \mathrm{id}_{\boldsymbol{P}}$. ∎

On that basis, we want to give a further proof of the nonequivalence of balls and polydisks (3.11). Both are *circular domains* in $\mathbb{C}^n$, i.e., domains G such that $e^{i\vartheta} z \in G$ for every $z \in G$, $\vartheta \in \mathbb{R}$.

4 A. 2 Corollary. *If $f: G \to H$ is a biholomorphic mapping between bounded circular domains, and if $0 \in G$ and $f(0) = 0$, then f is linear.*

Proof. For fixed $\theta \in \mathbb{R}$, the mapping $F: z \mapsto f^{-1}(e^{-i\theta} f(e^{i\theta} z))$ sends G holomorphically into itself, with 0 as a fixed point; by the Chain Rule, $\frac{\partial F}{\partial z}(0) = I_n$. According to 4 A. 1, F is the identity mapping; i.e., $f(e^{i\theta} z) = e^{i\theta} f(z)$. Expand f near 0 as a series $\sum_{k=0}^{\infty} P_k$ of mappings whose components P_{kj} are homogeneous polynomials of degree k; the fact that $f(0) = 0$ implies that $P_0 = 0$. It follows that

$$\sum_{k=1}^{\infty} e^{ik\theta} P_k(z) = \sum_{k=1}^{\infty} P_k(e^{i\theta} z) = \sum_{k=1}^{\infty} e^{i\theta} P_k(z) .$$

The series-representation with homogeneous polynomials is unique (E. 4e); consequently, $(e^{ik\theta} - e^{i\theta}) P_k = 0$ for each $\theta \in \mathbb{R}$, and $P_k = 0$ for $k \neq 1$. ■

By replacing 3.9 with the following, one can obtain a proof of 3.11:

4 A.3 Corollary. *Let G and H be circular domains in $\mathbb{C}^n$ such that both contain 0, and one is homogeneous and bounded. Then G and H are biholomorphically equivalent iff they are linearly equivalent.*

Proof. Only if. If $g: G \to H$ is biholomorphic and G is homogeneous, then there exists an $h \in \operatorname{Aut}(G)$ such that $h(0) = g^{-1}(0)$. The mapping $f := g \circ h$ is biholomorphic and fulfills the hypotheses of 4 A. 2, and is thus a linear equivalence. The direction "if" is trivial. ■

Incidentally, the assumption of homogeneity for one of the domains in 4A. 3 is superfluous [KaUp 3.5].

§ 5 Montel's Theorem

A fundamental result for the investigation of topological properties of subsets A of $\mathbb{R}^n$ is the *Heine-Borel Theorem*: A is compact iff A is closed and bounded. An analogous statement (under the name *Montel's Theorem*) holds for $\mathcal{O}(X)$. For $\mathcal{O}(\mathbb{C}^0) \cong \mathbb{C}$ (E. 1c), that is clear. For general regions X of arbitrary dimension, the concept "bounded in $\mathcal{O}(X)$" must be defined, and the concept "compact" must be made precise.

For a topology induced on a vector space V by means of a family of seminorms $(s_j)_{j \in J}$, a set $D \subset V$ is called *bounded* if $\|s_j\|_D < \infty$ for every $j \in J$. For example, $\{f \in \mathscr{C}(X); \|f\|_X < 1\}$ is bounded in $\mathscr{C}(X)$. Boundedness is preserved by passage to a linear subspace. A subset $D \subset V$ is called *sequentially compact* if each sequence in D has a convergent subsequence with a limit in D. For metrizable V, sequential compactness is equivalent to *compactness* (i.e., every open cover of D has a finite subcover; see [Bou GT IX § 2 Prop. 15]).

One direction of Montel's Theorem follows immediately from the following simple fact:

5.1 Lemma. *For a locally compact space T, every compact subset D of $\mathscr{C}(T)$ is closed and bounded.*

Proof. Since a compact subset of a Hausdorff space is necessarily closed, it suffices to prove that D is bounded. Let K be a compact subset of T; then $\|\cdot\|_K : \mathscr{C}(T) \to \mathbb{R}$ is a continuous function. Hence, $\{\|d\|_K; d \in D\}$ is a compact subset of $\mathbb{R}$, and $\sup\{\|d\|_K; d \in D\} < \infty$. ■

The converse of 5.1 does not hold for $\mathscr{C}(T)$ (see E. 5a), but it does hold for $\mathcal{O}(X)$:

5.2 Montel's Theorem. *A subset D of $\mathcal{O}(X)$ is compact iff it is closed and bounded.*

Proof. The "only if" direction follows from 5.1. *If.* Given a sequence $(f_j^1)_{j \in \mathbb{N}}$ in D, since D is metrizable (by E. 2e), we need only construct a convergent subsequence whose limit is in D; we reduce the problem to the Heine-Borel Theorem by means of a diagonal process. Since X has a representation of the form $X = \bigcup_{k=2}^{\infty} \boldsymbol{P}_k$ with polydisks $\boldsymbol{P}_k \subset\subset X$, we may induct on k: from the sequence $(f_j^{k-1})_{j \in \mathbb{N}}$ chosen at step $(k-1)$, we choose a subsequence $(f_j^k)_{j \in \mathbb{N}}$ that converges on $\boldsymbol{P}_k$; by E. 2g i), (f_k^k) converges on X. If we set $\boldsymbol{P}_1 = \emptyset$, then the procedure begins with (f_j^1). Thus, without loss of generality, we have this situation: (f_j) is a sequence of holomorphic functions on $\boldsymbol{P} = \boldsymbol{P}^n(1)$, and there exists an $M \in \mathbb{R}$ such that $\sup_j \|f_j\|_{\boldsymbol{P}} \leqq M$. For every coefficient ${}_j a_\nu$ of the Taylor series expansion $f_j = \Sigma_j a_\nu z^\nu$, we know that $|{}_j a_\nu| \leqq M$ by 4.3 ii). For fixed ν, the sequence $({}_j a_\nu)_j$ has a convergent subsequence, according to the Heine-Borel Theorem. Thus, with respect to a fixed ordering of $\mathbb{N}^n$, it is possible to construct a decreasing sequence of index sets J_ν in $\mathbb{N}$ such that $j_\nu := \min J_\nu \geqq |\nu|$ and $({}_j a_\nu)_{j \in J_\nu}$ converges to an $a_\nu \in \mathbb{C}$. Then the sequence $(f_{j_\mu})_{\mu \in \mathbb{N}^n}$ converges in $\mathcal{O}(\boldsymbol{P})$ to $f := \Sigma a_\nu z^\nu$: for $z \in \boldsymbol{P}$ with $|z| \leqq r < 1$ and $N \in \mathbb{N}$, we have that

$$|f(z) - f_{j_\mu}(z)| = |\sum_{\nu \in \mathbb{N}^n} (a_\nu - {}_{j_\mu}a_\nu) z^\nu| \leqq \sum_{|\nu| \leqq N} |a_\nu - {}_{j_\mu}a_\nu| + 2M \sum_{|\nu| > N} r^{|\nu|}.$$

For a fixed ε and sufficiently large N, the second summand is smaller than ε; for a fixed N and sufficiently large $\mu \geqq \mu_0(\varepsilon)$, the first summand is also smaller than ε: for each $\nu \in \mathbb{N}^n$, we have that $({}_{j_\mu}a_\nu)_{\mu \in \mathbb{N}^n} \to a_\nu$, due to the fact that $J_\mu \subset J_\nu$ for $\mu \geqq \nu$. ■

A Fréchet space in which the bounded sets are precisely the relatively compact sets is also called a *Montel space.* Thus 5.2 can be reformulated in the following way:

5.3 Corollary. *$\mathcal{O}(X)$ is a Montel space.* ■

A linear mapping $\varphi: E \to F$ between two topological vector spaces is called *compact* if there is a neighborhood U of zero in E such that $\varphi(U)$ is relatively compact in F.

5.4 Corollary. *If $i: X \to Y$ is the inclusion mapping between two regions in $\mathbb{C}^n$, and if X is relatively compact in Y, then $i^0: \mathcal{O}(Y) \to \mathcal{O}(X)$ is a compact mapping.*

Proof. The set $W(0, 1; \|\cdot\|_X) = \{f \in \mathcal{O}(Y); \|f\|_X < 1\}$ is a neighborhood of zero in $\mathcal{O}(Y)$, since $\bar{X}$ is compact by assumption. The set $i^0(W(0, 1; \|\cdot\|_X))$ is bounded in $\mathcal{O}(X)$; thus, by Montel's Theorem, it is relatively compact. ■

E. 5a. In $\mathscr{C}(\mathbb{R})$, the set $\{\sin jx; j \in \mathbb{N}\}$ of real analytic functions is closed and bounded, but not compact.

§ 6 The Identity Theorem, the Maximum Principle, and Runge's Theorem

A holomorphic function is completely determined on a domain by its values in a neighborhood of a single point:

6.1 Identity Theorem. *Let G be a domain, and suppose that f and g are in $\mathcal{O}(G)$. If there exists a nonempty open subset W of G such that $f|_W = g|_W$, then $f = g$.*

Proof. Suppose that $a \in W$. Since partial derivatives are formed locally, we see that $D^\nu f(a) = D^\nu(f|_W)(a) = D^\nu(g|_W)(a) = D^\nu g(a)$ for every $\nu \in \mathbb{N}^n$; the theorem follows from E. 4d. ■

6.2 Corollary. *$\mathcal{O}(X)$ is an integral domain (i.e., a commutative ring with a unit element and no zero-divisors) iff X is a domain.*

Proof. If X includes, say, the two components $X_1 \neq X_2$, then the characteristic functions χ_j of X_j are holomorphic on X; moreover, $\chi_1 \chi_2 = 0$, so $\mathcal{O}(X)$ has zero-divisors. If X is connected, and if $f, g \in \mathcal{O}(X)$ with $f \neq 0$, then, by virtue of continuity, there exists an open set W in X on which f has no zeros. Thus, if $fg = 0$, it follows that $g|_W = 0$; by the Identity Theorem, $g = 0$. ■

E. 6a. If a compact subset K of a domain G contains interior points, then $\|\cdot\|_K$ is a norm on $\mathcal{O}(G)$.

For the next theorem, we need the following terminology: a mapping is called *open* if it sends open sets to open sets; *closed* mappings are defined analogously.

6.3 Theorem. *If $f \in \mathcal{O}(X)$ is nonconstant on each connected component of X, then f is an open mapping.*

Proof. For $a \in X$, choose a $\boldsymbol{B}(a; r) \subset X$; by 6.1, there exists a $b \in \boldsymbol{B}(a; r)$ such that $f(a) \neq f(b)$. It follows that the assignment $\lambda \mapsto f((1-\lambda)a + \lambda b)$ determines a nonconstant holomorphic function g on $\overline{\boldsymbol{B}^1(1)}$; by the one-dimensional version of 6.3 (see [Co IV.7.5]), $g(\overline{\boldsymbol{B}^1(1)})$ is a neighborhood of $f(a)$ included in $f(X)$. ■

E. 6b. Show that the mapping $f \in \mathrm{Hol}(\mathbb{C}^2, \mathbb{C}^2)$, $(z, w) \mapsto (z, zw)$, is neither open nor closed (for § 7).

6.4 Maximum Principle. *Let G be a domain, and fix an $f \in \mathcal{O}(G)$. If there exists an $a \in G$ such that $\|f\|_G = |f(a)|$, then f is constant.*

Proof. If f were not constant, then, by 6.3, we would have that $f(G) \Subset \mathbb{C}$ and thus that $\{|f(z)|; z \in G\} \Subset \mathbb{R}_{\geq 0}$; therefore, $|f|$ would not assume a maximum value, in contradiction to the assumption. ■

6.5 Corollary. *If X is a bounded region, and if the restriction of an $f \in \mathscr{C}(\bar{X})$ to X is holomorphic, then the function $|f|$ assumes its maximum value on ∂X.*

Proof. The continuous function $|f|$ assumes its maximum value in $\bar{X}$, say at $a \in \bar{X}$. If a $\notin \partial X$, then, by the Maximum Principle, f is constant on the connected component Z of X containing a. Thus, due to the continuity of $|f|$ on $\bar{X}$, we know that $|f(b)| = \|f\|_X$ for any $b \in \partial Z \subset \partial X$. ∎

E. 6c. Show, with a domain $X \subset\subset \mathbb{C}$, that 6.5 does not hold for unbounded X even if f is bounded.

Every polydisk $\boldsymbol{P}$ has the property that each holomorphic function $f \in \mathcal{O}(\boldsymbol{P})$ can be approximated on $\boldsymbol{P}$ by polynomials (specifically, by the partial sums of the Taylor series of f). That does not hold for arbitrary domains $G \subset\subset \mathbb{C}^n$: for $n = 1$, *Runge's Theorem* states that the polynomials in z are dense in $\mathcal{O}(G)$ iff G is simply connected [Co VIII.3.2]. Approximability by polynomials is obviously equivalent to approximability by entire functions. That fact motivates the following definition, which later admits an immediate generalization to "Runge pairs":

6.6 Definition. *A domain (region) G in $\mathbb{C}^n$ is called a Runge domain (region) if the restriction-homomorphism $i^0 : \mathcal{O}(\mathbb{C}^n) \to \mathcal{O}(G)$ has a dense image.*

6.7 Proposition. *If $G = G_1 \times \ldots \times G_n$ is a product of simply connected one-dimensional domains, then G is a Runge domain.*

Proof. By the Riemann Mapping Theorem 3.7, there exists a biholomorphic mapping $\varphi = (\varphi_1, \ldots, \varphi_n): G \to W$ onto a product-domain W, each of whose factors is $\mathbb{C}$ or $\boldsymbol{B}^1(1)$. By 4.8, each $f \in \mathcal{O}(W)$ is represented by its Taylor series on all of W, and thus is approximable by partial sums of the form $\sum_0^{<\infty} c_{\nu_1 \ldots \nu_n} w_1^{\nu_1} \ldots w_n^{\nu_n}$.

By 3.4, $\varphi^0 : \mathcal{O}(W) \to \mathcal{O}(G)$ is an isomorphism of topological algebras, so $\varphi^0(f)$ can be approximated on G by finite sums of the form

$$\text{(6.7.1)} \quad \Sigma c_{\nu_1 \ldots \nu_n} \varphi_1^{\nu_1} \ldots \varphi_n^{\nu_n}.$$

By the one-dimensional version of Runge's Theorem, every $\varphi_j \in \mathcal{O}(G_j)$ is approximable by entire functions on G_j. Hence, every finite sum (6.7.1), and thereby every $\varphi^0(f)$, is approximable by entire functions on G. ∎

The property of being a Runge domain is not preserved under biholomorphic transformations, as the following example shows:

6.8 Example (Wermer). Consider the mapping $\varphi : \mathbb{C}^3 \to \mathbb{C}^3, z \mapsto (z_1, z_1 z_2 + z_3, z_1 z_2^2 - z_2 + 2z_2 z_3)$. By 8.4, for sufficiently small ε with $0 < \varepsilon < 1/2$, the restriction of φ (also denoted by φ) to the polydisk $\boldsymbol{P} := \boldsymbol{P}^3(0; 1 + \varepsilon, 1 + \varepsilon, \varepsilon)$ is a biholomorphic mapping onto a domain $G \subset\subset \mathbb{C}^3$. According to 6.7, $\boldsymbol{P}$ is a Runge domain; however, G is not a Runge domain: since $\varphi(z_1, z_1^{-1}, 0) =$

$(z_1, 1, 0)$ for $z_1 \neq 0$, it follows that $S := \{(z_1, 1, 0); |z_1| = 1\} \subset W := \{(z_1, 1, 0) \in G\}$. Now the restriction of $h := \mathrm{pr}_2 \circ \varphi^{-1}$ to W satisfies the equality

$$h(z_1, 1, 0) = \mathrm{pr}_2 \circ \varphi^{-1}(z_1, 1, 0) = \mathrm{pr}_2(z_1, z_1^{-1}, 0) = z_1^{-1}$$

(note that $(0, 1, 0) \notin W$); hence $\|h\|_S = 1$. If $|z_1| \leqq 1$ and $P \in \mathbb{C}[z_1, z_2, z_3]$, then $|P(z_1, 1, 0)| \leqq \leqq \|P\|_S$, since 6.5 can be applied to $X := \boldsymbol{P}^1(1)$ and the holomorphic function $P(z_1, 1, 0)$ of one variable. If G were a Runge domain, then h would be approximable by such polynomials P on G, and it would follow that $|h(z_1, 1, 0)| \leqq \|h\|_S$. For $z \in W$ such that $|z_1| < 1$, that would imply the "inequality" $1 < \dfrac{1}{|z_1|} = |h(z)| \leqq \|h\|_S = 1$. ↯ ■

E. 6d. For one-dimensional domains G, we have this fact: if $f \in \mathcal{O}(G)$ and if the set $\{z \in G; f(z) = 0\}$ has an accumulation point in G, then $f = 0$. Show that such a form of the Identity Theorem is false for $n \geqq 2$ (however, see the proof of 12 B. 1).

E. 6e. Let a set $K \subset\subset \mathbb{C}$ with nonempty interior $\mathring{K}$ be given. Prove that $\mathcal{O}(\mathbb{C})$ is not complete under the norm $\|\cdot\|_K$ (hint: $(z^j)_{j \in \mathbb{N}}$).

§ 7 The Riemann Removable Singularity Theorems

The one-dimensional Removable Singularity Theorem states that, for $G \subset\subset \mathbb{C}^1$ and $a \in G$, if $f \in \mathcal{O}(G \backslash a)$ is bounded near a, then f is extendible to G. In the generalization of that fundamental theorem, it turns out that the exceptional set $\{a\}$ – the zero-set of the function $z - a$ – can be replaced with an analytic set of codimension at least one (the fact that the codimension behaves in accordance with the intuitive notion of the concept is discussed in § 48). In the ensuing discussion, we use this notation: if $f_1, f_2, \ldots$ are mappings from a set U into an abelian group, then

$$N(U; f_1, f_2, \ldots) := N(f_1, f_2, \ldots) := \{u \in U; f_1(u) = f_2(u) = \ldots = 0\}$$

denotes their common *zero-set (Nullstellenmenge)*.

7.1 Definition. *A subset A of X is called an* <u>*analytic set*</u> *if for each $z \in X$ there exist a neighborhood U and holomorphic functions $f_1, \ldots, f_m \in \mathcal{O}(U)$ such that $A \cap U = N(U; f_1, \ldots, f_m)$. If, in addition, $A \neq X$, then A is called a proper analytic subset.*

Some examples of analytic sets in X are X, $\emptyset$, intersections of X with affine subspaces[12)] of $\mathbb{C}^n$ (in particular, each point of X), locally finite unions of analytic sets (locally represented by products of the corresponding functions), intersections of analytic sets, and inverse images of analytic sets under holomorphic mappings (proof?). Images of analytic sets are not in general analytic (E. 6b).

7.2 Remark. *Analytic subsets of X are closed in X.* ■

---

12) For a linear subspace V of $\mathbb{C}^n$, and an $a \in \mathbb{C}^n$, $a + V$ is called an *affine subspace* of $\mathbb{C}^n$.

E. 7a. A set $A \subset X$ is called *locally analytic* if, for each $a \in A$, there exists a neighborhood $U \subset X$ of a such that $A \cap U$ is analytic in U. Prove that *a subset A of X is analytic iff it is locally analytic and closed in X.*

Now we proceed to the promised generalization for n-dimensional domains, in which we use the following terminology: $f \in \mathcal{O}(G \setminus A)$ is called *locally bounded at A* if each $a \in A$ has a neighborhood U in G such that $\|f\|_{U \setminus A} < \infty$.

7.3 First Riemann Removable Singularity Theorem. *If $A \subset G$ is closed and contained in a proper analytic subset of G, then the inclusion $i: G \setminus A \to G$ induces an isomorphism i^o of topological algebras between $\mathcal{O}(G)$ and the subalgebra of $\mathcal{O}(G \setminus A)$ that consists of all functions in $\mathcal{O}(G \setminus A)$ that are locally bounded at A.*

Hence, every function in $\mathcal{O}(G \setminus A)$ that is locally bounded at A has a (unique) holomorphic extension to G, and a sequence of functions in $\mathcal{O}(G)$ converges on G iff it converges on $G \setminus A$. For the proof, we require some familiarity with analytic sets.

7.4 Proposition. *Let A be an analytic subset of the domain G in $\mathbb{C}^n$. Then either $A = G$, or A is nowhere dense in G and nowhere separating (i.e., if V is an open connected subset of G, then also $V \setminus A$ is connected). In particular, $A = G$ if $\mathring{A} \neq \emptyset$.*

Proof. Without loss of generality, suppose that $A = N(G; f_1, \ldots, f_m) \neq G$. It follows from the Identity Theorem that $\mathring{A}$ is empty. If $\boldsymbol{B} = \boldsymbol{B}(a; r) \subset G$, then $\boldsymbol{B} \setminus A$ is connected: for two arbitrarily chosen points $b, c \in \boldsymbol{B} \setminus A$, fix a complex line Γ joining them; then $\Gamma \cap \boldsymbol{B}$ is one-dimensional and the proper analytic subset $\Gamma \cap \boldsymbol{B} \cap A$ is discrete; it follows that b and c can be joined with a path in $\Gamma \cap \boldsymbol{B} \setminus A$. For an arbitrary connected region $V \Subset G$ and points $a, b \in V \setminus A$, join a and b in V with a chain of balls $\boldsymbol{B}_1, \ldots, \boldsymbol{B}_t$ such that $\boldsymbol{B}_\tau \cap \boldsymbol{B}_{\tau+1} \neq \emptyset$. Since $\mathring{A}$ is empty, $(\boldsymbol{B}_\tau \setminus A) \cap (\boldsymbol{B}_{\tau+1} \setminus A) \neq \emptyset$, so $\bigcup_\tau (\boldsymbol{B}_\tau \setminus A)$ is a connected subset of V that contains a and b; i.e., $U \setminus A$ is connected. ■

In analogy to linear algebra, one can measure the smallness of an analytic set A in comparison to the surrounding space.

7.5 Definition. *An analytic set $A \subset X$ has codimension s at $a \in A$ (in symbols, $s = \operatorname{codim}_a A$) if there exists an s-dimensional, but no $(s+1)$-dimensional, affine subspace Γ of $\mathbb{C}^n$ such that a is an isolated point of $\Gamma \cap A$. For nonempty A, we define*

$$\operatorname{codim} A := \min_{a \in A} \operatorname{codim}_a A.$$

7.6 Remark. *If A is a proper analytic subset of a domain G, then* $\operatorname{codim} A \geqq 1$.

Proof. Without loss of generality, let G be a ball $\boldsymbol{B}$ about $a \in A$. If Γ is a complex line joining a to some $b \in \boldsymbol{B} \setminus A$, then the proper analytic subset $A \cap \Gamma \cap \boldsymbol{B}$ of the one-dimensional set $\Gamma \cap \boldsymbol{B}$ is discrete; hence, $\operatorname{codim}_a A \geqq 1$. ■

Proof of 7.3. By 7.4, $G\backslash A$ is dense in G; thus, an $f \in \mathcal{O}(G\backslash A)$ can have at most one holomorphic extension to G, so the algebra-homomorphism i^0 is injective. Thus, in order to prove that i^0 is surjective, it suffices to construct an extension $\check{f}$ near a fixed point $a \in A$. For $f \in \mathcal{O}(G\backslash A)$, we may assume that the following statements hold (with respect to the decomposition $\mathbb{C}^n = \mathbb{C}^{n-1} \times \mathbb{C}$, $z = (z', z_n)$):

i) $a = 0$.

ii) 0 is an isolated point of $A \cap (0 \times \mathbb{C})$ (by 7.6).

iii) There is a polydisk $\boldsymbol{P} := \boldsymbol{P}^{n-1}(r) \times \boldsymbol{P}^1(r) \subset\subset G$ such that $A \cap (0 \times \overline{\boldsymbol{P}^1(r)}) = \{0\}$; in particular, the set $A \cap (0 \times \boldsymbol{T}^1(r))$ is empty.

iv) $\|f\|_{\boldsymbol{P}\backslash A} < \infty$.

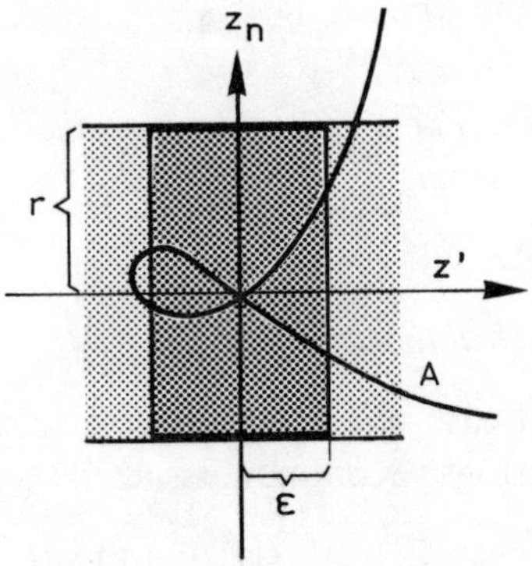

By iii), for $\boldsymbol{T}^1 := \boldsymbol{T}^1(r)$, the distance from the compact set $0 \times \boldsymbol{T}^1$ to the closed set A is positive; hence, there exists an $\varepsilon > 0$ such that $\boldsymbol{P}' \times \boldsymbol{T}^1 \subset \boldsymbol{P}\backslash A$, with $\boldsymbol{P}' := \boldsymbol{P}^{n-1}(\varepsilon)$. Set $\boldsymbol{P}^1 := \boldsymbol{P}^1(r)$; by the one-dimensional Identity Theorem, for each $b' \in \boldsymbol{P}'$, the intersection of A and $b' \times \boldsymbol{P}^1$ consits of isolated points. On $\boldsymbol{P}' \times \boldsymbol{P}^1$, the assignment

$$(z', z_n) \mapsto \frac{1}{2\pi i} \int_{|\zeta| = r} \frac{f(z', \zeta)}{\zeta - z_n} d\zeta$$

determines a continuous function $\check{f}$ that is partially holomorphic in (z', z_n), since integration and differentiation with respect to each $\bar{z}_j$ may be carried out in any order. For each $b' \in \boldsymbol{P}'$, the restriction $f|_{b' \times \boldsymbol{P}^1}$ has only removable singularities, by iv); thus, by the Cauchy Integral Formula, $\check{f}$ and f agree outside of A.

By 3.4, i^0 is continuous; by E. 2g i) and 4.5, the continuity of $(i^0)^{-1}$ follows from the local estimate

$$\|\check{f}\|_{\boldsymbol{P}' \times \boldsymbol{P}^1} \leqq \sup_{z' \in \boldsymbol{P}'} \|\check{f}\|_{z' \times \boldsymbol{P}^1} \leqq \sup_{z' \in \boldsymbol{P}'} \|f\|_{z' \times \boldsymbol{T}^1} \leqq \|f\|_{\boldsymbol{P}' \times \boldsymbol{P}^1 \backslash A}. \quad \blacksquare$$

In the following important consequence of 7.3, we encounter for the first time a phenomenon that is alien to the one-dimensional theory:

7.7 Second Riemann Removable Singularity Theorem. *If the analytic set A in X*

has codimension at least 2, *then the restriction-mapping* $\mathcal{O}(X) \to \mathcal{O}(X \setminus A)$ *is an isomorphism of topological algebras.*

Proof. By 7.3, it suffices, given $f \in \mathcal{O}(X \setminus A)$, to show that f is locally bounded at each $a \in A$. Without loss of generality, we may assume that $a = 0$, and that $A \cap [0 \times \overline{\boldsymbol{P}^2(r)}] = \{0\}$ for an appropriate polydisk $\boldsymbol{P} := \boldsymbol{P}^{n-2}(r) \times \boldsymbol{P}^2(r) \subset\subset X$.

Then $0 \times (\overline{\boldsymbol{P}^2(r) \setminus \boldsymbol{P}^2(r/2)}) \subset \bar{\boldsymbol{P}} \setminus A$; as in the proof of 7.3, there exists an ε such that $K := \overline{\boldsymbol{P}^{n-2}(\varepsilon)} \times (\overline{\boldsymbol{P}^2(r) \setminus \boldsymbol{P}^2(r/2)}) \subset \bar{\boldsymbol{P}} \setminus A$. In 7.8, we prove that for fixed $w \in \boldsymbol{P}^{n-2}(\varepsilon)$, the holomorphic function of two variables $f_w := f|_{w \times [\boldsymbol{P}^2(r) \setminus \overline{\boldsymbol{P}^2(r/2)}]}$ has a holomorphic extension $g_w \in \mathcal{O}(\boldsymbol{P}^2(r))$. Now 6.5 implies that $\| g_w \|_{\boldsymbol{P}^2(r)} \leqq \| f \|_K$; since we know that $g_w = f$ on $[w \times \boldsymbol{P}^2(r)] \setminus A$, we obtain the desired inequality

$$\| f \|_{[\boldsymbol{P}^{n-2}(\varepsilon) \times \boldsymbol{P}^2(r)] \setminus A} \leqq \| f \|_K < \infty . \quad \blacksquare$$

7.8 Proposition (Kugelsatz). *For* $n \geqq 2$ *and polyradii* $\varrho, \tilde{\varrho} \in \mathbb{R}^n_{>0}$ *such that* $\varrho_j > \tilde{\varrho}_j$, $j = 1, \ldots, n$, *the restriction-mapping*

$$\mathcal{O}(\boldsymbol{P}(\varrho)) \to \mathcal{O}(\boldsymbol{P}(\varrho) \setminus \overline{\boldsymbol{P}(\tilde{\varrho})})$$

is an isomorphism of topological algebras.

Proof. This simple special case of a much more general Kugelsatz 52.20 follows from an application of Hartogs's Kontinuitätssatz 7.9 to the decomposition

$$[(\boldsymbol{P}^{n-1}(\varrho') \setminus \overline{\boldsymbol{P}^{n-1}(\tilde{\varrho}')}) \times \boldsymbol{P}^1(\varrho_n)] \cup [\boldsymbol{P}^{n-1}(\varrho') \times (\boldsymbol{P}^1(\varrho_n) \setminus \overline{\boldsymbol{P}^1(\tilde{\varrho}_n)})] = \boldsymbol{P}(\varrho) \setminus \overline{\boldsymbol{P}(\tilde{\varrho})}$$

with $\varrho' := (\varrho_1, \ldots, \varrho_{n-1})$. $\blacksquare$

7.9 Hartogs's Kontinuitätssatz. *For* $n \geqq 2$, *let* Y *be a nonempty subregion of a domain* $H \subset\subset \mathbb{C}^{n-1}$, *and let* $r > \tilde{r} > 0$ *be real numbers. Then the restriction-mapping*

$$i^0 : \mathcal{O}(H \times \boldsymbol{P}^1(r)) \to \mathcal{O}([Y \times \boldsymbol{P}^1(r)] \cup [H \times (\boldsymbol{P}^1(r) \setminus \overline{\boldsymbol{P}^1(\tilde{r})})])$$

is an isomorphism of topological algebras.

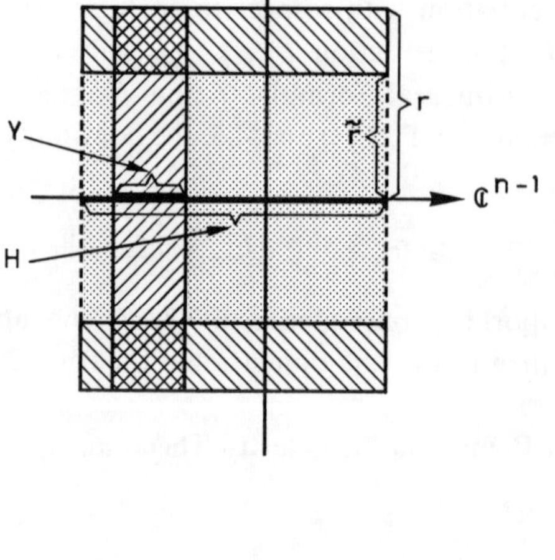

Proof. We argue as in the proof of 7.3: fix s so that $\tilde{r} < s < r$, and choose a function

$$f \in \mathcal{O}(Y \times \boldsymbol{P}^1(r) \cup H \times (\boldsymbol{P}^1(r) \setminus \overline{\boldsymbol{P}^1(\tilde{r})})) .$$

Then the function f_s defined by setting

$$f_s(z) := \frac{1}{2\pi i} \int_{|\zeta|=s} \frac{f(z_1, \ldots, z_{n-1}, \zeta)}{\zeta - z_n} d\zeta$$

is holomorphic on $H \times \boldsymbol{P}^1(s)$ and agrees with f on $Y \times \boldsymbol{P}^1(s)$; by the Identity Theorem, we have constructed a (unique) extension $\check{f}$. The continuity of $(i^0)^{-1}$ follows (see 7.3) from the inequality

$$\|\check{f}\|_{P^{n-1}(\varrho') \times P^1(s)} \leqq \|f\|_{P^{n-1}(\varrho') \times T^1(s)} . \quad \blacksquare\blacksquare$$

7.10 Corollary. *For $n \geqq 2$, holomorphic functions have neither isolated singularities nor isolated zeros.*

Proof. If $f \in \mathcal{O}(X \setminus a)$, then, by the Kugelsatz, f is holomorphically extendible to all of X. If $a \in X$ were an isolated zero of $f \in \mathcal{O}(X)$, then $\frac{1}{f} \in \mathcal{O}(X \setminus a)$ would not be holomorphically extendible at a. $\lightning$ $\blacksquare$

The next result is of fundamental importance for complex dimension theory.

7.11 Corollary. *If $G \subset \mathbb{C}^n$ is a domain and $0 \neq f \in \mathcal{O}(G)$ has a nonempty zero-set, then* $\operatorname{codim} N(f) = 1$.

Proof. By 7.6, it suffices to show that $\operatorname{codim} N(f) \leqq 1$. Suppose that $\operatorname{codim}_a N(f) \geqq 2$ for an $a \in N(f)$; that means that there exists a two-dimensional affine plane Γ in $\mathbb{C}^n$ such that a is an isolated point of $N(f) \cap \Gamma$. Then $f|_{\Gamma \cap G}$ has an isolated zero, in contradiction to 7.10. $\blacksquare$

E. 7b. Show by example that neither 7.10 nor 7.11 holds for every $f \in \mathbb{R}[x_1, \ldots, x_n]$.

E. 7c. *Kontinuitätssatz.* Let G be a domain in $\mathbb{C}^n = \mathbb{C}^{n-1} \times \mathbb{C}$ that includes $0 \times \boldsymbol{T}^1(r)$, and for which there exist arbitrarily small $\varepsilon > 0$ such that $G \cap [\boldsymbol{P}^{n-1}(\varepsilon) \times \boldsymbol{P}^1(r)]$ is connected. Suppose that there exists a sequence of points $({}_ja) = ({}_ja', {}_ja_n)_{j \in \mathbb{N}}$ converging to 0 such that each $\{{}_ja'\} \times \overline{\boldsymbol{P}^1({}_ja_n; r)}$ is included in G. Show that there exists an open neighborhood U of $0 \times \overline{\boldsymbol{P}^1(r)}$ such that the restriction-mapping $i^0: \mathcal{O}(G \cup U) \to \mathcal{O}(G)$ is an isomorphism.

E. 7d. If $\varphi: X_1 \to X_2$ is a holomorphic mapping, then its *graph*, $\Gamma(\varphi) := \{(z, \varphi(z)) \in X_1 \times X_2\}$, is an analytic set in $X_1 \times X_2$. Factor φ over a holomorphic bijection $\psi: X_1 \to \Gamma(\varphi)$. (For 32B. 8; in the language of 8.8, $\Gamma(\varphi)$ is a submanifold; ψ is actually biholomorphic (see §32).)

E. 7e. *Laurent series.* For domains $H \Subset \mathbb{C}^{n-1}$ and $\boldsymbol{R} := \{z \in \mathbb{C}^1; a < |z| < b\} \Subset \mathbb{C}^1$, show that every function $f \in \mathcal{O}(\boldsymbol{R} \times H)$ has a unique representation of the form $f = \sum_{j=-\infty}^{\infty} g_j z^j$ with $g_j \in \mathcal{O}(H)$ such that the series converges absolutely in $\mathscr{C}(\boldsymbol{R} \times H)$. (Hint: With the help of the one-dimensional Laurent series, construct a representation of the form

$$f(w,z) = \frac{1}{2\pi i}\int_{\gamma_1} \frac{f(w,u)}{u-z}\,du - \frac{1}{2\pi i}\int_{\gamma_2} \frac{f(w,u)}{u-z}\,du\,.)$$

§ 8 The Implicit Mapping Theorem

In the real differential calculus, linear approximation is one of the fundamental techniques. With its help, the investigation of nonlinear mappings is led back to the methods and results of linear algebra. In our discussion of the adaptation of that method to complex calculus, Y is a region in $\mathbb{C}^m$ and $f = (f_1, \dots, f_m): X \to Y$ is a holomorphic mapping.

8.1 Definition. *For* $f \in \mathrm{Hol}(X, Y)$, *let* $\dfrac{\partial f}{\partial z}$ *denote the mapping*

$$X \to \mathrm{Hom}(\mathbb{C}^n, \mathbb{C}^m),\ a \mapsto \frac{\partial f}{\partial z}(a) = \left(\frac{\partial f_j}{\partial z_k}(a)\right)_{\substack{1 \leqq j \leqq m \\ 1 \leqq k \leqq n}}.$$

The matrix $\dfrac{\partial f}{\partial z}(a)$ *is called the* <u>*holomorphic functional matrix*</u> *(or* <u>*Jacobian matrix*</u>*) of* f *at* a.

Note that, for such an f, the (*a priori* $\mathbb{R}$-linear) derivative $d_a f: \mathbb{C}^n \to \mathbb{C}^m$ is $\mathbb{C}$-linear, as it satisfies the equation

$$d_a f = \sum_{j=1}^{n} \frac{\partial f}{\partial z_j}(a)\, dz_j$$

(see E. 1a v)); hence, $\dfrac{\partial f}{\partial z}(a)$ is the matrix that describes $d_a f$ with respect to the canonical bases of $\mathbb{C}^n$ and $\mathbb{C}^m$. Thus, the usual computational rules for derivatives in real analysis remain valid for holomorphic functional matrices; in particular, we have the following fact:

8.2 Chain Rule. *If* $g \in \mathrm{Hol}(Y, Z)$, *then* $\dfrac{\partial (g \circ f)}{\partial z} = \left(\dfrac{\partial g}{\partial w} \circ f\right) \cdot \dfrac{\partial f}{\partial z}$. ■

For $n = m$, the determinant

$$J_f(a) := \det \frac{\partial f}{\partial z}(a)$$

exists; it is called the *complex functional determinant* (or *Jacobian determinant*) of f at a. With that notation, we have the following fact:

$$\frac{\partial f}{\partial z}(a) \text{ is an isomorphism} \Leftrightarrow J_f(a) \neq 0\,.$$

8.3 Inverse Mapping Theorem. *Suppose that $f \in \mathrm{Hol}(X, \mathbb{C}^n)$ and $a \in X \subset\subset \mathbb{C}^n$; then f is a biholomorphic mapping from an open neighborhood of a onto an open neighborhood of $f(a)$ iff $J_f(a) \neq 0$.*

Proof. Only if. If g is the inverse mapping near $f(a)$, then $J_g(f(a)) \cdot J_f(a) = J_{\mathrm{id}}(a) = \mathrm{id}$. *If.* The hypothesis $J_f(a) \neq 0$ implies that $d_a f$ is invertible; hence, by the real version of the Inverse Mapping Theorem, we may assume that the mapping $f: X \to f(X)$ has an inverse mapping $g: f(X) \to X$ with continuous *real* derivatives. In order to prove that g is holomorphic, we verify the Cauchy-Riemann differential equation $\frac{\partial g}{\partial \bar{w}} = 0$ (see E. 1a). The identity mapping $\mathrm{id}_{\mathbb{C}^n} = f \circ g$ is holomorphic, so the Chain Rule E. 1a implies that

$$0 = \frac{\partial (f \circ g)}{\partial \bar{w}} = \left(\frac{\partial f}{\partial z} \circ g\right) \cdot \frac{\partial g}{\partial \bar{w}};$$

since $\frac{\partial f}{\partial z} \circ g$ is invertible, we have that $\frac{\partial g}{\partial \bar{w}} = 0$. ■

E. 8a. Let the mapping $f: X \to \mathbb{C}^n$ have partial derivatives with respect to the real variables x_j and y_j, and let $f_j = g_j + ih_j$ denote the decomposition of the component f_j of f into real and imaginary parts. For

$$\Delta_f := \begin{pmatrix} \frac{\partial f_j}{\partial z_k} & \frac{\partial f_j}{\partial \bar{z}_k} \\ \frac{\partial \bar{f}_j}{\partial z_k} & \frac{\partial \bar{f}_j}{\partial \bar{z}_k} \end{pmatrix}_{1 \leq j,k \leq n} \quad \text{and} \quad D_f := \begin{pmatrix} \frac{\partial g_j}{\partial x_k} & \frac{\partial g_j}{\partial y_k} \\ \frac{\partial h_j}{\partial x_k} & \frac{\partial h_j}{\partial y_k} \end{pmatrix}_{1 \leq j,k \leq n},$$

show that
a) $\det \Delta_f = \det D_f$, and
b) if f is holomorphic, then $\det \Delta_f = |J_f|^2$.

E. 8b. Give an example showing that the mapping f in 8.3 is not necessarily biholomorphic on all of X.

8.4 Example. For the mapping $\varphi: \mathbb{C}^3 \to \mathbb{C}^3$, $z \mapsto (z_1, z_1 z_2 + z_3, z_1 z_2^2 - z_2 + 2 z_2 z_3)$, of 6.9, $J_\varphi(z) = 1 - 2z_3$; hence, for $0 < \varepsilon < 1/2$, φ is locally invertible on $\boldsymbol{P} := \boldsymbol{P}^3(1+\varepsilon, 1+\varepsilon, \varepsilon)$. If $\varphi: \boldsymbol{P} \to \varphi(\boldsymbol{P})$ were not biholomorphic for small ε, then there would exist two convergent sequences $({}_j a)$ and $({}_j b)$ in $\boldsymbol{P}$ with ${}_j a \neq {}_j b$, $\lim {}_j a_3 = 0 = \lim {}_j b_3$, and $\varphi({}_j a) = \varphi({}_j b)$. It is easy to see that the two sequences would have the same limit, so φ would not be bijective near a.

8.5 Theorem. *For $X \subset\subset \mathbb{C}^n$, if $f: X \to \mathbb{C}^n$ is a holomorphic injection, then $f(X)$ is open in $\mathbb{C}^n$, and f is a biholomorphic mapping from X to $f(X)$.*

Theorem 8.5 is equivalent to the following statement:

If $f \in \mathrm{Hol}(X, \mathbb{C}^n)$ is injective, then $J_f(z) \neq 0$ for every $z \in X$.

For, by 8.3, if that statement holds, then f, being injective, maps X biholomorphically onto $f(X)$; the reverse implication follows from 8.3. We forego the proof of 8.5 at this point, because a general version of the theorem will be proved in 46 A.1.

E. 8c. Give an example showing that 8.5 does not hold for real analytic functions.

Let $f\colon X \times Y \to \mathbb{C}^p$ be holomorphic. Then, for fixed $a \in X$, the mapping

$$f(a,\cdot)\colon Y \to \mathbb{C}^p,\ w \mapsto f(a,w),$$

is also holomorphic.

8.6 Implicit Mapping Theorem. *Suppose that $X \times Y \subset \mathbb{C}^n \times \mathbb{C}^m$, and that $f\colon X \times Y \to \mathbb{C}^m$ is holomorphic; let a point $(a, w_0) \in N(f)$ such that $\operatorname{rank} \dfrac{\partial f}{\partial w}(a, w_0) = m$ be given. Then there exists an open neighborhood $U \times W$ of (a, w_0) such that the (a priori set-valued) assignment*

$$u \mapsto \{w \in W; f(u,w) = 0\}$$

determines a holomorphic mapping $g\colon U \to W$.

Proof. After shrinking X and Y if necessary, we see by 8.3 that the holomorphic mapping

$$X \times Y \to \mathbb{C}^{n+m},\ (z,w) \mapsto (z, f(z,w)),$$

induces a biholomorphic mapping $F\colon X \times Y \to V$ with an inverse mapping $H = (\mathrm{pr}_n, H_Y)$. We may assume that V includes $X \times 0$; then the mapping $h := H_Y|_{X \times 0}$ is holomorphic, and we conclude that

$$f(z,w) = 0 \quad \text{iff} \quad w = h(z). \quad \blacksquare$$

As in the real case, those holomorphic functions whose functional matrices have constant rank admit a canonical representation:

8.7 Rank Theorem. *Let $f\colon X \to \mathbb{C}^m$ be a holomorphic mapping for which $\dfrac{\partial f}{\partial z}$ has constant rank r near a. Then there exist open neighborhoods U of a and V of $b = f(a)$, polydisks $\boldsymbol{P}^n \subset \mathbb{C}^n$ and $\boldsymbol{P}^m \subset \mathbb{C}^m$, each centered at 0, and biholomorphic mappings $\varphi\colon \boldsymbol{P}^n \to U$ and $\psi\colon V \to \boldsymbol{P}^m$ with $\varphi(0) = a$ and $\psi(b) = 0$ such that, with $\chi(z_1, \ldots, z_n) = (z_1, \ldots, z_r, 0, \ldots, 0)$, the diagram*

$$\begin{array}{ccc} U & \xrightarrow{\ f\ } & V \\ \uparrow{\scriptstyle\varphi} & & \downarrow{\scriptstyle\psi} \\ \boldsymbol{P}^n & \xrightarrow{\ \chi\ } & \boldsymbol{P}^m \end{array}$$

commutes.

Proof. Without loss of generality, set $a = b = 0$; moreover, let the coordinates of

$\mathbb{C}^n$ and $\mathbb{C}^m$ be chosen in such a manner that $\dfrac{\partial f}{\partial z}(0)$ has the matrix representation

$$\frac{\partial f}{\partial z}(0) = \begin{pmatrix} I_r & 0 \\ 0 & 0 \end{pmatrix}.$$

Then, for the mapping

$$g: X \to \mathbb{C}^n,\ z \mapsto (f_1(z), \ldots, f_r(z), z_{r+1}, \ldots, z_n),$$

we obviously have that $\dfrac{\partial g}{\partial z}(0) = I_n$. By 8.3, there exists an open neighborhood U of 0 in $\mathbb{C}^n$ that is mapped biholomorphically by g onto a polydisk $\boldsymbol{P}^n$; set $\varphi := (g|_U)^{-1}$. For $w \in \boldsymbol{P}^n$ and $z := \varphi(w)$, we have that

$$(f_1, \ldots, f_m)(z) = f(z) = f \circ \varphi(w) =: (w_1, \ldots, w_r, h_{r+1}(w), \ldots, h_m(w)),$$

where, in addition to f and φ, every h_j is holomorphic. The mapping $f \circ \varphi$ satisfies the inequality $\operatorname{rank} \dfrac{\partial(f \circ \varphi)}{\partial w} \geqq r$ on $\boldsymbol{P}^n$. We may assume that $\operatorname{rank} \dfrac{\partial f}{\partial z} = r$ on $\boldsymbol{P}^n$; thus, by 8.2,

$$\operatorname{rank} \frac{\partial(f \circ \varphi)}{\partial w} = r, \text{ so } \frac{\partial h_j}{\partial w_k} = 0 \ \forall j, k \geqq r+1.$$

Hence, the h_j's do not depend on the variables $w_{r+1}, \ldots, w_n$; their restrictions to the first r components determine a mapping $h: \boldsymbol{P}^r \to \mathbb{C}^{m-r}$. The bijective mapping

$$\gamma: \boldsymbol{P}^r \times \mathbb{C}^{m-r} \to \boldsymbol{P}^r \times \mathbb{C}^{m-r},\ (u, v) \mapsto (u, v - h(u)),$$

has the functional matrix (see E. 8d)

$$\begin{pmatrix} I_r & 0 \\ * & I_{m-r} \end{pmatrix};$$

by 8.3, γ is biholomorphic. Now choose a sufficiently large polydisk $\boldsymbol{P}^{m-r}$ in $\mathbb{C}^{m-r}$ such that

$$\gamma \circ f \circ \varphi(\boldsymbol{P}^n) \subset \boldsymbol{P}^r \times \boldsymbol{P}^{m-r} =: \boldsymbol{P}^m;$$

for $V := \gamma^{-1}(\boldsymbol{P}^m)$ and $\psi := \gamma|_V$, we conclude that

$$\psi \circ f \circ \varphi(w) = \gamma(w_1, \ldots, w_r, h_{r+1}(w), \ldots, h_m(w)) = (w_1, \ldots, w_r, 0, \ldots, 0) = \chi(w). \quad \blacksquare$$

E. 8d. If $f_j: X_j \to Y_j$, $j = 1,2$, are holomorphic mappings, then the mapping $f_1 \times f_2: X_1 \times X_2 \to Y_1 \times Y_2$ satisfies the equality

$$\frac{\partial(f_1 \times f_2)}{\partial(z_1, z_2)} = \begin{pmatrix} \dfrac{\partial f_1}{\partial z_1} & 0 \\ 0 & \dfrac{\partial f_2}{\partial z_2} \end{pmatrix}.$$

Finally, as an application of the Rank Theorem, we wish to discuss regularity statements for analytic sets. The central issue is the concept of a submanifold; here, that means an analytic set that is locally biholomorphically embeddable as a linear subspace in $\mathbb{C}^n$.

8.8 Definition. *A closed subset T of X is called a submanifold of X if, for every point $a \in T$, there exist an open neighborhood U of a in X and a biholomorphic mapping $\phi: U \to \boldsymbol{P}$ onto a polydisk $\boldsymbol{P}$ with center $0 = \phi(a)$ in $\mathbb{C}^n$, such that, with respect to a decomposition $\boldsymbol{P} = \boldsymbol{P}^s \times \boldsymbol{P}^{n-s}$,*

$$T \cap U = \phi^{-1}(\boldsymbol{P}^s \times 0).$$

By E. 8e, the number s is determined by $a \in T$; it is called the *dimension* of T at a, and is denoted by $\dim_a T$.

E. 8e. *If $f: X \to Y$ is biholomorphic, then* $\dim X = \operatorname{rank} \dfrac{\partial f}{\partial z} = \dim Y$.

Submanifolds are analytic sets:

8.9 Proposition. *Let $T \subset X$ be closed; then T is a submanifold of X iff, for each $a \in T$, there exist a neighborhood $U \subset\subset X$ of a and a mapping $f \in \operatorname{Hol}(U, \mathbb{C}^m)$ such that i) $U \cap T = N(U; f)$ and ii) $\dfrac{\partial f}{\partial z}$ has constant rank on U.*

If T is a submanifold of X, then $\dim_a T = n - \operatorname{rank} \dfrac{\partial f}{\partial z}(a)$.

Proof. Only if. For $a \in T$, choose a biholomorphic mapping $\phi: U \to \boldsymbol{P}^n$ in accordance with 8.8. Set $f := \pi \circ \phi$, where π denotes the canonical projection $\pi: \boldsymbol{P}^n = \boldsymbol{P}^s \times \boldsymbol{P}^{n-s} \to \boldsymbol{P}^{n-s}$; by 8.2,

$$\operatorname{rank} \frac{\partial f}{\partial z} = \operatorname{rank} \frac{\partial(\pi \circ \phi)}{\partial z} = \operatorname{rank} \frac{\partial \pi}{\partial w} = n - s$$

on U. Moreover, $N(U; f) = \phi^{-1}(N(\boldsymbol{P}^n; \pi)) = \phi^{-1}(\boldsymbol{P}^s \times 0)$.

If. Let r denote the rank of $\dfrac{\partial f}{\partial z}$ on the neighborhood U of a. According to 8.7, we may assume that a commutative diagram of the form

$$\begin{array}{ccc} U & \xrightarrow{\ f\ } & V \\ \varphi \uparrow \cong & & \cong \downarrow \psi \\ \boldsymbol{P}^n & \xrightarrow{\ \chi\ } & \boldsymbol{P}^m \end{array}$$

exists; then $\phi := \varphi^{-1}$ is biholomorphic, and

$$T \cap U = N(U; f) = N(U; \psi \circ f) = N(U; \chi \circ \phi) = \phi^{-1}(N(\boldsymbol{P}^n; \chi)) = \\ = \phi^{-1}(0 \times \boldsymbol{P}^{n-r}). \quad \blacksquare$$

It is possible to characterize submanifolds locally not only as zero-sets of holomorphic mappings, but also as images of holomorphic mappings.

8.10 Proposition. *Let $T \subset X$ be closed. Then T is a submanifold iff, for each $U \subset\subset T$ and each $a \in U$, there exist a polydisk $\boldsymbol{P}^k$ and a mapping $f \in \mathrm{Hol}(\boldsymbol{P}^k, U)$ such that*

i) $f(0) = a$,

ii) $U \cap T = f(\boldsymbol{P}^k)$, *and*

iii) $\dfrac{\partial f}{\partial z}$ *has constant rank.*

In that case, rank $\dfrac{\partial f}{\partial z}(a) = \dim_a T$*; that value is also the minimum of the k's that appear.*

Proof. Only if. For $a \in T$ and $s := \dim_a T$, choose, in accordance with 8.8, a biholomorphic mapping $\phi: U \to \boldsymbol{P}^n$ such that $T \cap U = \phi^{-1}(\boldsymbol{P}^s \times 0)$. Then $f := \phi^{-1}|_{\boldsymbol{P}^s}$ is holomorphic, and the functional matrix has constant rank s. *If.* Suppose that rank $\dfrac{\partial f}{\partial z} = r$ on U. By 8.7, we may assume that there is a commutative diagram

$$\begin{array}{ccc} \boldsymbol{P}^k & \xrightarrow{\;f\;} & X \\ \varphi \uparrow \cong & & \cong \downarrow \psi \\ \boldsymbol{P}' & \xrightarrow{\;\chi\;} & \boldsymbol{P}^n \end{array}$$

Then $T = f(\boldsymbol{P}^k) = f \circ \varphi(\boldsymbol{P}') = \psi^{-1}\chi(\boldsymbol{P}')$; due to the nature of the construction of χ, we have that $\chi(\boldsymbol{P}') = \boldsymbol{P}^r \times 0 \subset \boldsymbol{P}^n$. In particular, it follows that $r = \dim_a T$. ■

E. 8f. *Semicontinuity of the rank of holomorphic mappings.* Prove the following statements for $X \subset\subset \mathbb{C}^n$, $f \in \mathrm{Hol}(X, \mathbb{C}^m)$, and $k \in \mathbb{N}$:

i) $A_k := \{a \in X; \operatorname{rank} \dfrac{\partial f}{\partial z}(a) \leq k\}$ is analytic in X.

ii) If f is discrete (see § 33 B), then $A_{n-1} \neq X$.

iii) The converse of statement ii) is false.

(Hints: minors of $\dfrac{\partial f}{\partial z}$ in i), the Rank Theorem in ii), and $f(z_1, z_2) = (z_1, z_1 z_2)$ in iii); for 33.7.)

Chapter 1: Regions of Holomorphy

In the introductory chapter, we encountered a phenomenon that is unknown in the one-dimensional theory: every $f \in \mathcal{O}(\mathbb{C}^2 \backslash 0)$ can be extended holomorphically to $\mathbb{C}^2$; in other words, $\mathbb{C}^2 \backslash 0$ is not the *region of holomorphy* of any function. Before formulating and proving general versions of deeper one-dimensional results, such as the Mittag-Leffler Theorem and the Weierstrass Factorization Theorem, we consider a more elementary question: which regions $X \subset\subset \mathbb{C}^n$ are regions of holomorphy? By imparting an understanding of the essential differences from the one-dimensional theory, the answer should serve as a correction to the picture presented in Chapter 0.

We begin by investigating the domains of convergence of power series, which turn out to be those complete Reinhardt domains which have the additional property of being logarithmically convex. *Convexity* proves to be the essential property for the study of regions of holomorphy in general. We extend the concept of elementary convexity for both regions and functions. For the first generalization, we consider holomorphically convex regions, and obtain, as an answer to our question, the *Theorem of Cartan-Thullen*: "X is a region of holomorphy iff X is holomorphically convex." For the second generalization, we introduce plurisubharmonic functions, with the help of which the concept of a pseudoconvex domain can be defined. In that way, we obtain a second answer: "X is a domain of holomorphy iff X is pseudoconvex." We put off the proof that "pseudoconvexity implies holomorphic convexity", known as *Levi's problem*, until Chapter 6; by that point, we shall have prepared a method that makes possible the step from "local" to "global", namely, cohomology theory.

Extension theorems for holomorphic functions are closely related to the theory of domains of holomorphy. One such theorem in particular, the "Edge-of-the-wedge" Theorem, is relevant for physical applications; hence, we delve briefly into its mathematical aspects.

§ 11 Domains of convergence of power series and Reinhardt domains

The domain of convergence of a power series of one variable is a disk. If $P = \sum_{\nu \in \mathbb{N}^n} a_\nu z^\nu$ is a power series in n variables, then Abel's Lemma implies that the set A of points in $\mathbb{C}^n$ at which $\Sigma a_\nu z^\nu$ converges absolutely and the set B of points in $\mathbb{C}^n$ at which $\Sigma a_\nu z^\nu$ converges conditionally have the same interior X; we call X the *domain of convergence* of P. Hence, P is convergent iff X is nonempty; for nonconvergent P, both A and B may contain nonzero points.

E. 11a. If $\boldsymbol{P}$ is the domain of convergence of a one-dimensional series $\sum_{j=0}^{\infty} a_j z^j$, and if $t \notin \bar{\boldsymbol{P}}$, then $\sum_{j=0}^{\infty} a_j t^j$ diverges. Show that that statement does not hold for higher dimensions, using $\sum_{j=0}^{\infty} z_1^j z_2$.

Our first goal is a characterization of such domains of convergence X. Consider the mapping

$$\tau: \mathbb{C}^n \to \mathbb{R}^n_{\geq 0}, \; z \mapsto (|z_1|, \dots, |z_n|);$$

τ is continuous, proper (33 B. 1), open, and surjective. For each $\varrho \in \mathbb{R}^n_{>0}$, the inverse image $\tau^{-1}(\varrho) =: \boldsymbol{T}(\varrho)$ is a real n-dimensional torus.

11.1 Definition. *A domain $G \subset \mathbb{C}^n$ is called a* Reinhardt domain *if $G = \tau^{-1}\tau(G)$.*

A Reinhardt domain is called

i) proper if $G = \emptyset$ *or* $0 \in G$,

ii) complete if $\mathbf{P}(\tau(z)) \subset G$ *for each* $z \in G \cap \mathbb{C}^{*n}$,

iii) logarithmically convex if $\{(\log|z_1|, \ldots, \log|z_n|);\ z \in G \cap \mathbb{C}^{*n}\}$ *is a convex subset of* $\mathbb{R}^n$.

Hence, Reinhardt domains can be characterized in terms of their τ-images; in particular, such domains are circular. For $n = 2$, consider the following τ-images in the closed first quadrant, $\mathbb{R}^2_{\geqq 0}$:

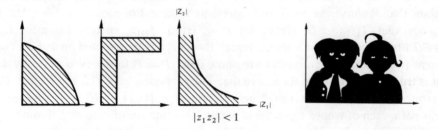

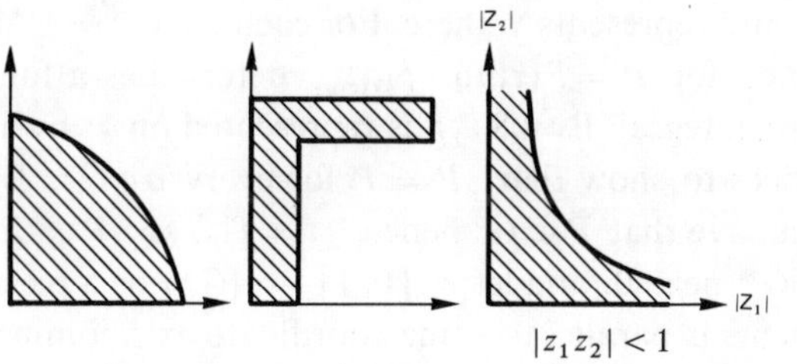

(Which of them represent complete (resp., logarithmically convex) Reinhardt domains?).

11.2 Proposition. *Domains of convergence of power series are logarithmically convex complete Reinhardt domains.*

Proof. If $X \subset\subset \mathbb{C}^n$ is the domain of convergence of $\Sigma a_\nu z^\nu$, then Abel's Lemma implies immediately that X is a complete Reinhardt domain. It remains to show that X is logarithmically convex. By Abel's Lemma, it suffices to prove that, for all $\varrho, \sigma, \chi \in \mathbb{R}^n_{>0}$ and $\alpha, \beta \in \mathbb{R}_{\geqq 0}$ such that $\alpha + \beta = 1$ and $\log \chi := (\log \chi_1, \ldots, \log \chi_n) = \alpha \log \varrho + \beta \log \sigma$, the following statement holds: if $|a_\nu| \varrho^\nu \leqq C$ and $|a_\nu| \sigma^\nu \leqq C$ for every ν, then $|a_\nu| \chi^\nu \leqq C$ for every ν. That statement follows from the inequality

$$\log \chi^\nu = \sum_{j=1}^{n} \nu_j \log \chi_j = \alpha \Sigma \nu_j \log \varrho_j + \beta \Sigma \nu_j \log \sigma_j = \alpha \log \varrho^\nu + \beta \log \sigma^\nu$$

$$\leqq \alpha \log \frac{C}{|a_\nu|} + \beta \log \frac{C}{|a_\nu|} = \log \frac{C}{|a_\nu|}. \quad \blacksquare$$

11.3 Remark. In 12.4 v), we prove that every logarithmically convex complete Reinhardt domain is the domain of convergence of a power series.

If (G_α) is a family of complete Reinhardt domains, then the interior H of $\bigcap_\alpha G_\alpha$ is obviously a complete Reinhardt domain. If, in addition, every G_α is logarithmically convex, then H is logarithmically convex, since $\log \tau(\bigcap_\alpha G_\alpha) = \bigcap_\alpha \log \tau(G_\alpha)$ and since an intersection of convex sets is convex. In particular, for each Reinhardt domain G, there exists a smallest logarithmically convex complete Reinhardt domain containing G; it is called the *hull* $\check{G}$ of G.

11.4 Proposition. *Let $G \subset\subset \mathbb{C}^n$ be a proper Reinhardt domain. Then each $f \in \mathcal{O}(G)$ is represented by a power series $\check{f}$ that converges on $\check{G}$ (thus $\check{f}$ extends f holomorphically to $\check{G}$); in particular, G is a Runge domain.*

Proof. Let P denote the Taylor series of $f \in \mathcal{O}(G)$ about the origin; by 11.2, it suffices to show that P converges on G and represents f there. For each $a \in G^* := \mathbb{C}^{*n} \cap G$, we know that $\boldsymbol{T}(\tau(a)) \subset G$. Hence, for $\boldsymbol{P} := \boldsymbol{P}(\tau(a))$, $f|_{\boldsymbol{T}(\tau(a))}$ determines a function ${}_af \in \mathcal{O}(\boldsymbol{P})$ by means of the Cauchy integral. By 4.8, ${}_af$ is represented on $\boldsymbol{P}$ by a convergent power series ${}_aP$; it suffices to show that ${}_aP = P$ for every $a \in G^*$. By 4.1, that is true for a near 0: we then have that $\bar{\boldsymbol{P}} \subset G$; hence, ${}_af = f|_{\boldsymbol{P}}$, so ${}_aP = P$. For an arbitrary $b \in G^*$, choose $a \in G^*$ near 0, and let $\sigma: [0, 1] \to \tau(G^*)$ be a piecewise linear path, each of whose segments is parallel to some coordinate axis, joining $\tau(a)$ and $\tau(b)$. Then ${}_{\sigma(t)}P = P$ for each $t \in [0,1]$, since only one variable changes on each segment, and since the one-dimensional Cauchy integral $\int_{|\zeta| = \varrho} \frac{h(\zeta)}{\zeta - z} d\zeta$ is independent of ϱ for holomorphic h. The fact that ${}_aP = P$ implies that ${}_bP = P$. ■

E. 11b. Determine $\check{G}$ for the following two domains:

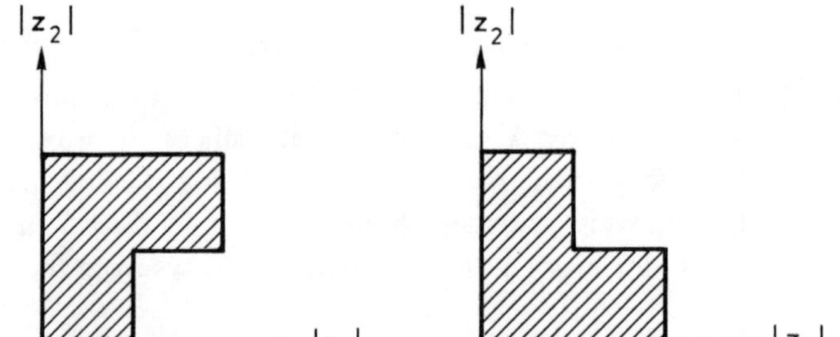

In general, a proper Reinhardt domain G is a proper subset of its hull, $\check{G}$. For $n \geqq 2$, a particularly important example is the n-dimensional *Hartogs figure*

$$H_n := \{z \in \boldsymbol{P}^n(1); |z_1| \leqq \tfrac{1}{2} \Rightarrow |z_j| < \tfrac{1}{2} \text{ for } j = 2, \dots, n\}$$
$$= (\overline{\boldsymbol{P}^1(\tfrac{1}{2}) \times \boldsymbol{P}^{n-1}(\tfrac{1}{2})}) \cup (\{\zeta \in \mathbb{C}; \tfrac{1}{2} < |\zeta| < 1\} \times \boldsymbol{P}^{n-1}(1)).$$

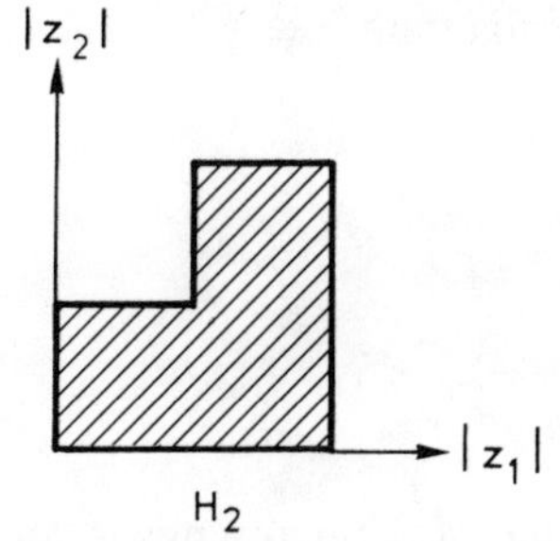

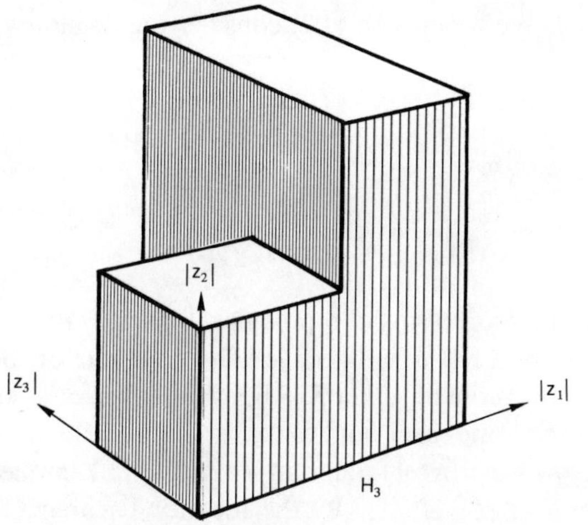

The pair $(\boldsymbol{P}^n(1), H_n)$ is called the n-dimensional *standard (Euclidean) Hartogs configuration*; it provides a model for the following more general concept:

11.5 Definition. *Let* $(\boldsymbol{P}, H)$ *be the n-dimensional standard Hartogs configuration. For each biholomorphic mapping g from* $\boldsymbol{P}$ *onto a domain* $\tilde{P} \Subset \mathbb{C}^n$, *the pair* $(\tilde{P}, \tilde{H}) := = (g(\boldsymbol{P}), g(H))$ *is called a (general) Hartogs configuration.*

11.6 Hartogs's Kontinuitätssatz. *Fix* $X \Subset \mathbb{C}^n$, *and suppose that* $(\tilde{P}, \tilde{H})$ *is a Hartogs configuration such that* $\tilde{P} \cap X$ *is connected and* $\tilde{H} \Subset X$. *Then the restriction-mapping* $\mathcal{O}(X \cup \tilde{P}) \to \mathcal{O}(X)$ *is an isomorphism of topological algebras. (Specifically, each* $f \in \mathcal{O}(X)$ *is holomorphically extendible to* $X \cup \tilde{P}$.)

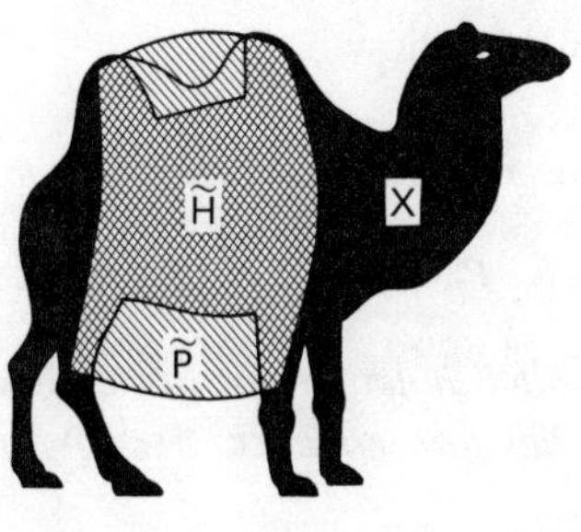

11.7 Corollary. *If* $(\tilde{P}, \tilde{H})$ *is a Hartogs configuration, then the restriction-mapping* $\mathcal{O}(\tilde{P}) \to \mathcal{O}(\tilde{H})$ *is an isomorphism of topological algebras.*

Proof. We begin with 11.7: consider the commutative diagram

$$\begin{array}{ccc} \mathcal{O}(\tilde{P}) & \xrightarrow{\quad \varrho \quad} & \mathcal{O}(\tilde{H}) \\ {\scriptstyle g^0} \Big\downarrow {\scriptstyle \cong} & & {\scriptstyle g^0} \Big\downarrow {\scriptstyle \cong} \\ \mathcal{O}(\mathbf{P}) & \xrightarrow{\quad \sigma \quad} & \mathcal{O}(H) \end{array}$$

of homomorphisms of topological algebras. By 3.4, each of the mappings is continuous; by 11.4, σ is bijective. The construction by means of the Cauchy integral, as in the proof of 7.3 (or 55.8 ii)), shows that σ is open. Thus σ is a homeomorphism, and the same must hold for ϱ.

Now we prove 11.6: for an $f \in \mathcal{O}(X)$, 11.7 implies the existence of a unique extension $\tilde{f} \in \mathcal{O}(\tilde{P})$ of $f|_{\tilde{H}}$. By the Identity Theorem, f and $\tilde{f}$ coincide on the domain $\tilde{P} \cap X$, so f determines a unique extension of f to $X \cup \tilde{P}$. It follows immediately from 11.7 that the restriction-mapping is open. ■

E. 11c. i) Show that $\mathbb{C} \times 0$ has a fundamental system of neighborhoods that are logarithmically convex complete Reinhardt domains in $\mathbb{C}^2$ (hint: in $\mathbb{R}^2_{\geqq 0}$, the set $(\mathbb{R}_{\geqq 0}) \times 0$ has a fundamental system of neighborhoods determined by decreasing step functions).

ii) Generalize i) to $\mathbb{C}^n \times 0$ in $\mathbb{C}^n \times \mathbb{C}^m$ (for E. 63e).

§ 12 Regions of Holomorphy and Holomorphic Convexity

In section 11, we investigated certain Reinhardt domains to see whether every holomorphic function on G could be extended to a domain that properly includes G. In this section, we consider the analogous question for arbitrary regions in $\mathbb{C}^n$; our discussion is couched in the following terms:

12.1 Definition. *A function $f \in \mathcal{O}(X)$ is called holomorphically extendible (from a point $a \in X$) to a polydisk $\mathbf{P}(a; \varrho)$ if its Taylor series, $\sum_{\nu \in \mathbb{N}^n} \frac{(D^\nu f)(a)}{\nu!}(z-a)^\nu$, converges on $\mathbf{P}(a; \varrho)$; it is called holomorphically extendible at a point $y \in \mathbb{C}^n \setminus X$ if, for some $a \in X$, the point y lies in a polydisk $\mathbf{P}(a; \varrho)$ to which f is holomorphically extendible.*

The fact that a function $f \in \mathcal{O}(X)$ is holomorphically extendible from $a \in X$ to $\mathbf{P} = \mathbf{P}(a; \varrho)$ does not necessarily imply that f is holomorphically extendible to $X \cup \mathbf{P}$, for the extension can depend on the given point a (that is the case, for example, with $f = \sqrt{\cdot} \in \mathcal{O}(\mathbb{C}^* \setminus \mathbb{R}_{<0})$); however, we can say the following:

E. 12a. If $X \cap \mathbf{P}$ is connected, then f is holomorphically extendible to $\mathbf{P}$ (in the sense of 12.1) iff f is holomorphically extendible to $X \cup \mathbf{P}$.

The principal result of this section is the equivalence, demonstrated in 12.8, of the following two concepts:

12.2 Definition. *A region (resp., domain) $X \subset\subset \mathbb{C}^n$ is called a region (resp., domain) of holomorphy if there exists an $f \in \mathcal{O}(X)$ that is not holomorphically extendible at any point y outside of X.*

12.3 Definition. *A region $X \subset\subset \mathbb{C}^n$ is called holomorphically convex if, for each compactum $K \subset X$, the "holomorphically convex hull of K in X",*

$$\hat{K}_{\mathcal{O}(X)} := \{x \in X;\, |f(x)| \leqq \|f\|_K \quad \forall f \in \mathcal{O}(X)\}\,,$$

is compact.

12.4 Remarks. i) The holomorphically convex hull $\hat{K}_{\mathcal{O}(X)}$ depends on the region X: for $K := S^1 \subset \mathbb{C}^* \subset \mathbb{C}$, we have that

$$\hat{K}_{\mathcal{O}(\mathbb{C}^*)} = K \underset{\neq}{\subseteq} \overline{\boldsymbol{P}^1(1)} = \hat{K}_{\mathcal{O}(\mathbb{C})}\,.$$

ii) $\hat{K}_{\mathcal{O}(X)}$ is closed in X (but not necessarily in $\mathbb{C}^n$) and bounded.

iii) For a one-dimensional region X and a compact set $K \subset X$, the hull $\hat{K}_{\mathcal{O}(X)}$ can be determined purely topologically: $\hat{K}_{\mathcal{O}(X)}$ is the union of K with those connected components of $X \backslash K$ which are relatively compact in X [Co VIII.1 Ex. 6].

iv) The concept "holomorphically convex" can be subsumed under the following general scheme, which illuminates the similarity to elementary convexity: for a set $D \subset X$, and a family $\mathfrak{F}$ of continuous functions on X,

$$\hat{D}_{\mathfrak{F}} := \{x \in X;\, |f(x)| \leqq \|f\|_D \forall f \in \mathfrak{F}\}$$

is called the *$\mathfrak{F}$-convex hull of D in X*, and X is called *$\mathfrak{F}$-convex* if $\hat{K}_{\mathfrak{F}}$ is compact for each compact $K \subset X$. In this book, the following types of convexity appear:

$\mathfrak{F} = \{f: \mathbb{C}^n \to \mathbb{R};\, f \text{ is real affine linear}\}$[1]	*elementary convexity if $X \subset\subset \mathbb{C}^n$* (proof?)
$\mathfrak{F} = \{m \in \mathbb{C}[z];\, m \text{ is a monomial}\}$	*monomial convexity*
$\mathfrak{F} = \mathbb{C}[z]$	*polynomial convexity*
$\mathfrak{F} = \mathcal{O}(X)$	*holomorphic convexity*

The pseudoconvexity in §14 can be described in a similar manner with the help of plurisubharmonic functions.

v) While elementary convexity is relatively easy to check, verification of the other types of convexity is usually difficult. However, the following case is simple:

For proper Reinhardt domains X the following statements are equivalent:

a) X is monomially convex.

b) X is polynomially convex.

c) X is holomorphically convex.

[1] That is, $f - f(0)$ is $\mathbb{R}$-linear.

d) X is a logarithmically convex complete Reinhardt domain.
e) X is the domain of convergence of a power series.

Proof. The implications "$a) \Rightarrow b) \Rightarrow c)$" are trivial.

$c) \Rightarrow e)$ By 12.8, since X is holomorphically convex, there exists an $f \in \mathcal{O}(X)$ that is at no $x \in \mathbb{C}^n \setminus X$ holomorphically extendible. According to 11.4, the Taylor series P of f at 0 converges on $\check{X}$; hence, $\check{X} = X$, and X is thus the domain of convergence of P.

The implication "$e) \Rightarrow d)$" was proved in 11.2. $d) \Rightarrow a)$ For compact $K \subset X$, we have to show that the monomially convex hull $\hat{K}_{M,X}$ is compact in X. Without loss of generality, let $K = \overline{\mathring{K}}$; then $K = \overline{K^*}$, where $K^* := K \cap \mathbb{C}^{*n}$. It is not difficult to verify that $\check{K}$ (the smallest logarithmically convex subset of $\mathbb{C}^n$ that includes both K and the sets $\overline{\boldsymbol{P}^n(a)}$ for $a \in \check{K}$) is a compact subset of $X = \check{X}$; hence, it suffices to prove the following statement:

(12.4.1) If $K = \overline{\mathring{K}} = \check{K} \subset \boldsymbol{P}^n(1)$ is compact, then $K = \hat{K}_{M,\mathbb{C}^n}$.

For then the compactness of $\check{K}$ implies that of

$$\hat{K}_{M,X} = \hat{K}_{M,\mathbb{C}^n} \cap X = K \cap X = K.$$

Proof of (12.4.1). It suffices to find for each $a \in \boldsymbol{P}^n(1) \setminus K$ a multi-index $\nu \in \mathbb{N}^n$ such that $\|z^\nu\|_K < |a^\nu|$. Set $L := \log \circ \tau$.

α) For $a \in \mathbb{C}^{*n}$, the point $L(a)$ does not lie in the convex set $L(K^*)$; hence, there exists a linear form $l = \sum_{j=1}^{n} \varrho_j x_j : \mathbb{R}^n \to \mathbb{R}$ such that $\sup l(L(K^*)) < l(L(a))$. Then, $l(L(a)) < 0$: for each $x \in L(K^*)$, there exists a positive integer k such that $k \cdot L(a) \in x + \mathbb{R}^n_{\leqq 0} \subset L(K^*)$, since $L(a)$ and $L(K^*)$ lie in $\mathbb{R}^n_{\leqq 0}$. Now $l|_{x + \mathbb{R}^n_{\leqq 0}} < l(L(a))$ for every $x \in L(K^*)$; it follows immediately that $l|_{\mathbb{R}^n_{\leqq 0}} \leqq 0$, and thus that $\varrho_j = l(0, \ldots, 0, 1, 0, \ldots, 0) \geqq 0$.

Since $L(K^*)$ is closed in $\mathbb{R}^n_{\leqq 0}$, all of the ϱ_j's can be chosen from $\mathbb{Q}_{\geqq 0}$. Finally, l can be replaced with $m \cdot l$ for any $m \in \mathbb{N}_{>0}$. Without loss of generality, set $\nu := (\varrho_1, \ldots, \varrho_n) \in \mathbb{N}^n$; for $z \in \mathbb{C}^{*n}$, then, $\log|z^\nu| = l(L(z))$, and it follows from the inequality

$$\log(\|z^\nu\|_{K^*}) = \sup_{b \in L(K^*)} l(b) < l(L(a)) = \log|a^\nu|$$

that $\|z^\nu\|_K = \|z^\nu\|_{K^*} < |a^\nu|$.

β) If $a \in \mathbb{C}^{*j} \times 0 \subset \mathbb{C}^n$, then, with the canonical projection $\mathrm{pr}: \mathbb{C}^n \to \mathbb{C}^j$, we have that $\mathrm{pr}(a) \notin \mathrm{pr}(K) = \widecheck{\mathrm{pr}(K)}$; as above, a multi-index $\mu \in \mathbb{N}^j$ can be determined so that $\|z^\mu\|_{\mathrm{pr}(K)} < |\mathrm{pr}(a)^\mu|$. For $\nu := (\mu, 0) \in \mathbb{N}^j \times 0 \subset \mathbb{N}^n$, then,

$$\|z^\nu\|_K = \|z^\mu\|_{\mathrm{pr}(K)} < |\mathrm{pr}(a)^\mu| = |a^\nu|. \quad \blacksquare$$

The next exercise explains why the extension of $f \in \mathcal{O}(X)$ at $y \in \mathbb{C}^n \setminus X$ was formulated in such a complicated manner. The phenomenon of a "many-valued" function, which appears in E. 12b, led to the definition of the (concrete) Riemann surface in the one-dimensional theory. The higher-dimensional generalization is called a *Riemann domain*; in the notation of § 33 B, that is a holomorphic covering $\pi: M \to G \Subset \mathbb{C}^n$ with a connected manifold M. In E. 30c and in § 37, we delve briefly into this extension of the construction of domains of holomorphic functions to include domains that do not lie in $\mathbb{C}^n$.

E. 12b. Consider the polydisks $\boldsymbol{P} := \boldsymbol{P}^1(1)$ and $\boldsymbol{Q} := \boldsymbol{P}^1(1; \frac{1}{4})$, and let $S \Subset \mathbb{C}$ be the simply connected domain depicted in the accompanying figure ($S \cap \partial \boldsymbol{Q} \cap \boldsymbol{P} = \emptyset$).

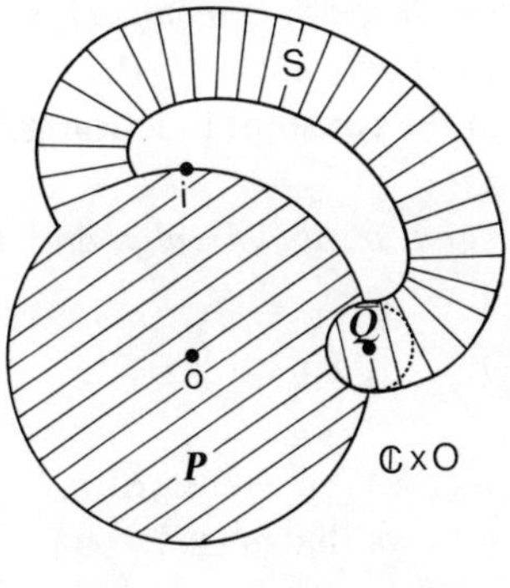

Prove the following statements for $X_1 := S \times \{|z_2| < \frac{1}{4}\}$, $X_2 := (\boldsymbol{P} \setminus \bar{\boldsymbol{Q}}) \times \{|z_2| = \frac{1}{4}\}$, $X_3 := \boldsymbol{P} \times \{\frac{1}{4} < |z_2| < 1\}$, and $X := X_1 \cup X_2 \cup X_3$:

i) For each $f \in \mathcal{O}(X)$, the restriction $f|_{\boldsymbol{P}^2(1) \setminus \boldsymbol{P}^2((1,0);\frac{1}{4})}$ has a unique holomorphic extension $\check{f} \in \mathcal{O}(\boldsymbol{P}^2(1))$.

ii) X is not a region of holomorphy.

iii) The function $f \in \mathcal{O}(X)$, $z \mapsto \sqrt{z_1 - i}$, is such that $f|_{X \cap \boldsymbol{P}^2(1)} \neq \check{f}|_{X \cap \boldsymbol{P}^2(1)}$.

With the help of the maximum norm, we define for $A, B \subset \mathbb{C}^n$

$$\operatorname{dist}(A, B) := \inf\{|a - b|;\ a \in A, b \in B\}$$

and the *boundary-distance function on* $X \subsetneqq \mathbb{C}^n$:

$$\delta_X : X \to \mathbb{R}_{\geqq 0},\ x \mapsto \operatorname{dist}(x, \partial X) = \sup\{r \in \mathbb{R};\ \boldsymbol{P}(x; r) \subset X\}.$$

The following fact is an essential aid for the proof of 12.8:

12.5 Thullen's Lemma. *For a region $X \subset\subset \mathbb{C}^n$ and a compact set $K \subset X$, let $g \in \mathcal{O}(X)$ be such that $|g| \leqq \delta_X$ on K. Then, for each $a \in \hat{K}_{\mathcal{O}(X)}$, every $f \in \mathcal{O}(X)$ is holomorphically extendible to $\boldsymbol{P}(a; |g(a)|)$.*

Proof. We may assume that $|g| < \delta_X$ on K, since g can be replaced by $(1-\varepsilon)g$ for any positive $\varepsilon < 1$. We want to show for $a \in \hat{K}_{\mathcal{O}(X)}$ that $\Sigma \dfrac{D^\nu f(a)}{\nu!}(z-a)^\nu$ converges on $\boldsymbol{P}(a; |g(a)|)$. By Abel's Lemma, it is sufficient to prove that the set $\left\{\dfrac{1}{\nu!}|D^\nu f(a) g^{|\nu|}(a)|;\ \nu \in \mathbb{N}^n\right\}$ is bounded. Now the fact that $K \subset\subset X$ implies that $L := \bigcup_{z \in K} \boldsymbol{P}(z; |g(z)|) \subset\subset X$; for $f \in \mathcal{O}(X)$ and $z \in K \setminus N(g)$, we have that

$$\left|\frac{D^\nu f(z)}{\nu!}\right| \underset{4.3\text{ ii)}}{\leqq} \frac{\|f\|_{\boldsymbol{P}(z;|g(z)|)}}{|g(z)|^{|\nu|}} \leqq \frac{\|f\|_L}{|g(z)|^{|\nu|}} < \infty,$$

and thus that $\|g^{|\nu|} D^\nu f\|_K \leqq \nu! \|f\|_L$. Since $g^{|\nu|} D^\nu f \in \mathcal{O}(X)$, the definition of $\hat{K}_{\mathcal{O}(X)}$ implies that, for $a \in \hat{K}_{\mathcal{O}(X)}$,

$$|(g^{|\nu|} D^\nu f)(a)| \leqq \|g^{|\nu|} D^\nu f\|_{\hat{K}} = \|g^{|\nu|} D^\nu f\|_K \leqq \nu! \|f\|_L. \quad \blacksquare$$

12.6 Corollary. *For a region $X \Subset \mathbb{C}^n$, a compact set $K \subset X$, and an $a \in \hat{K}_{\mathcal{O}(X)}$, every $f \in \mathcal{O}(X)$ is holomorphically extendible to $\boldsymbol{P}(a; \operatorname{dist}(K, \partial X))$.*

To prove that, apply 12.5 to the constant function $g := \operatorname{dist}(K, \partial X)$. ■

12.7 Corollary. *If X is a region of holomorphy, and K is a compact subset of X, then*

$$\operatorname{dist}(K, \partial X) = \operatorname{dist}(\hat{K}_{\mathcal{O}(X)}, \partial X).$$

Proof. Obviously, $d := \operatorname{dist}(K, \partial X) \geqq \hat{d} := \operatorname{dist}(\hat{K}, \partial X)$. For every $a \in \hat{K}$, we have that $\boldsymbol{P}(a; d) \subset X$ by 12.6; it follows that $\hat{d} \geqq d$. ■

E. 12c. Prove directly that $\operatorname{dist}(K, \partial X) = \operatorname{dist}(\hat{K}, \partial X)$ for every compact set $K \subset X \Subset \mathbb{C}$ $\left(\text{hint: for } a \in \partial X, \text{ use } \dfrac{1}{z-a}\right)$.

12.8 Theorem. *For $X \Subset \mathbb{C}^n$, the following statements are equivalent:*

i) X is holomorphically convex.

ii) For each sequence of points $(a_j)_{j \in \mathbb{N}}$ with no limit point in X, there exists an $f \in \mathcal{O}(X)$ such that $\sup_{j \in \mathbb{N}} |f(a_j)| = \infty$.

iii) X is a region of holomorphy.

iv) For each $z \in \mathbb{C}^n \setminus X$, there exists an $f \in \mathcal{O}(X)$ that is not holomorphically extendible at z.

v) For each $a \in X$ and each polydisk $\boldsymbol{P}(a; \varrho) \not\subset X$, there exists an $f \in \mathcal{O}(X)$ that is not holomorphically extendible to $\boldsymbol{P}(a; \varrho)$.

vi) There exists no connected $V \Subset \mathbb{C}^n$ with the following two properties:

α) $V \cap \partial X \neq \emptyset$.

β) There exists a connected component U of $X \cap V$ and a mapping $\varphi: \mathcal{O}(X) \to \mathcal{O}(V)$ such that $g|_U = (\varphi g)|_U$ for every $g \in \mathcal{O}(X)$.

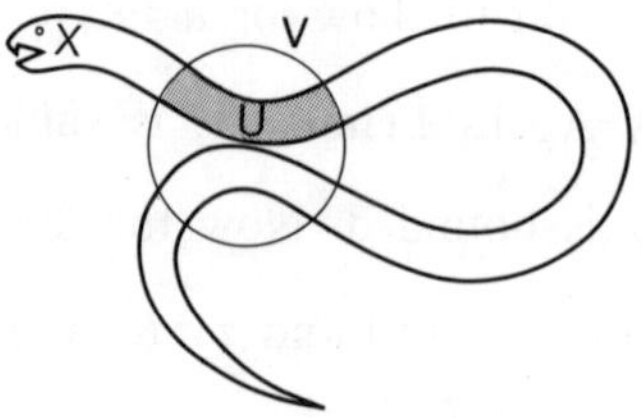

Proof. i) ⇒ ii) After replacing the sequences $(a_j)_{j \in \mathbb{N}}$ of ii) and $(K_j)_{j \in \mathbb{N}}$ of 12.9 with appropriate subsequences, we may assume that the hypotheses of 12.9 are satisfied; then 12.9 provides the desired function $f \in \mathcal{O}(X)$.

ii) ⇒ i) If X is not holomorphically convex, then there exists a compact set $K \subset X$ such that $\hat{K}$ contains a sequence (a_j) with no limit point in $\hat{K}$, and thus with no limit point in X (note that $\hat{K}$ is closed in X). For each $f \in \mathcal{O}(X)$, we have that $|f(a_j)| \leqq \|f\|_K < \infty$, in contradiction to ii).

i) $\Rightarrow$ iii) We may assume that $X \neq \mathbb{C}^n$. Then we want to construct a sequence in the setting of 12.9 that has every point of ∂X as a limit point: Let us call a point $x \in \mathbb{C}^n$ *rational*, if its components have rational real and imaginary parts. For $x \in X$, if we let $\boldsymbol{P}_X(x; r)$ denote that connected component of $X \cap \boldsymbol{P}^n(x; r)$ which contains x, then the set

$$\mathfrak{M} := \{\boldsymbol{P}_X(x; r);\ x \text{ and } r \text{ rational},\ \boldsymbol{P}(x; r) \not\subset X\}$$

is countable; hence, we can describe it as a sequence $(Q_j)_{j \in \mathbb{N}}$. Given an exhaustion[2] $(K_i)_{i \in \mathbb{N}}$ of X as in 12.9, we can choose recursively a sequence $(a_j, i_{j+1})_{j \in \mathbb{N}}$ in $X \times \mathbb{N}$ such that for every j,

$$a_j \in Q_j \cap (K_{i_{j+1}} \backslash K_{i_j})$$

(having chosen (a_{j-1}, i_j), we can fix an $a_j \in Q_j \backslash K_{i_j}$ and a number i_{j+1} such that $K_{i_{j+1}}$ contains a_j). Now 12.9, applied to the exhaustion $(K_{i_j})_{j \in \mathbb{N}}$ of X, provides an $f \in \mathcal{O}(X)$ such that $|f(a_j)| \geqq j$ for every j. We want to show that f is extendible at no point in $\mathbb{C}^n \backslash X$. Assume the contrary; then there exists a polydisk $\boldsymbol{P}(x_0; r_0) \not\subset X$ with $x_0 \in X$ on which the Taylor series T of f at x_0 converges. In particular, f and T coincide on $\boldsymbol{P}_X(x_0; r_0)$. We can construct recursively a sequence $\boldsymbol{P}_X(x_k; r_k)$ in $\mathfrak{M}$ with $x_k \in \boldsymbol{P}_X(x_0; r_0)$ and $r_k < 1/k$ such that $\boldsymbol{P}(x_{k+1}; r_{k+1}) \subset\subset \boldsymbol{P}(x_k; r_k)$. Then the corresponding points $a_{j(k)} \in \boldsymbol{P}_X(x_k; r_k)$ form a subsequence of (a_j) in $\boldsymbol{P}_X(x_0; r_0)$ that converges to a point $y \in \boldsymbol{P}(x_0; r_0)$. Although T converges on $\boldsymbol{P}(x_0; r_0)$, we have that $|T(y)| = |\lim_{k \to \infty} f(a_{j(k)})| = \infty$. ↯

The implications iii) $\Rightarrow$ iv) $\Rightarrow$ v) are trivial, since each involves only exchanging an existential quantifier with a universal quantifier.

v) $\Rightarrow$ i) Assume that there exists a compact $K \subset X$ such that $\hat{K}$ is not compact and thus not closed in $\mathbb{C}^n$; then $\operatorname{dist}(\hat{K}, \partial X) = 0$. In particular, there exists an $a \in \hat{K}$ such that $\delta_X(a) < \operatorname{dist}(K, \partial X)$. By 12.6, every $f \in \mathcal{O}(X)$ can be extended holomorphically to $\boldsymbol{P}(a; \operatorname{dist}(K, \partial X))$. ↯

iii) $\Rightarrow$ vi) Suppose that there is such a V, and consider a path $\gamma: [0,1] \to V$ such that $\gamma(0) \in U$ and $\gamma(1) \in V \backslash X$. The first point x at which γ intersects ∂U must lie in $\bar{X}$; hence, $x \in \partial X$, since U is a connected component of $X \cap V$. Thus it is possible to find an $a \in U$ and an $r > 0$ such that $x \in \boldsymbol{P}(a; r) \subset V$; since $g|_U = (\varphi g)|_U$, the functions g and φg must have the same Taylor series expansion about a, so g is holomorphically extendible at $x \in \partial X$. ↯

vi) $\Rightarrow$ v) If there exist a point $a \in X$ and a polydisk $\boldsymbol{P}(a; \varrho) \not\subset X$ to which each $g \in \mathcal{O}(X)$ is holomorphically extendible, then the set $V := \boldsymbol{P}(a; \varrho)$ has the properties described in vi) (take U to be that connected component of $V \cap X$ which contains a). ■

---

[2] An ascending union of subsets A_j of X is called an *exhaustion of* X, if $X = \bigcup A_j$.

The following fact remains to be proved:

12.9 Lemma. *Suppose that $X \Subset \mathbb{C}^n$ has a compact exhaustion $(K_j)_{j \in \mathbb{N}}$ such that $K_j = (\widehat{K_j})_{\mathcal{O}(X)} \subset \mathring{K}_{j+1}$. Then, for each sequence $(a_j)_{j \in \mathbb{N}}$ with $a_j \in K_{j+1} \backslash K_j$, there exists an $f \in \mathcal{O}(X)$ such that $|f(a_j)| \geqq j$ for every j. Moreover, every holomorphically convex region in $\mathbb{C}^n$ has such an exhaustion.*

Proof. For a sequence (a_j), we construct recursively a sequence (f_j) in $\mathcal{O}(X)$ such that

$$\text{i)} \quad \|f_j\|_{K_j} \leqq 2^{-j} \qquad \text{and} \qquad \text{ii)} \quad |f_j(a_j)| \geqq j + 1 + \Big|\sum_{i=0}^{j-1} f_i(a_j)\Big|:$$

set $K_0 := \emptyset$ and $f_0 := 1$, and suppose that $f_0, \ldots, f_{j-1}$ have been constructed; according to E. 12d, for $a_j \in K_{j+1} \backslash K_j$, it is possible to find an f_j that satisfies i) and ii). It follows from i) that $f := \sum_{j=0}^{\infty} f_j$ lies in $\mathcal{O}(X)$, and from i) and ii) that

$$|f(a_j)| \geqq \Big(|f_j(a_j)| - \Big|\sum_{i=0}^{j-1} f_i(a_j)\Big|\Big) - \Big|\sum_{i=j+1}^{\infty} f_i(a_j)\Big| \geqq (j+1) - 1 = j\,.$$

Now let X be holomorphically convex. By 51 A. 2, X has a compact exhaustion $X = \bigcup_{j=1}^{\infty} L_j$. Every $\hat{L}_j$ is compact; since $(\mathring{\hat{L}}_j)_{j \in \mathbb{N}}$ forms an open cover of X, an appropriate subsequence of $(\hat{L}_j)_{j \in \mathbb{N}}$ is an exhaustion with the desired properties. ■■

E. 12d. Prove that the following statements hold for compact sets $K \subset X$: For each $a \in X \backslash \hat{K}_{\mathcal{O}(X)}$, there exists an $f \in \mathcal{O}(X)$ such that $\|f\|_K < 1 < |f(a)|$. For an arbitrary $\varepsilon > 0$, it is possible to attain $\|f\|_K < \varepsilon$ and $|f(a)| > 1/\varepsilon$.

12.10 Examples. i) Every region $X \Subset \mathbb{C}$ is a region of holomorphy: for $a \in \mathbb{C} \backslash X$, the function $f := \dfrac{1}{z-a} \in \mathcal{O}(X)$ is not holomorphically extendible at a.

ii) $\mathbb{C}^2 \backslash 0$ is not a region of holomorphy (Kugelsatz).

iii) X is a region of holomorphy iff every connected component of X is a domain of holomorphy.

iv) *Every finite product of regions of holomorphy, and in particular every polydisk, is a region of holomorphy:* consider a product $X = X_1 \times X_2$ of holomorphically convex regions and a sequence of points $(a_j)_{j \in \mathbb{N}}$ with no limit point in X. We may assume that the sequence $(\mathrm{pr}_1 a_j)_{j \in \mathbb{N}}$ has no limit points in X_1. If $f_1 \in \mathcal{O}(X_1)$ is unbounded on that sequence, then $f_1 \circ \mathrm{pr}_1 \in \mathcal{O}(X)$ is unbounded on $\{a_j; j \in \mathbb{N}\}$.

v) If $\varphi : X \to Y$ is a *proper holomorphic mapping* between regions $X \Subset \mathbb{C}^n$ and $Y \subset \mathbb{C}^m$, then 33 B. 1 v) implies immediately that X is a region of holomorphy if Y is.

vi) *Analytic polyhedra are regions of holomorphy*: for $U \Subset X \Subset \mathbb{C}^n$, a product region $Z = \prod_{j=1}^{m} Z_j \Subset \mathbb{C}^m$, and $\varphi \in \mathrm{Hol}(X, \mathbb{C}^m)$, the set

$$W := U \cap \varphi^{-1}(Z)$$

is called an *analytic polyhedron* in X if W is a relatively compact subset of U. In that case, we deduce from v) that W is holomorphically convex: since $\partial W \cap \partial U = \emptyset$, it follows that $\partial W \subset \partial(\varphi^{-1}(Z)) \subset \varphi^{-1}(\partial Z)$; hence, by 53 A. 8 v), the mapping $\varphi: W \to Z$ is proper.

The set U in the definition of W ensures that each union of connected components of W is also an analytic polyhedron if it is at a positive distance from each other connected component of W.

In this book we apply only those analytic polyhedra for which Z is a polydisk.

vii) Convex domains in $\mathbb{C}^n$ (in particular, balls $\boldsymbol{B}^n(r)$) are domains of holomorphy (see E. 12f).

E. 12e. For $U = X = \mathbb{C}$, the analytic polyhedron determined by $Z = \boldsymbol{P}^1(1)$ and $\varphi(z) := z^2 - 1$ is not connected.

E. 12f. *If $G \Subset \mathbb{C}^n$ is convex, then G is holomorphically convex.* (Hint: assume that $0 \in \partial G$; then there exist $a_j, b_j \in \mathbb{R}$ such that $l := \sum_{j=1}^{n} (a_j x_j + b_j y_j)$ satisfies the conditions $l(0) = 0$ and $l|_G < 0$; use the function $f(z) := 1 \Big/ \sum_{j=1}^{n} (a_j - i b_j) z_j \in \mathcal{O}(G)$; for 14.9).

E. 12g. Show that $X := \{z \in \mathbb{C}^2; |z_1| < |z_2| < 1\}$ is a domain of holomorphy (hint: $f(z) := \dfrac{1}{e^{i\theta} z_1 - z_2} \in \mathcal{O}(X)$ for $\theta \in \mathbb{R}$; for E. 62b).

§ 12A Supplement: Further Extension Theorems for Holomorphic Functions

Here we discuss some more consequences of Thullen's Lemma; Proposition 12 A. 3 can serve as motivation for sections 13 and 14.

We understand a *complex disk D in $X \Subset \mathbb{C}^n$* to be a set of the form

$$D = d + \boldsymbol{B}^1(1) \cdot b \Subset X, \text{ with } b \neq 0 \text{ and } d \text{ in } \mathbb{C}^n;$$

we also write

$$\mathrm{bd}\, D := d + S^1 \cdot b \subset \bar{D}.$$

For $X \neq \mathbb{C}^n$, we let $h_D: \bar{D} \to \mathbb{R}$ denote the (uniquely determined [Co X. 2.4]) continuous extension of $\log \delta_X|_{\mathrm{bd}\, D}$ to $\bar{D}$, which, as a function of $\lambda \in \boldsymbol{B}^1(1)$, is harmonic (see 13.1)).

12 A. 1 Proposition *If D is a complex disk in X and if $a \in \bar{D}$, then every $f \in \mathcal{O}(X)$ is holomorphically extendible to the polydisk $\boldsymbol{P}(a; e^{h_D(a)})$.*

Proof. By 12 A. 2, for each $\varepsilon > 0$, there exists a polynomial $P \in \mathbb{C}[z]$ that satisfies the following inequality on $\operatorname{bd} D$:

$$h_D - \varepsilon < \operatorname{re} P < h_D.$$

Since h_D and $\operatorname{re} P$ are harmonic, and since D is a complex disk, the inequality is valid on all of $\bar{D}$; we apply it in the following form:

$$e^{h_D - \varepsilon} < e^{\operatorname{re} P} = |e^P| < e^{h_D}.$$

On $K := \operatorname{bd} D$ we have that $e^{h_D} = \delta_X$; by 12.5, then, every $f \in \mathcal{O}(X)$ is holomorphically extendible to $\boldsymbol{P}(a; |e^{P(a)}|) \supset \boldsymbol{P}(a; e^{h_D(a) - \varepsilon})$ for each $a \in \hat{K} \underset{6.5}{=} \bar{D}$. Since ε was arbitrary, $f \in \mathcal{O}(X)$ can even be extended to $\boldsymbol{P}(a; e^{h_D(a)})$. ■

12 A. 2 Lemma. *Let D be a complex disk in $\mathbb{C}^n$. For each continuous function $h: \operatorname{bd} D \to \mathbb{R}$, and each $\varepsilon > 0$, there exists a polynomial $P \in \mathbb{C}[z_1, \dots, z_n]$ such that $\|h - \operatorname{re} P\|_{\operatorname{bd} D} < \varepsilon$.*

Proof. In the special case in which $D = \boldsymbol{B}^1(1) \subset \mathbb{C}$, the point-separating real algebra

$$\left\{ e^{i\varphi} \mapsto \sum_{j=0}^{m} (a_j \cos j\varphi + b_j \sin j\varphi);\ a_j, b_j \in \mathbb{R}, m \in \mathbb{N} \right\}$$

is dense in $\mathscr{C}(\operatorname{bd} D; \mathbb{R})$ with respect to the norm $\|\cdot\|_{\operatorname{bd} D}$, according to the Stone-Weierstrass Theorem. For $P = \sum_{j=0}^{m} (a_j - i b_j) z^j$ and $z = e^{i\varphi} \in \operatorname{bd} D$, we have that

$$\operatorname{re} P(z) = \sum_{j=0}^{m} (a_j \cos j\varphi + b_j \sin j\varphi),$$

and the assertion follows for the special case.

For arbitrary n and $D = \boldsymbol{B}^1(1) \times 0 \subset \mathbb{C}^n$, then, there exists a $Q \in \mathbb{C}[z_1]$ such that $|\operatorname{re} Q(z_1) - h(z_1, 0)| < \varepsilon$ for $z \in \operatorname{bd} D$; set $P(z) := Q(z_1)$.

The general case follows by application of an affine automorphism $\varphi: \mathbb{C}^n \to \mathbb{C}^n$ such that $\varphi(D) = \boldsymbol{B}^1 \times 0$. ■■

For a complex disk D in a holomorphically convex region X, and a point $a \in \bar{D}$, we obtain that

$$\boldsymbol{P}(a; e^{h_D(a)}) \subset X,$$

by 12 A. 1. That leads to the next result:

12 A. 3 Corollary. *A region of holomorphy $X \subset\subset \mathbb{C}^n$ satisfies the inequality*

(12. A. 3.1) $h_D \leqq \log(\delta_X|_D)$

for each complex disk D in X. ■

By 13.11, condition (12 A. 3.1) is equivalent to $-\log \delta_X$ being plurisubharmonic, and thus to $X \subset\subset \mathbb{C}^n$ being *pseudoconvex* (see 14.1).

12 A. 4 Corollary. *Suppose that $X \subset\subset \mathbb{C}^n$ and $x \in \partial X$. If there is a complex disk D in X and a point $a \in D$ such that either*

i) $|x - a| < \operatorname{dist}(\operatorname{bd} D, \partial X)$, *or*

ii) $|x - a| = \operatorname{dist}(\operatorname{bd} D, \partial X)$ *and $\delta_X|_{\operatorname{bd} D}$ is nonconstant,*

then every $f \in \mathcal{O}(X)$ is holomorphically extendible at x.

Proof. By the Maximum Principle, $(\widehat{\operatorname{bd} D})_{\mathcal{O}(X)} = \bar{D}$; by 12.6, then, condition i) ensures that f is holomorphically extendible at a.

If δ_X is nonconstant on $\operatorname{bd} D$, then so is h_D; for $d := \operatorname{dist}(\partial X, \operatorname{bd} D)$, we have that

$$\log d = \min_{y \in \operatorname{bd} D} h_D(y) < h_D(a),$$

since the nonconstant harmonic function h_D cannot assume a minimum in D (in analogy to 13.3β)). Thus $d < e^{h_D(a)}$, and the inequality $|x-a| = d < e^{h_D(a)}$ yields that $x \in \boldsymbol{P}(a; e^{h_D(a)})$; the assertion follows from 12 A. 1. ■

E. 12 Aa. Show that 12 A.4 would not hold for $X = \boldsymbol{P}^2(1)$ if the stipulation "$\delta_X|_{\operatorname{bd} D}$ is nonconstant" were dropped from ii).

12 A. 5 Example. *For* $X := \mathbb{C}^2 \backslash \mathbb{R}^2$, *the restriction-mapping* $\varphi: \mathcal{O}(\mathbb{C}^2) \to \mathcal{O}(X)$ *is an isomorphism of topological algebras.*

Proof. Since X is dense in $\mathbb{C}^2$, it suffices, by 3.4 and 55.6 iii), to show that φ is surjective. For $x = (x_1, x_2) \in \mathbb{C}^2 \backslash X = \mathbb{R} \times \mathbb{R}$, $a := (x_1 + i, x_2)$, and the complex disk

$$D := a + \boldsymbol{B}^1(1) \cdot (0,2),$$

we have that $\operatorname{dist}(\operatorname{bd} D, \partial X) = 1$, and $\delta_X|_{\operatorname{bd} D}$ is nonconstant $(\partial_X(x_1 + i, x_2 + 2) = 1 < 2 = = \delta_X(x_1 + i, x_2 + 2i))$. Since

$$|x - a| = 1 = \operatorname{dist}(\operatorname{bd} D, \partial X),$$

12 A.4 yields that f is holomorphically extendible to a polydisk containing x. The assertion follows from E. 12a. ■

The set $\mathbb{R}^2$ is thus a real plane in $\mathbb{C}^2$ to which every holomorphic function defined on the complement is extendible. For real planes on which $\mathbb{C}^2$ induces the structure of a complex line, the analogous statement is of course false!

12 A. 6 Example. *For* $X := \mathbb{C}^2 \backslash (\mathbb{R} \times \overline{\boldsymbol{P}^1(1)})$, *the restriction-mapping* $\varphi: \mathcal{O}(\mathbb{C}^2) \to \mathcal{O}(X)$ *is an isomorphism of topological algebras.*

Proof. By E. 12a, for each fixed $x = (x_1, b) \in \mathbb{R} \times \overline{\boldsymbol{P}^1(1)}$, it is sufficient to find a polydisk $\boldsymbol{P}$ such that $\boldsymbol{P} \cap X$ is connected, and such that every $f \in \mathcal{O}(X)$ is holomorphically extendible to $\boldsymbol{P}$. For $a = (x_1 + i, 0) \in D := a + \boldsymbol{B}^1(1) \cdot (0,3)$, we have that $|x - a| = |(-i, b)| = 1 < 2 = \operatorname{dist}(\operatorname{bd} D, \partial X)$. The proof of 12 A.4 i) shows that each $f \in \mathcal{O}(X)$ is holomorphically extendible to $\boldsymbol{P}(a; 2)$; moreover, $X \cap \boldsymbol{P}(a; 2)$ is obviously connected. ■

E. 12 Ab. Prove 12A.6 for restriction-mappings of the form $\mathcal{O}(\boldsymbol{P}^2(\varrho)) \to \mathcal{O}(\boldsymbol{P}^2(\varrho) \backslash (\mathbb{R} \times K))$ with K compact.

In 7.8, 12 A. 5, and 12 A. 6, we have stated extension theorems for regions of the form

$$X = (G_1 \times G_2) \backslash (A_1 \times A_2) \subset\subset \mathbb{C}^2$$

with domains $G_j \subset\subset \mathbb{C}$ and nonempty relatively closed sets $A_j \not\subset G_j$. Every region X of that form is connected (since each $x \in X$ can be joined to a fixed point $(g_1, g_2) \in (G_1 \backslash A_1) \times (G_2 \backslash A_2)$ with a piecewise linear path, each of whose segments is parallel to a coordinate axis).

12 A. 7 Proposition. *Such a domain* X *is not a domain of holomorphy.*

Proof. Without loss of generality, we have that $\boldsymbol{P}^1(0; 4) \subset\subset G_j \backslash A_j$, $\operatorname{dist}(0, \partial G_j) > \operatorname{dist}(0, A_j) = 4$, and $4 \in A_j \subset \mathbb{C}$. For $x = (4,4)$, $D = \{2\} \times \boldsymbol{P}^1(2;1)$, and $a = (2,2)$, then, the hypotheses of 12A.4 ii) are satisfied: for each $y \in \boldsymbol{P}^1(2;1)$, we have that

$$1 = \operatorname{dist}(A_2, 3) \leqq \operatorname{dist}(A_2, y) \leqq \operatorname{dist}(A_2, 1) = 3\,,$$

and thus that

$$\operatorname{dist}(\operatorname{bd} D, \partial X) = \operatorname{dist}(\operatorname{bd} D, A_1 \times A_2) = \delta_X(2,3) = 2 = |x - a| < 3 = \delta_X(2,1)\,. \quad \blacksquare$$

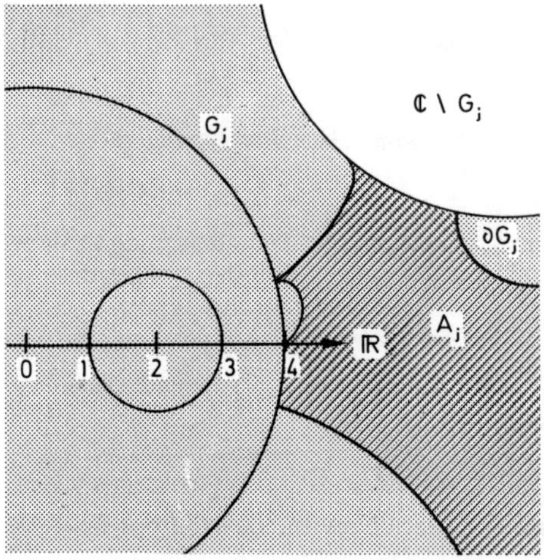

From 12A.7, it follows that the domains in $\mathbb{C}^2$ with the following τ-images (see 11.1) in the closed first quadrant of the plane are not holomorphically convex:

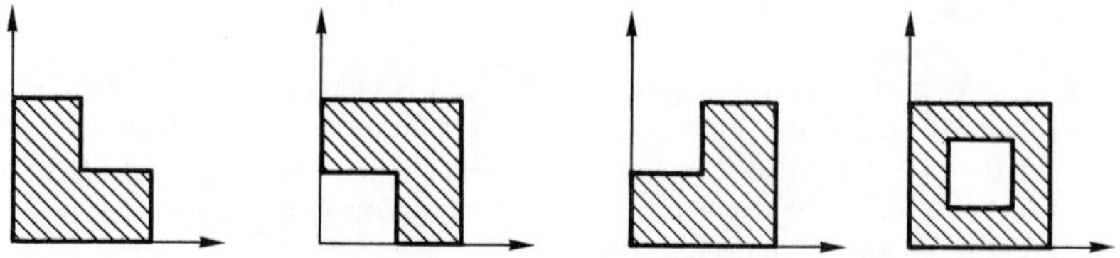

§ 12 B Supplement: The Edge-of-the-wedge Theorem

We turn now to the Edge-of-the-wedge Theorem, which is used in quantum field theory to show that the product of space inversion (parity) P, time reversal T, and charge conjugation C is a symmetry in the sense of local field theory (PCT-Theorem, see [StWi p. 16]).

Using the representation $\mathbb{C}^n = \mathbb{R}^n + i\mathbb{R}^n$, consider a set $V^0 \Subset \mathbb{R}^n$, a polyradius $\varrho \in \mathbb{R}^n_{>0}$, and an open real convex cone $C \Subset \mathbb{R}^n$ with vertex at the origin; set $W := C \cap \boldsymbol{P}^n(0;\varrho)$. Then the sets

$$V^+ := V^0 + iW \quad \text{and} \quad V^- := V^0 - iW$$

are called "wedges with edge V^0".

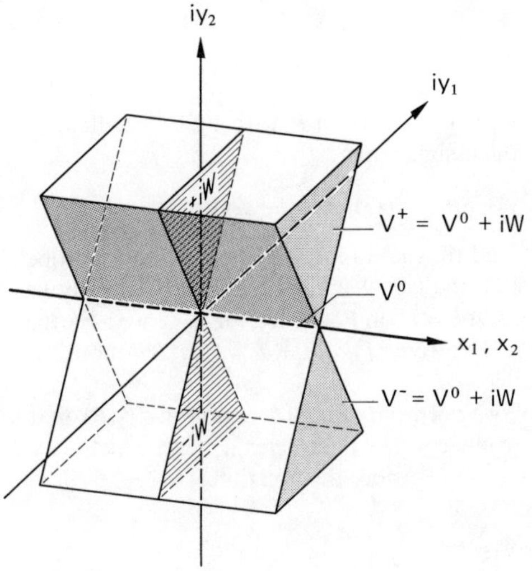

12 B. 1 Edge-of-the-wedge Theorem. *Let $V = V^+ \cup V^0 \cup V^-$ be given. Then there exists an open neighborhood X of V in $\mathbb{C}^n$ such that the restriction-mapping*

$$i^0 : \mathcal{O}(X) \to \mathcal{O}(V^+ \cup V^-) \cap \mathscr{C}(V)$$

is an isomorphism of topological algebras.

Proof. Without loss of generality, let V^0 be connected. It suffices to find for each $a \in V^0$ a connected neighborhood U_a of a in $\mathbb{C}^n$ such that every $f \in \mathcal{O}(V^+ \cup V^-) \cap \mathscr{C}(V)$ is holomorphically extendible to U_a. Then the domain $X := \bigcup U_a \cup V$ has the desired properties: the Identity Theorem implies that the continuous algebra-homomorphism i^0 is injective; that, in turn, implies that i^0 is surjective, since the local extensions patch together on X; it can be shown by means of the integrals appearing in the construction, as in 7.3 (or by 55.6 ii)), that i^0 is open.

Without loss of generality, suppose that $a = 0$. Choose a basis for $\mathbb{R}^n$ lying in the cone C, and let M be a real matrix that sends that basis to the canonical basis; then $\mathbb{R}^n_{>0} \subset M(C)$. Since M preserves the decomposition $\mathbb{C}^n = \mathbb{R}^n + i\mathbb{R}^n$, we may assume that

$$V^0 = \mathbb{R}^n \cap \boldsymbol{P}^n(7) \quad \text{and} \quad W = \mathbb{R}^n_{>0} \cap \boldsymbol{P}^n(7).$$

For $n = 1$, the proof is simple: the region V is bounded by a rectangle; without loss of generality, suppose that $f \in \mathcal{O}(V^+ \cup V^-)$ is continuous on $\bar{V}$. Then, by Cauchy's Integral Formula, the assignment

$$F(z) := \frac{1}{2\pi i} \int_{\partial V} \frac{f(\zeta)}{\zeta - z} d\zeta$$

determines a holomorphic function on V. For $a \in V^+$, we have that

$$f(a) = \frac{1}{2\pi i} \int_{\partial V^+} \frac{f(\zeta)}{\zeta - a} d\zeta \quad \text{and} \quad 0 = \frac{1}{2\pi i} \int_{\partial V^-} \frac{f(\zeta)}{\zeta - a} d\zeta .$$

Since $\partial V = \partial V^+ + \partial V^-$, it follows that $F|_{V^+} = f|_{V^+}$; similarly, $F|_{V^-} = f|_{V^-}$. We conclude immediately that $U_0 := V$ has the desired properties.

For general n, it suffices to consider $U_0 := \boldsymbol{P} := \boldsymbol{P}^n(1)$; we will find an $F \in \mathcal{O}(\boldsymbol{P})$ that satisfies the condition

(12 B. 1.1) $F|_{V^0 \cap P} = f|_{V^0 \cap P}$.

For such an F, fix a point $b = x + iy$ in $\boldsymbol{P}$ with $y \in \mathbb{R}^n_{>0}$ sufficiently small to ensure that the functions determined by the assignments

$$h(\lambda) := f(x + \lambda y) \quad \text{and} \quad H(\lambda) := F(x + \lambda y)$$

are well-defined near $\overline{\boldsymbol{P}^1}$ and thus holomorphic on $\overline{\boldsymbol{P}^1}$ (that y can be chosen so that h has the desired property depends on the fact that the case "$n = 1$" already has been established). Since $h|_{\boldsymbol{P}^1 \cap \mathbb{R}} = H|_{\boldsymbol{P}^1 \cap \mathbb{R}}$, it follows that $h = H$ on $\overline{\boldsymbol{P}^1}$; hence, for $\lambda = i$, we have that $f(b) = h(i) = H(i) = F(b)$. According to the Identity Theorem, $f|_{V^+ \cap \boldsymbol{P}} = F|_{V^+ \cap \boldsymbol{P}}$. By choosing $y \in \mathbb{R}^n_{<0}$, we ensure that $f|_{V^- \cap \boldsymbol{P}} = F|_{V^- \cap \boldsymbol{P}}$.

To complete the proof, we need only find a function $F \in \mathcal{O}(\boldsymbol{P})$ that satisfies (12 B. 1.1); in the construction of F, we introduce a new parameter λ, with which we can reduce the verification of (12 B. 1.1) to the case $n = 1$. Consider the function

$$\varphi \in \mathcal{O}(\overline{\boldsymbol{P}^2})^{3)}, \quad (w, \lambda) \mapsto \frac{2w + 4\lambda}{2 + \lambda w};$$

the following facts can be verified readily:

i) $\operatorname{im} \varphi(w, \lambda) = \dfrac{(4 - 4|\lambda|^2) \operatorname{im} w + (8 - 2|w|^2) \operatorname{im} \lambda}{|2 + \lambda w|^2}$.

ii) $\operatorname{im} \varphi(w, \lambda)$ and $\operatorname{im} \lambda$ have the same sign if $(1 - |\lambda|^2) \operatorname{im} w = 0$, i. e., if w is real or $|\lambda| = 1$.
iii) $\varphi(w, 0) = w$.
iv) $\|\varphi\|_{\boldsymbol{P}^2} \leqq 6$.

Now consider the holomorphic mapping

$$\Phi: \boldsymbol{P} \times \overline{\boldsymbol{P}^1} \to \boldsymbol{P}^n(7), \ (z, \lambda) \mapsto (\varphi(z_1, \lambda), \dots, \varphi(z_n, \lambda));$$

it follows from ii) and iv) that (with an obvious abuse of notation)

$$\Phi\left(\boldsymbol{P} \times (\boldsymbol{T}^1 \cap \{\operatorname{im} \begin{smallmatrix} > \\ = \\ < \end{smallmatrix} 0\})\right) \subset \begin{matrix} V^+ \\ V^0 \\ V^- \end{matrix}.$$

Then $f \circ \Phi$ is defined (and continuous) on $\boldsymbol{P} \times \boldsymbol{T}^1$. For $T_\varepsilon := \{z \in \boldsymbol{T}^1; |\operatorname{im} z| \geqq \varepsilon\}$, we define a function F_ε on $\boldsymbol{P}$ by setting

$$F_\varepsilon(z) := \frac{1}{2\pi i} \int_{T_\varepsilon} \frac{f \circ \Phi(z, \lambda)}{\lambda} d\lambda.$$

For $\varepsilon > 0$, F_ε is holomorphic; hence, the function $F := F_0 = \lim_{\varepsilon \to 0} F_\varepsilon$ is also in $\mathcal{O}(\boldsymbol{P})$. Now this function F satisfies (12 B.1.1): for fixed $x \in V^0 \cap \boldsymbol{P}$, it follows from ii) and iv) that, on $\overline{\boldsymbol{P}^1}$,

$$\Phi(x, \{\operatorname{im} \begin{smallmatrix} > \\ = \\ < \end{smallmatrix} 0\}) \subset \begin{matrix} V^+ \\ V^0 \\ V^- \end{matrix}.$$

---

3) $\mathcal{O}(\bar{X}) := \{f: X \to \mathbb{C};\ f \text{ is holomorphic near } \bar{X}\}$ for $X \subset \mathbb{C}^n$.

Thus, we obtain a continuous function

$$\overline{P^1} \to \mathbb{C}, \quad \lambda \mapsto f \circ \Phi(x, \lambda),$$

that is actually holomorphic on P^1, as we have seen in the case $n = 1$. Cauchy's Integral Formula yields that

$$F(x) = \frac{1}{2\pi i} \int_{T^1} \frac{f \circ \Phi(x, \lambda)}{\lambda} d\lambda = f \circ \Phi(x, 0) \underset{iii)}{=} f(x). \quad \blacksquare$$

12 B. 2 Remark. The algebra $\mathcal{O}(V^+ \cap V^-) \cap \mathscr{C}(V)$ in 12 B. 1 can be replaced with the (*a priori* larger) set of all functions $f \in \mathcal{O}(V^+ \cup V^-)$ for which

$$\lim_{\substack{y \to 0 \\ y \in \pm W}} \int_{V^0} f(x + iy) \psi(x) dx$$

exists for every $\psi \in \mathscr{C}^\infty(V^0)$ with compact support. For a pretty proof of this stronger version of 12 B. 1, we refer the reader to [Ru].

Additional literature: [Ru], [Vl].

§ 13 Plurisubharmonic Functions

We introduce yet another type of convexity in § 14, namely, pseudoconvexity; the concept is based on the theory of plurisubharmonic functions, which we develop here in analogy to the theory of convex functions. For each statement about plurisubharmonic functions, we recommend that the reader consider an analogous statement for convex functions.

We begin with the one-dimensional case:

For an interval $I \subset\subset \mathbb{R}$, the solutions $f \in \mathscr{C}^2(I, \mathbb{R})$ of the potential equation $\frac{d^2 f}{dx^2} = 0$ are precisely the affine linear functions $l = ax + b$. A continuous function $f: I \to \mathbb{R}$ is called *convex* or "subaffine linear", if, for every interval $J \subset\subset I$, and every affine linear function l, the following implication holds:

$$f|_{\partial J} \leqq l|_{\partial J} \Rightarrow f|_J \leqq l|_J.$$

Now a continuous function f is affine linear iff it satisfies the mean value equality

$$f(x) = \tfrac{1}{2}(f(x + \varepsilon) + f(x - \varepsilon))$$

for every x and ε such that $[x - \varepsilon, x + \varepsilon] \subset I$; it is convex if it satisfies the following mean value *in*equality (proof?)

$$f(x) \leqq \tfrac{1}{2}(f(x + \varepsilon) + f(x - \varepsilon))$$

for every x and ε such that $[x - \varepsilon, x + \varepsilon] \subset I$.

Let $G \subset\subset \mathbb{C} = \mathbb{R}^2$ be a domain. The solutions $h \in \mathscr{C}^2(G, \mathbb{R})$[4] of the potential equation $\frac{\partial^2 h}{\partial x^2} + \frac{\partial^2 h}{\partial y^2} = 0$ in G are the *harmonic functions* [CoX]; every such h is locally the real part of a

---

[4] $\mathscr{C}^p(G, \mathbb{R}) := \{f \in \mathscr{C}(G, \mathbb{R});\ f$ is p-fold continuously partially differentiable$\}$.

holomorphic function, and thus real analytic. Moreover, a function $h \in \mathscr{C}(G, \mathbb{R})$ is harmonic iff it satisfies the following *mean value equality* for every a and r such that $\overline{\boldsymbol{B}^1(a;r)} \subset G$:

$$h(a) = \frac{1}{2\pi} \int_0^{2\pi} h(a + re^{i\varphi}) d\varphi .$$

In the ensuing discussion, we replace $\mathbb{R}$ with $\mathbb{R} \cup \{-\infty\}$ provided with the canonical topology; that allows us to consider functions such as $\log|f|$ at the zeros of f. Such an enlargement of the range is superfluous for the treatment of convex functions f, since $f(I) \subset \mathbb{R}$ if $f \not\equiv -\infty$.

13.1 Definition. *A function $f \in \mathscr{C}(G, \mathbb{R} \cup \{-\infty\})$ is called* <u>*subharmonic*</u> *if, for every disk $\boldsymbol{B} \subset\subset G$, and every function $h \in \mathscr{C}(\bar{\boldsymbol{B}}, \mathbb{R})$ that is harmonic on $\boldsymbol{B}$, f satisfies the following condition:*

$$f|_{\partial \boldsymbol{B}} \leqq h|_{\partial \boldsymbol{B}} \Rightarrow f|_{\boldsymbol{B}} \leqq h|_{\boldsymbol{B}} .$$

The usual definition of subharmonic functions demands, instead of continuity, only upper semicontinuity, since that is what is necessary to ensure that the limit of a decreasing sequence of subharmonic functions be subharmonic (see 13.9). However, the subharmonic functions that we consider are all continuous; in fact, those of greatest interest to us stem from boundary-distance functions (see 13.5 iii)). The reader who is familiar with semicontinuity will be able to make the necessary modifications for a generalization to semicontinuous functions.

Our first characterization of subharmonic functions in 13.2 shows that the real parts $\operatorname{re} P$ of complex polynomials P suffice as test functions h in 13.1, that harmonic functions are subharmonic, and that being subharmonic is a local property (as is convexity).

A function $f \in \mathscr{C}(I, \mathbb{R})$ is convex iff each $x \in I$ is contained in a subinterval $J \subset I$ on which f is convex.

13.2 Lemma. *For a continuous function $f: G \to \mathbb{R} \cup \{-\infty\}$, (with $G \subset\subset \mathbb{C}$), the following statements are equivalent:*

i) f is subharmonic on G.

ii) For each ball $\boldsymbol{B} \subset\subset G$, and each polynomial $P \in \mathbb{C}[z]$, the following condition is satisfied:

$$f|_{\partial \boldsymbol{B}} \leqq \operatorname{re} P|_{\partial \boldsymbol{B}} \Rightarrow f|_{\boldsymbol{B}} \leqq \operatorname{re} P|_{\boldsymbol{B}} .$$

iii) For each ball $\boldsymbol{B}(a;r) \subset\subset G$, the mean value inequality

$$f(a) \leqq \frac{1}{2\pi} \int_0^{2\pi} f(a + re^{i\theta}) d\theta$$

holds (the existence of the integral in $\mathbb{R} \cup \{-\infty\}$ follows from (13.2.1)).

Furthermore, each of those statements is equivalent to its local version ($i)_{local}$ – $iii)_{local}$): for some function $\varepsilon: G \to \mathbb{R}_{>0}$, the conditions hold for every ball $\boldsymbol{B}(a;r) \subset\subset G$ such that $r \leqq \varepsilon(a)$.

Proof. Trivially, each of the first three statements implies its local version.

i) $\Rightarrow$ iii) Without loss of generality, suppose that $f(a) > -\infty$.

α) Suppose that $-\infty \notin f(G)$. For each disk $\boldsymbol{B} = \boldsymbol{B}(a; r) \subset\subset G$ and each continuous function on $\partial \boldsymbol{B}$, there exists a unique continuous extension to $\bar{\boldsymbol{B}}$ that is harmonic on $\boldsymbol{B}$ (solution of the Dirichlet problem [Co X.2.4.]); in particular, $f|_{\partial \boldsymbol{B}}$ has such an extension, say h, and iii) follows immediately:

$$f(a) \leqq h(a) = \frac{1}{2\pi} \int_0^{2\pi} h(a + re^{i\varphi})d\varphi = \frac{1}{2\pi} \int_0^{2\pi} f(a + re^{i\varphi})d\varphi .$$

β) Now suppose that $-\infty \in f(G)$. For each $j \in \mathbb{N}$, the function $f_j := \max(f, -j)$ is obviously subharmonic; moreover, $-\infty \notin f_j(G)$, so f_j satisfies condition iii) by α). Fix a j_0 such that $-j_0 \leqq f(a)$; then

$$f(a) = f_j(a) \leqq \frac{1}{2\pi} \int_0^{2\pi} f_j(a + re^{i\varphi})d\varphi$$

for every $j \geqq j_0$. Since the statement

(13.2.1) For each compact set $K \subset \mathbb{R}$, and each continuous function $g: K \to \mathbb{R} \cup \{-\infty\}$, the Lebesgue integral $\int_K g(x)dx \in \mathbb{R} \cup \{-\infty\}$ exists and satisfies the equation [WhZy (5.32)]

$$\int_K g(x)dx = \lim_{j \to \infty} \int_K \max(g, -j)(x)dx .$$

holds, it follows that the monotonically decreasing sequence $\left(\int_0^{2\pi} f_j(a + re^{i\varphi})d\varphi\right)_{j \in \mathbb{N}}$ converges to $\int_0^{2\pi} f(a + re^{i\varphi})d\varphi$, and iii) follows.

Before continuing with the proof, we mention some related facts:

13.3 Remarks. *α) For a family $(f_\alpha)_{\alpha \in A}$ of subharmonic functions on G, if the function $F := \sup f_\alpha : G \to \mathbb{R} \cup \{-\infty\}$ is continuous, then it is subharmonic.*

β) The Maximum Principle holds for each subharmonic function f on the domain G (i.e., f is constant if there exists a $b \in G$ such that $f \leqq f(b)$ on G).

γ) The set of subharmonic functions on G forms a real convex cone in $\mathscr{C}(G, \mathbb{R} \cup \{-\infty\})$; i.e., for $\lambda \in \mathbb{R}_{>0}$ and subharmonic functions f and g on G, the functions $f + g$ and λf are subharmonic.

Proof of 13.3. Statement α) is obviously true. For β), suppose that there exists a $b \in G$ such that $f \leqq f(b)$ on G; it suffices to show that the closed set $f^{-1}(f(b))$ is open. If b were not an interior point, then there would exist a $\boldsymbol{B}(b; r) \subset\subset G$ such that

$$f(b) \leqq \frac{1}{2\pi} \int_0^{2\pi} f(b + re^{i\varphi})d\varphi < \frac{1}{2\pi} \int_0^{2\pi} f(b)d\varphi = f(b) .$$

Finally, γ) follows easily from the equivalence of i) and iii) in 13.2 (we do not use γ) in the proof that iii) $\Rightarrow$ i)). ■

We return to the proof of 13.2:

iii)$_{\text{local}}$ ⇒ i) Let the function $h \in \mathscr{C}(\bar{\boldsymbol{B}}, \mathbb{R})$ be harmonic on $\bar{\boldsymbol{B}} \subset\subset G$ and such that $f \leqq h$ on $\partial \boldsymbol{B}$. We have to deduce the same inequality on the whole of $\boldsymbol{B}$, or, equivalently, the inequality $(f-h)|_{\boldsymbol{B}} \leqq 0$. By the proof of 13.3 β), the Maximum Principle holds for $f-h$ on $\boldsymbol{B}$ (as both h and $-h$ satisfy the mean value equality, iii)$_{\text{local}}$ holds for $f-h$); hence, $f-h$ attains its maximum on $\partial \boldsymbol{B}$.

i) ⇒ ii) This follows from the fact that $\operatorname{re} P$ is harmonic.

ii) ⇒ i) If the function $h \in \mathscr{C}(\bar{\boldsymbol{B}}, \mathbb{R})$ is harmonic on $\boldsymbol{B}$, then, for each $\varepsilon > 0$, there exists a polynomial $P \in \mathbb{C}[z]$ such that $h \leqq \operatorname{re} P \leqq h + \varepsilon$ on $\partial \boldsymbol{B}$, by the Stone-Weierstrass Theorem (see 12 A. 2). The constant function ε is harmonic; hence, $h+\varepsilon$ is harmonic. By the implication iii) ⇒ i) (which has been proved above), the harmonic functions h and $\operatorname{re} P$ are subharmonic, so $h \leqq \operatorname{re} P \leqq h + \varepsilon$ on $\bar{\boldsymbol{B}}$. If a function $f \in \mathscr{C}(G, \mathbb{R} \cup \{-\infty\})$ satisfies both the inequality $f \leqq h \leqq \operatorname{re} P \leqq h+\varepsilon$ on $\partial \boldsymbol{B}$ and condition ii), then $f \leqq h + \varepsilon$ on $\bar{\boldsymbol{B}}$. Since the choice of $\varepsilon > 0$ is arbitrary, it follows that $f \leqq h$ on $\bar{\boldsymbol{B}}$.

The local versions of statements i) – iii) obviously admit a treatment analogous to that used for the global versions above, and the equivalence of all of the statements follows. ■■

E. 13a. The harmonic functions on G form a real vector space; if both f and $-f$ are subharmonic, then f is harmonic.

A function $f \in \mathscr{C}^2(I, \mathbb{R})$ is convex iff $f'' \geqq 0$. That can be proved with the identity

$$\tfrac{1}{2}(f(x_0+h)+f(x_0-h)) - f(x_0) = \tfrac{1}{2}\int_0^h \left(\int_{x_0-t}^{x_0+t} f''(s)\,ds \right) dt\,.$$

That idea is imitated in 13.4, where the operator $\dfrac{d^2}{dx^2}$ is replaced by the *Laplace operator*

$$\Delta = \frac{\partial^2}{\partial x^2} + \frac{\partial^2}{\partial y^2} = 4\,\frac{\partial^2}{\partial z\, \partial \bar{z}}\,.$$

13.4 Proposition. *A function $f \in \mathscr{C}^2(G, \mathbb{R})$ is subharmonic iff $\Delta f \geqq 0$.*

Proof. Suppose that the following equality holds for each $a \in G$ and every sufficiently small $r > 0$:

$$\text{(13.4.1)} \quad F(r) := \frac{1}{2\pi}\int_0^{2\pi} f(a + re^{i\varphi})\,d\varphi - f(a) = \int_0^r \left(\frac{1}{2\pi\varrho} \int_{|z-a| \leqq \varrho} \Delta f(z)\,dx\,dy \right) d\varrho\,.$$

If $\Delta f \geqq 0$, then $H(\varrho) := \int_{|z-a| \leqq \varrho} \Delta f(z)\,dx\,dy$ is nonnegative for each a, and thus so is $F(r)$; by 13.2, f is subharmonic. On the other hand, if there exists an $a \in G$ such that $\Delta f(a) < 0$, then, for the corresponding H, we have that $H(\varrho) < 0$ for sufficiently small positive ϱ; consequently, $F(r) < 0$ for small positive r, and it follows that f is not subharmonic.

It remains to show that (13.4.1) holds, i.e., that

$$E(r) := F(r) - \int_0^r \frac{H(\varrho)}{2\pi\varrho}\, d\varrho = 0 .$$

Now $E(0) = F(0) = 0$; we will prove that $0 = \frac{dE}{dr}(r) = \frac{dF}{dr}(r) - \frac{H(r)}{2\pi r}$. By Green's Formula,

$$\int_B (u\Delta v - v\Delta u)\,dx\,dy = \int_{\partial B} \left(u\frac{\partial v}{\partial n} - v\frac{\partial u}{\partial n}\right) ds ,$$

where $\frac{\partial}{\partial n}$ is the normal derivative and ds is the line element of $\partial \boldsymbol{B}$, the following equality holds for $u := 1$, $v := f$, and $\boldsymbol{B} := \boldsymbol{B}(a; r)$:

$$H(r) = \int_B \Delta f\,dx\,dy = \int_{\partial B} \frac{\partial f}{\partial n}\, ds = \int_0^{2\pi} \frac{\partial}{\partial r} f(a + re^{i\varphi})\cdot r\,d\varphi = 2\pi r \frac{dF}{dr}(r) .$$

In particular, then, H is continuous at $r = 0$. ■

13.5 Examples. i) For $m \geqq 1$, the function $g \in \mathscr{C}(\mathbb{C})$, $z \mapsto (z\bar{z})^m$, is subharmonic but not harmonic, for we have that

$$\Delta g = 4g_{z\bar{z}}{}^{5)} = 4m^2(z\bar{z})^{m-1} \geqq 0 \quad \text{and} \quad \Delta g(z) \neq 0 \quad \text{for } z \neq 0 .$$

ii) For $f \in \mathcal{O}(G)$, the function $\log|f|$ is subharmonic on G and harmonic on $G \setminus N(f)$: for every zero of f, the mean value inequality is automatically satisfied; on $\boldsymbol{B}(a; r) \subset\subset G \setminus N(f)$, we have that $f = |f| e^{i \arg(f)}$, and thus that $\log f = \log|f| + i \arg(f)$; hence, $\log|f|$, as the real part of the holomorphic function $\log f$, is harmonic. Note that the proof makes use only of the fact that f is continuous on G and holomorphic on $G \setminus N(f)$ (by Rado's Theorem [Ns 4. Th. 1′], such an f is in fact holomorphic).

iii) The function $-\log \delta_G : G \to \mathbb{R} \cup \{-\infty\}$ is subharmonic: first, we show that the *boundary-distance function* δ_G *is continuous*: we may assume that $G \neq \mathbb{C}$; for $a, b \in G$, then, there exist $z_a, z_b \in \partial G$ such that $\delta_G(a) = |a - z_a|$ and $\delta_G(b) = |b - z_b|$; the triangle inequality provides that $|a - z_a| \leqq |a - z_b| \leqq |a - b| + |b - z_b|$, and, by virtue of symmetry, it follows that

$$|\delta_G(a) - \delta_G(b)| \leqq |a - b| .$$

As a consequence, the function $-\log \delta_G$ is continuous. By ii) and 13.3 α), the fact that

$$-\log \delta_G(z) = \sup_{a \in \partial G} (-\log|z - a|)$$

implies that $-\log \delta_G$ is subharmonic.

---

5) $g_z := \frac{\partial g}{\partial z}$, etc.

13.6 Definition. *A subharmonic function $f \in \mathscr{C}^2(G, \mathbb{R})$ is called strictly subharmonic if Δf has no zeros.*

Note the analogy to strictly convex functions: Example 13.5 i) shows that the condition

$$f(a) < \frac{1}{2\pi} \int_0^{2\pi} f(a + re^{i\varphi}) d\varphi, \quad \forall r > 0,$$

is not sufficiently strong to ensure that f be strictly subharmonic.

Convex functions can be characterized geometrically: for a function $f \in \mathscr{C}^1(I, \mathbb{R})$, the following conditions are equivalent:

i) f is convex.

ii) The set $\{(x, y) \in \mathbb{R}^2; x \in I, y > f(x)\}$ is convex.

iii) For each $x \in I$, the graph $\Gamma(f)$ lies above the line tangent to $\Gamma(f)$ at $(x, f(x))$.

iv) For every $x, y \in I$, f satisfies the condition $f(x) \geqq f(y) + (x - y) f'(y)$.

13.7 Proposition. *For $f \in \mathscr{C}^2(G, \mathbb{R})$ and $a \in G$, define*

$$P_{f,a} := 2(z - a) f_z(a) + (z - a)^2 f_{zz}(a) \in \mathbb{C}[z].$$

If f is strictly subharmonic, then f satisfies the following condition:
For each $a \in G$, there exists a neighborhood U of a on which

$$(13.7.1) \quad f \geqq f(a) + \operatorname{re} P_{f,a}.$$

Conversely, if (13.7.1) holds for each $a \in G$, then f is subharmonic.

Proof. Let f be in $\mathscr{C}^2(G, \mathbb{R})$. Since f is real-valued, we have by E. 1a ii) that

$$\operatorname{re} P_{f,a}(z) = ((z - a) f_z(a) + \overline{(z - a)} f_{\bar z}(a)) + \tfrac{1}{2}((z - a)^2 f_{zz}(a) + \overline{(z - a)}^2 f_{\bar z \bar z}(a)).$$

The Taylor expansion of f about a with respect to z and $\bar z$ has the following form (where the remainder $R_{f,a}(z)$ is such that $\lim_{z \to a} |z - a|^{-2} R_{f,a}(z) = 0$):

$$\begin{aligned} f(z) &= f(a) + (z - a) f_z(a) + \overline{(z - a)} f_{\bar z}(a) + \\ &\quad + \tfrac{1}{2}((z - a)^2 f_{zz}(a) + \overline{(z - a)}^2 f_{\bar z \bar z}(a) + 2|z - a|^2 f_{z \bar z}(a)) + R_{f,a}(z) \\ &= f(a) + \operatorname{re} P_{f,a}(z) + \tfrac{1}{4} \Delta f(a) |z - a|^2 + R_{f,a}(z). \end{aligned}$$

For $z \neq a$, define a function F by setting

$$F(z) := (f(z) - f(a) - \operatorname{re} P_{f,a}(z)) = (\tfrac{1}{4} \Delta f(a) + |z - a|^{-2} R_{f,a}(z)) |z - a|^2.$$

If f is strictly subharmonic, then $\Delta f(a) > 0$; hence, for z near a, we have that $F(z) > 0$, and (13.7.1) holds. On the other hand, if f is not subharmonic, 13.4 implies the existence of an $a \in G$ such that $\Delta f(a) < 0$; for $z \neq a$ near a, we then have that $F(z) < 0$, so (13.7.1) does not hold at a. ∎

A function $f \in \mathscr{C}^2(G, \mathbb{R})$ that satisfies (13.7.1) is not necessarily strictly subharmonic, as is shown, for example, by the constant functions. Moreover, subharmonic functions do not necessarily satisfy (13.7.1):

13.8 Example. The function f determined by

$$f(z) := -\operatorname{re} z^3 \in \mathscr{C}^2(\mathbb{C}, \mathbb{R})$$

is harmonic, and $P_{f,0} = 0$. Since $f(t) < 0$ for $t > 0$, it follows that (13.7.1) does not hold for $a = 0$.

13.9 Proposition. *If a monotonically decreasing sequence $(g_j)_{j \in \mathbb{N}}$ of subharmonic functions converges pointwise to a continuous function g, then g is subharmonic.*

Proof. For $\varepsilon > 0$ and a harmonic function $h: \bar{\boldsymbol{B}} \to \mathbb{R}$ such that $h|_{\partial \boldsymbol{B}} \geqq g|_{\partial \boldsymbol{B}}$, the intersection of the compact sets $M_j = \{z \in \partial \boldsymbol{B};\, g_j(z) \geqq h(z) + \varepsilon\}$ is empty; hence, there exists a k such that $M_k = \emptyset$. Then $g \leqq g_k \leqq h + \varepsilon$ on $\partial \boldsymbol{B}$ and thus also on $\boldsymbol{B}$. ∎

Now we have the necessary framework for the one-dimensional case. For the generalization to higher dimensions, we again proceed in analogy to the theory of convex functions (a function of several variables is convex iff its restriction to each real segment is a convex function of one variable):

13.10 Definition. *For $X \subset \mathbb{C}^n$, a continuous function $f: X \to \mathbb{R} \cup \{-\infty\}$ is called (strictly) plurisubharmonic if each restriction of f to a complex disk D in X is (strictly) subharmonic.*

By "the restriction of f to $D = d + \boldsymbol{B}^1(1) \cdot b$ is (strictly) subharmonic", we mean of course that the function $\lambda \mapsto f(d + \lambda b)$ is (strictly) subharmonic on the unit disk $\boldsymbol{B}^1(1)$.

13.11 Remark. For a continuous function $f: X \to \mathbb{R}$, let $h_D: \bar{D} \to \mathbb{R}$ denote the harmonic extension of $f|_{\operatorname{bd} D}$ to D (see [Co X. 2.4]). Then f is plurisubharmonic iff $f|_{\bar{D}} \leqq h_D$ for every complex disk D in X: if g is harmonic and $g|_{\operatorname{bd} D} \geqq f|_{\operatorname{bd} D} = h_D|_{\operatorname{bd} D}$, then $f|_{\bar{D}} \leqq h_D \leqq g|_{\bar{D}}$.

E. 13b. Show that the function $f: z \mapsto -\log \|z\|$ is not plurisubharmonic on $\mathbb{C}^n \setminus \{0\}$ for $n \geqq 2$ (hint: find a disk D in $\mathbb{C}^n \setminus \{0\}$ with center a such that $f(a) > f(\operatorname{bd} D)$; for E. 14c).

13.12 Example. For $X \subset \mathbb{C}^n$ and $f \in \mathcal{O}(X)$, the function $\log|f|$ is plurisubharmonic (13.5 ii)).

The Chain Rule provides a simple criterion:

13.13 Proposition. *For $X \subset \mathbb{C}^n$, a function $f \in \mathscr{C}^2(X, \mathbb{R})$ is (strictly) plurisubharmonic iff its Levi form*

$$L_{f,a}: \mathbb{C}^n \to \mathbb{C},\; h \mapsto \sum_{j,k=1}^{n} f_{z_j \bar{z}_k}(a) h_j \bar{h}_k,$$

is positive semidefinite (definite) for each $a \in X$ (in symbols, $L_{f,a} \geqq 0$ and $L_{f,a} > 0$, respectively).

Proof. It suffices to consider $L_{f,a}(h)$ for small h. Thus, we may assume that

$$g: \boldsymbol{B}^1 \to \mathbb{C}^n,\ \lambda \mapsto a + \lambda h,$$

maps $\boldsymbol{B}^1$ into X. Since $g_{j\lambda}$ is the constant function h_j, we see that $\bar{g}_{k\bar{\lambda}} = \bar{h}_k$; it follows that

$$(f \circ g)_{\lambda\bar{\lambda}} \underset{\text{E. 1a iv)}}{=} \sum_{j,k=1}^{n} (f_{z_j \bar{z}_k} \circ g) g_{j\lambda} \bar{g}_{k\bar{\lambda}} + 0 = L_{f,g(\cdot)}(h).$$

The proposition follows by 13.4 and 13.6. ■

For the description of open balls given in § 14, we need the following information:

13.14 Example. The function $f \in \mathscr{C}(\mathbb{C}^n)$, $z \mapsto \|z\|^2 - 1$, has the Levi form $L_{f,a}(h) = \sum_{j=1}^{n} |h_j|^2$ and is thus strictly plurisubharmonic.

E. 13c. Prove that, for $k \in \mathbb{N}$, the function determined by the assignment

$$f_k(z) = \frac{1}{n} \sum_{j=1}^{n} |z_j|^2 + \sum_{j=1}^{n} |z_j|^{2k} - 1$$

is strictly plurisubharmonic (for 62.2).

E. 13d. Show that, for $G \subset \mathbb{C}^n$, $a \in G$, and $f \in \mathscr{C}^2(G, \mathbb{R})$, there exists a polynomial $P_{f,a} \in \mathbb{C}[z_1, \dots, z_n]$ such that

$$f(z) = f(a) + \operatorname{re} P_{f,a}(z) + L_{f,a}(z-a) + R_{f,a}(z) \quad \text{near } a,$$

where $R_{f,a}$ is such that $\lim_{z \to a} \frac{R_{f,a}(z)}{|z-a|^2} = 0$. (Hint: with the notation $f[j] := f_{z_j}(a)(z_j - a_j)$, etc., we have the following equality:

$$f(z) - f(a) = \Sigma f[j] + \Sigma f[\bar{j}] + \tfrac{1}{2} \Sigma (f[j,k] + f[\bar{j},k] + f[j,\bar{k}] + f[\bar{j},\bar{k}]) + R_{f,a}(z);$$

set $P_{f,a} := \Sigma 2 f[j] + \Sigma f[j,k]$ and apply E. 1a ii); for 62.3).

§ 14 Pseudoconvex Domains

In this section, we introduce yet another type of convexity, called pseudoconvexity, with the help of plurisubharmonic functions. Domains of holomorphy are pseudoconvex (see 12 A. 4); whether, conversely, pseudoconvex domains must also be holomorphically convex, is a difficult question ("*Levi's problem*"), which we answer affirmatively in 63 A. 1 in connection with the proof of Theorem B.

A major advantage of pseudoconvexity is that it constitutes a local condition on the boundary; hence, it is often easier to investigate than holomorphic convexity (for example, see 14.5 and 14.7).

We use the following notation: for a norm β on $\mathbb{C}^n$,

$\boldsymbol{B}_\beta(a; r)$ is the (open) *β-ball* of radius $r > 0$ about a in $\mathbb{C}^n$, $\boldsymbol{B}_\beta(r) := \boldsymbol{B}_\beta(0; r)$,

and for a region $X \Subset \mathbb{C}^n$,

$$\delta_{X,\beta} : X \to \mathbb{R} \cup \{\infty\},\ x \mapsto \operatorname{dist}_\beta(x, \partial X) = \sup\{r;\ \boldsymbol{B}_\beta(x; r) \subset X\},$$

is the *β-boundary-distance function.*

Of course, we already have a notation for certain cases:

$$\boldsymbol{B}_{\|\cdot\|} = \boldsymbol{B},\ \boldsymbol{B}_{|\cdot|} = \boldsymbol{P}, \quad \text{and} \quad \delta_{X,|\cdot|} = \delta_X .$$

For $B \subset X \Subset \mathbb{C}^n$, we define the *pseudoconvex hull* of B in X (in analogy to the holomorphically convex hull $\hat{B}_{\mathcal{O}(X)}$) to be

$$\tilde{B} := \{x \in X;\ h(x) \leqq \sup_{z \in B} h(z)$$

for every plurisubharmonic $h : X \to \mathbb{R} \cup \{-\infty\}\}$.

We have that

$$\tilde{B} \subset \hat{B}_{\mathcal{O}(X)},$$

for, given an $a \in X \backslash \hat{B}$, we can find an $f \in \mathcal{O}(X)$ such that $\|f\|_B < |f(a)|$, and hence such that $\sup_{z \in B} \log|f(z)| < \log|f(a)|$; since $\log|f|$ is plurisubharmonic (by 13.12), a is not in $\tilde{B}$.

14.1 Definition and Proposition. *A region $X \Subset \mathbb{C}^n$ is called pseudoconvex if one of the following equivalent conditions is satisfied:*

i) For every norm β (on $\mathbb{C}^n$), the function $-\log \delta_{X,\beta} : X \to \mathbb{R}$ is plurisubharmonic.

ii) There exists a norm β such that $-\log \delta_{X,\beta}$ is plurisubharmonic.

iii) There exists a plurisubharmonic function $u : X \to \mathbb{R}$ that is proper and bounded from below, i.e., for every $r \in \mathbb{R}$,

$$X_r := \{x \in X;\ u(x) < r\} \Subset X .$$

iv) For every compact set $K \subset X$, the pseudoconvex hull $\tilde{K}$ is compact.

v) For every ball $\boldsymbol{B} \subset \mathbb{C}$, and every continuous mapping $\Phi : \boldsymbol{I} \times \mathbb{C} \to \mathbb{C}^n$ [6] *with $\Phi_t : \mathbb{C} \to \mathbb{C}^n$, $z \mapsto \Phi(t, z)$, is holomorphic and such that $\Phi_t(\partial \boldsymbol{B}) \subset X$ for each $t \in \boldsymbol{I}$, the following statement is true: the inclusion $\Phi_t(\boldsymbol{B}) \subset X$ holds either for every $t \in \boldsymbol{I}$ or for no $t \in \boldsymbol{I}$.*

The final condition precludes the following boundary behavior for X:

[6] $\boldsymbol{I} := [0,1]$.

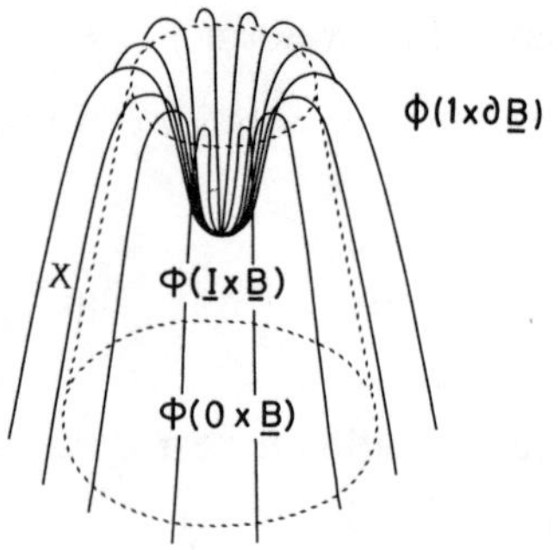

It follows from condition v) that one-dimensional regions are pseudoconvex; of course, that is also a consequence of 12.10 i) and the following fact (proved here for the second time):

14.2 Corollary. *Regions of holomorphy are pseudoconvex.*

Proof. If $X \Subset \mathbb{C}^n$ is a region of holomorphy, and if $K \subset X$ is compact, then $\tilde{K} \subset \hat{K}$ is also compact; hence, X satisfies condition iv). ■

Proof of 14.1. ii) ⇒ iii) For $X = \mathbb{C}^n$, set $u(z) = \|z\|^2$ (see 13.14). For $X \neq \mathbb{C}^n$, we may assume that $0 \notin X$; set $u(z) := \|z\|^2 - \log \delta_{X,\beta}(z)$. For an appropriate $c \in \mathbb{R}_{>0}$, we have that

$$u(z) \geqq \|z\|^2 - \delta_{X,\beta}(z) \geqq \|z\|^2 - \beta(z) \geqq \|z\|^2 - c\|z\| \geqq -c^2/4;$$

hence, u is bounded from below and proper (see 33 B. 1 v)).

iii) ⇒ iv) Each compact $K \subset X$ lies in some X_r; as a closed subset of the compact set $\bar{X}_r$, the hull $\tilde{K}$ is also compact.

iv) ⇒ v) It suffices to show that the set $T := \{t \in \boldsymbol{I};\ \Phi_t(\bar{\boldsymbol{B}}) \subset X\}$ is open and closed in $\boldsymbol{I}$. The distance from the compact set $\Phi_t(\bar{\boldsymbol{B}})$ to ∂X is positive for each $t \in T$; hence, T is open, since Φ is continuous. By 14.3, $\Phi_t(\bar{\boldsymbol{B}}) \subset \widetilde{\Phi_t(\partial \boldsymbol{B})}$ for every $t \in T$. For $K := \Phi(\boldsymbol{I} \times \partial \boldsymbol{B}) \subset X$, it follows that

$$\Phi(T \times \bar{\boldsymbol{B}}) \subset \widetilde{\Phi(T \times \partial \boldsymbol{B})} \subset \tilde{K} \Subset X;$$

passage to the closures yields that

$$\Phi(\bar{T} \times \bar{\boldsymbol{B}}) \subset \overline{\Phi(T \times \bar{\boldsymbol{B}})} \subset \bar{\tilde{K}} = \tilde{K} \subset X.$$

Then $\bar{T} \subset T$, so T is closed.

v) ⇒ i) Set $\delta := \delta_{X,\beta}$. By 13.2, it suffices to show that, for each complex disk $D = d + \boldsymbol{B}^1(1) \cdot b \Subset X$, and each polynomial $P \in \mathbb{C}[z_1]$, the inequality

$$(14.1.1) \quad -\log \delta(d + \lambda b) \leqq \operatorname{re} P(\lambda)$$

holds for each $\lambda \in \boldsymbol{B}^1 := \boldsymbol{B}^1(1)$ if it holds for each $\lambda \in \partial \boldsymbol{B}^1$. For fixed $\lambda \in \mathbb{C}$, (14.1.1) is equivalent to the inequality

$$\delta(d+\lambda b) \geqq e^{-\operatorname{re} P(\lambda)} = |e^{-P(\lambda)}| =: r(\lambda);$$

hence, we can replace (14.1.1) with the more geometric condition

(14.1.2) $d+\lambda b+\boldsymbol{B}^n_\beta(r(\lambda)) \subset X$.

Suppose that (14.1.2) holds for each $\lambda \in \partial\boldsymbol{B}^1$; we have to show that it holds for each $\lambda \in \boldsymbol{B}^1$ as well. To that end, fix $\zeta \in \boldsymbol{B}^1$ and $z \in \boldsymbol{B}^n_\beta(r(\zeta))$, and consider the mapping

$$\Phi: \boldsymbol{I} \times \mathbb{C} \to \mathbb{C}^n,\ (t,\lambda) \mapsto d+\lambda b+te^{P(\zeta)-P(\lambda)}z.$$

Now $\Phi_0(\boldsymbol{B}^1) = D \subset X$; if we can show that $\Phi(\boldsymbol{I} \times \partial\boldsymbol{B}^1) \subset X$, then we can conclude with the help of v) that $\Phi(\boldsymbol{I} \times \boldsymbol{B}^1) \subset X$ and $d+\zeta b+z = \Phi(1,\zeta) \in X$.

We have that $te^{P(\zeta)-P(\lambda)}z \in \boldsymbol{B}^n_\beta(r(\lambda))$, since

$$\beta(te^{P(\zeta)-P(\lambda)}z) \leqq 1 \cdot |e^{P(\zeta)}| \cdot |e^{-P(\lambda)}| \beta(z) < \frac{1}{r(\zeta)} \cdot r(\lambda) \cdot r(\zeta).$$

For $\lambda \in \partial\boldsymbol{B}^1$, it follows from (14.1.2) that $\Phi(t,\lambda) \in X$. ■

In the proof of 14.1, we used the following fact:

14.3 Lemma. *For $\boldsymbol{B} = \boldsymbol{B}^1(a;r)$ and $h \in \operatorname{Hol}(\bar{\boldsymbol{B}}, X)$,*

$$h(\bar{\boldsymbol{B}}) \subset \widehat{h(\partial\boldsymbol{B})}.$$

Proof. We need to show that, for each plurisubharmonic function $f: X \to \mathbb{R} \cup \{-\infty\}$, and each $w \in h(\bar{\boldsymbol{B}})$,

$$f(w) \leqq \sup\{f(z);\ z \in h(\partial\boldsymbol{B})\}.$$

By 14.4, that follows from the Maximum Principle 13.3 β). ■

14.4. Remark. *For every holomorphic mapping $h: Y \to X$, and every plurisubharmonic function f on X, the composition $f \circ h$ is plurisubharmonic on Y.*

Proof. Obviously, the function $g = f \circ h$ is continuous. Without loss of generality, suppose that $Y = \boldsymbol{P}^1(r)$ and $f \not\equiv -\infty$. If f is in $\mathscr{C}^2(X,\mathbb{R})$, then, by 13.4, we need only show that $\Delta g \geqq 0$. Since h is holomorphic, the Chain Rule and 13.13 yield that

$$\tfrac{1}{4}\Delta g = \frac{\partial^2 g}{\partial w\, \partial \bar{w}} = \sum_{j,k=1}^{n} f_{z_j \bar{z}_k} \circ g h_{jw} \bar{h}_{k\bar{w}} = L_{f,h}(h_w) \geqq 0 \text{ on } Y.$$

For $f \notin \mathscr{C}^2(X,\mathbb{R})$, suppose that, for each $a \in X$, there exists a monotonically decreasing sequence of plurisubharmonic functions $f_j \in \mathscr{C}^2(X,\mathbb{R})$ converging pointwise to f near a. Then $(f_j \circ h)_{j \in \mathbb{N}}$ must be a monotonically decreasing sequence of plurisubharmonic functions as well; by 13.9, g is plurisubharmonic. The construction of such a sequence (f_j) is based on standard techniques of real analysis, which we wish only to intimate: for sufficiently small fixed $s > 0$, choose a $\mathscr{C}^\infty$-function $\omega: \mathbb{R} \to \mathbb{R}_{\geqslant 0}$ that satisfies the equality $\int_{\mathbb{C}^n} \omega(\|z\|)d\lambda(z) = 1$ (with Lebesgue measure $d\lambda(z)$), and whose graph looks like this:

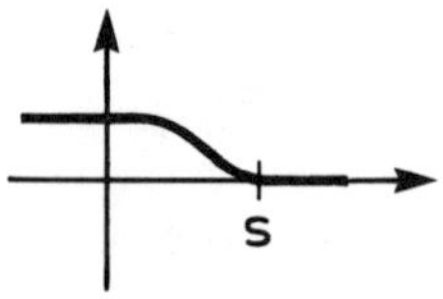

For $j \in \mathbb{N}$, define plurisubharmonic $\mathscr{C}^\infty$-functions f_j by setting

$$f_j(z) := \int_{\mathbb{C}^n} f(z + \zeta / j)\omega(\|\zeta\|)d\lambda(\zeta).$$

Then (f_j) is a monotonically decreasing sequence that convergers pointwise to f on $\{z \in X; \delta_X(z) > s\}$. Details may be found in [PfII. 1.4 Satz 13]. ■■

The pseudoconvexity of a domain is a local property of the boundary (as is convexity); more precisely, we have the following fact:

14.5 Proposition. *A region $X \subset\subset \mathbb{C}^n$ is pseudoconvex iff each $a \in \partial X$ has an open neighborhood U^a such that $U^a \cap X$ is pseudoconvex.*

Proof. Only if. This direction follows by application of 14.2 and E. 14a ii) to the pseudoconvex regions $\boldsymbol{B}^n(a; 1)$ and X. *If.* By E. 14a ii), each $X \cap \boldsymbol{B}^n(r)$ satisfies the same boundary condition that X satisfies, so, by E. 14a i), we may assume that X is bounded. For each $a \in \partial X$, choose a neighborhood V^a of a such that $V^a \subset\subset U^a$ and $\delta_X = \delta_{X \cap U^a}$ on $V^a \cap X$. Then the function $-\log \delta_X$ is plurisubharmonic on $V := X \cap (\bigcup_{a \in \partial X} V^a)$, and, by 13.3, the function $h := \max(-\log \delta_X, \|\log \delta_X\|_{X \setminus V} + 1)$ is plurisubharmonic on X. Obviously, h is proper and bounded from below, so condition 14.1 iii) is satisfied. ■

E. 14a. Prove the following statements, using 14.1 v):
i) Every increasing union of pseudoconvex regions is pseudoconvex.
ii) If the intersection of a family of pseudoconvex regions is open, then it is pseudoconvex.

A special kind of pseudoconvex domain is important for both the generalization to complex spaces in E. 62d and the solution of the Levi problem in Chapter 6:

14.6 Definition. *A bounded region $X \subset\subset \mathbb{C}^n$ is called strictly pseudoconvex if there exists on some neighborhood U of ∂X a strictly plurisubharmonic function φ such that $X \cap U = \{z \in U; \varphi(z) < 0\}$.*

By 13.14, balls are strictly pseudoconvex. It follows from 14.5 that strictly pseudoconvex regions are pseudoconvex: for each ball $\boldsymbol{B} \subset U$, the intersection $\boldsymbol{B} \cap X$ is pseudoconvex, since $\tilde{K} \subset \hat{K} \subset\subset \boldsymbol{B}$ for each compact $K \subset \boldsymbol{B} \cap X$; from the fact that $\varphi|_K < 0$, it follows that $\tilde{K} \subset\subset \boldsymbol{B} \cap X$, so 14.1 iv) is satisfied.

Pseudoconvex domains can be approximated by strictly pseudoconvex domains:

14.7 Proposition. *Each pseudoconvex region $X \subset\subset \mathbb{C}^n$ has a "strictly pseudoconvex"* *exhaustion* $X = \bigcup_{j=1}^{\infty} X_j$ *with strictly pseudoconvex regions* X_j *that satisfy the following two conditions for each* j:

i) $X_j \subset\subset X_{j+1}$.

ii) There exists a strictly plurisubharmonic function $\varphi_j \in \mathscr{C}^{\infty}(X_{j+1}, \mathbb{R})$ *such that the intersection of* X_j *with each connected component of* X *is a connected component of* $\{\varphi_j < 0\}$.

Proof. The connected components of X are countable in number; hence, we may write them as H_m, $m = 0, \ldots$. Let $f: X \to \mathbb{R}$ be a proper subharmonic function such that $\inf f|_{H_m} = m$ (see 14.1). Then it is easy to see how the following construction for $X := H_0$ generalizes for an arbitrary X. Let U_1 denote a nonempty connected component of $\{f < 1\}$; we obtain recursively an exhaustion of G with domains $U_j \subset\subset U_{j+1}$: given U_j, we let U_{j+1} be that connected component of $\{f < j+1\}$ which contains U_j. Now we construct, for fixed j, a strictly pseudoconvex domain G_j such that $U_{j-1} \subset\subset G_j \subset\subset U_j$: there exists a sequence of *strictly* plurisubharmonic $\mathscr{C}^{\infty}$-functions ψ_k on U_{j+1} such that $\psi_k \downarrow f|_{U_{j+1}}$ (for an appropriate sequence (f_k), such as the one in the proof of 14.4, set $\psi_k := f_k + \|z\|^2/k$). The increasing sequence $(U_j \cap \{\psi_k < j\})_{k \in \mathbb{N}}$ exhausts U_j, so there exists a k such that $\bar{U}_{j-1} \subset \subset U_j \cap \{\psi_k < j\}$. For sufficiently small $\varepsilon > 0$, the closure $\bar{U}_{j-1}$ lies in a connected component G_j of $U_j \cap \{\psi_k < j - \varepsilon\}$; by Sard's Theorem ([Na$_3$ § 2]), ε can be chosen so that $\frac{\partial \psi_k}{\partial z}$ has no zeros on $\{\psi_k = j - \varepsilon\}$ (specifically, G_j has a smooth boundary). Hence, for $\varphi_j := \psi_k - j + \varepsilon$, there exists a neighborhood W of $\bar{G}_j$ in U_j that does not intersect any other connected component of $\{\varphi_j < 0\}$; then $W \cap G_j = W \cap \{\varphi_j < 0\}$, and G_j is strictly pseudoconvex. ■

E. 14b. Prove that every annulus with positive radii is strictly pseudoconvex (hint: for $\alpha, \beta > 0$, the function $z \mapsto \alpha \bar{z} z - \log(\beta \bar{z} z)$ is strictly subharmonic; for E. 63Aa).

E. 14c. Show that $\mathbb{C}^n \setminus \{0\}$ is not pseudoconvex if $n \geqq 2$ (hint: E. 13b).

As an example of the applications of pseudoconvex domains, we give a characterization of holomorphically convex "tubes". These domains play a role in quantum physics; from a mathematical standpoint, they are remarkable because they are among the few domains G in $\mathbb{C}^n$ whose *envelope of holomorphy* in $\mathbb{C}^n$ (i.e., a domain of holomorphy that includes G and to which each function in $\mathcal{O}(G)$ extends holomorphically) not only exists, but can be given explicitly (see 57.9 ii)).

14.8 Definition. *If* B *is a domain in* $\mathbb{R}^n$, *then the domain* $T_B := B + i\mathbb{R}^n$ *is called the tube on* B *in* $\mathbb{C}^n = \mathbb{R}^n + i\mathbb{R}^n$.

14.9 Proposition. *For a domain* B *in* $\mathbb{R}^n$, *the following statements are equivalent:*

i) B *is convex.*

ii) T_B *is convex.*

iii) T_B *is holomorphically convex.*
iv) T_B *is pseudoconvex.*

Proof. That i) implies ii) is trivial; the implications "ii) ⇒ iii)" and "iii) ⇒ iv)" follow from E. 12f and 14.2, respectively. It remains to show that B is convex if T_B is pseudoconvex. Thus, it suffices to show, for each segment $A := b + [-1,1] \cdot d$ in B, that the restriction $\delta_B|_A$ assumes its minimum value at $b+d$ or $b-d$, or, equivalently, that $-\log \delta_B|_A$ assumes its maximum there.

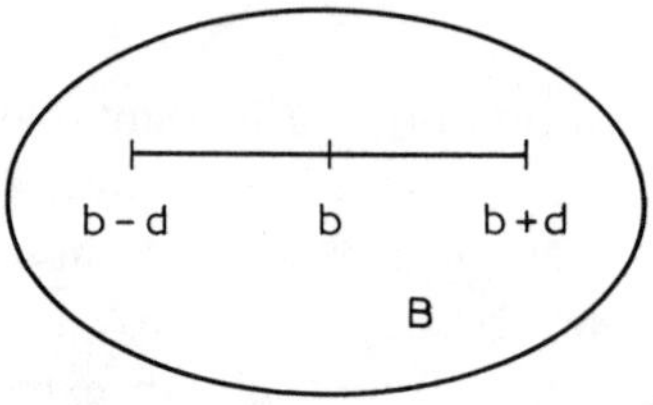

The function $\varphi: [-1,+1] \to \mathbb{R}, t \mapsto -\log \delta_B(b+td)$ has a subharmonic extension $\phi: \boldsymbol{B}^1(1) \to \mathbb{R}, \lambda \mapsto -\log \delta_{T_B}(b+\lambda d)$, and the equality $\delta_{T_B}(z) = \delta_B(\operatorname{re} z)$ for $z \in T_B$ implies that

$$\phi(\lambda) = \varphi(\operatorname{re}\lambda).$$

If φ assumes its maximum value at $t_0 \neq \pm 1$, then ϕ assumes its maximum value at $t_0 \in \boldsymbol{B}^1(1)$, and must be constant, by 13.3 β); hence, φ is constant. ■

Part Two: Function Theory on Analytic Sets

Whereas we were concerned with function theory on domains in complex number spaces in the first part of the book, we turn to the investigation of analytic sets in the second part. These solution sets of systems of holomorphic equations play a central role in complex analysis, a role similar to those played by solution sets of systems of linear equations and algebraic equations in linear algebra and algebraic geometry, respectively.

The theory of ringed spaces provides the formal framework within which we study analytic sets. By distinguishing "*structure sheaves*", we endow each analytic set with a complex structure that enables us to study holomorphic functions on it. This shift from embedded analytic sets to analytic sets with structure sheaves corresponds to that from concrete to abstract Riemann surfaces, the necessity of which was so clearly underscored by H. Weyl (see p. V).

One then extends the construction of holomorphic mappings to such abstract analytic sets; for example, that makes it possible to identify $\mathbb{C}$ with $N(\mathbb{C}^2; z_2)$ and $N(\mathbb{C}^3; z_2, z_3)$, but not with $N(\mathbb{C}^2; z_1^2 - z_2^3)$, even though there exists a holomorphic homeomorphism from $\mathbb{C}$ onto $N(\mathbb{C}^2; z_1^2 - z_2^3)$, namely, $z \mapsto (z^3, z^2)$. On the basis of the abstract construction, we are also in a position to provide the quotient space $\mathbb{C}^2/(z \sim -z)$ with a natural complex structure that makes it isomorphic to the embedded analytic set $A = N(\mathbb{C}^3;\ w_1 w_2 - w_3^2)$ (via the holomorphic mapping $\mathbb{C}^2 \to \mathbb{C}^3$, $(z_1, z_2) \mapsto (z_1^2, z_2^2, z_1 z_2)$), and thus isomorphic to the Riemann domain of $\sqrt{w_1 w_2}$.

In the study of a system of holomorphic equations $f_1 = \ldots = f_m = 0$, the principal object of interest may be the mere geometry of the solution set, in which case one analyzes the *zero-set* $N(f_1, \ldots, f_m)$; on the other hand, if algebraic properties of the system also are to be taken into account, then the "*variety*" $V(f_1, \ldots, f_m)$ is investigated. The zero-sets of the form $N(f_1, \ldots, f_m)$ serve as local models for the construction of "*reduced*" complex spaces, varieties $V(f_1, \ldots, f_m)$, for that of general complex spaces.

Chapter 3 contains the formal framework for the construction, independent of any embedding, of complex spaces (for some important classes of complex spaces, e.g., for the compact Riemann surfaces, there can be no embedding into a complex number space); specifically, Chapter 3 is devoted to ringed spaces. In addition, it is intended to provide motivation and to develop a geometric intuition; hence, in addition to standard examples of manifolds, we include the "*archsingularities*", which we use throughout the text to illustrate various phenomena: the *Achsenkreuz*, Neil's parabola, $\sqrt{uv}$, and the Whitney umbrella. Multiple points and the line with double point may be considered as the nonreduced arch-singularities; accordingly, they are treated mainly in Chapter 4, where a systematic theory of complex spaces is developed with the help of coherent sheaves.

The local study of complex spaces is based on the "Antiequivalence Theorem": *If* $x \in X$ *and* $y \in Y$ *are points in complex spaces such that the stalks of the structure sheaves* ${}_X\mathcal{O}_x$ *and* ${}_Y\mathcal{O}_y$ *are isomorphic algebras, then there exist neighborhoods* $U(x)$ *in* X *and* $U(y)$ *in* Y *that are isomorphic.* Therefore, in Chapter 2, we investigate first the "*analytic algebras*" ${}_X\mathcal{O}_x$ (i.e., quotients of power series algebras by ideals); from the Weierstrass Preparation Theorem, we derive the fundamental properties of power series algebras and their homomorphic images.

The bridge from this punctual theory to the local theory – the Antiequivalence Theorem – is built with the help of a coherence argument, as is the case with every step from "punctual" to "local" in complex analysis; a particularly illustrative example of that is the connection between the algebraic proposition 46.2 and the Representation Theorem for Prime Germs.

Chapter 2: The Weierstrass Preparation Theorem and its Applications

Although analytic sets are defined locally, we investigate them first from a punctual standpoint; a synopsis of the punctual statements will be possible in Chapter 3 after the introduction of the concept of a sheaf. The punctual theory is essentially formal and thus enables us to make good use of algebraic methods: we investigate many properties of convergent power series algebras and their homomorphisms at first from a formal algebraic standpoint, the necessary considerations of convergence being withheld for a second step (such an approach has proved to be useful in other mathematical disciplines as well). That is manifest in the fundamental result of the punctual theory, the *Weierstrass Preparation Theorem*, in which the principal ideals of (full) power series algebras are described in terms of Weierstrass polynomials. We give a proof of that theorem using ideas of Siegel-Stickelberger; the proof uses only elementary methods, and the verification of convergence statements requires only a slight additional effort.

The one-dimensional Weierstrass Preparation Theorem states that each power series f converging near 0 is of the form $z^{\text{order}(f)}$, up to multiplication by a unit, an assertion so elementary, that the fundamental significance of the representation is hardly noticed.

One consequence of the Preparation Theorem is the *Weierstrass Division Formula*, which is the analogue of the Euclidean Division Algorithm for power series algebras; the most important algebraic results follow from it: power series algebras are noetherian, local, factorial, and henselian.

Homomorphic images of power series algebras, called "*analytic algebras*", are not necessarily power series algebras themselves (that corresponds to the appearance of singularities in geometry). We investigate the basic algebraic properties of these analytic algebras and their homomorphisms, and explain their geometric significance (a characterization of complex space germs and their holomorphic mappings that is independent of embeddings) with the help of the "Antiequivalence Principle" in § 44.

§ 21 Power Series Algebras

According to 4.8, each function that is holomorphic at $0 \in \mathbb{C}^n$ can be represented near $0 \in \mathbb{C}^n$ by a convergent power series, namely, its Taylor series. Before investigating convergence, we consider the formal case:

21.1 Definition. *The set of formal power series in n indeterminates over* $\mathbb{C}$, *denoted by*

$$\mathbb{C}[\![X_1,\ldots,X_n]\!] = \mathbb{C}[\![X]\!] = {}_n\mathscr{F},$$

consists of all expressions of the form

$$P := \sum_{\nu \in \mathbb{N}^n} a_\nu X^\nu := \sum a_{\nu_1 \ldots \nu_n} X_1^{\nu_1} \ldots X_n^{\nu_n},$$

with $a_\nu \in \mathbb{C}$ *for each* $\nu \in \mathbb{N}^n$.

With the operations

addition, $\quad \Sigma a_\nu X^\nu + \Sigma b_\nu X^\nu := \Sigma (a_\nu + b_\nu) X^\nu,$

multiplication, $\quad (\Sigma a_\nu X^\nu)(\Sigma b_\mu X^\mu) := \sum_\lambda \left(\sum_{\nu+\mu=\lambda} a_\nu b_\mu \right) X^\lambda$, and

scalar multiplication, $\quad t(\Sigma a_\nu X^\nu) := \Sigma (t a_\nu) X^\nu,$

$\mathbb{C}[\![X]\!]$ is a ($\mathbb{C}$-)algebra that includes the polynomial algebra $\mathbb{C}[X] = \mathbb{C}[X_1,\ldots,X_n]$. Although we have not introduced any notion of convergence in $\mathbb{C}[\![X]\!]$ so far (see E. 21e), we can perform certain infinite sums: a family $(P_j)_{j \in J}$ of formal power series $P = \Sigma_j a_\nu X^\nu$ is called *summable* if, for each $q \in \mathbb{N}$, there are only finitely many indices $j \in J$ such that $o(P_j) \leq q$; in that case, for each $\nu \in \mathbb{N}^n$, the set $\{j \in J;\, {}_j a_\nu \neq 0\}$ is finite, and the sum

$$\sum_{j \in J} P_j := \sum_\nu \left(\sum_{j \in J} {}_j a_\nu \right) X^\nu$$

is well-defined (in particular, $P_j \neq 0$ for at most countably many j). For $P \in \mathbb{C}[\![X]\!]$ and summable families $(P_j)_{j \in J}$ and $(Q_j)_{j \in J}$, the following computational rules hold:

(21.1.1) $\quad \Sigma P_j + \Sigma Q_j = \Sigma (P_j + Q_j) \quad$ and $\quad P \cdot \left(\sum_{j \in J} P_j \right) = \sum_{j \in J} (P \cdot P_j).$

Every family $(P_j)_{j \in \mathbb{N}}$ of *homogeneous polynomials* $P_j = \sum_{|\nu|=j} a_\nu X^\nu$ *of degree* j is summable; each $P \in \mathbb{C}[\![X]\!]$ has a unique *representation by homogeneous polynomials* $P = \sum_{j=0}^{\infty} P_j$; for $P, Q \in \mathbb{C}[\![X]\!]$, that yields the equality

(21.1.2) $P \cdot Q = \sum_{l=0}^{\infty} \sum_{j+k=l} P_j \cdot Q_k$.

The only units of $\mathbb{C}[X]$ are the nonzero constant polynomials; in $\mathbb{C}[\![X]\!]$, however, every formal power series $P = \sum a_\nu X^\nu$ with nonzero "*value*" $P(0) := a_0$ is a unit:

21.2 Proposition. *$\mathbb{C}[\![X]\!]$ is a local algebra without zero-divisors; its maximal ideal is*

$$\mathfrak{m}_{[\![X]\!]} = \{P \in \mathbb{C}[\![X]\!];\ P(0) = 0\}.$$

A *ring*[1] R is called *local* if it has exactly one maximal ideal $\mathfrak{m} = \mathfrak{m}_R$ (an ideal $\mathfrak{m}$ is maximal iff $\mathfrak{m} \neq R$ and there exists no ideal $\mathfrak{a}$ such that $\mathfrak{m} \subsetneqq \mathfrak{a} \subsetneqq R$; that is the case iff $R/\mathfrak{m}$ is a field). Fields are local rings. An algebra R that is local as a ring is called a *local algebra* if the composition φ of the canonical mappings $\mathbb{C} \cdot 1_R \to R \twoheadrightarrow R/\mathfrak{m}$ is an isomorphism of fields; by identification of $\mathbb{C}$ with $\mathbb{C} \cdot 1_R$, that determines a vector-space-isomorphism $R \cong \mathbb{C} \oplus \mathfrak{m}$. Note that φ is not automatically an isomorphism, as the example $R = Q(\mathbb{C}[\![X]\!])$ shows (see § 23A).

E. 21a. Prove that a ring is local iff its nonunits form an ideal $\mathfrak{a}$ (in which case $\mathfrak{a}$ is the maximal ideal; for 21.2).

E. 21b. Prove that $\mathbb{C}[X]$ is not a local ring.

For the proof of 21.2, we need the concept "*order $o(P)$ of a power series P*", which, in the case of a polynomial P of one variable, is the usual multiplicity of the origin as a zero of P:

$$o\Big(\sum_{\nu \in \mathbb{N}^n} a_\nu X^\nu\Big) := \begin{cases} \infty, \text{ if } a_\nu = 0 \ \forall \nu \in \mathbb{N}^n \\ \min\{|\nu|;\ a_\nu \neq 0\}, \text{ otherwise.} \end{cases}$$

By (21.1.1) and (21.1.2), the order enjoys the following properties:

$$o(P+Q) \geqq \min\{o(P), o(Q)\},\ o(P \cdot Q) = o(P) + o(Q),$$
$$o(tP) = o(P)\ \forall t \in \mathbb{C}^*.$$

Proof of 21.2. The algebra $\mathbb{C}[\![X]\!]$ has no zero-divisors, since the product of two nonzero power series is of finite order: for $P, Q \neq 0$, $o(P \cdot Q) = o(P) + o(Q) < \infty$.

By (21.2.1), the set $\mathfrak{m}_{[\![X]\!]}$ is an ideal; it contains no unit, as we have that $1 \notin \mathfrak{m}_{[\![X]\!]}$. On the other hand, every $P \notin \mathfrak{m}_{[\![X]\!]}$ is a unit (thus by E. 21a, $\mathbb{C}[\![X]\!]$ is a local ring): We may assume that $P(0) = 1$. The family $((1-P)^j)_{j \in \mathbb{N}}$ is summable, since $o((1-P)^j) = j \cdot o(1-P) \geqq j$. From (21.2.1), we deduce that

$$P \cdot \sum_{j=0}^{\infty} (1-P)^j = (1-(1-P)) \cdot \sum_{j=0}^{\infty} (1-P)^j = \sum_{j=0}^{\infty} (1-P)^j - \sum_{j=0}^{\infty} (1-P)^{j+1} = 1;$$

hence, P is a unit with inverse $\sum_{j=0}^{\infty} (1-P)^j$.

---

[1] By "*ring*", we always mean "commutative ring with a unit element"; if S is a *subring* of R, then the unit element of S is that of R, by definition.

Finally, $\mathbb{C}[\![X]\!]$ is a local algebra, since every P admits a unique decomposition $P = P(0) + (P - P(0))$ with $P - P(0) \in \mathfrak{m}_{[\![X]\!]}$. ∎

Now we want to distinguish those formal power series which converge near 0. For each $r \in \mathbb{R}^n_{>0}$, the mapping

$$\|\cdot\|_r : \mathbb{C}[\![X]\!] \to \mathbb{R} \cup \{\infty\},\ \Sigma a_\nu X^\nu \mapsto \Sigma |a_\nu| r^\nu,$$

is (by 21.5) a *pseudonorm* (i.e., a "norm" that may assume the value ∞).

21.3 Definition. *A formal power series $P \in \mathbb{C}[\![X_1, \ldots, X_n]\!]$ is called convergent if there exists an $r \in \mathbb{R}^n_{>0}$ such that $\|P\|_r < \infty$. The set of convergent power series is denoted by $\mathbb{C}\{X_1, \ldots, X_n\} = \mathbb{C}\{X\} = {}_n\mathcal{O}_0$.*

For a fixed $a \in \mathbb{C}^n$, each convergent power series $P = \Sigma c_\nu X^\nu$ determines a holomorphic function on a polydisk $\boldsymbol{P}^n(a; r)$ by means of the assignment $z \mapsto \Sigma c_\nu (z-a)^\nu$ (and vice versa, see 4.8; of course, r depends on P). The set ${}_n\mathcal{O}_a$ of such "functions" obviously forms an algebra. In general, we consider only ${}_n\mathcal{O}_0$, since, by 3.4, the translation

$$\tau : \mathbb{C}^n \to \mathbb{C}^n,\ z \mapsto z - a,$$

determines an algebra-isomorphism

$$\tau^0 : {}_n\mathcal{O}_0 \to {}_n\mathcal{O}_a.$$

E. 21c. Let $\mathfrak{U} = (U_j)_{j \in J}$ be a family of open subsets of $\mathbb{C}^n$, and suppose that the order on J determined by

$$j \geqq k :\Leftrightarrow U_j \subset U_k$$

directs J *upward* (for every k, l there exists a j such that $j \geqq k$ and $j \geqq l$). Then, with the restriction-homomorphisms $\varrho_{jk} : \mathcal{O}(U_k) \to \mathcal{O}(U_j)$ for $U_j \subset U_k$, the system $(\mathcal{O}(U_j), \varrho_{jk})_{j \in J}$ is an *inductive* (or *direct*) *system*; in other words, the following two conditions hold:

i) $\varrho_{jj} = \mathrm{id}_{\mathcal{O}(U_j)}$,

ii) $\varrho_{jk} \circ \varrho_{kl} = \varrho_{jl}\ \forall j \geqq k \geqq l$.

We call functions $f \in \mathcal{O}(U_k)$ and $g \in \mathcal{O}(U_l)$ *equivalent* if there exists a neighborhood $U_j \subset U_k \cap U_l$ such that $\varrho_{jk}(f) = \varrho_{jl}(g)$. Then addition and multiplication of functions determine analogous operations for the equivalence classes; hence the latter form an algebra, which is called the *inductive (direct) limit* $\varinjlim_{U \in \mathfrak{U}} \mathcal{O}(U)$.

Prove that *if $\mathfrak{U}$ is a fundamental system of neighborhoods of $a \in \mathbb{C}^n$, then there exists an algebra-isomorphism ${}_n\mathcal{O}_a \cong \varinjlim_{U \in \mathfrak{U}} \mathcal{O}(U)$*. The equivalence class f_a of $f \in \mathcal{O}(U)$ is called the *germ of f at a* (for 23.11).

Analogous constructions can be carried out for a fundamental system of neighborhoods of an arbitrary set $A \subset \mathbb{C}^n$ to obtain an isomorphism $\mathcal{O}(A) \cong \varinjlim \mathcal{O}(U)$; more generally, $\mathcal{O}(U)$ can be replaced by systems of sets, abelian groups, rings, etc. (for § 30).

21.4 Proposition. *${}_n\mathcal{O}_0$ is a local algebra with maximal ideal ${}_n\mathfrak{m}_0 := {}_n\mathfrak{m} := \{P \in {}_n\mathcal{O}_0;$ $P(0) = 0\} = \mathfrak{m}_{[\![X]\!]} \cap {}_n\mathcal{O}_0 =: \mathfrak{m}_{\{X\}}$.*

Proof. By 21.5, ${}_n\mathcal{O}_0$ is a subalgebra of $\mathbb{C}[\![X]\!]$; hence, by 21.2, it remains only to show for each $P \in {}_n\mathcal{O}_0$ with nonzero value that the formal inverse $\sum_{j=0}^{\infty} (1-P)^j$ converges. We may suppose that $P(0) = 1$; if we choose an r such that $\|1-P\|_r < 1$ (see 21.5), then we find that $\|\Sigma(1-P)^j\|_r \leqq \Sigma \|1-P\|_r^j < \infty$. ■

21.5 Lemma. *The mapping* $\|\cdot\|_r$ *satisfies the following conditions:*

i) For each summable family $(P_j)_{j\in J}$,

$$\|\sum_{j\in J} P_j\|_r \leqq \sum_{j\in J} \|P_j\|_r ;$$

equality holds if each X^ν *appears in at most one* P_j.

ii) $\|P\cdot Q\|_r \leqq \|P\|_r \|Q\|_r$; *equality holds for* $P = a_\nu X^\nu$.

iii) If P *is convergent, then* $\lim_{r\to 0} \|P\|_r = |P(0)|$.

Proof. i) The sum of a series of positive terms is unaffected by a reordering of the terms; hence, we have the following inequality for $P_j = \Sigma_j a_\nu X^\nu$:

$$\|\sum_j P_j\|_r = \sum_\nu |\sum_j {}_j a_\nu| r^\nu \leqq \sum_\nu \sum_j |{}_j a_\nu| r^\nu = \sum_j \sum_\nu |{}_j a_\nu| r^\nu = \sum_j \|P_j\|_r .$$

ii) $$\|(\Sigma a_\nu X^\nu)\cdot(\Sigma b_\mu X^\mu)\|_r = \sum_\lambda |\sum_{\mu+\nu=\lambda} a_\nu b_\mu| r^\lambda \leqq \sum_\lambda \sum_{\mu+\nu=\lambda} |a_\nu b_\mu| r^\lambda$$
$$= (\Sigma|a_\nu| r^\nu)(\Sigma|b_\mu| r^\mu) = \|\Sigma a_\nu X^\nu\|_r \|\Sigma b_\mu X^\mu\|_r .$$

iii) Without loss of generality, suppose that $P(0) = 0$. It is possible to construct recursively a representation $P = \sum_{i=1}^{n} P_i X_i$ in such a manner that each monomial in P appears in exactly one $P_i X_i$. Then, by i) and ii), we have that $\|P\|_r = \sum_{i=1}^{n} \|P_i X_i\|_r = \sum_{i=1}^{n} \|P_i\|_r r_i$. In particular, $\|P_i\|_r < \infty$ if $\|P\|_r < \infty$, so the fact that $\|\cdot\|_r \leqq \|\cdot\|_s$ for $r \leqq s$ implies that $\lim_{r\to 0} \|P\|_r = 0$. ■

E. 21d. Find power series P and Q in ${}_1\mathcal{O}_0$ such that $\|P\cdot Q\|_1 < \|P\|_1 \|Q\|_1$.

Now we want to investigate homomorphisms of power series algebras; the convergent case reduces readily to the formal case:

21.6 Remark. *Every homomorphism* $\varphi: R \to S$ *of local algebras is local;* i.e., $\varphi(\mathfrak{m}_R) \subset \mathfrak{m}_S$.

Proof. The ideal $\mathfrak{a} := \varphi^{-1}(\mathfrak{m}_S)$ is proper, since it does not contain 1_R; hence, the induced injective homomorphism

$$\bar{\varphi}: R/\mathfrak{a} \to S/\mathfrak{m}_S \cong \mathbb{C}$$

of one-dimensional vector spaces is bijective. We conclude that $\mathfrak{a}$ is maximal (i.e., $\mathfrak{a} = \mathfrak{m}_R$), and that $\varphi(\mathfrak{m}_R) = \varphi(\mathfrak{a}) \subset \varphi(\mathfrak{m}_S)$. ■

Remark 21.6 does not hold for local *rings* R and S, as is shown by the inclusion $\mathbb{C}[\![X]\!] \hookrightarrow Q(\mathbb{C}[\![X]\!])$.

21.7 Proposition. *Let* $\varphi : {}_n\mathscr{F} \to {}_m\mathscr{F}$ *be an algebra-homomorphism.*

i) φ *is local, and sends summable families* $(P_j)_{j\in J}$ *to summable families* $(\varphi(P_j))_{j\in J}$, *and* $\varphi(\sum_{j\in J} P_j) = \sum_{j\in J} \varphi(P_j)$.

ii) For $Q_1, \ldots, Q_n \in \mathfrak{m}({}_m\mathscr{F})$, *there exists precisely one algebra-homomorphism* $\psi : {}_n\mathscr{F} \to {}_m\mathscr{F}$ *such that* $\psi(X_i) = Q_i$ *for* $i = 1, \ldots, n$; *moreover,* $\psi(\sum_\nu a_\nu X^\nu) = \sum_\nu a_\nu Q^\nu$.

iii) φ *has a restriction to* ${}_n\mathcal{O}_0 \to {}_m\mathcal{O}_0$ *iff each* $\varphi(X_i)$ *lies in* ${}_m\mathcal{O}_0$; *conversely, every algebra-homomorphism* ${}_n\mathcal{O}_0 \to {}_m\mathcal{O}_0$ *has a unique extension that is an algebra-homomorphism* ${}_n\mathscr{F} \to {}_m\mathscr{F}$.

Proof. Set ${}_k\mathfrak{m} := \mathfrak{m}({}_k\mathscr{F})$, then 21.6 implies that $\varphi({}_n\mathfrak{m}^i) \subset {}_m\mathfrak{m}^i$[2] for each $i \in \mathbb{N}$; hence, the family $(\varphi(P_j))_{j\in J}$ is summable, by 21.8 i).

ii) For $R := {}_n\mathscr{F}$, let us assume that the R-module $\mathfrak{m} := \mathfrak{m}_R$ is generated by $f_1, \ldots, f_l$. By inducting on $k \geqq 1$, we see that the condition

$$(21.7.1) \quad R = \sum_{|\nu| < k} \mathbb{C} f^\nu + \mathfrak{m}^k$$

holds for $f := (f_1, \ldots, f_l)$: using the representation $\mathfrak{m}^k = \sum_{|\nu| = k} R f^\nu$, the step "$k \Rightarrow k+1$" goes as follows:

$$\begin{aligned} R &= \sum_{|\nu|<k} \mathbb{C} f^\nu + \mathfrak{m}^k = \sum_{|\nu|<k} \mathbb{C} f^\nu + \sum_{|\nu|=k} R f^\nu \\ &= \sum_{|\nu|<k} \mathbb{C} f^\nu + \sum_{|\nu|=k} (\mathbb{C} f^\nu + \mathfrak{m} f^\nu) \subset \sum_{|\nu|<k+1} \mathbb{C} f^\nu + \mathfrak{m}^{k+1} \subset R. \end{aligned}$$

There exists at most one algebra-homomorphism that sends each f_j to Q_j: For two such homomorphisms φ and ψ, it suffices to demonstrate that the $\mathbb{C}$-linear map $\sigma := \varphi - \psi$ is the zero-mapping. Obviously, $\sigma(\sum_{|\nu|<k} \mathbb{C} f^\nu) = 0$; hence, (21.7.1) implies that $\sigma(R) \subset {}_m\mathfrak{m}^k$ for each k. Then 21.8 ii) implies that $\sigma = 0$.

Finally, the assignment $\varphi(\Sigma a_\nu X^\nu) := \Sigma a_\nu Q^\nu$ determines an algebra-homomorphism, since $(Q_\nu)_{\nu \in \mathbb{N}^n}$ is a summable family.

i) It remains to show that $\varphi(\Sigma P_j) = \Sigma \varphi(P_j)$. Following the idea of the proof of ii), it suffices to demonstrate that

$$\varphi(\Sigma P_j) - \Sigma \varphi(P_j) \in \bigcap_{k=1}^{\infty} {}_m\mathfrak{m}^k.$$

[2] $\mathfrak{m}^j = \mathfrak{m} \cdot \ldots \cdot \mathfrak{m}$ is not to be confused with $R^j = R \oplus \ldots \oplus R$.

For a fixed k, set $\Sigma' P_j := \sum\limits_{o(P_j)<k} P_j$ and $\Sigma'' P_j := \Sigma P_j - \Sigma' P_j$; then we have that

$$\begin{aligned}\varphi(\Sigma P_j) - \Sigma\varphi(P_j) &= \varphi(\Sigma' P_j + \Sigma'' P_j) - \Sigma'\varphi(P_j) - \Sigma''\varphi(P_j)\\ &= \varphi(\Sigma'' P_j) - \Sigma''\varphi(P_j) \in {}_m\mathfrak{m}^k .\end{aligned}$$

iii) It remains only to show that if $Q_j := \varphi(X_j) \in {}_m\mathcal{O}_0$ for $j = 1, \ldots, n$, then $\varphi({}_n\mathcal{O}_0) \subset {}_m\mathcal{O}_0$. For a fixed $P = \Sigma a_\nu X^\nu \in {}_n\mathcal{O}_0$, choose an r such that $\|P\|_r < \infty$. By 21.5, there exists an s such that $\|Q_j\|_s < r_j$, $j = 1, \ldots, n$; for such an s,

$$\|\varphi(P)\|_s = \|\Sigma a_\nu Q^\nu\|_s \leqq \Sigma |a_\nu| \, \|Q^\nu\|_s \leqq \Sigma |a_\nu| r^\nu = \|P\|_r < \infty . \quad ■$$

We have used the following special case of *Krull's Intersection Theorem* 23 A.5:

21.8 Lemma. *If* $\mathfrak{m}$ *is the maximal ideal of* $\mathbb{C}[\![X]\!]$ *or* ${}_n\mathcal{O}_0$, *then*

$$i)\ \mathfrak{m}^j = \{P; o(P) \geqq j\}, \quad \text{and} \quad ii)\ \bigcap_{j \in \mathbb{N}} \mathfrak{m}^j = 0 .$$

Proof. The inclusion "$\subset$" in i) follows by induction from 21.2, 21.4, and the fact that $o(P \cdot Q) = o(P) + o(Q)$. "$\supset$". If $o(P) \geqq j$, then P has a representation $P = \sum\limits_{|\nu|=j} P_\nu \cdot X^\nu$ in which each monomial of P appears exactly once. If P is convergent, then each P_ν is convergent, because $\|P\|_r = \sum\limits_{|\nu|=j} \|P_\nu\|_r \cdot r^\nu < \infty$. Since $X^\nu \in \mathfrak{m}^{|\nu|}$, we conclude that $P \in \mathfrak{m}^j$.

Condition ii) follows immediately from condition i). ■

21.9 Remark. *There is a canonical one-to-one correspondence between the algebra-homomorphisms* $\varphi: {}_n\mathcal{O}_0 \to {}_m\mathcal{O}_0$ *and germs of holomorphic mappings:* for $0 \in U \Subset \mathbb{C}^m$, if the mapping $h \in \mathrm{Hol}(U, \mathbb{C}^n)$ is such that $h(0) = 0$, then, by 3.4, h induces an algebra-homomorphism

$$h^0 : {}_n\mathcal{O}_0 \to {}_m\mathcal{O}_0, \ P \mapsto P \circ h .$$

Conversely, if $\varphi: {}_n\mathcal{O}_0 \to {}_m\mathcal{O}_0$ is an algebra-homomorphism, then, by 21.7 ii), the power series $h_j := \varphi(X_j)$ is in ${}_m\mathfrak{m}$; hence, $h = (h_1, \ldots, h_n): U \to \mathbb{C}^n$ is a holomorphic mapping with $h(0) = 0$ if $U \subset \mathbb{C}^m$ is chosen so that every h_j is holomorphic on U. Then $h^0(X_j) = h_j = \varphi(X_j)$, so $h^0 = \varphi$, by 21.7. ■

E. 21e. For $R = {}_n\mathcal{O}_0$ (analogously, for ${}_n\mathcal{F}$) and $\mathfrak{m} = \mathfrak{m}_R$, show that the following statements hold:

i) There is precisely one way to make R into a topological group such that $\{\mathfrak{m}^j; j \in \mathbb{N}\}$ is a fundamental system of neighborhoods of 0. (That topology is called the *Krull*, or $\mathfrak{m}$-*adic, topology*.)

ii) R is a Hausdorff topological algebra (hint: 21.8).

iii) Every homomorphism ${}_n\mathcal{O}_0 \to {}_m\mathcal{O}_0$ is continuous (hint: 21.7).

iv) If $(P_j)_{j \in J}$ is a summable family, then its sum $\sum\limits_j P_j$ can be interpreted as a convergent series in the topological group R.

§ 22 The Weierstrass Theorems

A nonzero power series $P = \sum_{\nu \in \mathbb{N}} a_\nu X^\nu$ in one variable admits a unique decomposition of the form

$$P = X^b(a_b + a_{b+1} X + \cdots), \quad a_b \neq 0,$$

i.e., a decomposition $P = X^b \cdot e$ with a unit $e \in \mathbb{C}[\![X]\!]$ that converges iff P converges. Hence, each nonzero ideal in $\mathbb{C}[\![X]\!]$ (or ${}_1\mathcal{O}_0$) is a principal ideal generated by a unique monomial X^b.

For $n \geqq 2$, the ring ${}_nR(= {}_n\mathcal{O}_0$ or ${}_n\mathcal{F})$ is no longer a principal ideal ring (see E. 22i); accordingly, the ideal theory is more complicated. Nonetheless, principal ideals play a central role – in proofs by induction for example; we analyze the deeper reason for that in dimension theory (see § 48). The *Weierstrass Preparation Theorem* says that, essentially (i.e., if necessary, after application of a linear coordinate-transformation), each principal ideal in ${}_nR$ is generated by a monic polynomial

$$X_n^b + a_1 X_n^{b-1} + \ldots + a_b \in {}_{n-1}R[X_n] \quad \text{with} \quad a_j(0) = 0 \quad \text{for} \quad j = 1, \ldots, b.$$

The significance of the Weierstrass Preparation Theorem is further emphasized by one of its corollaries, the *Weierstrass Division Formula*, which is an analogue of the Euclidean Division Theorem. With the notation for $g \in \mathbb{N}^n$,

$$R[X]_g := R[X_1, \ldots, X_n]_{(g_1, \ldots, g_n)} := \{P \in R[X]; \deg_{X_j} P \leqq g_j - 1\},$$

we have the following well-known result for $n = 1$:

Euclidean Division Theorem. *For a ring R, if $P \in R[X]$ is a monic polynomial of degree b, then the mapping*

$$R[X] \cdot P \oplus R[X]_b \to R[X], \ (q \cdot P, r) \mapsto qP + r,$$

is an isomorphism of R-modules. In other words, each $Q \in R[X]$ has a unique representation (division by P with remainder r) of the form

$$Q = q \cdot P + r \quad \text{with} \quad r \in R[X]_b. \quad \blacksquare$$

With the notation

$$\mathbb{C}[\![X, Y]\!] := \mathbb{C}[\![X_1, \ldots, X_n, Y]\!],\ \mathbb{C}\{X, Y\} := \mathbb{C}\{X_1, \ldots, X_n, Y\}, \quad \text{etc.},$$

we turn now to the characterization of power series that do not "vanish identically on the Y-axis":

22.1 Definition. *A power series $P \in \mathbb{C}[\![X, Y]\!]$ is called*

i) distinguished in Y with order b if $P(0, \ldots, 0, Y) = Y^b \cdot e$ for some unit $e \in \mathbb{C}[\![Y]\!]$,

ii) a Weierstrass polynomial of degree b in Y if $P = Y^b + \sum_{j=1}^{b} a_j Y^{b-j}$ with $a_j \in \mathfrak{m}_{[\![X]\!]}$, $j = 1, \ldots, b$.

Each monic polynomial $P \neq 0$ of degree b in $\mathbb{C}[\![X]\!][Y]$ is distinguished in Y

with order at most b; moreover, such a P is a Weierstrass polynomial iff $P(0, Y)$ has a zero of order b at $Y = 0$.

E. 22a. Let $P = P_1 \cdot \ldots \cdot P_m \in \mathbb{C}[\![X,Y]\!]$ be given; show that P is distinguished in Y iff every P_j is distinguished in Y. With the assumption that every P_j is a monic polynomial in $\mathbb{C}[\![X]\!][Y]$, show that P is a Weierstrass polynomial iff every P_j is a Weierstrass polynomial (for 23.6).

22.2 Weierstrass Preparation Theorem. *If $P \in \mathbb{C}[\![X, Y]\!]$ is distinguished in Y with order b, then there exist precisely one Weierstrass polynomial $\omega \in \mathbb{C}[\![X]\!][Y]$ of degree b and precisely one unit $e \in \mathbb{C}[\![X, Y]\!]$ such that $P = \omega \cdot e$; moreover, if P is in $\mathbb{C}\{X, Y\}$ (resp., $\mathbb{C}\{X\}[Y]$), then so are ω and e.*

22.3 Weierstrass Division Formula. *If $P \in \mathbb{C}[\![X, Y]\!]$ is distinguished in Y with order b, then the mapping*

$$\mathbb{C}[\![X, Y]\!] \cdot P \oplus \mathbb{C}[\![X]\!][Y]_b \to \mathbb{C}[\![X, Y]\!],\ (q \cdot P, r) \mapsto qP + r,$$

is an isomorphism of $\mathbb{C}[\![X]\!]$-modules. If P is in $\mathbb{C}\{X, Y\}$, then it induces a $\mathbb{C}\{X\}$-module-isomorphism

$$\mathbb{C}\{X, Y\} \cdot P \oplus \mathbb{C}\{X\}[Y]_b \cong \mathbb{C}\{X, Y\}.$$

If P is in $\mathbb{C}\{X\}[Y]$, then it induces a $\mathbb{C}\{X\}$-module-isomorphism

$$\mathbb{C}\{X\}[Y] \cdot P \oplus \mathbb{C}\{X\}[Y]_b \cong \mathbb{C}\{X\}[Y].$$

22.4 Corollary. *If the power series $P \in \mathbb{C}\{X, Y\}$ is distinguished in Y with order b, then the canonical mapping*

$$\mathbb{C}[\![X]\!]^b \to \mathbb{C}[\![X, Y]\!],\ (Q_0, \ldots, Q_{b-1}) \mapsto \sum_{j=0}^{b-1} Q_j Y^j,$$

induces $\mathbb{C}\{X\}$-module-isomorphisms

$$\mathbb{C}[\![X]\!]^b \to \mathbb{C}[\![X, Y]\!] / \mathbb{C}[\![X, Y]\!] \cdot P$$
$$\mathbb{C}\{X\}^b \to \mathbb{C}\{X, Y\} / \mathbb{C}\{X, Y\} \cdot P.$$

If P is in fact a Weierstrass polynomial, then it also induces a $\mathbb{C}\{X\}$-module-isomorphism

$$\mathbb{C}\{X\}^b \to \mathbb{C}\{X\}[Y] / \mathbb{C}\{X\}[Y] \cdot P. \quad \blacksquare$$

Let us first show that the Weierstrass Preparation Theorem implies the Implicit Mapping Theorem 8.6 and the Inverse Mapping Theorem 8.3. We use the following notation:

$$z = (z_1, \ldots, z_n),\ w = (w_1, \ldots, w_m),\ w' := (w_1, \ldots, w_{m-1}).$$

8.6_{alg} Implicit Mapping Theorem. *For a mapping $f = (f_1, \ldots, f_m)$ such that $f_j \in \mathfrak{m}_{\{z,w\}}$ and $\det \frac{\partial f}{\partial w}(0) \neq 0$, there exists a unique mapping $g = (g_1, \ldots, g_m)$, $g_j \in \mathfrak{m}_{\{z\}}$ such that, near $(0,0)$,*

$$f(z, w) = 0 \Leftrightarrow w = g(z).$$

Proof. After a linear change of the coordinates w, we may assume that $A := \frac{\partial f}{\partial w}(0)$ is the matrix I_m (since the condition "$f(z, A^{-1}w) = 0$ iff $w = g(z)$" implies "$f(z,w) = 0$ iff $w = (A^{-1}g)(z)$"; moreover, the Chain Rule implies that $\frac{\partial(f \circ A^{-1})}{\partial w}(0) = I_m$). In particular, then, f_m is distinguished with order 1 in w_m; hence, the Preparation Theorem provides a unique decomposition $f_m = e \cdot (w_m - g_0)$ with $g_0 \in \mathfrak{m}_{\{z,w'\}}$. It is easy to see that we may take e to be 1; i.e.,

$$f_m = w_m - g_0 \quad \text{with} \quad g_0 \in \mathfrak{m}_{\{z,w'\}}.$$

Now we induct on m. The case $m = 1$ is obvious. For "$m-1 \Rightarrow m$", let us define $h = (h_1, \ldots, h_{m-1})$ such that $h_j \in \mathfrak{m}_{\{z,w'\}}$ by setting

$$h(z,w') = f'(z,w',g_0(z,w')).$$

Then $\frac{\partial h}{\partial w'}(0) = I_{m-1}$, and the induction hypothesis yields the existence of a unique $g' = (g_1, \ldots, g_{m-1})$ such that $g_j \in \mathfrak{m}_{\{z\}}$ and

$$h(z,w') = 0 \Leftrightarrow w' = g'(z).$$

If we define the missing $g_m \in \mathfrak{m}_{\{z\}}$ by setting

$$g_m(z) := g_0(z, g'(z)),$$

then $g := (g_1, \ldots, g_m)$ has the desired property: $f(z,g(z)) = 0$ since $f_m(z,g(z)) = g_m(z) - g_0(z,g'(z)) = 0$ and $f'(z,g'(z),g_m(z)) = h(z,g'(z)) = 0$; on the other hand, if $f(z,w) = 0$, then the fact that $f_m(z,w) = 0$ implies that $w_m = g_0(z,w')$, and the equality $0 = f'(z,w',g_0(z,w')) = h(z,w')$ implies that $w' = g'(z)$; in particular $w_m = g_0(z,g'(z)) = g_m(z)$. ■

Proof of 8.3. For the nontrivial implication it obviously is sufficient to show that, for a mapping $f = (f_1, \ldots f_m)$ *with* $f_j \in \mathfrak{m}_{\{w\}}$, *if* $J_f(0) \neq 0$, *then* f *has a holomorphic inverse* $g \in {}_n\mathfrak{m}_{\{z\}}$. For the mapping

$$h: \mathbb{C}^n \times X \to \mathbb{C}^n, \ (z,w) \mapsto f(w) - z,$$

$\det \frac{\partial h}{\partial w}(0) = J_f(0) \neq 0$; hence, by 8.6_{alg}, there exists a unique $g = (g_1, \ldots, g_n)$ such that $g_j \in \mathfrak{m}_{\{z\}}$ and $0 = h(z,g(z)) = f(g(z)) - z$. Then $f \circ g = \mathrm{id}_{\mathbb{C}^n}$ near 0; in particular, near 0, f must be surjective and g, injective. Since the Chain Rule implies that $J_g(0) \neq 0$, we may apply the same argument to g in place of f in order to prove that the holomorphic mapping g is surjective and hence bijective. ■

Proof of 22.2 and 22.3. For $b = 0$, 22.2 and 22.3 are trivial, since the power series P is then a unit; for the remainder of this discussion, let $b \geqq 1$ be fixed. In the case of one variable (i.e., $n = 0$), the proof is simple: Y^b is *the* Weierstrass polynomial of degree b, and there is exactly one unit $e \in \mathbb{C}[\![Y]\!]$ such that $P = eY^b$; moreover, we have that

$$e \in \mathbb{C}\{Y\} \Leftrightarrow P \in \mathbb{C}\{Y\} \quad \text{and} \quad e \in \mathbb{C}[Y] \Leftrightarrow P \in \mathbb{C}[Y].$$

That settles 22.2 for $n = 0$; 22.3 follows just as easily, since P can be replaced by Y^b.

For $n \geqq 1$, we imitate the preceding argument, changing P by means of an appropriate injective transformation of variables

$$\varphi: \mathbb{C}[\![X,Y]\!] \to \mathbb{C}[\![S,T]\!]$$

into the form $\varphi(P) = e'T^b$ with a unit e'. The desired representation $P = e\omega$ of P in $\mathbb{C}[\![X,Y]\!]$ as the product of a unit and a Weierstrass polynomial stems from the representation $\varphi(P) = e'T^b$ in $\mathbb{C}[\![S,T]\!]$, which can be given elementarily and explicitly, thus allowing e and ω to be given explicitly as well. Since φ respects convergent power series and polynomials, the supplemental statements in 22.2 follow automatically. The proof of 22.3 runs along the same lines.

Fix $b \in \mathbb{N}_{>0}$, and let S represent $S_1, \ldots, S_n$. The mapping

$$\varphi: \mathbb{C}[\![X,Y]\!] \to \mathbb{C}[\![S,T]\!];\ X_j \mapsto S_j T^b,\ Y \mapsto T,$$

is an algebra-homomorphism whose essential properties we wish to describe by means of a *weight* γ: for a monomial $aS^\sigma T^\tau$ (with $a \neq 0$), we define

$$\gamma(aS^\sigma T^\tau) := \tau - b|\sigma| \in \mathbb{Z};$$

for a power series $P = \Sigma a_{\sigma\tau} S^\sigma T^\tau$ and $m \in \mathbb{Z}$, we write

$$\gamma(P) \geqq m$$

to indicate that only monomials of weight at least m appear in P. We also use the symbols $\leqq$, $<$, and $>$ correspondingly; in particular, $\gamma(0)$ satisfies all of those inequalities.

22.5 Lemma. *i)* φ *is injective.*

ii) $\operatorname{Im} \varphi = \{P \in \mathbb{C}[\![S,T]\!];\ \gamma(P) \geqq 0\}$.

iii) The following statements hold for $P \in \mathbb{C}[\![X,Y]\!]$:

a) P *is convergent iff* $\varphi(P)$ *is convergents; precisely stated,*

$$\|P\|_{(r,s)} = \|\varphi(P)\|_{(\tilde{r},s)} \quad \textit{for} \quad (r,s) \in \mathbb{R}^n_{>0} \times \mathbb{R}_{>0} \quad \textit{and} \quad \tilde{r} = rs^{-b}.$$

b) $P \in \mathbb{C}[\![X]\!][Y]_{d+1} \Leftrightarrow \gamma(\varphi(P)) < d+1$.

c) P *is distinguished in* Y *with order* b *iff* $\varphi(P) = eT^b$ *for some unit* $e \in \mathbb{C}[\![S,T]\!]$.

d) P *is a Weierstrass polynomial of degree* b *in* Y *iff* $\varphi(P) = (1+u)T^b$ *for some* $u \in \mathbb{C}[\![S,T]\!]$ *such that* $\gamma(u) < 0$ *(in particular,* $u \in \mathfrak{m}_{[\![S,T]\!]}$*).*

e) If P *is convergent, then* e *in c) and* u *in d) are convergent.*

iv) For each $m \in \mathbb{Z}$ *and each* $P \in \mathbb{C}[\![S,T]\!]$, *there exists a unique decomposition* $P = P_1 + P_2$ *in* $\mathbb{C}[\![S,T]\!]$ *such that* $\gamma(P_1) < m$ *and* $\gamma(P_2) \geqq m$; *if* P *is convergent, then so are* P_1 *and* P_2.

v) If the power series $P_1, P_2 \in \mathbb{C}[\![S,T]\!]$ *are such that* $\gamma(P_j) \geqq m_j \in \mathbb{Z}$, *then* $\gamma(P_1 \cdot P_2) \geqq m_1 + m_2$; *the analogous statement holds for "*$\leqq$*".*

vi) Each unit $e \in \mathbb{C}[\![S,T]\!]$ *has a unique decomposition of the form* $e = (1+u)\tilde{e}$ *such that* $\tilde{e}$ *is a unit,* $\gamma(\tilde{e}) \geqq 0$, *and* $\gamma(u) < 0$. *If* e *is convergent, then so are* u *and* $\tilde{e}$.

Before giving the proof of 22.5, which is simple up to vi), we use it to prove the Weierstrass Theorems (in the next two proofs, the numbers i) – vi) refer to 22.5).

Proof of 22.2. Let P be distinguished in Y with order b; then, by iii c), there exists precisely one unit $e' \in \mathbb{C}[\![S,T]\!]$ such that $\varphi(P) = e'T^b$. That implies the *existence* of the decomposition of P:

$$\varphi(P) = e'T^b \underset{\text{vi)}}{=} (1+u)\tilde{e}T^b \underset{\text{ii)}}{=} \varphi(e)(1+u)T^b,$$

where $e \in \mathbb{C}[\![X,Y]\!]$ is a unit, since φ is local; by iii d), the power series $\omega := e^{-1}P \in \mathbb{C}[\![X,Y]\!]$ is a Weierstrass polynomial of degree b.

For the *uniqueness*, suppose that P has a second decomposition $P = e^+\omega^+$; then

$$\varphi(e)(1+u)T^b = \varphi(P) = \varphi(e^+)\varphi(\omega^+) \underset{\text{iii d)}}{=} \varphi(e^+)(1+u^+)T^b,$$

so $\varphi(e) = \varphi(e^+)$ by vi), and i) implies that $e = e^+$.

By iii) and vi), both e and ω are convergent if P is.

If P is a polynomial in Y, then so is e: the divisions with remainder according to Euclid and Weierstrass agree on $\mathbb{C}\{X\}[Y]$, as we shall see in the proof of 22.3 (without using the fact that e is a polynomial). Since $P = e\omega + 0$ in $\mathbb{C}\{X,Y\}$, it follows that $e \in \mathbb{C}\{X\}[Y]$. ∎

Proof of 22.3. By 22.2, we may assume that P is a Weierstrass polynomial of degree b; according to iii) d), P determines a unit $e \in \mathbb{C}[\![S,T]\!]$ by means of $\varphi(P) = eT^b$.

Existence. By iv), for $Q \in \mathbb{C}[\![X,Y]\!]$, there exists a decomposition

(22.3.1) $e^{-1}\varphi(Q) = L\cdot T^b + M$ with $\gamma(M) < b$, $\gamma(L\cdot T^b) \geqq b$, and thus $\gamma(L) \geqq 0$.

Moreover, we can show that $\gamma(eM) \geqq 0$: since $eT^b = \varphi(P)$, ii) and v) imply that $\gamma(L\cdot eT^b) \geqq 0 + 0 = 0$ and hence that $\gamma(eM) = \gamma(\varphi(Q) - eLT^b) \geqq 0 + 0 = 0$. Thus by ii) there exist elements $q, r \in \mathbb{C}[\![X,Y]\!]$ such that $\varphi(q) = L$ and $\varphi(r) = eM$. It follows that

$$\varphi(qP + r) = L\cdot eT^b + eM = \varphi(Q), \text{ and thus that } Q = qP + r;$$

by iii b), we have that $r \in \mathbb{C}[\![X]\!][Y]_b$, since $\gamma(e\cdot M) \leqq 0 + b - 1 = b - 1$ by v).

If P and Q are convergent, then so are $L\cdot T^b$ and M by iv), and thus q and r are convergent, as well.

If Q is a polynomial in Y, then there exist divisions with remainder according to Euclid and Weierstrass; they must agree, since *the Weierstrass division is unique*: If $Q = qP + r$, then $\varphi(P) = eT^b$ provides a decomposition

$$e^{-1}\varphi(Q) = \varphi(q)T^b + e^{-1}\varphi(r).$$

The mapping φ is injective, so by iv), q and r are determined if $\gamma(\varphi(q)T^b) \geqq b$ and $\gamma(e^{-1}\varphi(r)) < b$. By v), we have that $\gamma(\varphi(q)T^b) \geqq 0 + b = b$. For $u := e - 1$ (see iii d)), we have that $e^{-1} = \sum_{j=0}^{\infty} (-u)^j$; since $\gamma(u^0) = 0$ and $\gamma(u) < 0$, it follows by v) and induction on j that $\gamma(u^j) < 0$, so $\gamma(e^{-1}) \leqq 0$. Thus iii b) implies that $\gamma(e^{-1}\varphi(r)) \leqq 0 + b - 1 = b - 1$. ∎

Proof of 22.5. ii)"$\subset$". We see that $\gamma \geqq 0$ on Im φ, since $\gamma(\varphi(X^\mu Y^\nu)) = \gamma(S^\mu T^{\nu+b|\mu|}) = (\nu + b|\mu|) - b|\mu| = \nu$. "$\supset$" If $\gamma(S^\sigma T^\tau) = \tau - b|\sigma| \geqq 0$, then $S^\sigma T^\tau = \varphi(X^\sigma Y^{\tau - b|\sigma|})$. The desired inclusion follows from the fact that $\varphi(\Sigma a_{\sigma\tau} X^\sigma Y^\tau) = \Sigma a_{\sigma\tau} \varphi(X^\sigma Y^\tau)$.

i) This is a consequence of the implication $\varphi(P) = \Sigma a_{\sigma\tau} S^\sigma T^\tau \Rightarrow P = \Sigma a_{\sigma\tau} X^\sigma Y^{\tau - b|\sigma|}$.

v) This follows from the equality $\gamma(S^\sigma T^\tau \cdot S^{\tilde\sigma} T^{\tilde\tau}) = \gamma(S^\sigma T^\tau) + \gamma(S^{\tilde\sigma} T^{\tilde\tau})$.

iii) For a), the following equality for monomials obviously is sufficient:

$$\|\varphi(X^\mu Y^\nu)\|_{(\tilde r, s)} = \|S^\mu T^{\nu+b|\mu|}\|_{(\tilde r, s)} = \tilde r^\mu s^{\nu + b|\mu|} = r^\mu s^\nu = \|X^\mu Y^\nu\|_{(r,s)}.$$

Part b) follows from the fact that $\gamma(\varphi(X^\mu Y^\nu)) = \nu$ (see proof of ii)).

c) We use the following fact: $P = \Sigma a_{\mu\nu} X^\mu Y^\nu$ is distinguished in Y with order b iff $a_{0\nu} = 0$ for $\nu < b$ and $a_{0b} \in \mathbb{C}^*$.

Only if. $\varphi(P) = a_{0b} T^b + \sum_{\mu \neq 0 \text{ or } \nu > b} a_{\mu\nu} S^\mu T^{\nu + b|\mu|} = \text{unit} \cdot T^b$.

If. If there exists a unit e such that

$$\Sigma a_{\mu\nu} S^\mu T^{\nu+b|\mu|} = \varphi(\Sigma a_{\mu\nu} X^\mu Y^\nu) = e \cdot T^b,$$

then $a_{0\nu} = 0$ for $\nu < b$, and $a_{0b} = e(0) \in \mathbb{C}^*$.

d) *Only if.* By c), for each Weierstrass polynomial $P = Y^b + f_1 Y^{b-1} + \ldots + f_b$, $f_j \in \mathfrak{m}_{[\![X]\!]}$, there exists a unit $e \in \mathbb{C}[\![S, T]\!]$ such that

$$eT^b = \varphi(P) = \varphi(Y^b) + \varphi\Big(\sum_{j=1}^{b} f_j Y^{b-j}\Big) = T^b + \sum_{j=1}^{b} \varphi(f_j) T^{b-j}.$$

Then $e(0) = 1$; hence, $u := e - 1 = T^{-b} \cdot \varphi(\Sigma f_j Y^{b-j})$ has value $u(0) = 0$. According to b), $\gamma(\varphi(\Sigma f_j Y^{b-j})) \leqq b - 1$; it follows from v) that $\gamma(u) \leqq (-b) + (b-1) < 0$.

If. By c), P is distinguished in Y with order b, and $\varphi(P - Y^b) = ((1+u) - 1) T^b$ with $\gamma(u) < 0$. According to v), $\gamma(uT^b) \leqq b - 1$, so $P - Y^b \in \mathbb{C}[\![X]\!][Y]_b$ by b). Therefore, P is a Weierstrass polynomial of degree b.

e) By 21.5 ii), a formal power series $Q \cdot T^b$ converges iff Q converges. Hence, the statement follows from a).

iv) Fix P and let P_1 consist of the monomials of weight less than m in P, and put $P_2 := P - P_1$. Then $\|P\|_r = \|P_1\|_r + \|P_2\|_r$, so P_1 and P_2 are convergent iff P is.

vi) Without loss of generality, suppose that $e(0) = 1$. For a decomposition $e = (1+u)\tilde e$ with $\gamma(u) < 0$ and $\gamma(\tilde e) \geqq 0$ (obviously, $\tilde e(0) = 1$), we use the logarithm to deduce the *uniqueness* of $\tilde e$ from iv). With the notation

$$\log(1+u) = u - \frac{u^2}{2} + \frac{u^3}{3} \pm \ldots, \quad \exp u := \sum_{j=0}^{\infty} \frac{u^j}{j!}$$

for $u \in \mathfrak{m}_{[\![S,T]\!]}$, the standard computational rules hold; in particular,

(22.5.1) $\log e = \log(1+u) + \log \tilde e$.

Since $\gamma(u) < 0$, we have that $\gamma(u^j) < 0$ for every $j > 0$, and thus that $\gamma(\log(1+u)) < 0$; analogously, it follows that $\gamma(\log \tilde e) \geqq 0$. By iv), with $m = 0$, the decomposition

(22.5.1) is unique, and the uniqueness of $\log \tilde{e}$ and $\tilde{e} = \exp \log \tilde{e}$ follows immediately.

For the *existence*, consider the decomposition

$$\log e = P_1 + P_2 \quad \text{with} \quad \gamma(P_1) < 0,\ \gamma(P_2) \geqq 0$$

in accordance with iv). Set $\tilde{e} := \exp P_2 := \exp(P_2(0)) \cdot \exp(P_2 - P_2(0))$ and $u := \exp(P_1) - 1$; then $\gamma(\tilde{e}) \geqq 0$ and $\gamma(u) < 0$. Thus

$$e = \exp \log e = \exp(P_1 + P_2) = \exp(P_1) \cdot \exp(P_2) = (1+u)\tilde{e}$$

is the desired decomposition. ■■

Application of a simple convergence estimate to the *punctual* Weierstrass Formula 22.3 yields a *local* formula for the algebra $B(\boldsymbol{P}^{n+1}(s)) := \{\Sigma a_{\mu\nu} X^\mu Y^\nu;\ \|\Sigma a_{\mu\nu} X^\mu Y^\nu\|_s < < \infty\}$. Note that $\|P\|_s < \infty$ *iff* P *converges absolutely on the closed polydisk* $\overline{\boldsymbol{P}(s)}$ (see the proof of 4.7).

22.6 Weierstrass Formula for Polydisks. *For a Weierstrass polynomial* $\omega \in \mathbb{C}\{X\}[Y]$ *of degree* b*, and every sufficiently small* $s = (s', s_{n+1}) \in \mathbb{R}^n_{>0} \times \mathbb{R}_{>0}$*, the Weierstrass Formula induces an isomorphism*

$$B(\boldsymbol{P}^{n+1}(s)) \cong B(\boldsymbol{P}^{n+1}(s)) \cdot \omega \oplus B(\boldsymbol{P}^n(s'))[Y]_b$$

of $B(\boldsymbol{P}^n(s'))$*-modules.*

Proof. For the Weierstrass decomposition $q\omega + r = Q \in B(\boldsymbol{P}^{n+1}(s))$, it suffices to show that q and r converge absolutely on $\boldsymbol{P}^{n+1}(s)$ if ω does. To that end, we prove the following fact:

22.7 Lemma. *For a Weierstrass polynomial* $\omega \in \mathbb{C}[\![X]\!][Y]$ *of degree* b*, and* $s \in \mathbb{R}^n_{>0} \times \mathbb{R}_{>0}$*, there exist constants* $C_1, C_2 \in \mathbb{R} \cup \{\infty\}$ *such that, for each Weierstrass decomposition* $q\omega + r = Q$ *in* $\mathbb{C}[\![X, Y]\!]$,

$$\|q\|_s \leqq s_{n+1}^{-b} C_1 \|Q\|_s \quad \text{and} \quad \|r\|_s \leqq C_1 C_2 \|Q\|_s .$$

Define $e = 1 + u$ according to 22.5 iii d) by setting $\varphi(\omega) =: eT^b$; for $s := (s_{n+1}^{-b} s', s_{n+}\ )$, put $C_1 := \|e^{-1}\|_{\tilde{s}}$ and $C_2 := \|e\|_{\tilde{s}}$.

Proof of 22.7. In the decomposition $e^{-1}\varphi(Q) = T^b L + M$ of (22.3.1), we see that

$$\varphi(q) = L,\ \varphi(r) = eM, \quad \text{and} \quad \|T^b L + M\| = \|T^b\| \cdot \|L\| + \|M\|;$$

hence, 22.5 iii a) implies that

$$\|q\|_s = \|\varphi(q)\|_{\tilde{s}} = \|L\|_{\tilde{s}} \leqq \|T^b\|_{s_{n+1}}^{-1} \|e^{-1}\|_{\tilde{s}} \|\varphi(Q)\|_{\tilde{s}} = s_{n+1}^{-b} C_1 |Q\|_s$$

and

$$\|r\|_s = \|\varphi(r)\|_{\tilde{s}} \leqq \|e\|_{\tilde{s}} \|M\|_{\tilde{s}} \leqq C_2 C_1 \|Q\|_s . \quad ■$$

Proof of 22.6 (completion). It remains to show that, for sufficiently small s, the constants C_1 and C_2 belong to $\mathbb{R}$. It suffices to show that $\varepsilon := \|u\|_{\tilde{s}} < 1$, for then

$$C_1 = \|e^{-1}\|_{\tilde{s}} = \|\sum_{j=0}^{\infty} (-u)^j\|_{\tilde{s}} \leqq \frac{1}{1-\varepsilon}, \quad \text{and} \quad C_2 = \|e\|_{\tilde{s}} = 1 + \varepsilon .$$

Now, using a representation $\omega = Y^b + \sum_{j=0}^{b-1} a_j Y^j$, we obtain that

$$\|u\|_{\tilde{s}} = \|uT^b\|_{\tilde{s}} s_{n+1}^{-b} = \|\varphi(\omega - Y^b)\|_{\tilde{s}} s_{n+1}^{-b} = \|\omega - Y^b\|_s s_{n+1}^{-b}$$

$$= \|\sum_{j=0}^{b-1} a_j Y^j\|_s s_{n+1}^{-b} = \sum_{j=0}^{b-1} \|a_j\|_{s'} s_{n+1}^{j-b} .$$

Thus, for fixed s_{n+1}, we see that $\|u\|_{\tilde{s}} < 1$ for sufficiently small s', by 21.5 iii). ■■

For the interested reader, we note the following consequence of the proof:

22.8 Remark. If $\varepsilon < 1$ and $\|\omega - Y^b\|_s \leqq \varepsilon s_{n+1}^b$, then

$$\|q\|_s \leqq \frac{1}{(1-\varepsilon) s_{n+1}^b} \|Q\|_s \quad \text{and} \quad \|r\|_s \leqq \frac{1+\varepsilon}{1-\varepsilon} \|Q\|_s . \quad ■$$

Whereas for $n = 0$ every nonzero $f \in \mathbb{C}[\![Y]\!]$ is distinguished in Y, in general, that holds for $n \geqq 1$ only after application of a "*shearing*" σ_c that is defined for $c = (c_1, \ldots, c_n) \in \mathbb{C}^n$ as follows:

$$\sigma_c : \mathbb{C}[\![X, Y]\!] \to \mathbb{C}[\![X, Y]\!], \; X_j \mapsto X_j + c_j Y, \; Y \mapsto Y .$$

Obviously, σ_c is an automorphism with inverse σ_{-c}.

22.9 Proposition. *If $P \in \mathbb{C}[\![X, Y]\!]$ is represented as a sum of homogeneous polynomials $P = \sum_{j=b}^{\infty} P_j$ with $P_b \neq 0$, then $q(X) := P_b(X, 1) \in \mathbb{C}[X]$ is not the zero polynomial. For $c \in \mathbb{C}^n$, $\sigma_c(P)$ is distinguished in Y with order b iff $q(c) \neq 0$.*

Proof. For $P = \sum_{\nu} a_\nu X^{\nu'} Y^{\nu_{n+1}}$ with $\nu = (\nu', \nu_{n+1})$, we note that

$$\sigma_c(P) = \sum_{\nu} a_\nu (X + cY)^{\nu'} Y^{\nu_{n+1}} ;$$

in particular,

$$\sigma_c(P)(0, \ldots, 0, Y) = \sum_{\nu} a_\nu c^{\nu'} Y^{|\nu|} = \sum_{j=0}^{\infty} \Big(\sum_{|\nu| = j} a_\nu c^{\nu'} \Big) Y^j = \sum_{j=b}^{\infty} P_j(c, 1) Y^j .$$

Thus $\sigma_c(P)$ is distinguished in Y with order b iff $P_b(c, 1) = q(c) \neq 0$. It remains to show that $q \neq 0$. If there exists a $z \in \mathbb{C}^{n+1}$ such that $z_{n+1} \neq 0 \neq P_b(z)$, then

$$0 \neq z_{n+1}^{-b} P_b(z_1, \ldots, z_{n+1}) = P_b\left(\frac{z_1}{z_{n+1}}, \ldots, \frac{z_n}{z_{n+1}}, 1\right) = q\left(\frac{z_1}{z_{n+1}}, \ldots, \frac{z_n}{z_{n+1}}\right).$$

Such a z must exist, for if $P_b(z) = 0$ held for every z such that $z_{n+1} \neq 0$, then P_b, as a continuous function, would have to be the zero polynomial. ∎

E. 22 b. For $n = 1$, determine for which $c \in \mathbb{C}$ the polynomial $\sigma_c(XY)$ is distinguished in Y.

E. 22 c. Show that, for finitely many $P_1, \ldots, P_m \in \mathbb{C}[[X,Y]] \setminus \{0\}$, there exists a shearing σ_c that makes all of the P_j distinguished in Y (for 42.1).

For Proposition 23.11, we need the following theorem:

22.10 Generalized Weierstrass Formula. *For ${}_n\mathcal{O}_0 := \mathbb{C}\{X_1, \ldots, X_n\}$ and $j = 1, \ldots, m$, let $P_j \in {}_n\mathcal{O}_0\{Y_1, \ldots, Y_j\}$ be distinguished in Y_j with order g_j. Then there exists a canonical isomorphism*

$${}_n\mathcal{O}_0\{Y_1, \ldots, Y_m\} \cong {}_n\mathcal{O}_0\{Y_1, \ldots, Y_m\} \cdot (P_1, \ldots, P_m) \oplus {}_n\mathcal{O}_0[Y_1, \ldots, Y_m]_{(g_1, \ldots, g_m)}$$

of ${}_n\mathcal{O}_0$-modules.

Proof. (by induction) The case $m = 1$ is just the Weierstrass Formula. "$m - 1 \Rightarrow m$". Without loss of generality, suppose that P_m is a Weierstrass polynomial of degree g_m. With the abbreviation $R_j := {}_n\mathcal{O}_0\{Y_1, \ldots, Y_j\}$, the Weierstrass Formula states that

$$R_m \cong R_m \cdot P_m \oplus R_{m-1}[Y_m]_{g_m}.$$

The induction hypothesis

$$R_{m-1} \cong R_{m-1} \cdot (P_1, \ldots, P_{m-1}) \oplus {}_n\mathcal{O}_0[Y_1, \ldots, Y_{m-1}]_{(g_1, \ldots, g_{m-1})},$$

implies that

$$R_m \cong R_m \cdot P_m \oplus R_{m-1} \cdot (P_1, \ldots, P_{m-1})[Y_m]_{g_m} \oplus {}_n\mathcal{O}_0[Y_1, \ldots, Y_m]_{(g_1, \ldots, g_m)}.$$

Finally, the assertion follows from the fact that

$$\begin{aligned} R_m \cdot (P_1, \ldots, P_m) &= R_m \cdot P_m + (R_m \cdot P_m \oplus R_{m-1}[Y_m]_{g_m}) \cdot (P_1, \ldots, P_{m-1}) \\ &= R_m \cdot P_m \oplus R_{m-1}[Y_m]_{g_m} \cdot (P_1, \ldots, P_{m-1}). \end{aligned}$$ ∎

E. 22 d. In the notation of 22.10, show that ${}_n\mathcal{O}_0\{Y_1, \ldots, Y_m\} / (P_1, \ldots, P_m)$ is a finitely-generated ${}_n\mathcal{O}_0$-module.

E. 22 e. Let $P \in \mathbb{C}[[X,Y]]$ be distinguished in Y with order b; show that $b \geqq o(P)$, and find a P such that $b > o(P)$.

E. 22 f. Show that the isomorphism in 22.6 determines an isomorphism of Banach spaces.

E. 22 g. Show by means of examples that the inclusions $\mathcal{O}(\bar{\boldsymbol{P}}) \subset B(\boldsymbol{P}) \subset \mathcal{O}(\boldsymbol{P})$ are proper.

E. 22 h. Discuss 22.2 and 22.3 for $P(X,Y) = X_1 Y$.

E. 22 i. Show that ${}_2\mathfrak{m}$ is not a principal ideal in ${}_2\mathcal{O}_0$ (hint: 7.10).

§ 23 Algebraic Properties of ${}_n\mathcal{O}_0$ and ${}_n\mathcal{F}$

With the help of the Weierstrass Theorems, the investigation of the ring-structure of power series algebras can be led back to polynomial algebras. For $n > 1$, ${}_n\mathcal{O}_0$ and ${}_n\mathcal{F}$ are no longer principal ideal domains, but all of their ideals are finitely-generated.

A *ring* R is called *noetherian* if every ideal in R is finitely-generated. In particular, principal ideal rings are noetherian. A *module*[3] M over a ring R is called *noetherian* if every R-submodule of M is finitely-generated (see E. 23d).

E. 23a. If $0 \to M' \xrightarrow{\alpha} M \xrightarrow{\beta} M'' \to 0$ is an *exact sequence* of R-modules (i.e., α is injective, β is surjective, and $\operatorname{Ker}\beta = \operatorname{Im}\alpha$), then M is noetherian iff M' and M'' are noetherian. In particular, finite direct sums, submodules, and residue class modules of noetherian modules are noetherian (for 23.1).

In analogy to *Hilbert's Basis Theorem* for polynomial rings (if R is noetherian, then so is $R[X]$; see [La VI. §2]), we have the following fact:

23.1 Proposition. *The rings ${}_n\mathcal{O}_0$ and ${}_n\mathcal{F}$ are noetherian.*

Proof. (for ${}_n\mathcal{O}_0$; the proof for ${}_n\mathcal{F}$ runs analogously).

The field ${}_0\mathcal{O}_0 = \mathbb{C}$ is certainly noetherian. We proceed by induction on n; suppose that ${}_{n-1}\mathcal{O}_0$ is noetherian, and let $\mathfrak{a} \neq 0$ be an ideal in ${}_n\mathcal{O}_0$. First we consider the case in which there exists a polynomial $f \in \mathfrak{a}$ that is distinguished with respect to X_n. By 22.4, $R := {}_n\mathcal{O}_0 / {}_n\mathcal{O}_0 \cdot f$ is a finitely-generated free ${}_{n-1}\mathcal{O}_0$-module, so, by E. 23a and the induction hypothesis, R is a noetherian ${}_{n-1}\mathcal{O}_0$-module. The residue class ideal $\bar{\mathfrak{a}}$ of $\mathfrak{a}$ in R is generated by some finite collection of elements $\bar{f}_1, \ldots, \bar{f}_s \in \bar{\mathfrak{a}}$ over ${}_{n-1}\mathcal{O}_0$, and thus also over ${}_n\mathcal{O}_0$. Hence, $\mathfrak{a}$ is generated over ${}_n\mathcal{O}_0$ by $\{f, f_1, \ldots, f_s\}$. Now suppose that $\mathfrak{a}$ contains no distinguished polynomial with respect to X_n. According to 22.9, for $0 \neq f \in \mathfrak{a}$, there exists a shearing σ such that $\sigma(f)$ is a distinguished polynomial with respect to X_n; then $\sigma(\mathfrak{a})$ is finitely-generated, so $\mathfrak{a}$ must be finitely-generated as well. ∎

Divisibility theory is of fundamental importance for the investigation of certain subsets of rings. In particular, for $R[X]_{\text{mon}} := \{P \in R[X];\, P \text{ monic}\}$, divisibility is essentially an aspect of the multiplicative structure; we therefore consider *monoids* M (i.e., commutative semigroups with 1 such that $ab = ac \Leftrightarrow b = c$):

23.2 Definition. *Let M be a monoid and $a, b \in M$ be nonunits. We call* a *a divisor of b (in symbols, $a|b$) iff $b = ac$ for some $c \in M$, a proper divisor of b iff $b = ac$ for some nonunit $c \in M$, irreducible iff a has no proper divisor, and prime iff the following implication holds: $a|bc \Rightarrow a|b$ or $a|c$.*

Prime elements are irreducible: if $a = bc$, and if $a|b$, say, then $b = ad$ for some d; hence, $a = adc$, and $dc = 1$, so c is a unit.

[3] Modules are understood to be unitary.

In $\mathbb{Z}$ (more generally, in every principal ideal domain), the Unique Factorization Theorem holds; such rings are "factorial":

23.3 Definition. *A monoid M is called factorial if each nonunit in M is the product of finitely many prime elements. A ring R is called factorial if $R \setminus \{0\}$ is a factorial monoid.*

Note that factorial rings have no (nontrivial) zero-divisors. In analogy to rings, we have the following fact (for a proof, see [ReScVe Satz 160]):

23.4 Proposition. *The following statements are equivalent for a monoid M:*

i) M is factorial.

ii) Every nonunit of M is a product of finitely many irreducible elements, and such a factoriza- is essentially unique (i.e., up to units and order). ■

For polynomial rings we have the following (see [La V §6 Thm. 10]):

23.5 Gauss's Theorem. *For a ring R and $n \in \mathbb{N}$, R is factorial iff $R[X_1, \ldots, X_n]$ is factorial.* ■

We prove the corresponding statement for power series algebras:

23.6 Proposition. *The rings ${}_n\mathcal{O}_0$ and ${}_n\mathcal{F}$ are factorial.*

The proof is based on this fact:

23.7 Lemma. *A Weierstrass polynomial $\omega \in {}_n\mathcal{O}_0[X_{n+1}]$ is prime in ${}_n\mathcal{O}_0[X_{n+1}]$ iff it is prime in ${}_{n+1}\mathcal{O}_0$.*

Proof of 23.6 (by induction on n for ${}_n\mathcal{O}_0$; the proof for ${}_n\mathcal{F}$ runs similarly). As a field, ${}_0\mathcal{O}_0 = \mathbb{C}$ is factorial. Suppose that ${}_n\mathcal{O}_0$ is factorial, and let $f \neq 0$ be a nonunit in ${}_{n+1}\mathcal{O}_0$. By 22.9 and the Preparation Theorem, since automorphisms preserve properties of divisibility, we may assume that f is a Weierstrass polynomial $\omega \in {}_n\mathcal{O}_0[X_{n+1}]$. By assumption, ${}_n\mathcal{O}_0$ is factorial; according to Gauss's Theorem, then, so is ${}_n\mathcal{O}_0[X_{n+1}]$. There is a factorization $\omega = \omega_1 \cdot \ldots \cdot \omega_r$ with prime factors $\omega_j \in {}_n\mathcal{O}_0[X_{n+1}]$ such that each ω_j is monic; by E. 22a, every ω_j is a Weierstrass polynomial in ${}_n\mathcal{O}_0[X_{n+1}]$, and, by 23.7, every ω_j is prime in ${}_{n+1}\mathcal{O}_0$. Therefore, $\omega = \omega_1 \cdot \ldots \cdot \omega_r$ is a prime factorization. ■

Proof of 23.7. We use the simple fact that a nonunit r in a ring R is prime iff $R/R \cdot r$ is an integral domain. By 22.4, the ring-homomorphism ${}_n\mathcal{O}_0[X_{n+1}] \hookrightarrow {}_{n+1}\mathcal{O}_0$ induces an isomorphism ${}_n\mathcal{O}_0[X_{n+1}] / {}_n\mathcal{O}_0[X_{n+1}] \cdot \omega \to {}_{n+1}\mathcal{O}_0 / {}_{n+1}\mathcal{O}_0 \cdot \omega$. Hence, ${}_n\mathcal{O}_0[X_{n+1}] / {}_n\mathcal{O}_0[X_{n+1}] \cdot \omega$ is an integral domain iff ${}_{n+1}\mathcal{O}_0 / {}_{n+1}\mathcal{O}_0 \cdot \omega$ is an integral domain. ■■

The question, whether meromorphic solutions of holomorphic polynomial equations are holomorphic, is of importance in the investigation of singularities. A partial answer follows from 23.6 and 23 A.9:

23.8 Corollary. *The rings ${}_n\mathcal{O}_0$ and ${}_n\mathcal{F}$ are normal.* ■

For a domain $G \subset \mathbb{C}^n$, even for $n = 1$, the ring $\mathcal{O}(G)$ is not factorial (see E. 23e); hence, neither Gauss's Theorem nor 23 A.9 are applicable. However, a result of interest for its applications to divisibility follows from 23.8:

23.9 Corollary. *If $G \subset\subset \mathbb{C}^n$ is a domain, then the ring $\mathcal{O}(G)$ is normal; thus $\mathcal{O}(G)[T]_{\mathrm{mon}}$ is a factorial monoid, and the factorizations into primes over $\mathcal{O}(G)$ and over the field of fractions $Q(\mathcal{O}(G))$ coincide.*

Proof. By 6.2, $\mathcal{O}(G)$ has no zero-divisors. If $g/h \in Q(\mathcal{O}(G))$ is integral over $\mathcal{O}(G)$, then, for each $x \in G$, the germ $g_x/h_x \in Q({}_G\mathcal{O}_x)$ is integral over ${}_G\mathcal{O}_x$; hence, $g_x/h_x \in {}_G\mathcal{O}_x$, and thus $g/h \in \mathcal{O}(G)$. That implies that $\mathcal{O}(G)$ is normal; the rest follows from 23 A.10. ■

Before proceeding, we need to generalize some of our terminology. Fix $a \in \mathbb{C}^n$, and let $\tau^0: {}_n\mathcal{O}_a \xrightarrow{\cong} {}_n\mathcal{O}_0$ be the isomorphism induced by the translation $\tau: \mathbb{C}^n \to \mathbb{C}^n$, $z \mapsto z + a$. A power series $f \in {}_n\mathcal{O}_a$ is called *distinguished in z_n at the point a* if $\tau^0 f$ is distinguished in z_n; f is called a *Weierstrass polynomial at a* if $\tau^0 f$ is a Weierstrass polynomial.

E. 23b. *Weierstrass factorization.* For $P \in {}_n\mathcal{O}_0[Y]_{\mathrm{mon}}$ and $(0, c) \in \mathbb{C}^n \times \mathbb{C}$, show that there exists a unique factorization $P = e \cdot \omega$ in ${}_n\mathcal{O}_0[Y]_{\mathrm{mon}}$ such that ω is a Weierstrass polynomial at $(0, c)$ and e is a unit in ${}_{n+1}\mathcal{O}_{(0,c)}$ (for 23.10).

A monic polynomial $P \in {}_n\mathcal{O}_0[Y]$ has a factorization $P(0, Y) = \prod_{j=1}^{m} (Y - c_j)^{g_j}$ at $X = 0$ with pairwise different zeros c_j; that factorization induces a factorization $P = \prod_{j=1}^{m} P_j$ with Weierstrass polynomials P_j at $(0, c_j)$:

23.10 Hensel's Lemma. *If $P \in {}_n\mathcal{O}_0[Y]$ is a monic polynomial, and if $P(0, Y) = \prod_{\tau=1}^{m} (Y - c_j)^{g_j}$ is the factorization into powers of distinct linear factors, then there exist monic polynomial $P_j \in {}_n\mathcal{O}_0[Y]$ of degree g_j such that*

$$\text{i) } P = \prod_{j=1}^{m} P_j \quad \text{and} \quad \text{ii) } P_j(0, Y) = (Y - c_j)^{g_j}.$$

Those properties determine the P_j.

Proof. We proceed by induction: For $m = 1$ there is nothing to show. For the step "$m - 1 \Rightarrow m$", let $P = e\omega$ denote the unique Weierstrass factorization at the point $(0, c_m)$ according to E. 23b. We see that $e(0, Y) = \prod_{j=1}^{m-1} (Y - c_j)^{g_j}$; hence, for the existence proof, it suffices to apply the induction hypothesis to e.

The uniqueness follows analogously, since ii) ensures that, in a factorization $P = \prod_{j=1}^{m} P_j$, the product $\prod_{j=1}^{m-1} P_j$ is a unit at $(0, c_m)$, and P_m is a Weierstrass polynomial; consequently, we can apply E. 23b and the induction hypothesis. ■

We want to present a special application of Hensel's Lemma that provides the key to the proof of the Finite Coherence Theorem 45.1; it can be viewed as a generalization of Corollary 22.4 of the Weierstrass Formula.

For each $a \in \mathbb{C}^m$, there exist canonical inclusions of algebras (with $(T_1, \dots, T_m)$ denoted by T)

(23.11.1) ${}_n\mathcal{O}_0[T_j] \hookrightarrow {}_n\mathcal{O}_0[T] \hookrightarrow {}_{n+m}\mathcal{O}_{(0,a)}$

such that the second inclusion sends each T_k to the $(n+k)$–th coordinate projection of $\mathbb{C}^{n+m}$. Let polynomials $P_j \in {}_n\mathcal{O}_0[T_j]_{\text{mon}} \subset {}_n\mathcal{O}_0[T]$ of degree g_j be given, and put $g := (g_1, \dots, g_m)$. Then the set $F := \{a \in \mathbb{C}^m;\ P_j(0, a_j) = 0,\ j = 1, \dots, m\}$ is *finite*. Via formation of residue classes and restriction, (23.11.1) yields a canonical commutative diagram

(23.11.2)
$$\begin{array}{ccc} & {}_n\mathcal{O}_0[T]_g & \\ \alpha \swarrow & & \searrow \psi = (\psi_a)_{a\in F} \\ {}_n\mathcal{O}_0[T]/(P_1,\dots,P_m) & \xrightarrow[\varphi = (\varphi_a)_{a\in F}]{} & \bigoplus_{a\in F} ({}_{n+m}\mathcal{O}_{(0,a)}/(P_1,\dots,P_m)) \end{array}$$

in which α, φ_a, ψ_a, and thus φ and ψ as well, are ${}_n\mathcal{O}_0$-linear. The φ_a are even homomorphisms of ${}_n\mathcal{O}_0$-algebras[4]; viewing $(\bigoplus_{a\in F} {}_{m+n}\mathcal{O}_{(0,a)}/(P_1, \dots, P_m))$ as an ${}_n\mathcal{O}_0$-algebra with componentwise operations (ring-direct product), we also see that φ is a homomorphism of ${}_n\mathcal{O}_0$-algebras, as well.

Although the morphism

$$\varphi_a : {}_n\mathcal{O}_0[T]/(P_1, \dots, P_m) \to {}_{n+m}\mathcal{O}_{(0,a)}/(P_1, \dots, P_m)$$

is induced by an inclusion, it is not necessarily injective:

E. 23c. Determine $\operatorname{Ker} \varphi_{(0,0)}$ for $n = 0, m = 2, P_1 = T_1$, and $P_2 = T_2^2 - T_2$.

23.11 Theorem. *The following statements hold for the diagram (23.11.2):*

i) α and ψ are isomorphisms of ${}_n\mathcal{O}_0$-modules.

ii) φ is an isomorphism of ${}_n\mathcal{O}_0$-algebras.

Proof. Obviously, ii) follows from i). That α is bijective follows directly from an m-fold application of the Euclidean Division Theorem. For ψ, we proceed as follows:

a) If there exists an $a \in \mathbb{C}^m$ such that every P_j is a Weierstrass polynomial at $(0, a)$, then $F = \{a\}$, and a translation reduces the problem to the case in which $a = 0$; the assertion follows from the Generalized Weierstrass Formula 22.10.

b) In the general case, by Hensel's Lemma, each P_j has a factorization $P_j = P_{j1} \cdot \ldots \cdot P_{js_j}$ in ${}_n\mathcal{O}_0[T_j]$ with Weierstrass polynomials P_{ji} at $(0, a_{ji})$; moreover,

---

[4] For the definition of an *R-algebra*, replace $\mathbb{C}$ with R in the definition of an algebra in § 1.

$a_{ji} \neq a_{jk}$ for $i \neq k$, and $g_j = \sum_{i=1}^{s_j} g_{ji}$, g_j and g_{ji} being the degrees of P_j and P_{ji}, respectively, as polynomials in T_j. To simplify the notation, we consider the following bijection onto the "index set" B:

$$F \to B := \{\sigma \in \mathbb{N}^m;\ 1 \leqq \sigma_j \leqq s_j, \forall j\},\ a = (a_{1\sigma_1}, \ldots, a_{m\sigma_m}) =: a(\sigma) \mapsto \sigma = (\sigma_1, \ldots, \sigma_m).$$

We want to apply the following consequence of the Hensel factorization:

(23.11.3) $\left\{\right.$ In ${}_{n+m}\mathcal{O}_{(0,a(\sigma))}$, the polynomials P_j and $P_{j\sigma_j}$ generate the same ideal; in other words, their germs in ${}_{n+m}\mathcal{O}_{(0,a(\sigma))}$ differ only by a unit. In particular, $Q_\sigma := \prod_{j=1}^{m} P_j / P_{jk_j}$ is a unit.

(Proof: $P_j = P_{j\sigma_j} \cdot (\prod_{i \neq \sigma_j} P_{ji})$, $P_{ji}(0, T) = (T_j - a_{ji})^{g_{ji}}$, and $P_{ji}(0, a(\sigma)) = (a_{j\sigma_j} - a_{ji})^{g_{ji}} \neq$ $\neq 0$ if $i \neq \sigma_j$; hence, $\prod_{i \neq \sigma_j} P_{ji}$ is a unit in ${}_{n+m}\mathcal{O}_{(0,a(\sigma))}$.)

With $g(\sigma) := (g_{1\sigma_1}, \ldots, g_{m\sigma_m})$, put

$$M_\sigma := {}_n\mathcal{O}_0[T]_{g(\sigma)} \cdot Q_\sigma \subset {}_n\mathcal{O}_0[T]_g$$

and

$$\psi_\sigma := \psi_{a(\sigma)}|_{M_\sigma} : M_\sigma \to {}_{n+m}\mathcal{O}_{(0,a(\sigma))} / (P_1, \ldots, P_m) =: R_\sigma\,;$$

to show that ψ is an isomorphism, it suffices to prove the following three statements:

$b_1)$ $\quad {}_n\mathcal{O}_0[T]_g = \bigoplus_{\sigma \in B} M_\sigma$.

$b_2)$ $\quad \psi = \bigoplus_{\sigma \in B} \psi_\sigma : \bigoplus_{\sigma \in B} M_\sigma \to \bigoplus_{\sigma \in B} R_\sigma$.

$b_3)$ $\quad$ Each $\psi_\sigma : M_\sigma \to R_\sigma$ is an ${}_n\mathcal{O}_0$-module-isomorphism.

$b_1)$ To show that the morphism

$$\bigoplus_{\sigma \in B} M_\sigma \to {}_n\mathcal{O}_0[T]_g =: M$$

determined by the inclusions is an isomorphism, we observe first that the (obviously free) ${}_n\mathcal{O}_0$-modules have the same rank:

$$\operatorname{rank}(\bigoplus_{\sigma \in B} M_\sigma) = \sum_{\sigma \in B} \prod_{j=1}^{m} g_{j\sigma_j} = \sum_{\sigma_1=1}^{s_1} \cdots \sum_{\sigma_m=1}^{s_m} \prod_{j=1}^{m} g_{j\sigma_j}$$
$$= \prod_{j=1}^{m} \sum_{k=1}^{s_j} g_{jk} = \prod_{j=1}^{m} g_j = \operatorname{rank} M\,.$$

By E. 24b (with $R = {}_n\mathcal{O}_0$), it suffices to show that $\oplus M_\sigma \to M$ is bijective for $n = 0$, i.e., for ${}_0\mathcal{O}_0$. Then the set

$$\{(T - a(\sigma))^\beta \cdot Q_\sigma;\ \beta \in \mathbb{N}^m, \sigma \in B, 0 \leqq \beta < g(\sigma)\}$$

is a $\mathbb{C}$-basis of $\bigoplus_{\sigma\in B} M_\sigma$; it suffices to show that its elements are linearly independent in $M = \mathbb{C}[T]_g$. To that end, let a linear combination

$$\sum_{\substack{0\leq\beta<g(\sigma)\\ \sigma\in B}} \lambda_{\beta\sigma}(T-a(\sigma))^\beta Q_\sigma = 0, \quad \lambda_{\beta\sigma}\in\mathbb{C}$$

be given. For a fixed $\tau\in B$ (without loss of generality, suppose that $a(\tau)=0$), define

$$S := \sum_{0\leq\beta<g(\tau)} \lambda_{\beta\tau}T^\beta Q_\tau = -\sum_{\substack{0\leq\beta<g(\sigma)\\ \sigma\neq\tau}} \lambda_{\beta\sigma}(T-a(\sigma))^\beta Q_\sigma =: U;$$

then $S=U$ lies in $\mathfrak{a} := \mathbb{C}\{T\}\cdot(T_1^{g_1\tau_1},\ldots,T_m^{g_m\tau_m})$, since, for each $\sigma\neq\tau$, there exists a j for which $\sigma_j\neq\tau_j$, so that $T^{g_j\tau_j}_j = P_{j\tau_j}$ is a divisor of Q_σ. Since Q_τ is a unit in $\mathbb{C}\{T\}$, it follows that $Q_\tau^{-1}\cdot S = \sum_{0\leq\beta<g(\tau)} \lambda_{\beta\tau}T^\beta \in \mathfrak{a}\cap\mathbb{C}[T]_{g(\tau)} = 0$; thus $\lambda_{\beta\tau}=0$ for every β. The choice of τ was arbitrary, so it follows that $\lambda_{\beta\sigma}=0$ for all β,σ.

b_2) By the definition of ψ_τ, we need to show that if $\tau\neq\sigma$, then $\psi_{a(\tau)}|M_\sigma = 0$; that is certainly true if $\psi_{a(\tau)}(Q_\sigma)=0$. That means that the germ generated by Q_σ at $(0,a(\tau))$ lies in ${}_{n+m}\mathcal{O}_{(0,a(\tau))}\cdot(P_1,\ldots,P_m) \underset{(23.11.3)}{=} {}_{n+m}\mathcal{O}_{(0,a(\tau))}\cdot(P_{1\tau_1},\ldots,P_{m\tau_m})$. There exists a j such that $\tau_j\neq\sigma_j$, so $P_{j\tau_j}$ is a divisor of Q_σ.

b_3) For a fixed $\sigma\in B$, each polynomial $P_{j\sigma_j}\in{}_n\mathcal{O}_0[T_j]$ is a Weierstrass polynomial of degree $g_{j\sigma_j}$ at $(0,a(\sigma))$; by a), then, the morphism

$$\tilde{\psi}_\sigma: {}_n\mathcal{O}_0[T]_{g(\sigma)} \to {}_{n+m}\mathcal{O}_{(0,a(\sigma))}/(P_{1\sigma_1},\ldots,P_{m\sigma_m})$$

determined by $P_{1\sigma_1},\ldots,P_{m\sigma_m}$ is an isomorphism of ${}_n\mathcal{O}_0$-modules, and there is a commutative diagram

$$\begin{array}{ccc}
{}_n\mathcal{O}_0[T]_{g(\sigma)} & \xrightarrow{\tilde{\psi}_\sigma} & {}_{n+m}\mathcal{O}_{(0,a(\sigma))}/(P_{1\sigma_1},\ldots,P_{m\sigma_m}) \\
\downarrow \alpha_\sigma & & \downarrow \beta_\sigma \\
{}_n\mathcal{O}_0[T]_g \supset {}_n\mathcal{O}_0[T]_{g(\sigma)}\cdot Q_\sigma = M_\sigma & \xrightarrow{\psi_\sigma} & R_\sigma = {}_{n+m}\mathcal{O}_{(0,a(\sigma))}/(P_1,\ldots,P_m),
\end{array}$$

in which α_σ and β_σ represent multiplication by Q_σ and by the germ of Q_σ at $(0,a(\sigma))$, respectively. Since ${}_{n+m}\mathcal{O}_0$ is an integral domain, α_σ is an isomorphism; moreover, β_σ is an isomorphism, since (23.11.3) tells us that it is just multiplication in the ring R_σ by the unit Q_σ. ■

E. 23d. Let M be an R-module; then the following statements are equivalent:

i) M is noetherian.

ii) Every ascending chain of submodules $M_1\subset M_2\subset\ldots$ in M becomes stationary after finitely many steps (for 44.9).

iii) With respect to inclusion, every nonempty system $\mathfrak{M}$ of submodules of M has a maximal element in $\mathfrak{M}$ (for 23 A.5).

E. 23e. Show that $\mathcal{O}(\mathbb{C})$ is neither noetherian nor factorial.

§ 23 A Supplement: Finite Ring-extensions, Normality

Here we collect some simple results from commutative algebra that are helpful at various places in the text.

23 A. 1 Dedekind's Lemma. *For an R-module M generated by p elements, and an ideal $\mathfrak{a} \subset R$, let $\varphi: M \to M$ be an endomorphism such that $\varphi(M) \subset \mathfrak{a}M$. Then φ satisfies an equation of the form*

$$\varphi^p + a_1 \varphi^{p-1} + \ldots + a_p = 0$$

with $a_j \in \mathfrak{a}$ (of course, $\varphi^p = \varphi \circ \ldots \circ \varphi$).

Proof. If $m_1, \ldots, m_p$ are generators of M, then each $\varphi(m_j) \in \mathfrak{a}M$ has a representation of the form $\varphi(m_j) = \sum_{k=1}^{p} a_{jk} m_k$ with $a_{jk} \in \mathfrak{a}$; with the abbreviation $m := (m_1, \ldots, m_p)$, that can be reformulated as

(23A.1.1) $A \cdot (m)^t = 0$ with $A = \varphi \circ I_p - (a_{jk})$.

As a polynomial in φ, the determinant of A provides the desired equation: for the adjoint of A, formed from its cofactors, $\tilde{A} \cdot A = (\det A) \cdot I_p$ (see [La XIII § 4 Prop. 8]); it follows from (23 A. 1.1) that

$$0 = \tilde{A} \cdot A \cdot (m)^t = (\det A) \cdot (m)^t, \quad \text{and} \quad \det A = 0. \quad \blacksquare$$

23 A. 2 Corollary. *For a finitely-generated R-module M, if $\mathfrak{a}$ is an ideal in R such that $M \subset \mathfrak{a}M$, then there exists an $a \in \mathfrak{a}$ such that $(1 + a)M = 0$.*

Proof. The mapping $\varphi = \mathrm{id}_M$ satisfies the hypotheses of 23 A. 1; set $a = a_1 + \ldots + a_p$. ■

23 A. 3 Corollary (Nakayama's Lemma). *Consider a finitely-generated module M over a local ring R, a submodule N of M, and a proper ideal $\mathfrak{a}$ of R. If $M \subset \mathfrak{a}M + N$, then $M = N$.*

Proof. In the special case $N = 0$, we have that $M \subset \mathfrak{a}M$. For an a as in 23 A. 2, $1 + a$ is a unit in R, since R is local; it follows that $M = (1 + a)^{-1}(1 + a)M = 0$. For general N, the mapping $\pi: M \to M/N$ is surjective. By assumption, $\pi(M) \subset \pi(\mathfrak{a}M) = \mathfrak{a}\pi(M)$; thus, by the special case, $\pi(M) = 0$; i.e., $M = N$. ■

E. 23 A a. Prove the following generalization of 23 A. 3: for submodules N and N' of a module M over a local ring R, with N' finitely-generated, if $N' \subset \mathfrak{a}N' + N$, then $N' \subset N$.

23 A. 4 Corollary. *Let M be a finitely-generated module over a local ring R. If the residue classes of $x_1, \ldots, x_n \in M$ generate the $R/\mathfrak{m}$-vector space $M/\mathfrak{m}M$, then $x_1, \ldots, x_n$ generate the R-module M.*

Proof. If $N = Rx_1 + \ldots + Rx_n$, then the composition $N \hookrightarrow M \twoheadrightarrow M/\mathfrak{m}M$ sends the module N onto $M/\mathfrak{m}M$; i.e., $M \subset \mathfrak{m}M + N$. By 23 A. 3, it follows that $M = N$. ■

E. 23 A b. Show that, for a finitely-generated module M over a local ring R, $\operatorname{corank}_R M = \dim_{R/\mathfrak{m}} M/\mathfrak{m}M$, where $\operatorname{corank}_R M := \min\{d; \exists \text{ epimorphism } R^d \to M\}$ is the minimal length of a system of generators of M (for 52.13).

23 A. 5 Krull's Intersection Theorem. *For each submodule N of a finitely-generated module M* *over a noetherian local ring R,*

$$\bigcap_{j=1}^{\infty} (N + \mathfrak{m}^j M) = N .$$

Proof. a) We may assume that $N = 0$ (that is, it suffices to prove that $D := \bigcap_{j=1}^{\infty} \mathfrak{m}^j M = 0$): since M is finitely generated, so is $\bar{M} := M/N$; if we can prove that $\bigcap_{j=1}^{\infty} \mathfrak{m}^j \bar{M} = 0$, then, with the canonical projection $\pi: M \to \bar{M}$, it will follow that

$$\bigcap_{j=1}^{\infty} (N + \mathfrak{m}^j M) = \bigcap_{j=1}^{\infty} \pi^{-1}(\mathfrak{m}^j \bar{M}) = \pi^{-1}\left(\bigcap_{j=1}^{\infty} \mathfrak{m}^j \bar{M}\right) = \pi^{-1}(0) = N .$$

b) We may assume similarly that $\mathfrak{m}D = 0$: set $\bar{\bar{M}} := M/\mathfrak{m}D$; if we can prove that $\bigcap_{j=1}^{\infty} \mathfrak{m}^j \bar{\bar{M}} = 0$, then $D \subset \bigcap_{j=1}^{\infty} (\mathfrak{m}D + \mathfrak{m}^j M) \subset \mathfrak{m}D$, and, by Nakayama's Lemma, we conclude that $D = 0$.

c) It suffices to find an R-module $F \subset M$ and a $k \in \mathbb{N}$ such that $F \cap D = 0$ and $\mathfrak{m}^k M \subset F$: in that case, $D \subset \mathfrak{m}^k M \subset F$, and thus $D = 0$.[5] We find F and k as follows: since M is noetherian (E. 23a), there exists, among the submodules E of M such that $E \cap D = 0$, a maximal one (E. 23d); call it F. Since $\mathfrak{m}$ is finitely-generated, it remains only to show that, for each fixed $r \in \mathfrak{m}$, there exists a k such that $r^k M \subset F$ or, equivalently (due to the maximality of F), such that $(r^k M + F) \cap D = 0$. Consider the ascending chain of submodules of M

$$M_j = \{m \in M; r^j m \in F\} \quad \text{for} \quad j \in \mathbb{N} .$$

Since M is noetherian, the chain becomes stationary; if $M_k = M_{k+1}$, say, then $(r^k M + F) \cap D = 0$: for $x = r^k m + f \in (r^k M + F) \cap D$, we have that $rx = r^{k+1} m + rf \in \mathfrak{m}D = 0$, and thus that $r^{k+1} m = -rf \in F$; hence, $m \in M_{k+1} = M_k$, so $r^k m \in F$, and $x = r^k m + f \in F \cap D = 0$. ∎

For the investigation of singularities, we need the following information for subrings R of a ring S (see 24.4):

23 A. 6 Proposition and Definition. *An element $s \in S$ is called integral over R[6] if one of the following equivalent conditions is satisfied:*

i) s satisfies an "equation of integral dependence"

$$s^p + a_1 s^{p-1} + \ldots + a_p = 0 \quad \textit{with} \quad a_j \in R .$$

ii) $R[s]$ is a finitely-generated R-module.

iii) There exists a subring T of S containing R and s that is finitely-generated as an R-module.

Proof. As usual, let $R[s]$ denote the subring $\left\{\sum_{j=0}^{<\infty} r_j s^j; r_j \in R\right\}$ of S. The implications "i) ⇒ ii) ⇒ iii)" are trivial; for the implication "iii) ⇒ i)", apply Dedekind's Lemma to $\varphi: T \xrightarrow{\cdot s} T$ and $\mathfrak{a} = R$, and consider $\varphi(1)$. ∎

---

[5] This approach is motivated by the special case in which R is a local algebra; M then has a $R/\mathfrak{m}$-vector-space decomposition $M = D \oplus F$. Assume that F is an R-module; then we have that $\mathfrak{m}M \subset \mathfrak{m}D + \mathfrak{m}F = \mathfrak{m}F \subset F$, so we may set $k = 1$.

[6] That generalizes the concept of an integer: since $\mathbb{Z}$ is normal (by 23 A. 9), an element r of $\mathbb{Q} = Q(\mathbb{Z})$ is integral over $\mathbb{Z}$ iff $r \in \mathbb{Z}$.

23 A. 7 Corollary. *i) Elements s and t of S are integral over R iff $R[s,t]$ is a finitely-generated R-module.*

ii) The set $\tilde{R} := \{s \in S;\ s$ is integral over $R\}$ is a subring of S containing R (and is called the <u>*integral closure of R in S*</u>*).*

Proof. i) *Only if.* This follows from 23 A. 6 and the fact that "finitely-generated over finitely-generated is finitely-generated". *If.* This follows from 23 A. 6.

ii) The inclusion $R \subset \tilde{R}$ is clear. If s and t are in $\tilde{R}$, then, by i), $R[s,t]$ is a finitely-generated R-module; it follows from 23 A. 6 that $R[s,t] \subset \tilde{R}$, so $\tilde{R}$ is a ring. ■

23 A. 8 Definition. *The set $\hat{R} := \{s \in Q(R);\ s$ is integral over $R\}$ is called the* <u>*integral closure*</u> *of R. If R has no zero-divisors, and if $R = \hat{R}$, then R is called a* <u>*normal ring*</u>*.*

If R is a reduced analytic algebra (see 24.1 and §31), then $\hat{R}$ is also called the *normalization*[7] of R; in that case, if $R = \hat{R}$, then R has no zero-divisors, and is thus normal (see 53 A. 6). Examples of normal rings are provided by the following fact:

23 A. 9 Lemma. *Every factorial ring R is normal.*

Proof. Let $s \in Q(R)$ be integral over R, say

$$s^b = \sum_{j=0}^{b-1} r_j s^j \quad \text{with} \quad r_j \in R.$$

Since R is factorial, we can write $s = f/g$, with relatively prime $f, g \in R$. It follows that

$$f^b = g\left(\sum_{j=0}^{b-1} r_j f^j g^{b-1-j}\right);$$

so g is a divisor of f^b. Since f and g are relatively prime, g must be a unit in R; it follows that $s = fg^{-1} \in R$. ■

23 A. 10 Lemma. *If R is a normal ring of characteristic 0, then $R[T]_{\text{mon}}$ is a factorial monoid, and the factorization into monic prime factors over R coincides with that over the field of fractions $Q(R)$.*

Proof. Let P be in $R[T]_{\text{mon}}$, and let $P = \prod_{k=1}^{r} P_k$ be the factorization of P over $Q(R)$ into monic irreducible factors. Each zero a_{kj} of P_k in the splitting field of P over $Q(R)$ (see [La VII § 3]) is integral over R, since $P(a_{kj}) = P_k(a_{kj}) = 0$ (23 A. 6); thus, the coefficients of P_k in $Q(R)$, being elementary symmetric functions in the a_{kj} [La V §9], are integral over R. Since R is normal, it follows that P_k is in $R[T]$. ■

For the treatment of prime ideals in 44.9, we need the following information:

23 A. 11 Proposition. *Let $\mathfrak{a}$ and $\mathfrak{a}_j$ be ideals in R, and let $\mathfrak{p}$ and $\mathfrak{p}_j$ be prime ideals in R for $j = 1, \dots, s$.*

i) If $\mathfrak{a} \subset \bigcup_{j=1}^{s} \mathfrak{p}_j$, then there exists a k such that $\mathfrak{a} \subset \mathfrak{p}_k$.

ii) If $\bigcap_{j=1}^{s} \mathfrak{a}_j \subset \mathfrak{p}$, then there exists a k such that $\mathfrak{a}_k \subset \mathfrak{p}$.

[7] In analogy to the geometric situation; see 71.8.

Proof. i) We prove by induction on s that, if $\mathfrak{a} \not\subset \mathfrak{p}_j$ for $j = 1, \ldots, s$, then $\mathfrak{a} \not\subset \bigcup_{j=1}^{s} \mathfrak{p}_j$.

(Note that the set-theoretic union $\bigcup \mathfrak{p}_j$ is not necessarily an ideal!) The case in which $s = 1$ is trivial. "$s-1 \Rightarrow s$". By the induction hypothesis, there exists for each j an $a_j \in \mathfrak{a} \setminus (\bigcup_{i \neq j} \mathfrak{p}_i)$. If some a_j does not lie in $\mathfrak{p}_j$, then the assertion follows. Otherwise every a_j lies in $\mathfrak{p}_j$. If

$$a := \sum_{j=1}^{s} a_1 a_2 \ldots \hat{a}_j \ldots a_s \in \mathfrak{a}$$ [8]

were to lie in some $\mathfrak{p}_j$, then $a_1 \ldots a_j \ldots a_s$ would be in $\mathfrak{p}_j$, although $\mathfrak{p}_j$ is prime! It follows that $\mathfrak{a} \not\subset \bigcup_{j=1}^{s} \mathfrak{p}_j$. ↯

ii) If no $\mathfrak{a}_j$ is included in $\mathfrak{p}$, then there exist elements $a_j \in \mathfrak{a}_j \setminus \mathfrak{p}$. Hence, $a_1 \ldots a_s$ lies in $\bigcap \mathfrak{a}_j$, but not in $\mathfrak{p}$, since $\mathfrak{p}$ is prime. ↯ ∎

Further literature: [AtMac]

§ 24 Analytic Algebras

For the investigation of "singular points" in the solution sets of systems of holomorphic equations, we use homomorphic images of algebras of convergent power series:

24.1 Definition. *An algebra is called an* <u>*analytic algebra*</u> *if it is isomorphic to the residue class algebra* ${}_n\mathcal{O}_0/\mathfrak{a}$ *of* ${}_n\mathcal{O}_0$ *over an ideal* $\mathfrak{a}$.

Note that for $n \geqq 1$ the algebra $\mathbb{C}[\![X_1, \ldots, X_n]\!]$ is not an analytic algebra (see 45.6 and E. 45b).

We summarize here the fundamental properties of analytic algebras (bearing in mind that algebra-homomorphisms φ satisfy the equality $\varphi(1) = 1$):

24.2 Proposition. *The following statements hold for analytic algebras R and S:*

i) Every homomorphic image of R is an analytic algebra.

ii) R is a local algebra.

iii) R is noetherian.

iv) $\bigcap_{j=1}^{\infty} \mathfrak{m}_R^j = (0)$.

v) Each algebra-homomorphism $\varphi: R \to S$ is local; it is determined by its values on a system of generators of $\mathfrak{m}_R$ over R.

vi) For arbitrary elements $f_1, \ldots, f_n$ of $\mathfrak{m}_S$, there exists a unique algebra-homomorphism $\varphi: {}_n\mathcal{O}_0 \to S$ such that $\varphi(X_j) = f_j, j = 1, \ldots, n$.

---

8) $\hat{a}_j$ means that a_j is to be omitted.

vii) For each diagram of algebra-homomorphisms of the form

$$\begin{array}{ccc} {}_n\mathcal{O}_0 & \overset{\hat{\varphi}}{\dashrightarrow} & {}_m\mathcal{O}_0 \\ {\scriptstyle \pi_R}\big\downarrow & & \big\downarrow{\scriptstyle \pi_S} \\ R & \xrightarrow{\ \varphi\ } & S \end{array},$$

there exists a homomorphism $\hat{\varphi}$ *such that the diagram commutes.*

Proof. Fix a homomorphism $\pi: {}_n\mathcal{O}_0 \to R \cong: {}_n\mathcal{O}_0/\mathfrak{a}$.

i) This is trivial.

ii) The maximal ideal $\mathfrak{m}_R$ of R is just $\pi({}_n\mathfrak{m})$. The vector space decomposition ${}_n\mathcal{O}_0 = \mathbb{C}\cdot 1 \oplus {}_n\mathfrak{m}$ induces immediately the isomorphism ${}_n\mathcal{O}_0/\mathfrak{a} \cong (\mathbb{C} \oplus {}_n\mathfrak{m}) / (0 \oplus \mathfrak{a}) \cong \cong \mathbb{C} \oplus {}_n\mathfrak{m}/\mathfrak{a}$.

iii) As a residue class ring of ${}_n\mathcal{O}_0$, R is noetherian by E. 23a.

iv) By 23 A. 5, this follows from ii) and iii).

v) By 21.6, φ is local. The proof parallels that of 21.7 ii) using iv) instead of 21.8 ii).

vi) By vii), φ exists; by v), φ is unique.

vii) Choose elements $g_j \in {}_m\mathcal{O}_0$ such that $\varphi\pi_R(X_j) = \pi_S(g_j)$, $j = i, \dots, n$. By 21.7, the assignment $\hat{\varphi}(X_j) := g_j$ determines an algebra-homomorphism $\hat{\varphi}$, and by v), the diagram commutes. ■

A ring-homomorphism $\varphi: R \to S$ is called *finite* if S is a finitely-generated module over the subring $\varphi(R)$; with that terminology, we have the following generalization of 22.4:

24.3 Noether's Normalization Theorem. *For each analytic algebra R, there exists a finite injective algebra-homomorphism ${}_d\mathcal{O}_0 \hookrightarrow R$.*

Proof. Surjective algebra-homomorphisms are clearly finite; hence, there exist finite homomorphisms $\sigma_n: {}_n\mathcal{O}_0 \to R$. Let d be the smallest n that appears; we shall show that $\sigma_d: {}_d\mathcal{O}_0 \to R$ is injective. If there exists an $f \neq 0$ in Ker σ_d, then we choose appropriate coordinates and apply 22.4, according to which ${}_d\mathcal{O}_0/(f)$ is a finitely-generated ${}_{d-1}\mathcal{O}_0$-module. It follows that the finitely-generated ${}_d\mathcal{O}_0/(f)$-module R is finitely-generated over ${}_{d-1}\mathcal{O}_0$. ↯ ■

In 48.8_{alg} we shall see that there is a unique such d – the dimension of R.

With the sharpened version 46.1 of Noether's Normalization Theorem for an analytic algebra R without zero-divisors, it follows from 46 B. 4 and 46 A. 3 that the normalization $R \hookrightarrow \hat{R}$ satisfies all of the hypotheses of the following result:

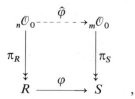

24.4 Theorem. *Let $\varphi: R \to S$ be a finite algebra-homomorphism. If R is an analytic algebra, then S is the ring-direct product of finitely many analytic algebras (which*

are unique up to isomorphism). If, additionally, S has no zero-divisors, then S is an analytic algebra.

Proof. Suppose, without loss of generality, that $R = {}_n\mathcal{O}_0$ (if $R = {}_n\mathcal{O}_0/\mathfrak{a}$, then ${}_n\mathcal{O}_0 \twoheadrightarrow R \xrightarrow{\varphi} S$ is finite). If $s_1, \ldots, s_m$ generate S over $\varphi({}_n\mathcal{O}_0)$, then

$${}_n\mathcal{O}_0[T_1, \ldots, T_m] \to S,\; P \mapsto P(s_1, \ldots, s_m),$$

is a surjective homomorphism of ${}_n\mathcal{O}_0$-algebras; moreover, by 23 A. 6, each s_j satisfies an equation of integral dependence $P_j(s_j) = 0$ with $P_j \in {}_n\mathcal{O}_0[T_j]_{\text{mon}}$, $j = 1, \ldots, m$. Hence, there exists a factorization

$$\begin{array}{ccc} {}_n\mathcal{O}_0[T] & \twoheadrightarrow & S \\ & \searrow \quad \nearrow_{\chi} & \\ & {}_n\mathcal{O}_0[T]\,/\,(P_1, \ldots, P_m). & \end{array}$$

According to 23.11, ${}_n\mathcal{O}_0[T]\,/\,(P_1, \ldots, P_m)$ is the ring-direct product of finitely many analytic algebras, so the assertion follows from parts ii) and iii) of the following simple algebraic exercise:

E. 24 a. If $R = R_1 \oplus \ldots \oplus R_m$ is a ring-direct product of rings R_j (i.e., componentwise operations), then the following statements hold:

i) There exist unique elements $e_j \in R_j$ such that $1 = e_1 + \ldots + e_m$. They satisfy the conditions $e_k e_j = \delta_{kj} e_j$ and $R_j = Re_j$; moreover, $e_j = 1_{R_j}$.

ii) Each ideal $\mathfrak{a}$ in R has a unique decomposition of the form $\mathfrak{a} = \mathfrak{a}_1 \oplus \ldots \oplus \mathfrak{a}_m$ with each $\mathfrak{a}_j$ being an ideal in R_j (of course $\mathfrak{a}_j = R_j\mathfrak{a}$).

iii) Each residue class ring $R/\mathfrak{a}$ is isomorphic to $\bigoplus_j R_j/\mathfrak{a}_j$.

iv) If $\mathfrak{m}_j$ is a maximal ideal in R_j, then

$$\mathfrak{m} = R_1 \oplus \ldots \oplus R_{j-1} \oplus \mathfrak{m}_j \oplus R_{j+1} \oplus \ldots \oplus R_m$$

is a maximal ideal in R; moreover, every maximal ideal in R is of that type.

v) Let $\varphi: R_1 \oplus \ldots \oplus R_m \to S_1 \oplus \ldots \oplus S_n$ be a ring-homomorphism of ring-direct products. If every S_i is a local ring, then the induced mappings $\varphi_{ij}: R_j \to S_i$ are such that, for each i, there exists precisely one $j = j(i)$ with $\varphi_{ij} \neq 0$. (Hint: For $n = 1$, apply the decomposition of $\varphi^{-1}(\mathfrak{m}_S)$ according to ii); for 45.15.)

E. 24 b. Let R be an analytic algebra, and choose $\varphi \in \operatorname{Hom}_R(R^b, R^b)$. Prove that φ is an isomorphism iff the mapping $\bar{\varphi} \in \operatorname{Hom}_{\mathbb{C}}(\mathbb{C}^b, \mathbb{C}^b)$ determined by reduction mod $\mathfrak{m}_R$ is an isomorphism. (Hint: for $f \in R$, set $f(0) := \operatorname{pr}_{\mathbb{C}}(f)$ with the "*evaluation*" $\operatorname{pr}_{\mathbb{C}}: R = \mathbb{C} \oplus \mathfrak{m}_R \to \mathbb{C}$, and apply the determinant of $\bar{\varphi} = \varphi(0)$; for 23.11.)

E. 24 c. Let R be an analytic algebra, let M' be a finitely-generated R-module, and choose $\varphi \in \operatorname{Hom}_R(M, M')$. Then φ is surjective iff $\varphi: M/\mathfrak{m}M \to M'/\mathfrak{m}M'$ is surjective (hint: 23 A. 3; for 41.7 iv)).

E. 24 d. "*Strukturausdünnung*" *in analytic algebras.* Let S denote a subalgebra of an analytic algebra R. Show that S is an analytic algebra if there exists a $k \in \mathbb{N}_{>0}$ such that $\mathfrak{m}_R^k \subset S$. (Hints:

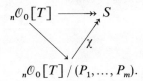

i) For $R = {}_n\mathcal{O}_0$, $S = \mathbb{C} \oplus {}_n\mathfrak{m}^k$: the set $J := \{\nu \in \mathbb{N}^n; k \leqq |\nu| \leqq 2k-1\}$ contains $m = \sum_{j=k}^{2k-1} \binom{j+n-1}{n-1}$ elements; the image of the algebra-homomorphism

$$\varphi: {}_m\mathcal{O}_0 \to {}_n\mathcal{O}_0, \; w_\nu \mapsto z^\nu \quad \text{for} \quad \nu \in J,$$

lies in S. Moreover, $S \subset \operatorname{Im}\varphi$: for every $\mu \in \mathbb{N}^n$ such that $|\mu| \geqq k$, fix a monomial $P_\mu \in \mathbb{C}[w]$ such that $\varphi(P_\mu) = z^\mu$. For $f = \sum_{|\mu| \geqq k} a_\mu z^\mu$, $\hat{f} := \sum a_\mu P_\mu$ is in $\mathbb{C}[\![w]\!]$, and $\varphi(\hat{f}) = f$. The series $\hat{f}$ converges: for $r \in \mathbb{R}^n_{>0}$ and $s := (r^\nu)_{\nu \in J} \in \mathbb{R}^m_{>0}$, we have that $\|f\|_r = \|\hat{f}\|_s$ (for E. 49Ai).
ii) For $R = {}_n\mathcal{O}_0$ and $\mathbb{C} \oplus {}_n\mathfrak{m}^k \subset S$, apply 24.4.)

E. 24e. Introduce the Krull topology for analytic algebras (in analogy to E. 21e).

Further literature: [Ab], [GrRe].

Chapter 3: Complex Manifolds and the Elementary Theory of Complex Spaces

In this chapter, we describe, in a manner independent of embeddings, submanifolds of domains in $\mathbb{C}^n$, which were introduced in 8.8. In so doing, we obtain the latitude necessary for a lucid and systematic presentation of function theory on analytic sets, proceeding along lines similar to the construction of Riemann surfaces, differentiable manifolds, etc. In complex analysis, in contrast to the real case, one quickly encounters manifolds that permit no embedding in domains of complex number spaces; as an example, we mention the Riemann sphere $\mathbb{P}_1 = \mathbb{C} \cup \{\infty\}$.

Chapter 3 may be viewed as a paradigmatic introduction to the theory of complex spaces. It is true that the formal framework of ringed spaces, which we use to discuss manifolds and complex spaces, is rather abstract; however, we present it in an elementary fashion with just enough generality for it to be used in various constructions, such as the formation of quotients, which cannot be carried out in the smaller category of complex spaces (ringed spaces play a similar role in algebraic geometry, p-adic analysis, and the theory of real-analytic spaces).

A ringed space is a topological space with a distinguished "*structure sheaf*" of algebras that unites punctual objects (such as the analytic algebras treated in Chapter 2) into a global object. It is a formalization of the concept of a sheaf of structure-preserving functions. In § 30, we collect the necessary elements of sheaf theory. In the context of the theory of ringed spaces, we present in § 31 two canonical constructions that provide an analytic set with a complex structure. The first one is motivated geometrically, and thus is grasped more easily by geometers: it is characterized purely in terms of the solution set of a system of equations $f_1 = \ldots = f_m = 0$; such a "*Nullstellenmenge*" is denoted by $N(G; f_1, \ldots, f_m)$. The second construction is algebraic in nature, and includes a finiteness condition, so that it can be treated easily with the methods of the algebraist: the objects so constructed are denoted by $V(G; f_1, \dots, f_m)$, in order to evoke the word "variety". We shall adopt both approaches for the construction of local models of complex spaces; the relationship between them is clarified in Cartan's Coherence Theorem 47.1, according to which the first construction is a special case of the second.

In § 32 and its supplements, we present some simple properties of manifolds and reduced complex spaces, and discuss in particular the reduced "*archsingularities*": the *Achsenkreuz*, Neil's parabola, Whitney's umbrella, and $\sqrt{uv}$.

We investigate the behaviour of "*Nullstellengebilde*" on monic polynomials in $\mathcal{O}(G)[T]$ as branched coverings in § 33. The results there generalize the Theorem on the Continuity of the Roots of a Polynomial; in particular, they provide a prototype for finite mappings, and facilitate a geometric understanding of the decomposition into irreducible components that is helpful in the abstract theory (e.g., for the Representation Theorem for Prime Germs).

At certain places in this chapter, we deliberately write "complex space" instead of "reduced complex space", in order to indicate that the statement at hand can easily be carried over to the class of general complex spaces, as treated in Chapter 4.

§ 30 Bringing in the Sheaves

In the description of the holomorphic structure of complex number spaces, we make use of the concept of the "*structure sheaf*" of $\mathbb{C}^n$, which is the sheaf of functions ${}_n\mathcal{O}$ on $\mathbb{C}^n$ determined by the family $(\mathcal{O}(X))_{X \subset\subset \mathbb{C}^n}$ with the restriction-homomorphisms of functions.

In order to clarify the concepts involved, we present here notation and results in the broader context of general sheaf theory (throughout this discussion, T denotes a topological space, and R, a ring):

A *presheaf* $(\mathcal{G}, \varrho)$ *of R-modules*[1] on T is determined by the provision of an R-module $\mathcal{G}(U)$ for each $U \subset\subset T$, and of "*restriction-homomorphisms*" $\varrho_V^U: \mathcal{G}(U) \to \mathcal{G}(V)$ for $V \subset\subset U$ satisfying the condition $\varrho_W^U = \varrho_W^V \circ \varrho_V^U$ for $W \subset\subset V$. In general, a presheaf $(\mathcal{G}, \varrho)$ is abbreviated as $\mathcal{G}$ and $\varrho_V^U(f)$, as $f|_V$. Note that the "restriction-homomorphism" $f \mapsto f|_V$ need not be the usual restriction of functions, even if the elements of $\mathcal{G}(U)$ and $\mathcal{G}(V)$ are functions on U and V, respectively.

A presheaf $\mathcal{G}$ is called a *sheaf* if the following two conditions are satisfied for each family of open sets $(U_j)_{j \in J}$ with $U := \bigcup_j U_j$:

i) *(Uniqueness of sections)*. Two "sections" $f, g \in \mathcal{G}(U)$ are equal iff $f|_{U_j} = g|_{U_j}$ for every $j \in J$.

ii) *(Existence of sections)*. For each family of sections $s_j \in \mathcal{G}(U_j)$ such that $s_i|_{U_{ij}} = s_j|_{U_{ij}}$[2] for all $i, j \in J$, there exists a section $s \in \mathcal{G}(U)$ such that $s|_{U_j} = s_j$ for every j.

Note that $\mathcal{G}(\emptyset) = 0$ for every sheaf $\mathcal{G}$.

---

[1] Presheaves of R-algebras, etc., are defined analogously; for $R = \mathbb{Z}$, those are just the presheaves of abelian groups, rings, etc.

[2] $U_{i_0, \dots, i_p} := U_{i_0} \cap \dots \cap U_{i_p}$.

E. 30 a. Let $\mathscr{G}$ be a presheaf of R-modules; then $\mathscr{G}$ is a sheaf iff, for each family of open sets $(U_i)_{i\in I}$ with $U := \bigcup_I U_i$, the sequence

$$\text{(E. 30a.1)}\quad 0 \to \mathscr{G}(U) \xrightarrow{\alpha} \prod_{i\in I} \mathscr{G}(U_i) \xrightarrow{\beta} \prod_{(i,j)\in I^2} \mathscr{G}(U_{ij}),$$

with $\alpha(f) := (f|_{U_i})_{i\in I}$ and $\beta((f_i)_{i\in I}) := (f_i|_{U_{ij}} - f_j|_{U_{ij}})_{(i,j)\in I^2}$, is exact (for 55.5).

For each subset $U \Subset X$, the *restriction* of the (pre-)sheaf $\mathscr{G}$ to U is, by definition, the (pre-) sheaf $\mathscr{G}|_U := (\mathscr{G}(V))_{V\Subset U}$.

A *(pre-)sheaf-morphism* $\varphi: \mathscr{F} \to \mathscr{G}$ consists of a family of morphisms

$$\varphi = (\varphi(U): \mathscr{F}(U) \to \mathscr{G}(U))_{U\Subset T}$$

such that the diagram

$$\begin{array}{ccc} \mathscr{F}(U) & \xrightarrow{\varphi(U)} & \mathscr{G}(U) \\ \Big\downarrow \varrho_V^U(\mathscr{F}) & & \Big\downarrow \varrho_V^U(\mathscr{G}) \\ \mathscr{F}(V) & \xrightarrow{\varphi(V)} & \mathscr{G}(V) \end{array}$$

commutes for all $V \Subset U \Subset T$ (in categorical terms, φ can be interpreted as a natural transformation of contravariant functors $\mathscr{F}$ and $\mathscr{G}$). The set of morphisms $\varphi: \mathscr{F} \to \mathscr{G}$ and the corresponding (pre-)sheaf will be denoted by $\operatorname{Hom}(\mathscr{F}, \mathscr{G})$ and $\mathscr{H}om(\mathscr{F}, \mathscr{G}) := (\operatorname{Hom}(\mathscr{F}|_U, \mathscr{G}|_U))_{U\Subset X}$, respectively.

For each presheaf $\mathscr{G}$, there exists a presheaf-morphism $\varphi: \mathscr{G} \to \overline{\mathscr{G}}$ into a sheaf $\overline{\mathscr{G}}$ that has the following universal property: each presheaf-morphism $\psi: \mathscr{G} \to \mathscr{F}$ to a sheaf $\mathscr{F}$ factors uniquely over φ. The construction of $\overline{\mathscr{G}}$ and φ runs as follows: For an open cover $\mathfrak{U} = (U_i)_{i\in I}$ of $U := \bigcup_I U_i \Subset T$, set

$$\Gamma(\mathfrak{U}, \mathscr{G}) := \{f \in \prod_i \mathscr{G}(U_i);\ \ f_i|_{U_{ij}} = f_j|_{U_{ij}}\ \forall i,j \in I\}.$$

For each refinement $\mathfrak{V} = (V_k)_{k\in K}$ of $\mathfrak{U}$ with refinement-mapping $\tau: K \to I$ (see §51A), define

$$\Gamma(\mathfrak{U}, \mathscr{G}) \to \Gamma(\mathfrak{V}, \mathscr{G}), \quad (f_i) \mapsto (f_{\tau(k)}|_{V_k}).$$

Set $\overline{\mathscr{G}}(U) := \varinjlim_{\mathfrak{U}} \Gamma(\mathfrak{U}, \mathscr{G})$, and let $\varphi(U): \mathscr{G}(U) \to \overline{\mathscr{G}}(U)$ be the mapping that sends $f \in \mathscr{G}(U) = \Gamma(\{U\}, \mathscr{G})$ to its class in $\overline{\mathscr{G}}(U)$. Then $\overline{\mathscr{G}}$ is a sheaf called the *sheaf associated to* $\mathscr{G}$, and is unique up ot isomorphism. It is sometimes useful to know that the construction of the associated sheaf $\overline{\mathscr{G}}$ depends only on the presheaf $\mathscr{G}$ on a basis of open sets of T.

For a presheaf $\mathscr{G}$ and $t \in T$, the direct limit

$$\mathscr{G}_t := \varinjlim_{U\ni t} \mathscr{G}(U)$$

is called the *stalk* of $\mathscr{G}$ at t; for each $t \in T$, the morphism $\varphi: \mathscr{G} \to \overline{\mathscr{G}}$ obviously induces an isomorphism $\varphi_t: \mathscr{G}_t \xrightarrow{\cong} \overline{\mathscr{G}}_t$.

We understand a (pre-)*sheaf of functions* to be a (pre-)sheaf $\mathscr{G}$ for which each $\mathscr{G}(U)$ is a space of functions and each ϱ_V^U is the usual restriction of functions; for such sheaves the sequence (E. 30a. 1) is automatically exact at $\mathscr{G}(U)$.

One can assign to each (pre-)sheaf $\mathscr{G}$ on T a unique topological space $|\mathscr{G}|$[3)] over T, called its *sheaf-space* (or *espace étalé* = displayed space), with which some constructions of sheaf theory

[3)] In the sequel, we usually write $\mathscr{G}$ for $|\mathscr{G}|$.

are easier to interpret geometrically: if $\mathscr{G}$ is a presheaf of R-modules on T, then one provides each $\mathscr{G}(U)$ with the discrete topology and defines an equivalence relation on the disjoint union

$$V = \dot{\bigcup_{U \Subset T}} U \times \mathscr{G}(U)$$

(viewed as a topological sum):

$$(t,f) \sim (u,g) :\Leftrightarrow t = u, \text{ and } \; f = g \quad \text{near} \quad t.$$

The quotient space $|\mathscr{G}| := V/\sim$, with the projection $\pi: |\mathscr{G}| \to T$ onto the first component, has as fibers $\pi^{-1}(t)$ the stalks $\mathscr{G}_t$ (with the discrete topology); furthermore, the following statements hold:

30.1 Remarks. i) *π is surjective and is a local homeomorphism.*

ii) *Each stalk $\mathscr{G}_t = \pi^{-1}(t)$ is an R-module.*

iii) *The algebraic structure of $\mathscr{G}_t$ depends continuously on t; i.e., for each $U \Subset T$, the set $\Gamma(U, |\mathscr{G}|) := \{\sigma: U \to |\mathscr{G}| \text{ continuous}; \ \pi \circ \sigma = \mathrm{id}_U\}$ of (continuous) sections with stalkwise-defined algebraic operations forms an R-module.*

Conversely, we call every topological space S over T (with projection $\pi: S \to T$) that satisfies conditions i)–iii) a *sheaf space over T*. The *section-functor* Γ provides a one-to-one correspondence between such spaces S and the sheaves of section $U \mapsto \Gamma(U, S)$; for a presheaf $\mathscr{G}$ the assignment $U \mapsto \Gamma(U, |\mathscr{G}|)$ yields the associated sheaf $\overline{\mathscr{G}}$. If $\mathscr{G}$ is a sheaf, then it follows that $\Gamma(U, |\mathscr{G}|) \cong \mathscr{G}(U)$.

The essential properties of a sheaf space $\pi: S \to T$ are as follows:

iv) *The projection π and all sections $\sigma \in \Gamma(U, S)$ are open mappings.*

v) *The set $\{\sigma(U); \sigma \in \Gamma(U, S), U \Subset T\}$ is a basis for the topology of S.*

vi) *For $f \in \Gamma(U, S)$ and $g \in \Gamma(V, S)$, the set $\{t \in U \cap V; f(t) = g(t)\}$ is open in T.*

To each sheaf-morphism $\varphi: \mathscr{F} \to \mathscr{G}$ there corresponds a unique continuous mapping $|\varphi|: |\mathscr{F}| \to |\mathscr{G}|$ that induces homomorphisms of R-modules between the corresponding stalks; conversely, each such mapping corresponds to a sheaf-morphism. Thus we have established an "equivalence" between the category of sheaf-spaces over T and that of sheaves of R-modules over T.

30.2 Example. We can illustrate those concepts with the sheaf ${}_n\mathcal{O}$: For $a \in U \Subset \mathbb{C}^n$ and $f \in {}_n\mathcal{O}(U)$, let $f_a \in {}_n\mathcal{O}_a$ denote the equivalence class of $(a,f) \in U \times {}_n\mathcal{O}(U)$ in ${}_n\mathcal{O}_a$ (the attentive reader will not mistake f_a for a partial derivative); f_a is called the *germ* of f *at* a (E. 21c). Each stalk ${}_n\mathcal{O}_a$ can be identified with ${}_n\mathcal{O}_0 = \mathbb{C}\{z_1, \ldots, z_n\}$: in §21, we obtained, by means of translation, a canonical isomorphism

$$\tau^0: {}_n\mathcal{O}_a \to {}_n\mathcal{O}_0, f_a \mapsto \sum_\nu \frac{D^\nu f(a)}{\nu!} z^\nu .$$

Two functions $f, g \in \mathcal{O}(U)$ define the same germ at $a \in U$ iff their Taylor expansions $\sum_\nu \frac{D^\nu f(a)}{\nu!}(z-a)^\nu$ and $\sum_\nu \frac{D^\nu g(a)}{\nu!}(z-a)^\nu$ coincide.

A basis for the topology of $|{}_n\mathcal{O}|$ is given by all sets of the form $W := \{f_b; b \in U\}$ for $U \Subset \mathbb{C}^n$ and $f \in \mathcal{O}(U)$. The mapping $\pi: W \to U, f_b \mapsto b$, is a homeomorphism.

For $f \in \mathrm{Hol}(U, \mathbb{C})$ and $a \in U \Subset \mathbb{C}^n$, $f(a)$ is a complex number. If we interpret f as a section $U \to |{}_n\mathcal{O}|$, then the image of a will be denoted as above by f_a, and not by $f(a)$! Then we abbreviate the value $f_a(a) \in \mathbb{C}$ of this germ as $f(a)$.

E. 30 b. Prove that the Identity Theorem 6.1 says merely that $|{}_n\mathcal{O}|$ is a Hausdorff space.

E. 30 c. For $n = 1$, let Z denote a connected component of $|{}_1\mathcal{O}|$, and $\pi: {}_1\mathcal{O} \to \mathbb{C}^1$, the sheaf-projection. For $\pi: Z \to \pi(Z) \subset\subset \mathbb{C}^1$, interpret Z as Riemann surface over $\pi(Z)$.

Accordingly, for general n, one calls the pair $(Z, \pi: Z \to \pi(Z))$ a *Riemann domain* over $\pi(Z)$; by E. 30b, Z is Hausdorff. Interpret Z as the "maximal domain of definition" of a holomorphic function.

The fundamental constructions of module theory generalize directly to the formation of the corresponding presheaves or sheaves: there are sheaves of rings $\mathcal{R}$ (with continuous unit-section), sheaves of $\mathcal{R}$-modules, sheaves of ideals, direct sums or products of sheaves (in particular, free sheaves of modules), quotient-sheaves, sheaves of homomorphisms, tensor products, etc. In the following, we shall frequently refer to sheaves of rings, modules, ideals, etc., simply as *rings*, *modules*, *ideals*, etc.

A sequence of sheaves

$$\mathcal{F} \xrightarrow{\varphi} \mathcal{G} \xrightarrow{\psi} \mathcal{H}$$

is an *exact sequence* iff, for every $t \in T$, the sequence of stalks

$$\mathcal{F}_t \xrightarrow{\varphi_t} \mathcal{G}_t \xrightarrow{\psi_t} \mathcal{H}_t$$

is exact (i.e., $\operatorname{Ker} \psi_t = \operatorname{Im} \varphi_t$). That is weaker than the requirement that the sequence

$$\mathcal{F}(U) \xrightarrow{\varphi(U)} \mathcal{G}(U) \xrightarrow{\psi(U)} \mathcal{H}(U)$$

be exact for each $U \subset\subset X$, for *the section-functor Γ is left-exact, but not in general exact* (that fact later leads to the construction of cohomology groups). We present an example, typical for our situation, in which Γ is not right-exact:

30.3 Example. For $T := \mathbb{C}^{2*}, \mathcal{O} := {}_2\mathcal{O}|_T$, and $A := \mathbb{C}^* \times 0$, define an ideal $\mathcal{I}$ in $\mathcal{O}$ by setting $\mathcal{I}(U) := \{f \in \mathcal{O}(U); f|_A = 0\}$; then the sequence of sheaves

$$0 \to \mathcal{I} \to \mathcal{O} \to \mathcal{O}/\mathcal{I} \to 0$$

is obviously exact, but the induced sequence of sections

$$0 \to \Gamma(T, \mathcal{I}) \to \Gamma(T, \mathcal{O}) \to \Gamma(T, \mathcal{O}/\mathcal{I}) \to 0$$

is not: letting

$$\operatorname{supp} \mathcal{G} := \{t \in T; \mathcal{G}_t \neq 0\}$$

denote the *support* of a sheaf $\mathcal{G}$, we have here that $\operatorname{supp} \mathcal{O}/\mathcal{I} = A$. It is easy to see that ${}_{\mathbb{C}^*}\mathcal{O}$ and $\mathcal{O}/\mathcal{I}$ can be identified in a canonical manner (see 31.5 i)). We have that $\Gamma(T, \mathcal{O}/\mathcal{I}) \neq \Gamma(T, \mathcal{O})/\Gamma(T, \mathcal{I})$: the function $1/z_1 \in {}_{\mathbb{C}^*}\mathcal{O}(\mathbb{C}^*)$ certainly can be extended holomorphically to the neighborhood $\mathbb{C}^* \times \mathbb{C}$ of A in T, but not to T (7.10); thus, $1/z_1$ does not determine a class in $\Gamma(T, \mathcal{O})/\Gamma(T, \mathcal{I})$, but it does determine a section in the quotient-sheaf $\mathcal{O}/\mathcal{I}$, since the support of the latter is A.

E. 30 d. Let $\mathcal{R}$ be a sheaf of rings on T, and fix $U \subset\subset T$ and $r \in \mathcal{R}(U)$. Prove the following statement: r is a unit in $\mathcal{R}(U)$ iff r_t is a unit in $\mathcal{R}_t$ for each $t \in U$ (for 31.1).

E. 30 e. ${}_n\mathcal{O}$ is a sheaf of algebras whose stalks are local algebras. Prove that, for $n \geq 1$ and $U \subset\subset \mathbb{C}^n$, the algebra ${}_n\mathcal{O}(U)$ is not local.

If $\varphi: S \to T$ is a continuous mapping of topological spaces, and if $\mathcal{G}$ is a sheaf on T, then the presheaf on S

$$U \mapsto (\varphi^{-1}\mathcal{G})(U) = \{\sigma: U \to |\mathcal{G}| \text{ continuous}; \pi_{\mathcal{G}} \circ \sigma = \varphi\}$$

is a sheaf, called the *(topological) inverse image sheaf* $\varphi^{-1}\mathscr{G}$ *of* $\mathscr{G}$ *with respect to* φ. For every $s \in S$, there is a canonical isomorphism $(\varphi^{-1}\mathscr{G})_s \cong \mathscr{G}_{\varphi(s)}$. The sheaf space of $\varphi^{-1}(\mathscr{G})$ is the projection onto S of the topological subspace $\{(s, f_t); \varphi(s) = t\}$ of $S \times |\mathscr{G}|$.

In particular, if $i: S \to T$ is the inclusion of a topological subspace S of T, then $i^{-1}\mathscr{G} =: \mathscr{G}|_S$ is the *restriction of* $\mathscr{G}$ *to* S. If $\mathscr{G}$ is an $\mathscr{R}$-module, then $\varphi^{-1}\mathscr{G}$ is a module over the sheaf of rings $\varphi^{-1}\mathscr{R}$; in fact, φ^{-1} can be extended to an exact covariant functor from the category of $\mathscr{R}$-modules into the category of $(\varphi^{-1}\mathscr{R})$-modules.

E. 30 f. i) Prove that $\varphi^{-1}\mathscr{G}$ is associated to the presheaf

$$U \mapsto \varinjlim_{\varphi^{-1}(V) \supset U} \mathscr{G}(V).$$

ii) For $G \subset\subset \mathbb{C}^n$, $H \subset\subset \mathbb{C}^m$, and $\varphi = \mathrm{pr}_G: G \times H \to G$, prove that $\varphi^{-1}({}_G\mathcal{O}) \subset {}_{G\times H}\mathcal{O}$ is the subsheaf of those functions which are locally independent of $z_{n+1}, \ldots, z_{n+m}$.

Supplementary literature: [Ku], [Br], [Go], [Te].

§ 31 Ringed Spaces and Local Models of Complex Spaces

If a Hausdorff space X has the structure of a complex manifold (32.1), then one can approach a characterization of that structure in different ways:

i) by defining a complex atlas on X (i.e., a cover of X with "holomorphically compatible" charts);

ii) by distinguishing, in the sheaf ${}_X\mathscr{C}$ of continuous functions on X, a subsheaf ${}_X\mathcal{O}$ with the following property: each $x \in X$ admits an open neighborhood $U \subset\subset X$ and a homeomorphism $\varphi: U \to U' \subset\subset \mathbb{C}^n$ that induces an isomorphism between ${}_X\mathcal{O}|_U$ and ${}_n\mathcal{O}|_{U'}$.

We will choose primarily the second way, since that approach can more naturally be extended to a theory of analytic sets that is independent of specific embeddings. Apart from that consideration, the concept of a sheaf plays such a central role in multidimensional function theory that we would have to apply it anyway.

In this section, we collect the formal part of the theory; for that purpose, we consider topological spaces T with "structure sheaves":

31.1 Definition. *A* ringed space *is a pair* $(T, \mathscr{A})$ *in which* T *is a (not necessarily Hausdorff) topological space and* $\mathscr{A}$ *is a sheaf of algebras on* T *such that*

i) Every stalk $\mathscr{A}_t$ *of* $\mathscr{A}$ *is a local algebra (see § 21).*

ii) For every $U \subset\subset T$ *and* $f \in \mathscr{A}(U)$, *the function*

$$\mathrm{Red}\, f: U \to \mathbb{C},\ t \mapsto f(t) := \varepsilon_t(f_t),$$

(with the "evaluation" $\varepsilon_t = \mathrm{pr}_{\mathbb{C}}: \mathscr{A}_t = \mathbb{C} \oplus \mathfrak{m}_t \to \mathbb{C}$*) is continuous.* $\mathrm{Red}\, f$ *is called the "reduction" of* f.

The sheaf $\mathscr{A}$ *is called the* structure sheaf; *it is also denoted by* ${}_T\mathscr{A}$. *Instead of* $(T, \mathscr{A})$, *one often writes simply* T; *if it is necessary to distinguish the ringed space from the underlying topological space, then the latter is denoted by* $|T|$.

The algebra $\mathscr{A}(T)$ includes in a natural way the field $\mathbb{C}$, since it contains the global section 1, and since $1_t \neq 0_t$ in every local algebra $\mathscr{A}_t$.

The ringed spaces represent a generalization of local algebras to "continuous families of local algebras with parameter-space $|T|$". Accordingly, we have an analogue of the following criterion: an algebra R is a local algebra iff there exists an algebra-homomorphism $\varphi: R \to \mathbb{C}$ such that the elements of R satisfying the condition "$\varphi(r) \neq 0$" are precisely the units of R (then φ necessarily coincides with the evaluation $\varepsilon: R = \mathbb{C} \oplus \mathfrak{m}_R \to \mathbb{C}$):

31.2 Lemma. *Let $\mathscr{A}$ be a sheaf of algebras on T, and let $\varphi: \mathscr{A} \to {}_T\mathscr{C}$ be a homomorphism of algebras. Suppose that, for each $U \subset T$ and each $f \in \mathscr{A}(U)$, f is a unit of $\mathscr{A}(U)$ iff $(\varphi f)(t) \neq 0$ for every $t \in U$. Then $(T, \mathscr{A})$ is a ringed space (every $\mathscr{A}_t$ is a local algebra), and φ coincides with the sheaf-homomorphism*

$$\mathrm{Red}: \mathscr{A} \to {}_T\mathscr{C}, \; f \mapsto \mathrm{Red}\, f.$$

Proof. It is easy to see that Red is in fact a sheaf-homomorphism. By the preceding remark on local algebras, we need only show for every $t \in T$ that an $f_t \in \mathscr{A}_t$ is a unit in $\mathscr{A}_t$ iff $\varphi(f_t)(t) \neq 0 \in \mathbb{C}$. *If.* Let $f \in \mathscr{A}(U)$ be a representative of f_t. By virtue of continuity, we may assume that $\varphi(f)$ has no zeros in U. By the assumption of the lemma, f is a unit in $\mathscr{A}(U)$, so f_t is a unit in $\mathscr{A}_t$. The direction "only if" is trivial. ■

E. 31a. Prove the following statements: i) $(T, {}_T\mathscr{C})$ is a ringed space.
ii) The sheaf $\mathscr{F}$ of all functions on $\mathbb{C}$ is a sheaf of algebras such that no stalk is a local algebra. *A fortiori*, $(\mathbb{C}, \mathscr{F})$ is not a ringed space.

31.3 Remark. In the literature, ringed spaces usually are defined without the requirement 31.1 ii) that the reduction be continuous. However, the ringed spaces of interest to us satisfy that condition anyway; since it can facilitate an intuitive interpretation, we have included it in the definition. Moreover, almost all propositions presented in this book concerning ringed spaces hold also without that assumption (even with unaltered proofs).

31.4 Examples. i) For $T \subset \mathbb{R}^n$ and $p \in \mathbb{N} \cup \{\infty, \omega\}$, $(T, {}_T\mathscr{C}^p)$ is a ringed space (where $\mathscr{C}^\omega$ denotes the sheaf of real-analytic functions).

ii) $(\mathbb{C}^n, {}_n\mathscr{O})$ is a ringed space.

iii) *Open subspaces.* For a ringed space $(T, \mathscr{A})$ and $U \subset T$, the ringed space $(U, \mathscr{A}|_U)$ is called an open subspace of $(T, \mathscr{A})$, and is usually denoted by $U \subset T$. In particular, for $X \subset \mathbb{C}^n$ and ${}_X\mathscr{O} := {}_n\mathscr{O}|_X$, $(X, {}_X\mathscr{O})$ is a ringed space.

iv) *Multiple points.* For $T = \mathbb{C}$, $j \geqq 1$ and $\mathscr{I} := {}_1\mathscr{O} \cdot z^j \subset {}_1\mathscr{O}$, the associated ringed subspace $A = (\{0\}, {}_1\mathscr{O}_0/\mathscr{I}_0)$ is called a *j-fold (multiple) point* (for $j = 2$, also *double point*). We shall make use of the following geometric symbols:

simple point double point *j*-fold point.

We have that

$$_A\mathcal{O} \cong \{a_0 + a_1\varepsilon + \ldots + a_{j-1}\varepsilon^{j-1};\ a_k \in \mathbb{C}\},$$

with $\varepsilon^j = 0$, so $_A\mathcal{O}$ is a sheaf of functions only for $j = 1$. (Many other ringed structures may be introduced on $\{0\}$; see 48.2_{alg}.)

v) *The reduction of a space.* The structure sheaf $\mathscr{A}$ of a ringed space T need not be a sheaf of functions; however, by means of the homomorphism Red, $\mathscr{A}$ is comparable to the sheaf of functions $\mathrm{Red}\,\mathscr{A} \subset {}_T\mathscr{C}$ generated by the presheaf $U \mapsto \mathrm{Red}\,\mathscr{A}(U)$ (in other words, $\mathrm{Red}: \mathscr{A} \to {}_T\mathscr{C}$ is not in general injective, so it is impossible to characterize the sections of $\mathscr{A}$ by means of their functional values). Since $\mathrm{Red}\,\mathscr{A}$ is a subalgebra of $_T\mathscr{C}$, it is easy to see that

$$\mathrm{Red}\,T := (|T|, \mathrm{Red}\,\mathscr{A})$$

is a ringed space; it is called the reduction of T (in 31.10 we extend the reduction to a covariant functor). The space $(T, \mathscr{A})$ is called a *reduced ringed space* if $\mathrm{Red}: \mathscr{A} \to {}_T\mathscr{C}$ is injective, in which case $\mathscr{A}$ may be identified with $\mathrm{Red}\,\mathscr{A}$.

vi) *Closed subspaces.* Our goal is to be able to assign ringed spaces to analytic sets. To that end, we introduce the following more general procedure:

Let $(T, \mathscr{A})$ be a ringed space, and let $\mathscr{I} \subset \mathscr{A}$ be a sheaf of ideals. By defining

$$A := \{t \in T;\ \mathscr{I}_t \neq \mathscr{A}_t\} = \{t \in T;\ 1_t \notin \mathscr{I}_t\} = \{t \in T;\ \mathscr{A}_t/\mathscr{I}_t \neq 0\} = \mathrm{supp}(\mathscr{A}/\mathscr{I})$$

and $_A\mathscr{A} := (\mathscr{A}/\mathscr{I})|_A$, we have that *$(A, {}_A\mathscr{A})$ is a ringed space*: every $_A\mathscr{A}_t$ is a local algebra, and the composition

$$_T\mathscr{A}|_A \xrightarrow{\mathrm{Red}_T|_A} {}_T\mathscr{C}|_A \to {}_A\mathscr{C}$$

provides the homomorphism $\mathrm{Red}_A: {}_A\mathscr{A} \to {}_A\mathscr{C}$ that is necessary for 31.2. Since $A = \{t \in T;\ 1_t \notin \mathscr{I}_t\}$ is closed in T, the space

$$V(T; \mathscr{I}) := (A, {}_A\mathscr{A})$$

is called a closed (ringed) subspace of T, written as $A \hookrightarrow T$. Examples are provided by the multiple points, which are closed subspaces of $\mathbb{C}$, and by Red T, which is the closed subspace $V(T;\ \mathscr{K}er\ \mathrm{Red})$ of T. A *locally closed subspace* of T is understood to be a closed subspace of an open subspace of T.

We are particularly interested in three special cases of the construction described in vi):

31.5 Examples. i) *Nullstellen ideals.* Fix $X \subset\!\!\!\!\subset \mathbb{C}^n$, suppose that $A \subset X$ is an analytic set, and let $_A\mathscr{I} \subset {}_X\mathcal{O}$ be the sheaf of all holomorphic functions that vanish on A (called the *nullstellen ideal of* A). *Then $(A, {}_A\mathcal{O}) := V(X; {}_A\mathscr{I})$ is a reduced ringed space*: we shall show that $\mathrm{Red}: {}_A\mathcal{O} \to {}_A\mathscr{C}$ is (stalkwise) injective. For $f \in {}_A\mathcal{O}_a$ such that $\mathrm{Red}\,f = 0$, let $F \in {}_X\mathcal{O}(U)$ be a representative of f. Then, near a, we have that

$$F|_{U\cap A} = (\mathrm{Red}\, F)|_{U\cap A} = (\mathrm{Red}\, f)|_{U\cap A} = 0;$$

it follows that $F \in {}_A\mathscr{I}_a$, and hence that $f = 0 \in {}_A\mathcal{O}_a$.

We note the following fact from the proof: *for $V \Subset A$, a function $f: V \to \mathbb{C}$ is in ${}_A\mathcal{O}(V)$ iff, for each $a \in V$, there exists a neighborhood $U \Subset X$ and an $F \in {}_X\mathcal{O}(U)$ such that $F|_{U\cap V} = f|_{U\cap V}$.*

ii) *Locally finitely generated ideals.* Let $A = N(X; f_1, \ldots, f_m) \subset X \Subset \mathbb{C}^n$ be the analytic set determined by $f_1, \ldots, f_m \in \mathcal{O}(X)$, and let $\mathscr{I} := {}_X\mathcal{O}\cdot(f_1, \ldots, f_m) \subset {}_X\mathcal{O}$ be the *ideal generated by* $f_1, \ldots, f_m$. Then, for ${}_A\mathcal{O} := ({}_X\mathcal{O}/\mathscr{I})|_A$, the ringed space $V(X; f_1, \ldots, f_m) := V(X; \mathscr{I}) = (A, {}_A\mathcal{O})$ is not necessarily reduced.

iii) *Characteristic ideals.* Let $(T, \mathscr{A})$ be a ringed space, let $A \subset T$ be a closed subset, and let $\mathscr{I}$ be the *characteristic ideal of* A in $\mathscr{A}$, i.e., $\mathscr{I}|_{T\setminus A} = \mathscr{A}|_{T\setminus A}$ and $\mathscr{I}|_A = 0$. Then $(\mathscr{A}/\mathscr{I})|_A$ is the (topological) restriction $\mathscr{A}|_A$ of the sheaf $\mathscr{A}$ to A; the ringed space $(A, \mathscr{A}|_A)$ is not in general reduced.

The construction given in i) leads to the local model of a reduced complex space as it is treated in §32; the sections of the structure sheaf are obtained, in accord with the geometric presentation, through the restriction of holomorphic functions to A. The second construction leads to the local model of a complex space as it is viewed in §43. The structure sheaf is not in general a sheaf of functions (see 31.4 v)), but it preserves the essential information concerning the defining system of equations.

The stalks ${}_A\mathscr{I}_a$ of the nullstellen ideal of an analytic set A are finitely-generated, because the stalks ${}_X\mathcal{O}_a$ of the structure sheaf are noetherian rings. That this *punctual* statement also holds *locally* is an essential and by no means trivial result (see 47.1): the ${}_X\mathcal{O}$-ideal ${}_A\mathscr{I}$ is *locally finitely generated*; thus, X can be covered with open sets U such that ${}_A\mathscr{I}|_U = ({}_X\mathcal{O}|_U)(f_1, \ldots, f_m)$ for appropriate functions $f_1, \ldots, f_m \in {}_A\mathscr{I}(U)$. The analogous punctual statement for the zero locus of real analytic functions holds also, but the local version is false [Ns_2 V.3]!

Construction iii) can lead to a better understanding of reduction:

E. 31 b. For an analytic set $A \subset X \Subset \mathbb{C}^n$ and $\mathcal{O} = {}_X\mathcal{O}$, show that $\mathrm{Red}\,(A, \mathcal{O}|_A) = (A, (\mathcal{O}/{}_A\mathscr{I})|_A)$. (Hint: show that the diagram

$$\begin{array}{ccc} \mathcal{O}|_A & \longrightarrow & \mathrm{Red}\,(\mathcal{O}|_A) \\ \downarrow & & \downarrow \\ (\mathcal{O}|_A)/({}_A\mathscr{I}|_A) & \rightarrow & (\mathcal{O}/{}_A\mathscr{I})|_A \end{array}$$

commutes. For 31.6.)

31.6 Remark. *If $A \subset X \Subset \mathbb{C}^n$ is an analytic set, then ${}_X\mathcal{O}/{}_A\mathscr{I}$ is the unique sheaf $\mathscr{A}$ on X that makes $(A, \mathscr{A}|_A)$ a reduced closed subspace of $(X, {}_X\mathcal{O})$.*

Proof. If $\mathscr{A} = {}_X\mathcal{O}/\mathscr{I}$, then, for the characteristic ideal sheaf $\mathscr{J}$ of A, we obviously have that $\mathscr{J} \subset \mathscr{I} \subset {}_A\mathscr{I}$. Thus it follows from E. 31 b that $\mathrm{Red}\,\mathscr{A} = {}_X\mathcal{O}/{}_A\mathscr{I}$. ■

A corresponding statement for an arbitrary closed subspace of a ringed space T can be proved with the help of the *nullstellen ideal* of A

$${}_A\mathscr{I} := \{f \in {}_T\mathscr{A}; \mathrm{Red}\, f|_A = 0\}.$$

E. 31c. For a fixed integer $j \geqq 1$, set $A = \mathbb{C} \times \{0\}$ and $\mathscr{I} := {}_2\mathcal{O} \cdot z_2^j$. Prove the following statements for ${}_A\mathcal{O} := ({}_2\mathcal{O}/\mathscr{I})|_A$:

i) There exists a canonical isomorphism of ${}_1\mathcal{O}$-modules ${}_A\mathcal{O} \cong {}_1\mathcal{O}[z_2]_j$; that implies that $\operatorname{Red}\left(\sum_{k=0}^{j-1} f_k(z_1) z_2^k\right) = f_0(z_1)$.

ii) $(A, {}_A\mathcal{O})$ is reduced iff $j = 1$.

Having introduced ringed spaces, we turn now to their morphisms, for which we mimic the characterization of holomorphic mappings $\varphi: X \to Y$ in 3.2: the lifting of holomorphic functions defines a "comorphism"

$$\varphi^0: {}_Y\mathcal{O} \to {}_X\mathcal{O},\ f \mapsto f \circ \varphi|_{\varphi^{-1}(W)},\ W \Subset Y,\ f \in {}_Y\mathcal{O}(W).$$

31.7 Definition. *A* <u>*morphism*</u> *$\varphi: S \to T$ of ringed spaces is a pair $(|\varphi|, \varphi^0)$ consisting of a continuous mapping $|\varphi|: |S| \to |T|$ and a $(|\varphi|$-$)$*<u>*comorphism*</u> *$\varphi^0: {}_T\mathscr{A} \to {}_S\mathscr{A}$ of algebras, i.e., a family of algebra-homomorphisms*

$$\varphi^0 := (\varphi^0(V): {}_T\mathscr{A}(V) \to {}_S\mathscr{A}(\varphi^{-1}(V)))_{V \Subset T}$$

that is compatible with the restrictions in ${}_T\mathscr{A}$ and ${}_S\mathscr{A}$.

For two morphisms $\varphi: S \to T$ and $\psi: T \to U$ of ringed spaces, we define

$$\psi \circ \varphi := (|\psi| \circ |\varphi|,\ \varphi^0 \circ \psi^0).$$

Frequently, we use "φ" instead of "$|\varphi|$", although the comorphism φ^0 is not determined by $|\varphi|$ (however, see 31.11):

E. 31d. i) Give two different morphisms $\varphi, \psi: \bullet \to (\mathbb{C}, {}_1\mathcal{O})$ such that $|\varphi| = 0 = |\psi|$.

ii) For $j \geqq 2$, give two different morphisms $(|\varphi|, \varphi^0): (A, {}_A\mathcal{O}) \to (A, \operatorname{Red}_A \mathcal{O})$ with $|\varphi| = \mathrm{id}_A$ (in the notation of E. 31c).

For each $s \in S$, the comorphism φ^0 induces a canonical local (21.6) algebra-homomorphism

$$\varphi_s^0: {}_T\mathscr{A}_{\varphi(s)} \to {}_S\mathscr{A}_s.$$

E. 31e. Prove that, for a morphism $(|\varphi|, \varphi^0): S \to T$ of ringed spaces, φ^0 is determined by $(\varphi_s^0)_{s \in S}$ (for 31.9).

E. 31f. *Canonical factorization of morphisms.* Let S and T be ringed spaces, and let $\tau: |S| \to |T|$ be a continuous mapping. Show that the following statements hold:

i) $(|S|, \tau^{-1}{}_T\mathscr{A})$ is a ringed space.

ii) The mappings $\Gamma(V, {}_T\mathscr{A}) \to (\tau^{-1}{}_T\mathscr{A})(\tau^{-1}(V))$, $\gamma \mapsto \gamma \circ \tau$, for $V \Subset T$, determine a *canonical comorphism* ${}_T\tau^0: {}_T\mathscr{A} \to \tau^{-1}{}_T\mathscr{A}$.

iii) For each morphism $(\tau, \varphi^0): S \to T$ of ringed spaces there is exactly one morphism ${}_S\varphi^0: \tau^{-1}{}_T\mathscr{A} \to {}_S\mathscr{A}$ of algebras on $|S|$ such that $\varphi^0 = {}_S\varphi^0 \circ {}_T\tau^0$; moreover, the assignment $\varphi \mapsto {}_S\varphi^0$ is bijective (for E. 31o).

31.8 Examples. i) If $\varphi: S \to T$ is a continuous mapping of topological spaces, then, with φ^0 denoting the lifting of functions, $(\varphi, \varphi^0): (S, {}_S\mathscr{C}) \to (T, {}_T\mathscr{C})$ is a morphism.

ii) For $X \Subset \mathbb{C}^n$, let $i: {}_X\mathcal{O} \to {}_X\mathscr{C}$ denote the canonical inclusion; then $(\mathrm{id}_X, i): (X, {}_X\mathscr{C}) \to (X, {}_X\mathcal{O})$ is a morphism.

iii) For a closed subspace $A := V(T; \mathscr{I})$ of a ringed space $(T, \mathscr{A})$, we may interpret "$A \hookrightarrow T$" as a morphism of ringed spaces

$$\hookrightarrow := (\text{set-inclusion}, \mathscr{A} \twoheadrightarrow (\mathscr{A}/\mathscr{I})|_A).$$

In particular, Red $T \hookrightarrow T$ is a morphism.

iv) *Inverse images.* If $\varphi: (S, {}_S\mathscr{A}) \to (T, {}_T\mathscr{A})$ is a morphism, and $\mathscr{J}$, an $\mathscr{A}_T$-ideal, then the *(ringed) inverse image sheaf* $\varphi^*\mathscr{J}$ of $\mathscr{J}$ is the ${}_S\mathscr{A}$-ideal generated by $\varphi^0(\mathscr{J}(W)) \subset {}_S\mathscr{A}(\varphi^{-1}(W))$; in particular, we have that

$$(\varphi^*\mathscr{J})_s = \varphi_s^0(\mathscr{J}_{\varphi(s)}) \cdot {}_S\mathscr{A}_s.$$

For the subspace $B := V(T; \mathscr{J}) \hookrightarrow T$, the subspace

$$\varphi^{-1}(B) := V(S; \varphi^*\mathscr{J}) \hookrightarrow S$$

is called the *inverse image of B under* φ. In particular, the nullstellen ideal ${}_{\{t\}}\mathscr{I}$ of a point t determines a ringed structure on the *fiber* $\varphi^{-1}(t)$ *of the morphism* φ.

E. 31 g. Determine $\varphi^{-1}(0)$ for i) $\varphi: \mathbb{C}^n = \mathbb{C}^n \times \{0\} \hookrightarrow \mathbb{C}^n \times \mathbb{C}^m$ and ii) $\varphi = \mathrm{pr}_n: \mathbb{C}^n \times \mathbb{C}^m \to \mathbb{C}^n$.

E. 31 h. For $A := V(\mathbb{C}^2; z_1^2 - z_2)$ and $f: A \overset{i}{\hookrightarrow} \mathbb{C}^2 \overset{\mathrm{pr}_2}{\to} \mathbb{C}$, prove that the fiber $f^{-1}(z)$ is reduced iff $z \neq 0$.

E. 31 i. Determine the canonical structure on the fibers of the holomorphic mapping $\mathbb{C}^2 \to \mathbb{C}^2, (z_1, z_2) \mapsto (z_1^j, z_2^k)$.

We can summarize the essential properties of closed subspaces as follows:

31.9 Proposition. *Let $\varphi: S \to T$ be a morphism of ringed spaces, and let $A := V(S; \mathscr{I})$ and $B := V(T; \mathscr{J})$ be subspaces of S and T, respectively. Then, for the inclusions $i: A \hookrightarrow S$ and $j: B \hookrightarrow T$, the following statements hold:*

i) Two morphisms $\psi, \chi: Z \to A$ are equal iff $i\psi = i\chi$ (i.e., i is a "monomorphism").

ii) "Restriction Lemma". A morphism ψ such that the diagram

$$\begin{array}{ccc} S & \xrightarrow{\ \varphi\ } & T \\ {\scriptstyle i}\uparrow & & \uparrow{\scriptstyle j} \\ A & \overset{\psi}{\dashrightarrow} & B \end{array}$$

commutes, exists iff $\varphi^*\mathscr{J} \subset \mathscr{I}$. In that case, ψ is unique and is called the restriction $\varphi|_A: A \to B$ of φ.

iii) There exists a restriction $\psi = \varphi|_{\varphi^{-1}(B)}: \varphi^{-1}(B) \to B$ *with the following universal property: every commutative diagram*

(31.9.1)

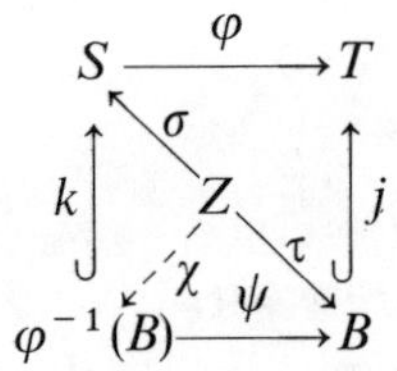

can be completed commutatively with a unique morphism $\chi: Z \to \varphi^{-1}(B)$ *("fiber product").*

Proof. i) *If.* Since $|i|$ is injective, $|\psi| = |\chi|$. Since $i^0_{\psi(z)}$ is surjective for each $z \in Z$, it follows from the equality

$$\psi^0_z \circ i^0_{\psi(z)} = (i\psi)^0_z = (i\chi)^0_z = \chi^0_z \circ i_{\psi(z)}$$

that $\psi^0_z = \chi^0_z$, and hence that $\psi^0 = \chi^0$ (see E. 31e).

ii) We may assume that $\mathscr{I} = 0$ and thus that $A = S$ (since we have that $\varphi^*\mathscr{J} \subset \mathscr{I} \Leftrightarrow \varphi^0\mathscr{J} \subset \mathscr{I} = \mathscr{Ker}\ i^0 \Leftrightarrow i^0\varphi^0\mathscr{J} = 0 \Leftrightarrow i^*\varphi^*\mathscr{J} = 0$, we may replace φ with $\varphi \circ i$). Then the direction "*only if*" follows from the equality

$$\varphi^*\mathscr{J} = (j\psi)^*\mathscr{J} = \psi^*(j^*\mathscr{J}) = \psi^*(0) = 0\,.$$

If. We certainly have that $\varphi(S) \subset B$, for, if $\varphi(s)$ were in $T \backslash B$, then it would follow that $1 \in \mathscr{J}_{\varphi(s)} = {}_T\mathscr{A}_{\varphi(s)}$, and hence that $(\varphi^*\mathscr{J})_s \neq 0$. Thus, $|\varphi|$ factors over a continuous mapping $|\psi|: S \to B$. For each $V \subset\subset T$, we have that $\varphi^0(\mathscr{J}(V)) = 0$; hence, φ^0 induces algebra-homomorphisms $\bar{\varphi}^0: {}_T\mathscr{A}(V)/\mathscr{J}(V) \to {}_S\mathscr{A}(\varphi^{-1}(V))$, and thus induces a $\psi^0: {}_B\mathscr{A} \to {}_S\mathscr{A}$. Finally, $\psi = (|\psi|, \psi^0)$ is unique, since j is a monomorphism.

iii) By ii), the restriction $\varphi|_{\varphi^{-1}(B)}$ exists, since $\varphi^{-1}(B) = V(S; \varphi^*\mathscr{J})$. In (31.9.1), $\varphi\sigma$ factors over j, so $\sigma^*(\varphi^*\mathscr{J}) = 0$; by ii), then, σ factors over k with a unique χ. It follows that $j\psi\chi = \varphi k\chi = \varphi\sigma = j\tau$, and i) implies that $\psi\chi = \tau$. ∎

Now we want to show that *reduction* is a *covariant functor* from the category of ringed spaces to the category of reduced ringed spaces:

31.10 Proposition. *If* $\varphi: S \to T$ *is a morphism of ringed spaces, then there is precisely one morphism* $\operatorname{Red}\varphi: \operatorname{Red} S \to \operatorname{Red} T$ *such that the diagram*

$$\begin{array}{ccc} (S, {}_S\mathscr{A}) & \xrightarrow{\quad\varphi\quad} & (T, {}_T\mathscr{A}) \\ \uparrow{\scriptstyle \operatorname{Red}_S} & & \uparrow{\scriptstyle \operatorname{Red}_T} \\ (S, \operatorname{Red}{}_S\mathscr{A}) & \xrightarrow{\ \operatorname{Red}\varphi\ } & (T, \operatorname{Red}{}_T\mathscr{A}) \end{array}$$

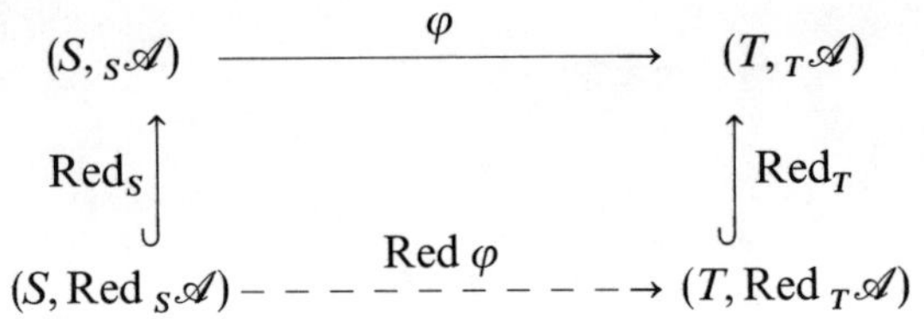

commutes; moreover, $|\mathrm{Red}\,\varphi| = |\varphi|$ *and*

(31.10.1) $(\mathrm{Red}\,\varphi)^0(f) = f \circ |\varphi|$ *for every* $f \in \mathrm{Red}\,{}_T\mathscr{A}$.

Proof. We have that $\mathrm{Red}\,T = V(T;\ \mathscr{K}er\ \mathrm{Red}_T)$, and that

$$\mathscr{K}er\ \mathrm{Red}_T(V) = \{f \in {}_T\mathscr{A}(V);\ f_t \in \mathfrak{m}({}_T\mathscr{A}_t),\ \forall t \in V\},\ V \mathrel{\subset\!\!\!\!\circ} T.$$

Since every $\varphi_s^0\colon {}_T\mathscr{A}_{\varphi(s)} \to {}_S\mathscr{A}_s$ is local, it follows that $\varphi^*(\mathscr{K}er\ \mathrm{Red}_T) \subset \mathscr{K}er\ \mathrm{Red}_S$; by the Restriction Lemma, then, $\mathrm{Red}\,\varphi$ exists and is unique. Fix $f \in (\mathrm{Red}\,{}_T\mathscr{A})(V)$ and $s \in \varphi^{-1}(V)$. Since (31.10.1) is a local statement, we may suppose that $f = \mathrm{Red}\,g$ with $g \in {}_T\mathscr{A}(V)$. Then it follows from the commutativity of the diagram that

$$((\mathrm{Red}\,\varphi)^0(f))(s) = (\mathrm{Red}\,(\varphi^0 g))(s) = (\mathrm{Red}\,(\varphi_s^0\, g_{\varphi(s)}))(s)$$
$$\underset{\mathrm{E.31j}}{=} (\mathrm{Red}\,g_{\varphi(s)})(\varphi(s)) = f(\varphi(s)) = (f \circ |\varphi|)(s),$$

and thus that (31.10.1) holds. ■

31.11 Corollary. *If* $(|\varphi|, \varphi^0)\colon S \to T$ *is a morphism of ringed spaces, and if* S *is reduced, then* φ^0 *is determined by* $|\varphi|$:

$$\varphi^0(f) = (\mathrm{Red}\,f) \circ |\varphi| \quad \textit{for} \quad f \in {}_T\mathscr{A}.\ ■$$

That justifies the following standard terminology for reduced ringed spaces S and T: a continuous mapping $\tau\colon |S| \to |T|$ is called a *morphism of ringed spaces* S and T if the inverse image $f \circ \tau$ of each f in ${}_T\mathscr{A}$ lies in ${}_S\mathscr{A}$.

E. 31j. Show that, for a homomorphism $\varphi^0\colon R \to S$ of local algebras, and the evaluation-homomorphisms, the following diagram commutes:

$$\begin{array}{ccc} R & \xrightarrow{\varphi^0} & S \\ & {\scriptstyle \varepsilon_R} \searrow \quad \swarrow {\scriptstyle \varepsilon_S} & \\ & \mathbb{C} & \end{array}$$

E. 31k. Let $\varphi\colon S \to T$ be a morphism of ringed spaces, and suppose that $A \hookrightarrow S$ and $B \hookrightarrow T$. Prove the following statements:

i) If A is reduced, then the restriction $\varphi|_A\colon A \to B$ exists iff $|\varphi|(A) \subset B$ (hint: 31.6; for 32 B.2).

ii) Construct for $\varphi = \mathrm{id}_{\mathbb{C}}$ an example of closed subspaces $A, B \hookrightarrow \mathbb{C}$ with $|\varphi|(A) \subset B$ such that no restriction ψ exists.

E. 31l. Prove a *Restriction Lemma* for *open* subspaces: there exists a ψ such that the diagram

$$\begin{array}{ccc} S & \xrightarrow{\varphi} & T \\ \cup & & \cup \\ A & \overset{\psi}{\dashrightarrow} & B \end{array}$$

commutes iff $|\varphi|(A) \subset B$.

In general, it is difficult to determine explicitly the reduction of a ringed space, since it cannot be done punctually (i.e., without local information).

31.12 Example. For $A := \{0\} \subset \mathbb{C}$, the ringed space $(A, {}_1\mathcal{O}|_A)$ is not reduced, nevertheless it has no "nilpotent" elements; its structure sheaf is coherent in the sense of § 41; the stalks ${}_A\mathcal{O}_0$ and ${}_1\mathcal{O}_0$ coincide, but locally, A and $\mathbb{C}$ are different.

If $\mathfrak{a}$ is an ideal in the ring R, then

$$\sqrt{\mathfrak{a}} := \{r \in R; \exists j \in \mathbb{N} \quad \text{with} \quad r^j \in \mathfrak{a}\}$$

is called the *radical of* $\mathfrak{a}$ *in* R; it is an ideal (Binomial Theorem). The radical $\mathfrak{n}_R := \sqrt{0}$ is called the *nilradical* of R, since it consists of the *nilpotent elements*. If $\mathfrak{n}_R = (0)$, then R is called *(algebraically) reduced*; that always holds for $R/\mathfrak{n}_R$. These concepts can be extended to sheaves of ideals $\mathcal{I}$ in sheaves of rings $\mathcal{R}$. One easily sees that $\sqrt{\mathcal{I}_t} = (\sqrt{\mathcal{I}})_t$.

Let $(T, \mathcal{A})$ be a ringed space; since the sheaf ${}_T\mathcal{C}$ has no nilpotent elements, the $\mathcal{A}$-ideal ${}_T\mathcal{N}$ of nilpotent elements in $\mathcal{A}$ always lies in the kernel of Red_T. By 31.4 vi), $(T, \mathcal{A}/{}_T\mathcal{N})$ is also a ringed space, called the *algebraic reduction* of $(T, \mathcal{A})$. An essential theorem concerning the "complex spaces" to be introduced in § 43 states that algebraic reduction is the same as reduction for those spaces (see 47.2).

E. 31 m. *Gluing ringed spaces.* Let T be a topological space with an open cover $\mathfrak{U} = (U_j)_{j \in J}$, and suppose that there are sheaves $\mathcal{A}_j$ on U_j and sheaf-isomorphisms $\vartheta_{jk}: \mathcal{A}_k|_{U_{jk}} \to \mathcal{A}_j|_{U_{jk}}$ for all $j,k \in J$. Suppose further that $\vartheta_{jj} = \mathrm{id}_{\mathcal{A}_j}$ and $\vartheta_{ij} \circ \vartheta_{jk} = \vartheta_{ik}$ for all $i,j,k \in J$. Prove the following statements:

i) There exist on T a unique sheaf $\mathcal{A}$ (up to isomorphism) and isomorphisms $\eta_j: \mathcal{A}|_{U_j} \to \mathcal{A}_j$ such that $\vartheta_{jk} = \eta_j \circ \eta_k^{-1}$ on U_{jk} for all $j, k \in J$.

ii) If every ϑ_{jk} determines an isomorphism of ringed spaces, then $(T, \mathcal{A})$ is a ringed space (for E. 32a).

iii) The natural mapping

$$\mathcal{A}(T) \to \{(s_j) \in \prod_J \mathcal{A}_j(U_j); s_j = \vartheta_{jk}(s_k) \text{ on } U_{jk}, \forall j,k\}, s \mapsto (\eta_j(s|_{U_j}))_{j \in J},$$

is a bijection (for 54.7).

E. 31 n. Let $\varphi: S \to T$ be a morphism of ringed spaces. Prove the following:

i) If $|\varphi|$ is injective and φ_s^0 is surjective for every $s \in S$, then φ is a *monomorphism* (example: $A \hookrightarrow T$; see E. 45n; for 44.3).

ii) If $|\varphi|$ is surjective and φ_s^0 is injective for every $s \in S$, then φ is an *epimorphism* (i.e., $\sigma\varphi = \tau\varphi \Rightarrow \sigma = \tau$, $\forall \sigma, \tau$; the converse does not hold in the category of complex spaces, as the inclusion $\boldsymbol{P}^1(1) \not\subset \mathbb{C}$ shows; see 49.11).

iii) φ is an *isomorphism* iff $|\varphi|$ is a homeomorphism and φ_s^0 is an isomorphism for every $s \in S$.

E. 31 o. *Description of the ringed inverse image sheaf.* Let $\varphi: S \to T$ be a morphism of ringed spaces, and let $\mathcal{J}$ be an ${}_T\mathcal{A}$-ideal. Prove the following:

i) The ${}_S\mathcal{A}$-ideal $\varphi^*\mathcal{J}$ is generated by the topological inverse image sheaf $\varphi^{-1}\mathcal{J} \subset \varphi^{-1}{}_T\mathcal{A}$ (use ${}_S\varphi^0$ from E. 31f. in order to define a surjective homomorphism ${}_S\mathcal{A} \oplus_{\varphi^{-1}{}_T\mathcal{A}} \varphi^{-1}\mathcal{J} \twoheadrightarrow \varphi^*\mathcal{J}$).

ii) If $\mathcal{J}$ is finitely-generated over ${}_T\mathcal{A}$, then so is $\varphi^*\mathcal{J}$ over ${}_S\mathcal{A}$ (for 43.5).

§ 32 Complex Manifolds and Reduced Complex Spaces

The submanifolds introduced in §8 lead to the definition of a class of ringed spaces that are of particular significance due to their simple local behavior:

32.1 Definition. *A ringed space $(X, \mathcal{O})$ is called a (complex) manifold if it satisfies these two conditions:*

i) X is a Hausdorff space.

ii) X has a cover consisting of open subspaces U, each of which is isomorphic (as a ringed space) to an open subspace $(V, {}_n\mathcal{O}|_V)$ of $(\mathbb{C}^n, {}_n\mathcal{O})$.

If $a \in X$ lies in an open subspace $U \cong V \subset\subset \mathbb{C}^n$, then n is called the *dimension* $\dim_a X$ of X at a. It is well-defined (see E. 8e); the function $X \to \mathbb{Z}$, $x \mapsto \dim_x X$, is continuous, and thus constant on each connected component of X.

E. 32a. *Charts on manifolds.* Let T be a Hausdorff space with an open cover $\mathfrak{U} = (U_j)_{j \in J}$, and suppose that, for each $j \in J$, there exists a homeomorphism $\varphi_j: U_j \to B_j \subset\subset \mathbb{C}^{n_j}$ such that the "coordinate-transformation" $\varphi_j \circ \varphi_i^{-1}: \varphi_i(U_{ij}) \to \varphi_j(U_{ij})$ is biholomorphic for each nonempty $U_{ij} := U_i \cap U_j$. Using E. 31m, show that T has the structure of a manifold in a canonical fashion, with ${}_T\mathcal{O}|_{U_j} = \varphi_j^{-1}({}_{B_j}\mathcal{O})$. The pair (U_j, φ_j) is called a chart on T (for 32.2).

32.2 Examples. i) For $X \subset\subset \mathbb{C}^n$, the ringed space $(X, {}_n\mathcal{O}|_X)$ is a complex manifold.

ii) If A is a submanifold of $X \subset\subset \mathbb{C}^n$, then, by 8.8 and E. 32a, A is a manifold.

iii) As a domain in $\mathbb{C}^{n \times n}$, $GL(n, \mathbb{C})$ is a manifold: for $A = (a_{ij}) \in \mathbb{C}^{n \times n}$, the determinant $\det A$ is a polynomial in the a_{ij}; by 7.4, then, $GL(n, \mathbb{C}) = \mathbb{C}^{n \times n} \setminus N(\det)$ is a domain in $\mathbb{C}^{n \times n}$.

iv) Riemann surfaces are connected one-dimensional complex manifolds.

In order to investigate analytic sets together with manifolds, we generalize Definition 32.1. We understand a *reduced local model* to be a ringed space that is isomorphic to a space $(A, ({}_U\mathcal{O}/{}_A\mathcal{I})|_A)$, where $A \subset U \subset\subset \mathbb{C}^n$ is an analytic set and ${}_A\mathcal{I} \subset {}_U\mathcal{O}$ is its nullstellen ideal.

32.3 Definition. *A ringed space $(X, {}_X\mathcal{O})$ is called a reduced complex space if it satisfies these two conditions:*

i) X is a Hausdorff space.

ii) $(X, {}_X\mathcal{O})$ has an open cover consisting of reduced local models.

The sections of the structure sheaf ${}_X\mathcal{O}$ are called holomorphic functions. A continuous mapping $\varphi: X \to Y$ of reduced complex spaces is called holomorphic if it satisfies the following condition:

$$W \subset\subset Y, f \in {}_Y\mathcal{O}(W) \Rightarrow f \circ \varphi \in {}_X\mathcal{O}(\varphi^{-1}(W)).$$

The set of holomorphic mappings $\varphi: X \to Y$ is denoted by $\mathrm{Hol}(X, Y)$.

By 31.11, then, for reduced complex spaces, the holomorphic mappings coincide with the morphisms of ringed spaces (so that the reduced complex spaces form a "full subcategory" of the ringed spaces, as do the manifolds).

32.4 Examples. i) Manifolds are reduced complex spaces.

ii) If $A \subset X \Subset \mathbb{C}^n$ is an analytic set, then $(A, {}_X\mathcal{O}/{}_A\mathcal{I})$ is a reduced complex space.

iii) For a reduced complex space X,

$$\mathrm{Hol}(X, \mathbb{C}) = \mathcal{O}(X).$$

For, if f is in $\mathrm{Hol}(X, \mathbb{C})$, then $f = \mathrm{id}_{\mathbb{C}} \circ f \in \mathcal{O}(X)$; conversely, if f is in $\mathcal{O}(X)$, then it suffices to show locally that f determines a holomorphic mapping $f: X \to \mathbb{C}$. That is a consequence of 31.5 i); we may assume that $X \hookrightarrow G \Subset \mathbb{C}^n$ and that there exists an $F \in \mathcal{O}(G) = \mathrm{Hol}(G, \mathbb{C})$ such that $f = F|_X$.

E. 32 b. Prove the following statements for a connected manifold $(X, \mathcal{O})$:

i) The *Identity Theorem* holds for X (hint: the set $\{x \in X; f = g \text{ near } x\}$ is closed).

ii) *Maximum Principle:* If the absolute value of $f \in \mathcal{O}(X)$ assumes its maximum value on X, then f is constant.

iii) If X is compact, then $\mathcal{O}(X) = \mathbb{C}$; i.e., every holomorphic function on a compact manifold X is constant.

iv) $\mathbb{C}^n$ has no compact submanifolds of positive dimension.

iv) The *(complex) projective space* $\mathbb{P}_n$. Up to now, we have restricted our explicit examples of reduced complex spaces to subspaces of $\mathbb{C}^n$; of these, only the finite point sets are compact (Maximum Principle). For compact spaces, the projective space $\mathbb{P}_n$ plays a fundamental role similar to that of $\mathbb{C}^n$ for the noncompact spaces. Its points are the lines passing through the origin in $\mathbb{C}^{n+1}$: for $z, w \in (\mathbb{C}^{n+1})^* := \mathbb{C}^{n+1} \setminus 0$, we define

$$z \sim w :\Leftrightarrow \mathbb{C} \cdot z = \mathbb{C} \cdot w \Leftrightarrow \exists \lambda \in \mathbb{C}^* \quad \text{such that} \quad z = \lambda w.$$

Then $|\mathbb{P}_n| = (\mathbb{C}^{n+1})^*/\sim$. With the quotient topology, $\mathbb{P}_n$ is *compact*; to prove that, we represent $\mathbb{P}_n$ as a quotient space of the compact sphere $S^{2n+1} := \{z \in \mathbb{C}^{n+1}; \|z\| = 1\}$. Consider the mapping

$$\tau: (\mathbb{C}^{n+1})^* \twoheadrightarrow S^{2n+1} \twoheadrightarrow \mathbb{P}_n, \ z \mapsto z/\|z\| \mapsto [z/\|z\|],$$

which induces the following equivalence relation R on S^{2n+1}:

$$z \sim w \Leftrightarrow \exists \lambda \in S^1 \quad \text{with} \quad z = \lambda w.$$

Now R is proper, since for each compact $K \subset S^{2n+1}$, the set

$$R(K) = S^1 \cdot K = \mathrm{Im}[S^1 \times K \to S^{2n+1}, (\lambda, z) \mapsto \lambda \cdot z]$$

is compact. Hence, $\mathbb{P}_n$ is compact by 33. B. 4.

By E. 32a, to show that $\mathbb{P}_n$ is a manifold, it suffices to construct charts U_j. For $U_j^0 := \{(z_0, \ldots, z_n) \in \mathbb{C}^{n+1}; z_j \neq 0\}$, the family $\mathfrak{U}^0 = (U_j^0)_{j=0,\ldots,n}$ is an open cover of $(\mathbb{C}^{n+1})^*$; with the open mapping $\tau: (\mathbb{C}^{n+1})^* \to |\mathbb{P}_n|$ and $U_j := \tau(U_j^0)$, the family $\mathfrak{U} := (U_j)_{j=0,\ldots,n}$ is an open cover of $|\mathbb{P}_n|$. The compositions

$$\gamma_j: \mathbb{C}^n \cong \{z \in \mathbb{C}^{n+1}; z_j = 1\} \hookrightarrow U_j^0 \xrightarrow{\tau} U_j$$

are homeomorphisms for $j = 0, \ldots, n$. For $p \in U_j$, one refers to

$(z_0, \ldots, \hat{z}_j, \ldots, z_n) := \gamma_j^{-1}(p)$ as *nonhomogeneous* or *affine coordinates* of p and to $(z_0, \ldots, z_n) \in \tau^{-1}(p)$ as *homogeneous coordinates*. The latter are unique up to a factor $\lambda \in \mathbb{C}^*$; thus it is usual to write

$$p = [z_0, \ldots, z_n]$$

if $\tau(z_0, \ldots, z_n) = p$. With that notation, it follows that $\gamma_j^{-1}([z_0, \ldots, z_n]) = \frac{1}{z_j}(z_0, \ldots, \hat{z}_j, \ldots, z_n)$ for $z_j \neq 0$. It remains to show that the mapping

$$\gamma_j^{-1}\gamma_i : \mathbb{C}^n \setminus \{z_j = 0\} \to \mathbb{C}^n \setminus \{z_i = 0\}$$

is biholomorphic for $U_{ij} = U_i \setminus \{z_j = 0\} = U_j \setminus \{z_i = 0\}$. That is easy to see: for $j < i$, we have that

$$\gamma_j^{-1}\gamma_i : (z_1, \ldots, z_n) \mapsto \frac{1}{z_j}(z_1, \ldots, z_j, \ldots, z_{i-1}, 1, z_i, \ldots, z_n).$$

With the subsheaf ${}_{\mathbb{P}_n}\mathcal{O} \subset {}_{\mathbb{P}_n}\mathcal{C}$ constructed as in E. 32a, $(\mathbb{P}_n, {}_{\mathbb{P}_n}\mathcal{O})$ is a manifold.

E. 32c. Prove that, in the notation used above, $\mathbb{P}_n \setminus U_j \cong \mathbb{P}_{n-1}$.

E. 32d. Prove that $\mathbb{P}_1$ is isomorphic to S^2 (the "Riemann sphere"). (Hint: one point compactification or explicit construction of

$\mathbb{P}_1 \overset{\varphi}{\underset{\psi}{\leftrightarrows}} S^2 = \{(z,t) \in \mathbb{C} \times \mathbb{R};\, z\bar{z} + t^2 = 1\}$ with $\varphi([w_0, w_1]) := (1/(w_0\bar{w}_0 + w_1\bar{w}_1)) \cdot (2\bar{w}_0 w_1, w_1\bar{w}_1 - w_0\bar{w}_0)$, $\psi(z,t) := [1-t, z]$ for $t \neq 1$, and $\psi(z,t) = [\bar{z}, 1+t]$ for $t \neq -1$.)

E. 32e. *Projective linear transformations.* Prove the following statements:

i) Every $A \in GL(n+1, \mathbb{C})$ induces an automorphism $P(A)$ on $\mathbb{P}_n$ such that $P(A) \circ \tau = \tau \circ A$.

ii) The assignment $P: A \mapsto P(A)$ determines a homomorphism from $GL(n+1, \mathbb{C})$ into the automorphism group $\operatorname{Aut} \mathbb{P}_n$ with $\operatorname{Ker} P = \mathbb{C}^* \cdot I_{n+1}$; the image group is called the *projective linear group* $\mathbb{P}GL(\mathbb{C}^{n+1})$ (by E. 53Ah, $\mathbb{P}GL(\mathbb{C}^{n+1}) = \operatorname{Aut} \mathbb{P}_n$).

iii) Every projective linear transformation $P(A)$ can be represented as a linear fractional transformation

$$z \mapsto \left(\frac{a_1 z + b_1}{a_0 z + b_0}, \frac{a_2 z + b_2}{a_0 z + b_0}, \ldots, \frac{a_n z + b_n}{a_0 z + b_0}\right)$$

in the nonhomogeneous coordinates $\gamma_0 : \mathbb{C}^n \xrightarrow{\cong} U_0 \subset\!\!\!\subset \mathbb{P}_n$ with $a_0, \ldots, a_n \in \operatorname{Hom}(\mathbb{C}^n, \mathbb{C}), b_0, \ldots, b_n \in \mathbb{C}$, and $a_0 \neq 0$ or $b_0 \neq 0$.

E. 32f. For each hyperplane $H \hookrightarrow \mathbb{P}_n$, there exists a biholomorphic mapping $\mathbb{P}_n \setminus H \to \mathbb{C}^n$ (hint: E. 32e i)).

E. 32g. Show, using the following example, that the Riemann Removable Singularity Theorems do not hold for holomorphic mappings: the projection $\tau : (\mathbb{C}^{n+1})^* \to \mathbb{P}_n$ is a holomorphic mapping that has no holomorphic extension (not even a continuous one) to $\mathbb{C}^{n+1}$. (Hint: $\tau(\boldsymbol{B}^{n+1}(\varepsilon) \setminus 0) = \mathbb{P}_n$ for all $\varepsilon > 0$.) Show that the closure of the graph $\Gamma(\tau) = \{(z, \tau(z)) \in (\mathbb{C}^{n+1})^* \times \mathbb{P}_n\}$ in $\mathbb{C}^{n+1} \times \mathbb{P}_n$ is not the graph of a continuous mapping.)

With an obvious generalization of 7.1, we construct new reduced complex spaces:

32.5 Definition. *A subset A of a complex space X is called an* analytic set *if, for each $z \in X$, there exist a neighborhood U of z and functions $f_1, \ldots, f_r \in {}_X\mathcal{O}(U)$ such that $A \cap U = N(U; f_1, \ldots, f_r)$.*

By 31.6, such an *analytic set can be made into a reduced complex subspace of X*: if ${}_A\mathcal{I}$ is the nullstellen ideal of A, then ${}_A\mathcal{O} := ({}_X\mathcal{O}/{}_A\mathcal{I})|_A$ is the associated structure sheaf.

E. 32 h. Generalize E. 7a to analytic sets in complex spaces.

32.6 Example. *Projective algebraic varieties.* Let $P \in \mathbb{C}[z_0, \ldots, z_n]$ be a homogeneous polynomial of degree $q \geqq 1$. Then $P(\lambda z) = \lambda^q P(z)$ for every $\lambda \in \mathbb{C}^*$, so $A := N((\mathbb{C}^{n+1})^*; P)$ is a (punctured) "cone"; in other words, it is saturated with respect to the equivalence relation $z \sim \lambda z$ (see §33B). Now $|H| := \tau(A)$ is an analytic set in $\mathbb{P}_n$, since we have for each j that

$$H \cap U_j = \{[z_0, \ldots, \overset{\overset{j}{\downarrow}}{1}, \ldots, z_n] \in \mathbb{P}_n;\ P(z_0, \ldots, 1, \ldots, z_n) = 0\}.$$

The space $(H, {}_{\mathbb{P}_n}\mathcal{O}/{}_H\mathcal{I})$ is called a reduced *hypersurface* in $\mathbb{P}_n$, or, for $q = 1$, a *hyperplane.*

Each set $\{P_j\}$ of homogeneous polynomials generates an ideal in $\mathbb{C}[z_0, \ldots, z_n]$; by Hilbert's Basis Theorem, that ideal is generated by finitely many P_j. Their common zero-set in $(\mathbb{C}^{n+1})^*$ is a cone, the image of which determines an analytic set A in $\mathbb{P}_n$. The complex space $(A, {}_{\mathbb{P}_n}\mathcal{O}/{}_A\mathcal{I})$ is called a reduced *projective algebraic variety.* As a closed subset of $\mathbb{P}_n$, A is compact. Incidentally, one can show that every closed complex subspace of $\mathbb{P}_n$ is projective algebraic (*Chow's Theorem* [Mu Cor. (4.6)]).

32.7 Definition. *If $(X, \mathcal{O})$ is a complex space, then $S(X) := \{x \in X;$ no neighborhood of x is a complex manifold$\}$ is called the* set of singularities *of X.*

For the characterization of singular points in the succeeding exercises, we make the following observation:

32.8 Remark. *For a manifold X, a nonzero function $f \in \mathcal{O}(X)$, and an $a \in N(f)$, the point a is not a singularity of the reduced complex space $N(f)$ iff f has a prime factorization in ${}_X\mathcal{O}_a$ of the form $f = g^s$ such that $\frac{\partial g}{\partial z}(a) \neq 0$ in local coordinates z at the point a.*

Proof. Without loss of generality, choose $X \Subset \mathbb{C}^n$. *If.* By virtue of continuity, $\frac{\partial g}{\partial z}$ has rank 1 near a; by 8.9, $N(f) = N(g)$ is a manifold near a.

Only if. This direction follows from E. 47e. ∎

E. 32 i. Determine $S(A)$ for the reduced spaces $A = N(\mathbb{C}^n; f)$ determined by the following equations:

i) $n = 2, f = z_1 z_2$ *(Achsenkreuz)*.
ii) $n = 2, f = z_1^2 - z_2^3$ *(Neil's parabola)*.
iii) $n = 3, f = z_1^2 - z_2 z_3$ ($\sqrt{z_2 z_3}$).
iv) $n = 3, f = z_1^2 z_2 - z_3^2$ *(Whitney's umbrella)*.

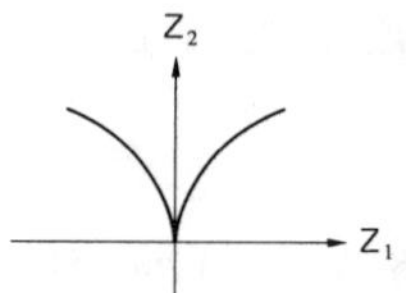

$N(\mathbb{C}^2; z_1^2 - z_2^3) \cap \mathbb{R}^2$

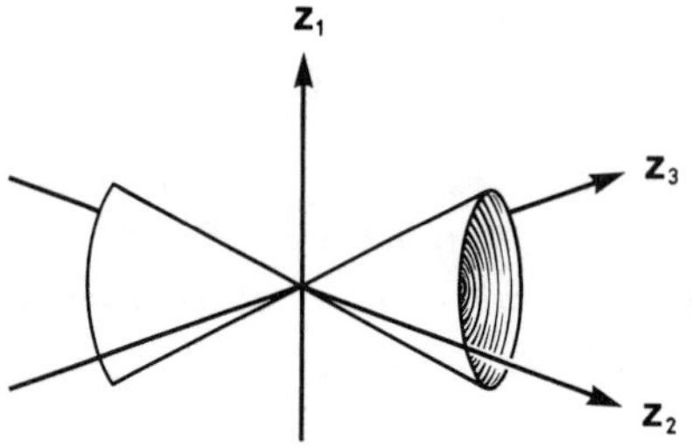

$N(\mathbb{C}^3; z_1^2 - z_2 z_3) \cap \mathbb{R}^3$

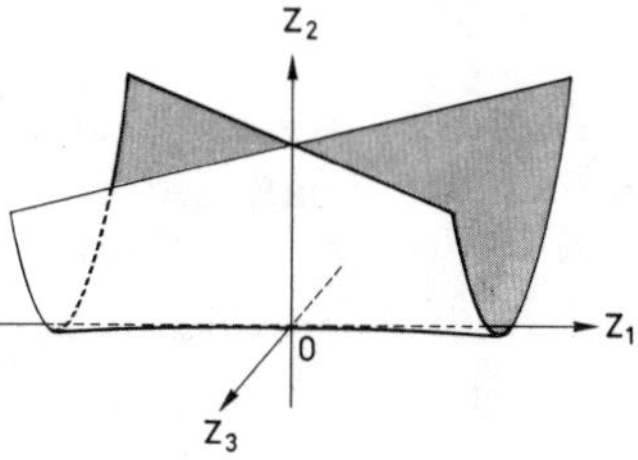

$N(\mathbb{C}^3; z_1^2 z_2 - z_3^2) \cap \mathbb{R}^3$

E. 32 j. Prove the following statements: i) $z_1^2 - z_2^2 - z_2^3$ is irreducible in $\mathbb{C}[z_1, z_2]$, but reducible in ${}_2\mathcal{O}_0$.

ii) A homogeneous polynomial $P \in \mathbb{C}[z_1, \ldots, z_n]$ is irreducible in $\mathbb{C}[z]$ iff P is irreducible in ${}_n\mathcal{O}_0$ (hint: $g \in {}_n\mathcal{O}_0$ can be represented as a sum of homogeneous polynomials).

E. 32 k. For $P = z_0^q + \ldots + z_r^q \in \mathbb{C}[z_0, \ldots, z_n]$, determine $S(N(\mathbb{P}_n; P))$. For $n = r = 2$, $N(\mathbb{P}_n; P)$ is a Riemann surface; the number $\frac{1}{2}(q-1)(q-2)$ is called its genus.

For the description of complex structures in the next exercises, it is helpful to apply holomorphic mappings:

32.9 Remark. *Let $\varphi: X \to Y$ be a holomorphic mapping of reduced complex spaces, suppose that $A := \varphi(X)$ is an analytic set in Y and that $\varphi: X \to A$ is an open mapping. Then, for every $x \in X$, we have that* $\operatorname{Ker} \varphi_x^0 = {}_A\mathcal{I}_x$, *and in particular that*

$${}_A\mathcal{O}_{\varphi(x)} \cong \varphi^0({}_Y\mathcal{O}_{\varphi(x)}) \subset {}_X\mathcal{O}_x . \quad \blacksquare$$

E. 32 l. Prove the following statements: a) The local algebras ${}_A\mathcal{O}_0$ in the examples of E. 32 i have the following form (use that algebraic reduction is the same as reduction, by 47.2):

i) *Achsenkreuz:* ${}_A\mathcal{O}_0 \cong \{a + \sum_{j \geq 1} b_j z_1^j + \sum_{k \geq 1} c_k z_2^k \in {}_2\mathcal{O}_0\}$ provided with the canonical multiplication modulo $z_1 z_2$; i.e., $z_1 z_2 = 0$.

ii) Neil's parabola: ${}_A\mathcal{O}_0 \cong \{\sum_{j \geq 0} a_j t^j \in {}_1\mathcal{O}_0;\ a_1 = 0\}$ (hint: $\varphi: \mathbb{C} \to \mathbb{C}^2, t \mapsto (t^3, t^2)$).

iii) $\sqrt{z_2 z_3}$: ${}_A\mathcal{O}_0 = \{\sum_{j \equiv k(2)} a_{jk} t^j u^k \in {}_2\mathcal{O}_0\}$ (hint: $\varphi: \mathbb{C}^2 \to \mathbb{C}^3, (t, u) \mapsto (tu, t^2, u^2)$).

iv) Whitney's umbrella: ${}_A\mathcal{O}_0 = \{\sum_{(i,j) \in I} a_{ij} t^i u^j \in {}_2\mathcal{O}_0\}$ where $I = \{(i,j);\ i \geq 1 \text{ or } j \equiv 0 \pmod 2\}$ (hint: $\varphi: \mathbb{C}^2 \to \mathbb{C}^3, (t, u) \mapsto (t, u^2, tu)$).

b) In i), ii), and iv), ${}_A\mathcal{O}_0$ is *not a normal ring* (hint: $\dfrac{z_1}{z_1 + z_2} = \dfrac{z_1^2}{z_1^2 + z_2^2}$, t, resp., u is integral over ${}_A\mathcal{O}_0$).

c) *None* of the rings ${}_A\mathcal{O}_0$ above is a *factorial ring* (hint: factor each of 0, t^6 and t^2u^2 (in iii) and iv)) as a product of irreducible elements in two different ways).

32.10 Definition. *For* $\{P_j; j\in J\} \subset \mathbb{C}[z_1,\dots,z_n]$, *the set* $N(\mathbb{C}^n; \{P_j\})$ *is called an* *affine algebraic variety;* $N(\mathbb{C}^n; P_{j_0})$ *is called an* *affine algebraic hypersurface* *(or hyperplane if* P_{j_0} *is affine linear).*

E. 32 m. Show that $\mathbb{C}^*$ is isomorphic to an affine algebraic hypersurface (hint: consider $\varphi: \mathbb{C}^* \to \mathbb{C}^2, z \mapsto (z, 1/z)$).

Every affine algebraic variety $A = N(\mathbb{C}^n; \{P_j\})$ can be compactified to a projective algebraic variety: if P_j has (total) degree q, then

(32.10.1) $\bar{P}_j := z_0^q \cdot P_j(z_1/z_0, \dots, z_n/z_0) \in \mathbb{C}[z_0, \dots, z_n]$

is homogeneous of degree q. The projective algebraic variety $B := N(\mathbb{P}_n; \{\bar{P}_j\})$ is such that $B \setminus \{z_0 = 0\} = A$. This "projective closure" coincides with the topological closure of A in $\mathbb{P}_n$ [Mu 2.33, 2.5b].

E. 32 n. Prove that, for $q > \deg(P)$, in the notations used in (32.10.1), $\mathbb{P}_{n-1} = \{z_0 = 0\} \subset N(\mathbb{P}_n; \bar{P})$; thus $N(\mathbb{P}_n; \bar{P})$ is "reducible" (49.4).

E. 32 o. Determine $S(N(\mathbb{P}_n; \bar{P}))$ for $P(z) = \sum_{j=1}^{n} z_j^j$ (for $\bar{P}$, see (32.10.1)).

E. 32 p. *Extension of holomorphic mappings to* $\mathbb{P}_n$. Consider $\mathbb{C}^n = \{[z_0, \dots, z_n] \in \mathbb{P}_n; z_0 \neq 0\}$ and prove the following:

i) For $n = 1$, $P_j \in \mathbb{C}[z]$, and $a_j \in \mathbb{C}$ (for $j = 0, \dots, s$), the mapping

$$\mathbb{C} \setminus \{a_1, \dots, a_s\} \to \mathbb{C},\ z \mapsto P_0(z) + \sum_{j=1}^{s} P_j\left(\frac{1}{z - a_j}\right),$$

has a holomorphic extension $\mathbb{P}_1 \to \mathbb{P}_1$. (Conversely, every holomorphic mapping $\varphi: \mathbb{P}_1 \to \mathbb{P}_1$ has such a representation: φ can be viewed as a meromorphic function on $\mathbb{C} \setminus \varphi^{-1}(\infty)$; by 53.1, it is thus a quotient of two polynomials.)

ii) For $P \in \mathbb{C}[z_1, \dots, z_n]$, there exists an analytic set $A \subset \mathbb{P}_n$ of codimension two such that the mapping $\mathbb{C}^n \to \mathbb{C}$, $z \mapsto P(z)$, has a holomorphic extension $\mathbb{P}_n \setminus A \to \mathbb{P}_1$ (hint: choose a linear function $l(z_0, \dots, z_n)$ and consider $(\bar{P}, l^{\deg(P)})$; for $\bar{P}$, see (32.10.1)).

iii) The "meromorphic function" $\mathbb{C}^2 \to \mathbb{C}, (z_1, z_2) \mapsto z_1/z_2$, has no continuous extension $\mathbb{P}_2 \to \mathbb{P}_1$.

We conclude this section with some remarks on complex differential calculus. In real analysis, tangent and cotangent spaces are used to linearize differentiable mappings. That concept is just as important in complex analysis. We introduce it in sufficient generality for our purposes:

For an analytic algebra $R = \mathbb{C} \oplus \mathfrak{m}$, the space $\mathfrak{m}/\mathfrak{m}^2$ is called the *cotangent space*, and $TR := (\mathfrak{m}/\mathfrak{m}^2)'$ [4], the *tangent space of* R.

---

[4] M' denotes the dual vector space $\operatorname{Hom}(M, \mathbb{C})$.

For $R = {}_n\mathcal{O}_0$ and the basis $\{\bar{z}_1, \dots, \bar{z}_n\}$[5)] of $\mathfrak{m}/\mathfrak{m}^2$, the dual basis for $T_n\mathcal{O}_0 = (\mathfrak{m}/\mathfrak{m}^2)'$ is obviously $\{\frac{\partial}{\partial z_1}\Big|_0, \dots, \frac{\partial}{\partial z_n}\Big|_0\}$. Thus, if X is a manifold, then $\dim_a X = \dim T_X\mathcal{O}_a$.

By 21.6, a homomorphism $\varphi: R \to S$ of analytic algebras induces linear mappings

$$\bar{\varphi}: \mathfrak{m}_R/\mathfrak{m}_R^2 \to \mathfrak{m}_S/\mathfrak{m}_S^2 \quad \text{and} \quad T\varphi: TS \to TR.$$

For $R = {}_m\mathcal{O}_0$ and $S = {}_n\mathcal{O}_0$, if $\varphi = f_0^0$ is given by a holomorphic mapping $f: X \to Y$ with $0 \in X \subset\subset \mathbb{C}^n$, $0 \in Y \subset\subset \mathbb{C}^m$ and $f(0) = 0$, then, for the canonical bases,

$$\text{(32.11.1)} \quad \bar{\varphi} = \left(\frac{\partial f}{\partial z}(0)\right)^t \quad \text{and} \quad T\varphi = \frac{\partial f}{\partial z}(0),$$

for we have that $f_0^0(w_j) = f_j \equiv \sum_{k=1}^{n} \frac{\partial f_j}{\partial z_k}(0) \cdot z_k (\mathrm{mod}_n \mathfrak{m}^2)$, and thus that $\overline{f_0^0}(\overline{w}_j) = \sum_{k=1}^{n} \frac{\partial f_j}{\partial z_k}(0) \cdot \bar{z}_k$.

For a holomorphic mapping $f: X \to Y$ of complex spaces and $a \in X$, we use the notation

$$T_a X := T_X\mathcal{O}_a \quad \text{and} \quad T_a f := \acute{T} f_a^0 : T_a X \to T_{fa} Y$$

for the *tangent space* of X at a and the *tangent mapping* (or *differential*) for f, respectively. The construction of tangent spaces is thus a covariant functor from the category of complex spaces (with "base point") to the category of finite-dimensional vector spaces.

32.11 Proposition. *If $\varphi: R_1 \to R_2$ is a homomorphism of analytic algebras with maximal ideals $\mathfrak{m}_1$ and $\mathfrak{m}_2$, then, with the projection $\pi: R_2 \to R_2 / R_2 \cdot \varphi(\mathfrak{m}_1) =: R$, the sequence*

$$\mathfrak{m}_1/\mathfrak{m}_1^2 \xrightarrow{\bar{\varphi}} \mathfrak{m}_2/\mathfrak{m}_2^2 \xrightarrow{\bar{\pi}} \mathfrak{m}/\mathfrak{m}^2 \to 0$$

is exact.

Let us note that, by 44.3 and 31.8 iv), R may be interpreted as a stalk of the structure sheaf of a fiber.

Proof. We show only that $\operatorname{Ker} \bar{\pi} \subset \operatorname{Im} \bar{\varphi}$, since the rest is trivial. For each $r \in \mathfrak{m}_2$ such that $\bar{\pi}(\bar{r}) = 0$, we see that $\pi(r) \in \mathfrak{m}^2 = \pi(\mathfrak{m}_2)^2$, so

$$r \in \mathfrak{m}_2^2 + R_2 \cdot \varphi(\mathfrak{m}_1) = \mathfrak{m}_2^2 + (\mathbb{C} \oplus \mathfrak{m}_2)\varphi(\mathfrak{m}_1) = \mathfrak{m}_2^2 + \varphi(\mathfrak{m}_1);$$

i.e., $r \in \varphi(\mathfrak{m}_1) \bmod \mathfrak{m}_2^2$. ■

2

5) The attentive reader will not confuse the notation of equivalence classes with that of conjugation!

32.12 Corollary. *For* $R = {}_n\mathcal{O}_0 / {}_n\mathcal{O}_0 \cdot (f_1, \dots, f_m)$,

$$\dim \mathfrak{m}_R/\mathfrak{m}_R^2 = n - \operatorname{rank}\left(\frac{\partial f}{\partial z}(0)\right).$$

Proof. By 32.11, the mapping $\varphi : {}_m\mathcal{O}_0 \to {}_n\mathcal{O}_0$, $z_j \mapsto f_j$, yields an exact sequence

$${}_m\mathfrak{m}/{}_m\mathfrak{m}^2 \xrightarrow{\bar\varphi} {}_n\mathfrak{m}/{}_n\mathfrak{m}^2 \xrightarrow{\bar\pi} \mathfrak{m}_R/\mathfrak{m}_R^2 \to 0$$

of vector spaces; it follows that

$$\dim \mathfrak{m}_R/\mathfrak{m}_R^2 = n - \dim(\operatorname{Im}\bar\varphi) = n - \operatorname{rank}\bar\varphi \underset{(32.11.1)}{=} n - \operatorname{rank}\left(\frac{\partial f}{\partial z}(0)\right). \quad \blacksquare$$

The following exercise deals with the explicit computation of tangent spaces:

E. 32 q. For an analytic algebra R, a $\delta \in R'$ is called a *derivation of* R if, for the evaluation $\varepsilon_0 : R \to \mathbb{C}, f \mapsto f(0)$,

$$\delta(f \cdot g) = f(0)\delta(g) + g(0)\delta(f) \quad \forall f, g \in R.$$

Prove the following statements:

i) The set $\operatorname{Der} R$ of all derivations of R is a linear subspace of R'.

ii) Each $\delta \in \operatorname{Der} R$ satisfies the condition $\delta(\mathbb{C}) = 0 = \delta(\mathfrak{m}^2)$. Thus there exists a canonical $\mathbb{C}$-linear mapping

$$\operatorname{Der} R \to (\mathfrak{m}/\mathfrak{m}^2)' = TR;$$

moreover, that mapping is bijective.

iii) In the situation of 32.11, the sequence

$$0 \to \operatorname{Der} R \xrightarrow{T\pi} \operatorname{Der} R_2 \xrightarrow{T\varphi} \operatorname{Der} R_1$$

is exact.

iv) For each ideal $\mathfrak{a} = (f_1, \dots, f_m) \subset R$,

$$\operatorname{Der} R/\mathfrak{a} \cong \{\delta \in \operatorname{Der} R;\ \delta(\mathfrak{a}) = 0\} = \{\delta \in \operatorname{Der} R;\ \delta f_1 = \dots = \delta f_m = 0\}.$$

E. 32 r. Determine $T_0 j$ for the following embeddings $j : X \hookrightarrow \mathbb{C}^2$:
i) $X = N(z_1 z_2)$; ii) $X = N(z_1^2 - z_2^3)$; iii) $X = N(z_1 z_2, z_2^2)$.

E. 32 s. For an analytic algebra R and a vector space homomorphism $\chi : TR \to T_n\mathcal{O}_0$, find an algebra-homomorphism $\varphi : {}_n\mathcal{O}_0 \to R$ such that $T\varphi = \chi$ (for 45.10).

32.13 Remark. *For an algebra-homomorphism* $\varphi : {}_m\mathcal{O}_0 \to {}_n\mathcal{O}_0$, *the following statements are equivalent:*

i) φ *is an isomorphism.*

ii) $\bar\varphi : {}_m\mathfrak{m}/{}_m\mathfrak{m}^2 \to {}_n\mathfrak{m}/{}_n\mathfrak{m}^2$ *is an isomorphism.*

iii) $T\varphi : T_n\mathcal{O}_0 \to T_m\mathcal{O}_0$ *is an isomorphism.*

Proof. ii) $\Rightarrow$ i) First we observe that $m = \dim {}_m\mathfrak{m}/{}_m\mathfrak{m}^2 = \dim {}_n\mathfrak{m}/{}_n\mathfrak{m}^2 = n$. By

21.9, there exists a holomorphic mapping f between neighborhoods $U, V \subset\subset \mathbb{C}^n$ of the origin such that $f(0) = 0$ and $\varphi = f_0^0$. The equality $\operatorname{rank} \bar{\varphi} \underset{(32.11.1)}{=} \operatorname{rank}\left(\frac{\partial f}{\partial z}(0)\right)$ implies that $\operatorname{rank} \frac{\partial f}{\partial z} = n$ near 0 (E. 8f); thus it follows from the Inverse Mapping Theorem that f is biholomorphic near 0, so φ is an isomorphism. ■

E. 32t. Give an example showing that 32.13 does not hold for $\varphi : {}_m\mathcal{O}_0 \to R$ with an arbitrary analytic algebra R (however, see 48.16).

§ 32A Supplement: Submersions and Immersions

For the construction of manifolds in §32B, we use primarily two particularly simple types of holomorphic mappings, which we characterize here with the help of differential calculus.

32 A. 1 Definition. *A holomorphic mapping $f: X \to Y$ of manifolds is called a submersion at $a \in X$ if f can be represented near a and $f(a)$ as a projection, i.e., if there exists a commutative diagram*

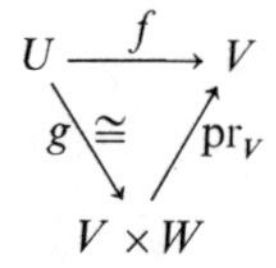

of holomorphic mappings such that $a \in U \subset\subset X$, $f(U) \subset V \subset\subset Y$, and $W \subset\subset \mathbb{C}^d$ for $d = \dim_a X - \dim_{f(a)} Y$.

32 A. 2$_{\text{geo}}$ Proposition. *For a holomorphic mapping $f: X \to Y$ of manifolds, and an $a \in X$, the following statements are equivalent:*

i) f is a submersion at a.

ii) Existence of a "local section": There exist open neighborhoods U of a and V of $f(U)$, and an $h \in \operatorname{Hol}(V, U)$, such that $f \circ h = \mathrm{id}_V$.

iii) $T_a f$ is surjective.

Instead of a direct geometric proof of 32 A. 2$_{\text{geo}}$, we show that an (obviously equivalent, by 21.9) algebraic formulation holds:

32 A. 2$_{\text{alg}}$ Proposition. *For an algebra-homomorphism $\varphi : {}_m\mathcal{O}_0 \to {}_n\mathcal{O}_0$, the following statements are equivalent:*

i) For $d := n - m$ and the canonical inclusion j, there exists an isomorphism ψ such that the diagram

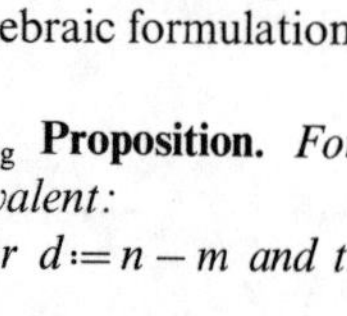

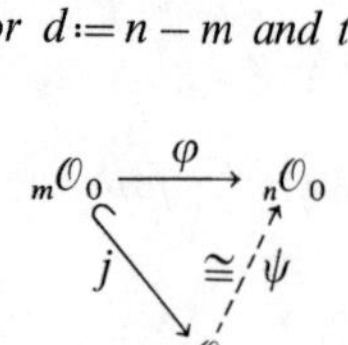

commutes. The algebra ${}_n\mathcal{O}_0 / {}_n\mathcal{O}_0 \cdot \varphi({}_m\mathfrak{m})$ is isomorphic to ${}_d\mathcal{O}_0$.

ii) There exists an algebra-homomorphism $\chi : {}_n\mathcal{O}_0 \to {}_m\mathcal{O}_0$ such that $\chi \circ \varphi = \mathrm{id}_{{}_m\mathcal{O}_0}$.

iii) $\bar{\varphi} : {}_m\mathfrak{m} / {}_m\mathfrak{m}^2 \to {}_n\mathfrak{m} / {}_n\mathfrak{m}^2$ is injective.

iv) For $f_j := \varphi(w_j)$, rank $\left(\dfrac{\partial f_j}{\partial z_k}(0)\right)_{\substack{j=1,\dots,m\\ k=1,\dots,n}} = m$.

Proof. i) $\Rightarrow$ ii) For $\pi: {}_{m+d}\mathcal{O}_0 \twoheadrightarrow {}_{m+d}\mathcal{O}_0/(w_{m+1},\dots,w_{m+d}) = {}_m\mathcal{O}_0$ and $\chi = \pi \circ \psi^{-1}$, we have the equality $\chi \circ \varphi = \pi \circ \psi^{-1} \circ \psi \circ j = \mathrm{id}_{{}_m\mathcal{O}_0}$.

ii) $\Rightarrow$ iv) Since $\mathrm{id}_{{}_m\mathcal{O}_0} = \bar{\chi} \circ \bar{\varphi}$ has rank m, it follows that

$$m \leqq \operatorname{rank} \bar{\varphi} \underset{(32.11.1)}{=} \operatorname{rank}\left(\frac{\partial f_j}{\partial z_k}(0)\right)^t = \operatorname{rank}\left(\frac{\partial f_j}{\partial z_k}(0)\right) \leqq m .$$

iv) $\Rightarrow$ iii) This follows from (32.11.1).

iii) $\Rightarrow$ i) Since $\dim \bar{\varphi}({}_m\mathfrak{m}/{}_m\mathfrak{m}^2) = m$, it follows that there exist $f_1,\dots,f_d \in {}_n\mathfrak{m}$ such that ${}_n\mathfrak{m}/{}_n\mathfrak{m}^2 = \operatorname{Im}\bar{\varphi} \oplus \mathbb{C}\bar{f}_1 \oplus \dots \oplus \mathbb{C}\bar{f}_d$. By 21.7, the assignment

$$w_j \mapsto \begin{cases} \varphi(w_j) & j = 1,\dots,m \\ f_{j-m} & j = m+1,\dots,m+d = n \end{cases}$$

determines a homomorphism $\psi: {}_{m+d}\mathcal{O}_0 \to {}_n\mathcal{O}_0$, which, by 32.13, is an isomorphism. Moreover, we have that ${}_n\mathcal{O}_0/{}_n\mathcal{O}_0 \cdot \varphi({}_m\mathfrak{m}) \cong {}_{m+d}\mathcal{O}_0/{}_{m+d}\mathcal{O}_0 \cdot {}_m\mathfrak{m} \cong {}_d\mathcal{O}_0$. ■

32 A. 3 Corollary. *The following statements hold for a holomorphic mapping $f: X \to Y$ of manifolds:*

i) The set $\{a \in X;\, f$ is a submersion at $a\}$ is open in X.

ii) If f is a submersion at every $a \in X$ (in which case f is called a submersion), then

a) f is an open mapping, and the induced topology on $f(X) \subset\!\!\!\!\!\!\;\subset Y$ is the quotient topology;

b) a mapping $g: f(X) \to Z$ of manifolds is holomorphic iff $g \circ f: X \to Z$ is holomorphic; and

c) the manifold-structure on $f(X)$ is determined by that of X and by the equivalence relation R_f (see 33 B. 3 i)).

Proof. If f is a submersion at a, then f can be represented near a as a projection; that implies i) and ii a). Statements ii b) and c) follow from the existence of local sections. ■

"Local embeddings" between manifolds can be characterized in a dual manner:

32 A. 4$_{\text{geo}}$ Proposition and Definition. *For a holomorphic mapping $f: X \to Y$ of manifolds, and a point $a \in X$, the following statements are equivalent:*

i) f is an immersion at a; i.e., there exist neighborhoods $a \in U \subset\!\!\!\!\!\!\;\subset X$, $f(U) \subset V \subset\!\!\!\!\!\!\;\subset Y$, and $0 \in W \subset\!\!\!\!\!\!\;\subset \mathbb{C}^d$ (with $d := \dim_{f(a)} Y - \dim_a X$), and a holomorphic mapping g such that the following diagram commutes:

$$\begin{array}{ccc} u & \xrightarrow{\;f\;} & v \\ \downarrow{\cong} & & {\scriptstyle g}\downarrow{\cong} \\ U \times \{0\} & \hookrightarrow & U \times W . \end{array}$$

ii) Existence of a left inverse: There exist open neighborhoods U of a and V of $f(U)$, and an $h \in \operatorname{Hol}(V, U)$ such that $h \circ f = \mathrm{id}_U$.

iii) $T_a f$ is injective.

Again we give an equivalent (by 21.9) algebraic version:

32 A. 4$_{\text{alg}}$ Proposition. *For an algebra-homomorphism $\varphi: {}_m\mathcal{O}_0 \to {}_n\mathcal{O}_0$, the following statements are equivalent:*

i) For the canonical mapping $\pi: {}_{n+d}\mathcal{O}_0 \twoheadrightarrow {}_{n+d}\mathcal{O}_0/(z_{n+1},\dots,z_{n+d})$ to residue classes, with $d := m - n$, there exists an isomorphism ψ such that the diagram

$$\begin{array}{ccc} {}_m\mathcal{O}_0 & \xrightarrow{\varphi} & {}_n\mathcal{O}_0 \\ & \nwarrow^{\psi}\ \cong \quad \nearrow^{\pi} & \\ & {}_{n+d}\mathcal{O}_0 & \end{array}$$

commutes. In particular, the algebra ${}_m\mathcal{O}_0/\mathrm{Ker}\,\varphi$ *is isomorphic to* ${}_n\mathcal{O}_0$.

ii) *There exists an algebra-homomorphism* $\chi: {}_n\mathcal{O}_0 \to {}_m\mathcal{O}_0$ *such that* $\varphi \circ \chi = \mathrm{id}_{{}_n\mathcal{O}_0}$.

iii) $\bar{\varphi}: {}_m\mathfrak{m}/{}_m\mathfrak{m}^2 \to {}_n\mathfrak{m}/{}_n\mathfrak{m}^2$ *is surjective.*

iv) *For* $f_j := \varphi(w_j)$, $\operatorname{rank}\left(\frac{\partial f_j}{\partial z_k}(0)\right)_{\substack{j=1,\dots,m\\ k=1,\dots,n}} = n$.

Proof. iii) ⇒ i) Let the generators of ${}_m\mathfrak{m}$ and ${}_{n+d}\mathfrak{m}$ be denoted by $w_1,\dots,w_m$ and $z_1,\dots,z_{n+d}$, respectively. By a suitable choice of $w_1,\dots,w_m$, we have for $f_j := \varphi(w_j) \in {}_n\mathcal{O}_0 \hookrightarrow {}_{n+d}\mathcal{O}_0$ that ${}_n\mathfrak{m}/{}_n\mathfrak{m}^2 = \bigoplus_{j=1}^{n} \mathbb{C}\cdot\bar{f}_j$ and $\mathrm{Ker}\,\bar{\varphi} = \bigoplus_{j=n+1}^{m} \mathbb{C}\cdot\bar{w}_j$.

By 21.7, an algebra-homomorphism $\vartheta: {}_m\mathcal{O}_0 \to {}_{n+d}\mathcal{O}_0$ is determined by the assignments $w_j \mapsto f_j$ for $j = 1,\dots,n$ and $w_j \mapsto z_j$ for $j = n+1,\dots,d$. Since $\bar{\vartheta}: {}_m\mathfrak{m}/{}_m\mathfrak{m}^2 \to {}_n\mathfrak{m}/{}_n\mathfrak{m}^2 \oplus {}_d\mathfrak{m}/{}_d\mathfrak{m}^2 = {}_{n+d}\mathfrak{m}/{}_{n+d}\mathfrak{m}^2$ is an isomorphism, it follows from 32.13 that ϑ is an isomorphism as well. Thus, $\psi := \vartheta^{-1}$ has the desired properties. ■

E. 32 A a. Prove the implications "i) ⇒ ii) ⇒ iv) ⇒ iii)" in 32A.4_{alg}. ■■

32 A. 5 Corollary. *Every holomorphic mapping* $f: X \to Y$ *of manifolds satisfies the following conditions:*

i) *The set* $\{a \in X;\ f$ *is an immersion at* $a\}$ *is open in* X.

ii) *If* f *is an immersion at every* $a \in X$ *(in which case* f *is called an immersion), then*

a) *a continuous mapping* $g: Z \to X$ *of manifolds is holomorphic iff* $f \circ g: Z \to Y$ *is holomorphic;*

b) *the manifold-structure on* X *is determined by that of* Y *and the continuous mapping* f.

Proof. The first statement follows directly from the definition. The second follows from 32 A. 4_{geo} ii). ■

32 A. 6 Example. If $f: X \to Y$ and $g: Y \to X$ are holomorphic mappings such that $g \circ f = \mathrm{id}_X$, then f is an injective immersion and g is a submersion; moreover, $f(X)$ is a closed submanifold, and $g|_{f(X)}$ is biholomorphic. ■

E. 32 Ab. For manifolds X and Y,

$$T_{(a,b)} X \times Y = T_a X \oplus T_b Y.$$

E. 32 Ac. Let $f: X \to Y$ be a submersion of manifolds. If $M \hookrightarrow Y$ is a submanifold, then $f^{-1}(M) \hookrightarrow X$ is a submanifold; moreover, $\dim_x f^{-1}(fx) = \dim_x X - \dim_{fx} Y$ (for 32 B. 2).

E. 32 Ad. Show that the following statements holds for holomorphic mappings $X \xrightarrow{f} Y \xrightarrow{g} Z$ of manifolds and $a \in X$: i) If $g \circ f$ is a submersion at a, then g is a submersion at $f(a)$ (for 32 B. 4).

ii) If $g \circ f$ is an immersion at a, then f is an immersion at a.

E. 32 Ae. *If* $X \hookrightarrow \mathbb{P}_n$ *is a projective algebraic variety, and if* H *is a hypersurface in* $\mathbb{P}_n$, *then* $X \setminus H$ *is biholomorphic to an affine algebraic variety.* (Hint: Let q be the total degree of a homogeneous equation $f \in \mathbb{C}[z_0,\dots,z_n]$ defining H. For $z = [z_0,\dots,z_n]$ and $\nu \in \mathbb{N}^{n+1}$ with $|\nu| = q$

(there are $N+1 := \binom{q+n}{q}$ such ν's), set $w_\nu := z^\nu$, and prove the following statements:

i) A fixed ordering of the ν's yields a holomorphic mapping $\varphi_q\colon \mathbb{P}_n \to \mathbb{P}_N$, $[z_0, \dots, z_n] \mapsto [(w_\nu)]$ *(Veronese mapping)*.

ii) φ_q is a homeomorphism $\mathbb{P}_n \to \varphi_q(\mathbb{P}_n) \subset \mathbb{P}_N$; moreover, it is an immersion, and $\varphi_q(\mathbb{P}_n)$ is a compact submanifold of $\mathbb{P}_N$.

iii) $|\varphi_q(\mathbb{P}_n)| = N(\mathbb{P}_N; w_\nu w_\mu - w_{\nu'} w_{\mu'},\ \nu+\mu = \nu'+\mu')$; describe the associated cone Y in $\mathbb{C}^{N+1}$ and show that $S(Y) = \{0\}$ for $q \geqq 2$ and $n \geqq 1$.

iv) $\varphi_q(N(\mathbb{P}_n; f))$ is the intersection of $\varphi_q(\mathbb{P}_n)$ with a hyperplane in $\mathbb{P}_N$.)

E. 32 Af. *Every finite product of projective algebraic varieties is projective algebraic.* (Hint: $\psi\colon \mathbb{P}_m \times \mathbb{P}_n \to \mathbb{P}_{mn+m+n}$, $([z_0, \dots, z_m], [w_0, \dots, w_n]) \mapsto [x_{00}, x_{01}, \dots, x_{mn}]$, with $x_{jk} := z_j w_k$, is a closed immersion (*Segre embedding;* see also 45.9).) Moreover,

$$\psi(\mathbb{P}_m \times \mathbb{P}_n) = N(\mathbb{P}_{mn+m+n}; (x_{ij}x_{kl} - x_{il}x_{kj})_{i,k;j,l}),$$

and the associated punctured *Segre cone* $\tau^{-1}(\psi(\mathbb{P}_m \times \mathbb{P}_n)) \hookrightarrow (\mathbb{C}^{(m+1)(n+1)})^*$ has no singularities.

Further literature: [Se]

§ 32 B Examples of Complex Manifolds

The manifolds $\mathbb{C}^n$ and $GL(n, \mathbb{C})$ also have a group-structure that is compatible with the complex structure; they are thus Lie groups:

32 B. 1 Definition. *A (complex) manifold G for which a group structure is given is called a (complex) Lie group if the multiplication $G \times G \to G$, $(g,h) \mapsto gh$, and the inversion $G \to G$, $g \mapsto g^{-1}$, are holomorphic mappings.*

32 B. 2 Examples. i) The additive group $\mathbb{C}^{n \times m}$ of $n \times m$-matrices over $\mathbb{C}$ is an mn-dimensional Lie group.

ii) *The general linear group.* With respect to matrix multiplication, $GL(n, \mathbb{C})$ is an n^2-dimensional Lie group: multiplication is bilinear, and the determinants appearing in

$$(a_{ij}) \mapsto (a_{ij})^{-1} = (\tilde{a}_{kl}) / \det(a_{ij})$$

with the cofactors $\tilde{a}_{kl} = (-1)^{kl} \det((a_{ij})_{i \neq l, j \neq k})$ are polynomials in the a_{ij}.

iii) Every group can be made into a 0-dimensional Lie group by providing it with the discrete topology.

iv) *Lie subgroups.* If G is a Lie group and $H \hookrightarrow G$ is both a (closed) submanifold and a subgroup of G, then H is a *Lie subgroup* of G (E. 31 k or 32 A. 5 ii)). As a special case in $G = GL(n, \mathbb{C})$, we mention the *special linear group* $SL(n, \mathbb{C}) := \{A \in GL(n, \mathbb{C}); \det A = 1\}$, which is an (n^2-1)-dimensional Lie group, since the mapping $\det\colon GL(n, \mathbb{C}) \to \mathbb{C}^*$ is a group-homomorphism and a submersion (E. 32 Ac): it suffices to verify that near I_n; there, $f\colon \mathbf{P}^1(1;1) \to GL(n, \mathbb{C})$, $\lambda \mapsto \sqrt[n]{\lambda}\, I_n$ (with $\sqrt[n]{1} = 1$), is a section for det.

The subgroup of $GL(n, \mathbb{C})$ consisting of the diagonal matrices, the subgroup consisting of the upper-triangular matrices, and the subgroup $GL(k, n-k)$ of matrices of the form $\begin{pmatrix} A & B \\ 0 & C \end{pmatrix}$ with $A \in GL(k, \mathbb{C})$, $C \in GL(n-k, \mathbb{C})$, and $B \in \mathbb{C}^{k \times (n-k)}$ are Lie groups of dimension $n, n(n+1)/2$, and $n^2 - k(n-k)$, respectively, since they can be described by means of linear equations.

A closed additive subgroup Γ of $G = \mathbb{C}^n$ that is isomorphic to $\mathbb{Z}^{2n}$ is called a *lattice*. As a discrete subgroup, Γ is a Lie group.

v) If G and H are Lie groups, then the direct product $G \times H$ is a Lie group.

E. 32 Ba. *$GL(n, \mathbb{C})$ is affine algebraic: $GL(n, \mathbb{C}) \cong \{(A, d) \in \mathbb{C}^{n \times n} \times \mathbb{C}; d \cdot \det A = 1\}$.*

For the investigation of quotient spaces, we make use of the following result:

32 B. 3 Proposition. *Let X be a manifold, and let $R \subset X \times X$ be an equivalence relation. The following two statements are equivalent:*

i) R is a closed[6] *submanifold of $X \times X$, and the canonical projections $\mathrm{pr}_1, \mathrm{pr}_2 : R \to X$ are submersions.*[7]

ii) The quotient space X/R has a manifold-structure for which the quotient mapping $\pi : X \to X/R$ is a submersion.

Before turning to the proof of 32 B. 3, we look at some applications:

32 B. 4 Corollary. *If G is a Lie group with a Lie subgroup $H \hookrightarrow G$, then there exists a unique manifold-structure (of dimension $\dim G - \dim H$) on the "homogeneous space" G/H of left cosets gH such that the mapping $\pi : G \to G/H$ is a submersion. Moreover, $G \times G/H \to G/H$, $(a, \bar{b}) \mapsto \overline{ab}$, is a submersion; in particular, G acts "holomorphically" on G/H.*

Proof. By 32 A. 3, there exists at most one manifold-structure such that π is a submersion. The existence follows from 32 B. 3 with $R = \{(gh, g) \in G \times G; g \in G, h \in H\}$: in the commutative diagram

$$\begin{array}{ccc} G \times G & \xrightarrow{(g,h) \mapsto (gh,g)} & G \times G \\ \uparrow & & \cup \\ G \times H & \longrightarrow & R \\ \downarrow \mathrm{pr}_G & & \downarrow \mathrm{pr}_2 \\ G & \xrightarrow{\mathrm{id}} & G \end{array},$$

the upper mapping is biholomorphic with inverse $(u, v) \mapsto (v, v^{-1}u)$, and the middle mapping is bijective. Thus, R is a closed submanifold and $\mathrm{pr}_2 : R \to G$ is a submersion.

The multiplication-diagram

$$(D) \qquad \begin{array}{ccc} G \times G & \xrightarrow{\bullet} & G \\ \downarrow \mathrm{id} \times \pi & & \downarrow \pi \\ G \times G/H & \longrightarrow & G/H \end{array}$$

commutes. The mapping $\mathrm{id} \times \pi$ is, as a direct product of submersions, a submersion. Thus, the action $G \times G/H \to G/H$ is holomorphic by 32 A. 3 ii b), and is a submersion by E. 32Ad. ■

32 B. 5 Remark. *If H in 32 B. 4 is a normal subgroup, then G/H is a Lie group.*

That follows immediately from diagrams similar to (D). ■

32 B. 6 Examples. i) *Complex tori.* If $\Gamma \subset \mathbb{C}^n$ is a lattice, then the manifold $T := \mathbb{C}^n/\Gamma$ is a complex torus, since T is obviously homeomorphic to the real torus $(S^1)^{2n}$; in particular, T is a compact

[6] For locally closed R see E. 32 Bf.

[7] Since R is symmetric, pr_1 is a submersion iff pr_2 is.

Lie group. The complex structure of T depends on the "periods" ω_j; we mention only that, for $n=1$, the systems $\{1,\omega_2\}$ and $\{1,\tilde{\omega}_2\}$ determine two equivalent holomorphic structures iff there exists an integral fractional linear transformation with determinant ± 1 such that the relation $\tilde{\omega}_2 = \dfrac{a\omega_2+b}{c\omega_2+d}$ holds [Gu I. 6].

ii) The *Grassmann manifold* $G_k(n) := GL(n,\mathbb{C})\,/\,GL(k,n-k;\mathbb{C})$ can be viewed as the space of k-dimensional linear subspaces of $\mathbb{C}^n$ (and is thus a generalization of $\mathbb{P}_{n-1} = G_1(n)$): for $\mathbb{C}^k := \mathbb{C}^k \times 0 \hookrightarrow \mathbb{C}^n$,

$$GL(k,n-k;\mathbb{C}) = \{A \in GL(n,\mathbb{C});\, A(\mathbb{C}^k) = \mathbb{C}^k\}.$$

For each k-dimensional linear subspace W of $\mathbb{C}^n$, there exists an $A \in GL(n,\mathbb{C})$ such that $A(W) = \mathbb{C}^k$. If A_1 is another such matrix, then $A_1 A^{-1}(\mathbb{C}^k) = \mathbb{C}^k$; i.e., $A_1 A^{-1} \in GL(k,n-k;\mathbb{C})$. Hence, the class of A in $G_k(n)$ can be identified with W.

$G_k(n)$ is compact: the *unitary group* $U(n) \subset GL(n,\mathbb{C})$ is determined by the equation $A\cdot\bar{A}^t = I_n$, with $\bar{A}$ denoting the conjugate matrix of A ($U(n)$ is not a complex Lie group). The columns of A are all bounded by 1, so $U(n)$ is compact. Since $U(n)$ acts transitively on the set of k-dimensional linear subspaces of $\mathbb{C}^n$ (existence of orthonormal bases), the composition

$$U(n) \subset GL(n,\mathbb{C}) \twoheadrightarrow G_k(n)$$

is surjective, and it follows that $G_k(n)$ is compact.

E. 32 Bb. Prove the following for $G_k(n)$: i) *Local coordinates:* Let $E \subset \mathbb{C}^n$ be a k-dimensional linear subspace, and fix an F such that $\mathbb{C}^n = E \oplus F$. With the canonical projection $\pi: \mathbb{C}^n \to \mathbb{C}^n/F$, the mapping

$$\{\varphi \in \operatorname{Hom}(\mathbb{C}^n/F,\mathbb{C}^n);\, \varphi\circ\pi = \mathrm{id}_{\mathbb{C}^n/F}\} \to \{V \subset \mathbb{C}^n;\, V \oplus F = \mathbb{C}^n\},\ \varphi \mapsto \operatorname{Im}\varphi,$$

is biholomorphic (use $\operatorname{Hom}(\mathbb{C}^n/F,\mathbb{C}^n) = \mathbb{C}^{n\times k}$). Thus, we obtain a local chart in E.

ii) *Correlation:* For $E \in G_k(n)$, set

$$E^0 := \{z \in \mathbb{C}^n;\, \langle E,z\rangle = 0\}$$

with $\langle z,w\rangle = \sum_{j=1}^{n} z_j w_j$. The correlation

$$G_k(n) \to G_{n-k}(n),\ E \mapsto E^0,$$

is a biholomorphic mapping. Thus, in particular, we have found an automorphism of order 2 of $G_k(2k)$.

32 B. 6. iii) The *flag manifold* $F_n := GL(n,\mathbb{C})\,/$ (triangular matrices) has as points the "flags" $(V_0,V_1,\dots,V_n)$, where V_j is a j-dimensional linear subspace of $\mathbb{C}^n$ and $V_j \subset V_{j+1}$ for each j: the triangular matrices are distinguished by the fact that they map the standard flag ($V_j = \sum_{k=1}^{j} \mathbb{C}\cdot e_k$ with the canonical basis $e_1,\dots,e_n$ of $\mathbb{C}^n$) into itself. Since $GL(n,\mathbb{C})$ acts transitively on the flags, the assertion follows. The manifold F_n is compact, since $U(n)$ acts transitively, as well.

iv) The *projective linear group* $\mathbb{P}GL(\mathbb{C}^n) = GL(n,\mathbb{C})/\mathbb{C}^*\cdot I_n$ is an (n^2-1)-dimensional Lie group, according to 32 B. 5, for $\mathbb{C}^*\cdot I_n$ obviously is a normal subgroup and a closed submanifold of $GL(n,\mathbb{C})$. Clearly, we also have that $\mathbb{P}GL(\mathbb{C}^n) \cong SL(n,\mathbb{C})/\{\zeta I_n;\, \zeta^n = 1\}$.

If X is a complex space and G is a subgroup of the *automorphism group* $\operatorname{Aut}(X) := \{f: X \to X$ biholomorphically$\}$, then G is called a *transformation group* of X.

32 B. 7 Definition. *A transformation group G is said to act on the complex space X*

i) freely, if no $g \in G$ other than id_X has a fixed point in X, and
ii) properly discontinuously, if, for each compact $K \subset X$, there exist only finitely many $g \in G$ such that $K \cap g(K) \neq \emptyset$.

32 B. 8 Corollary. *For a manifold X, if the transformation group G acts freely and properly discontinuously on X, then the orbit space X/G (see 33 B. 3) is a manifold, and the quotient-mapping $\pi : X \to X/G$ is a submersion.*

Proof. First, we show that the equivalence relation $R_G := \{(x,gx); x \in X, g \in G\}$ (see 33 B. 3) is a submanifold of $X \times X$: since G acts freely, $R_G = \bigcup_{g \in G} \Gamma(g)$ is the disjoint union of the graphs $\Gamma(g)$; it is locally finite, since the action of G is properly discontinuous; every $\Gamma(g)$ is a submanifold (E. 7d), so R must also be a submanifold. Second, we note that pr_1 is a submersion, since that holds on each $\Gamma(g)$. Now the assertion follows from those two facts by 32 B. 3. ∎

E. 32 Bc. Show that the mapping $\pi : X \to X/G$ in 32 B. 8 is a covering (§ 33 B).

32 B. 9 Examples. i) *Hopf surfaces:* The free cyclic group

$$G := \{z \mapsto 2^j z; j \in \mathbb{Z}\} \subset \mathrm{Aut}(\mathbb{C}^{2*})$$

obviously acts freely and properly discontinuously on $\mathbb{C}^{2*}$. Thus the Hopf surface $\mathbb{C}^{2*}/G$ is a manifold. It is homeomorphic to $S^1 \times S^3$ and hence compact: the proof rests on the homeomorphism

$$\mathbb{R} \times \mathbb{C}^2 \supset \mathbb{R} \times S^3 \xrightarrow{\varphi} \mathbb{C}^{2*}, \quad (t, z_1, z_2) \mapsto 2^t (z_1, z_2);$$

with respect to the operation of $\mathbb{Z}$ on $\mathbb{R} \times S^3$ defined by $m \cdot (t,z) := (t+m, z)$, that homeomorphism is *equivariant* (i. e., $\varphi(m \cdot a) = 2^m \varphi(a)$), and the assertion follows from the commutative diagram

$$\begin{array}{ccc} \mathbb{R} \times S^3 & \xrightarrow{\varphi} & \mathbb{C}^{2*} \\ \downarrow & & \downarrow \\ S^1 \times S^3 \cong (\mathbb{R} \times S^3)/\mathbb{Z} & \xrightarrow{\bar{\varphi}} & \mathbb{C}^{2*}/G. \end{array}$$

Incidentally, every manifold-structure on $S^1 \times S^3$ can be represented analogously in the form $\mathbb{C}^{2*}/G$ for an appropriate G [Ko, Thm. 1].

It can be shown that *the Hopf surface is not projective-algebraic* (otherwise, as a compact Kähler manifold, its "first Betti number" b_1 would be even [We V. 4.2], whereas $b_1(S^1 \times S^3) = 1$; see also E. 53 Aj).

ii) *Godeaux surface.* Fix $\zeta := e^{2\pi i/5}$, and let g be the automorphism of order 5 in $\mathbb{P}_3$ defined by the assignment $[z_j] \mapsto [\zeta^j z_j]$, with fixed points $[1,0,0,0]$, $[0,1,0,0]$, $[0,0,1,0]$, and $[0,0,0,1]$. For the two-dimensional manifold $M := N(\mathbb{P}_3; \sum_{j=0}^{3} z_j^5)$, we have that $g(M) \subset M$, so $G := \{g^j; 0 \leqq j \leqq 4\}$ acts freely and properly discontinuously on M. The compact manifold M/G is called a Godeaux surface.

E. 32 Bd. *Grassmann manifolds are projective-algebraic.* Verify the following statements:

i) *Charts on $G_k(n)$*: The set

$$M_k(n) := \{A \in \mathbb{C}^{k \times n}; \ \mathrm{rank}\, A = k\} \subset\!\!\!\!\!\!\;\; \mathbb{C}^{k \times n}$$

is $GL(k, \mathbb{C})$-saturated (see § 33 B). Let A_j and A_j^t denote the j-th column and the j-th row of $A \in M_k(n)$, respectively, and set $A_\nu := (A_{\nu_1} \ldots A_{\nu_k})$ for each $\nu = (\nu_1, \ldots, \nu_k) \in \mathbb{N}^k$ such that

$1 \leqq \nu_1 < \ldots < \nu_k \leqq n$; then $V_\nu := \{A \in M_k(n);\ \text{rank}\, A_\nu = k\} \subset M_k(n)$ is $GL(k,\mathbb{C})$-saturated, and $M_k(n) = \bigcup_\nu V_\nu$. The mapping

$$\psi: M_k(n) \longrightarrow G_k(n),\ A \mapsto \bigoplus_{j=1}^{k} \mathbb{C}\cdot A_j^t,$$

maps V_ν onto $V_\nu / GL(k,\mathbb{C}) \subset G_k(n) = \bigcup_\nu \psi(V_\nu)$; moreover,

$$V_\nu \cong V := V_{(1,\ldots,k)} = GL(k,\mathbb{C}) \times \mathbb{C}^{k\times(n-k)}$$

and

$$W := \psi(V) \cong \mathbb{C}^{k\times(n-k)} = \{(I_k B);\ B \in \mathbb{C}^{k\times(n-k)}\}.$$

ψ is a submersion; if $\psi(A) = \psi(\tilde{A})$, then there exists a $g \in GL(k,\mathbb{C})$ such that $g\cdot A = \tilde{A}$, and it follows that $G_k(n) \cong M_k(n) / GL(k,\mathbb{C})$.

ii) *Plücker embedding:* There exists a commutative diagram

$$\begin{array}{ccc} M_k(n) & \xrightarrow{\ \varphi\ } & \Lambda^k\mathbb{C}^n\setminus 0 \\ \psi\downarrow & & \downarrow\pi \\ G_k(n) & \xrightarrow{\ \Phi\ } & \mathbb{P}_{N-1} \end{array}$$

of holomorphic mappings such that $\varphi(A) := A_1^t \wedge \ldots \wedge A_k^t$ (see §58A), $N = \binom{n}{k}$, and π is the canonical projection (hint: $\varphi(g\cdot A) = (\det g)\cdot\varphi(A)$ for $g \in GL(k,\mathbb{C})$).

The mapping Φ is injective, since if $\Phi A = \Phi\tilde{A}$, then there exists a $\mu \in \mathbb{C}$ such that $\varphi A = \mu^k \varphi\tilde{A}$, and thus also a $g \in GL(k,\mathbb{C})$ such that $A = (\mu g)\tilde{A}$.

In addition, Φ is an immersion: with the canonical basis $\{e_1,\ldots,e_n\}$ of $\mathbb{C}^n$, we have that

$$\varphi(A) = (\det A_\nu)\, e_{\nu_1} \wedge \ldots \wedge e_{\nu_k};$$

for the chart $U \cong \mathbb{C}^{N-1}$ in $\mathbb{P}_{N-1}$ distinguished by the coefficient 1 for $e_1 \wedge \ldots \wedge e_k$, the mapping

$$\Phi: W \cong \mathbb{C}^{k\times(n-k)} \longrightarrow U \cong \mathbb{C}^{N-1}$$

has components Φ_ν with $\Phi_\nu(A) = \det A_\nu$; for $j > k$,

$$\Phi_{(1,\ldots,\hat{i},\ldots,k,j)}(A) = \det \begin{pmatrix} 1 & & & & & & b_{1j} \\ 0 & \ddots & & 0 & & & \vdots \\ & \ddots & 1 & 0 & & & \vdots \\ & & 0 & 0 & & & \vdots \\ & & & 1 & \ddots & 0 & \vdots \\ & 0 & & i & \ddots & & \vdots \\ & & & & & 1 & b_{kj} \end{pmatrix} = (-1)^{k-i-1} b_{ij},$$

and it follows that $\text{rank}\, T\Phi = k(n-k)$.

Explicit quadratic equations that describe $G_k(n)$ in $\mathbb{P}_{N-1}$ are to be found in [Wh, Chap. 5, (13.7)].

E. 32 Be. Generalize the construction of Hopf surfaces to n-dimensional manifolds X_n and show that

i) X_n is homeomorphic to $S^1 \times S^{2n-1}$, and

ii) the quotient mapping $\mathbb{C}^{n*} \to X_n$ cannot be extended holomorphically to $\mathbb{C}^n$. (Here, in contrast to E. 32 g, it is not even possible to define a meromorphic extension.)

E. 32 Bf. *If the action of G on X is not properly discontinuous, then X/G need not be Hausdorff:* with $\mathbb{C}^*$ acting on the vector space $\mathbb{C}^2$ by componentwise multiplication, $\mathbb{C}^2 / \mathbb{C}^*$ is the only open set in $\mathbb{C}^2 / \mathbb{C}^*$ that contains $\overline{0}$. For the action of $\mathbb{C}^*$ on $X = \mathbb{C}^{2*}$ defined by setting $\lambda \cdot z := (\lambda z_1, \lambda^{-1} z_2)$, the points $\overline{(1,0)}$ and $\overline{(0,1)}$ in $X/\mathbb{C}^*$ cannot be separated by disjoint neighborhoods. (Of course, $X/\mathbb{C}^*$ satisfies condition 32.3 ii); such examples have led some authors to drop the condition "Hausdorff" from the definition of complex spaces. Then, 32 B. 3, for example, can be generalized to locally closed submanifolds R of $X \times X$ and not necessarily Hausdorff "manifolds" X/R.)

Proof of 32 B. 3. i) $\Rightarrow$ ii) First we note that the mapping π is open with respect to the quotient topology: the submersion pr_2 is open (32 A. 3); hence, for $U \overset{\circ}{\subset} Y$, the set $R(U) := \mathrm{pr}_2 \mathrm{pr}_1^{-1}(U)$ is open in X. It follows that $\bar{X} := X/R$ ist Hausdorff: since R is closed in $X \times X$, there exists for each $(x_1, x_2) \in (X \times X) \setminus R$ an open neighborhood $U_1 \times U_2 \overset{\circ}{\subset} (X_1 \times X_2) \setminus R$; thus, $\pi(U_1)$ and $\pi(U_2)$ are disjoint neighborhoods of the distinct points $\pi(x_1)$ and $\pi(x_2)$. If π is a submersion, then, according to 32 A. 2_{geo}, there must exist local sections $\sigma : \pi(U) \to U$ for π; then, by 32 A. 6, $M := \sigma(\pi(U))$ and $\tau := \sigma \circ \pi$ satisfy the following condition:

iii) *Each $a \in X$ lies in a locally closed submanifold $M \hookrightarrow U \overset{\circ}{\subset} X$ for which the following condition holds:*

a) There exists a surjective submersion $\tau : U \to M$ such that

$$\tau^2 = \tau \quad \text{and} \quad R_\tau{}^{8)} = R|_U := R \cap (U \times U).$$

(In particular, $\tau^2 = \tau$ implies that $M \cap R(u) = \{\tau(u)\}$ for each $u \in U$). As a first step in our proof, we want to show that i) and iii a) imply ii). From iii a), we obtain the commutative diagram

$$\begin{array}{ccc} U & \xrightarrow{\ \pi|_U\ } & \bar{U} := U/R|_U \overset{\circ}{\subset} \bar{X} \\ & {\scriptstyle\tau}\searrow \quad \swarrow{\scriptstyle\bar{\tau}} & \\ & M & \end{array},$$

in which $\bar{\tau}$ is a homeomorphism; hence, $(\bar{U}, \bar{\tau}^{-1}({}_M\mathcal{O}))$ is a manifold, and $\pi|_U$ is a submersion. We can use $\bar{\tau} : \bar{U} \to M$ as a chart for $\bar{X}$ once we have shown that the structure sheaf $\bar{\tau}^{-1}({}_M\mathcal{O})$ does not depend on M or τ; by 32 A. 3 ii c), we need only verify that $\pi : R(U) \to \bar{U}$ is a submersion. In the commutative diagram

$$\begin{array}{ccc} & R \cap (U \times R(U)) & \\ {\scriptstyle \mathrm{pr}_1}\swarrow & & \searrow{\scriptstyle \mathrm{pr}_2} \\ U & & R(U) \\ {\scriptstyle \pi|_U}\searrow & & \swarrow{\scriptstyle \pi|_{R(U)}} \\ & \bar{U} & \end{array},$$

since $\pi|_U \circ \mathrm{pr}_1$ is a submersion, $\pi|_{R(U)} \circ \mathrm{pr}_2$ is a submersion as well; by 32 A. 3, since pr_2 is a submersion, $\pi|_{R(U)}$ must be holomorphic, and by E. 32 Ad, it also is a submersion.

Now it suffices to show that i) implies iii a). To that end, we replace iii a) with the condition

b) For $R|_U \xrightarrow{\mathrm{pr}_2} U$, there exists a holomorphic section σ through (a, a) such that $\sigma(U) = R \cap (M \times U) =: Y$;

that can be obtained from a) with $\sigma = (\tau, \mathrm{id}_U)$. By E. 32 Ac, $Y = R|_U \cap \mathrm{pr}_1^{-1}(M)$ is a sub-

---

8) $u \sim_{R_\tau} v$ iff $\tau(u) = \tau(v)$.

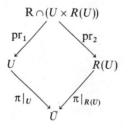

manifold of $R|_U$, since pr_1 is a submersion. To see that b) implies a), note that $\sigma = (\tau, \mathrm{id}_U)$ with $\tau \in \mathrm{Hol}(U, M)$; since pr_1 is a submersion and σ is biholomorphic (by 32 A.6), it follows that $\tau = \mathrm{pr}_1 \circ \sigma$ is a submersion; moreover, τ is surjective (since $(u, u) \in Y$ for each $u \in M$), and with the equality $\sigma \circ \mathrm{pr}_2|_Y = \mathrm{id}_Y$, we see that $\tau^2 = \tau$: since

$$(u, u) = \sigma \circ \mathrm{pr}_2(u, u) = \sigma(u) = (\tau(u), u), \ u \in M,$$

it follows that $\tau|_M = \mathrm{id}_M$; to verify that $R_\tau = R|_U$, we show that, for $u, v \in U$,

$$u \sim_R v \underset{\alpha)}{\Leftrightarrow} \tau(u) \sim_R \tau(v) \underset{\beta)}{\Leftrightarrow} \tau(u) = \tau(v):$$

α) follows from the fact that $u \sim_R \tau(u)$, which holds because $\sigma(u) = (\tau(u), u) \in Y \subset R$; for β), we observe that $\tau(U) \subset M \subset U$, and thus that

$$\tau u \sim_R \tau v \Rightarrow (\tau u, \tau v) \in Y \Rightarrow (\tau u, \tau v) = \sigma \circ \mathrm{pr}_2(\tau u, \tau v) = \sigma(\tau v) = (\tau^2 v, \tau v) = (\tau v, \tau v).$$

By 32 A.6, b) is equivalent to

c) $\mathrm{pr}_2 : Y \to U$ *is biholomorphic*; by 21.9, after sufficient shrinking of U, c) can be replaced with

d) ${}_X\mathcal{O}_a \xrightarrow{\mathrm{pr}_2^0} {}_Y\mathcal{O}_{(a,a)}$ *is an isomorphism*; finally, by 32.13, d) is equivalent to

e) $T_{(a,a)}\mathrm{pr}_2 : T_{(a,a)}Y \to T_aX$ *is an isomorphism.*

It remains to show that i) implies iii e); to that end, we first choose U and M: for $S := R(a)$, it follows from E.32 Ac that $\mathrm{pr}_1^{-1}(a) = a \times S \hookrightarrow R \hookrightarrow X \times X$ is a submanifold, and thus that $S \hookrightarrow X$ is a submanifold; hence, there exists a locally closed manifold M such that $a \in M \hookrightarrow U \Subset X$ and $U \cong M \times (S \cap U)$ (see 8.8). With $\dot{X}$, $\dot{\mathrm{pr}}_j$, etc., denoting T_aX, $T_{(a,a)}\mathrm{pr}_j$, etc., and $\dot{\Delta} \hookrightarrow \dot{M} \times \dot{M}$ denoting the diagonal, pr_2 clearly induces isomorphisms

$$0 \times \dot{S} \to \dot{S} \quad \text{and} \quad \dot{\Delta} \to \dot{M}.$$

By construction, $\dot{M} \oplus \dot{S} = \dot{X}$, so the sum $0 \times \dot{S} + \dot{\Delta}$ is direct in $\dot{R}$, and pr_2 induces an isomorphism

$$\dot{\mathrm{pr}}_2 : (0 \times \dot{S}) \oplus \dot{\Delta} \to \dot{X}.$$

To obtain e), it remains only to verify that

$$\dot{Y} = (0 \times \dot{S}) \oplus \dot{\Delta};$$

since $0 \times \dot{S} = \mathrm{Ker}\,\dot{\mathrm{pr}}_1$ that follows from the fact that $Y = \mathrm{pr}_1^{-1}(M)$ by transfer to the tangent spaces.

ii) $\Rightarrow$ i) By E. 32Ac, $R = (\pi \times \pi)^{-1}(\Delta_{\bar{X}})$ (see 49 A. 6 v)) is a submanifold of $X \times X$, for $\Delta_{\bar{X}}$ is a submanifold and $\pi \times \pi : X \times X \to \bar{X} \times \bar{X}$ is a submersion, since π is. By 32 A. 2_{geo}, $\mathrm{pr}_1 : R \to X$ is a submersion: for $(x_1, x_2) \in R$, there exists a neighborhood $\bar{U} \Subset \bar{X}$ of $\pi(x_1) = \pi(x_2)$ with a section $h : \bar{U} \to \pi^{-1}(\bar{U})$ through x_2. Then $\pi^{-1}(\bar{U}) \to R, x \mapsto (x, h(\pi(x)))$, is a holomorphic section for pr_1 through (x_1, x_2). ■

Now we carry over from topology a method of construction of complex structures that is called "*surgery on complex spaces*": let complex spaces $A \hookrightarrow X$, $A' \hookrightarrow U'$, an open neighborhood U of A in X, and a biholomorphic mapping $\varphi : U \backslash A \to U' \backslash A'$ be given; then we define a topological space

$$X' := (X \backslash A) \cup_\varphi U' := [(X \backslash A) \dot{\cup} U'] / \sim$$

by means of the equivalence relation $U \backslash A \ni u \sim \varphi(u) \in U' \backslash A'$. There clearly exists exactly one complex structure on X' such that U' and $X \backslash A$ are open subspaces in a canonical fashion (thus X' is formed from X by "replacing" A with A').

32 B. 10 Examples of surgery. i) *Changing the complex structure of a curve*: Fix $A = \{0\} \subset$

$U \Subset \mathbb{C} = X$, and let $\varphi: U \to U'$ be a holomorphic homeomorphism onto an analytic set $U' \hookrightarrow G \subset \mathbb{C}^n$ with $A' = S(U') = \varphi(0) = 0$ (e.g., $U' = N(\mathbb{C}^2; z_1^2 - z_2^3)$ (*Neil's parabola*) and $\varphi(t) = (t^3, t^2)$ (E. 321 ii)), or $U' = N(\mathbb{C}^3; z_1 z_3 - z_2^2, z_1^4 - z_2^3, z_1^5 - z_3^3)$ and $\varphi(t) = (t^3, t^4, t^5)$ (46.3)). Then surgery on X yields a structure sheaf ${}_{X'}\mathcal{O}$ with ${}_{X'}\mathcal{O}|_{X\setminus\{0\}} \cong {}_1\mathcal{O}|_{X\setminus\{0\}}$ and ${}_{X'}\mathcal{O}_0 = {}_{U'}\mathcal{O}_0$ ("bending" a singularity into the curve X); conversely, if we interpret that operation as surgery on U', then it represents a "smoothing" of the singularity of U'. A combination of these two processes makes possible an "exchange" of singularities of curves.

ii) *Resolution of the remaining singularities of* E. 32i:

α) For $N(z_1 z_2)$, see E. 32 Bhvγ), or 71.14 and 74.4.

β) For *Whitney's umbrella*, see E. 49 Am ii).

γ) For $\sqrt{z_1 z_2}$, consider $X = U = N(\mathbb{C}^3; z_1 z_2 - z_3^2)$ and $A := \{0\} = S(X)$. The coordinate transformation

$$\chi: \mathbb{C}^2 \setminus \{v = 0\} \to \mathbb{C}^2 \setminus \{y = 0\}, (u, v) \mapsto (uv^2, 1/v) =: (x, y)$$

determines a manifold $U' = \mathbb{C}^2 \cup_\chi \mathbb{C}^2$. The assignments $\pi(u, v) = (u, uv^2, uv)$ and $\pi(x, y) = (xy^2, x, xy)$ determine a proper holomorphic mapping $\pi: U' \to X$ whose restriction $\pi: U' \setminus \pi^{-1}(0) \to X \setminus \{0\}$ is biholomorphic. Thus $X' := (X \setminus \{0\}) \cup_{\pi^{-1}} U' = U'$ is obtained from X by cutting out the singularity and inserting the "exceptional set" $A' = \pi^{-1}(0) \cong \mathbb{P}_1$.

iii) *The Hirzebruch surfaces* Σ_m. Let X be $\mathbb{P}_1 \times \mathbb{P}_1$. In homogeneous coordinates $(z, \zeta) = ([z_0, z_1], [\zeta_0, \zeta_1])$, set

$$A := A' := \mathrm{N}(\mathbb{P}_1; z_1) \times \mathbb{P}_1 \quad \text{and} \quad U := U' := \{(z, \zeta) \in X; z_0 = 1\}.$$

For $m \in \mathbb{N}$, the mapping

$$\varphi_m: U \setminus A \to U' \setminus A', (z, \zeta) \mapsto (z, [\zeta_0 z_1^m, \zeta_1 z_0^m]),$$

is biholomorphic. Thus $\Sigma_m := (X \setminus A) \cup_{\varphi_m} U'$ is, in a natural manner, a nonsingular compact manifold. It is shown in [Hi] that Σ_m is homeomorphic to $\Sigma_0 = \mathbb{P}_1 \times \mathbb{P}_1$ for m even, and to Σ_1 for m odd (computation of "intersection-numbers" proves that Σ_0 and Σ_1 are not homeomorphic); for $l \neq m$, the surfaces Σ_l and Σ_m are not biholomorphically equivalent.

E. 32 Bg. Prove the following statements for Σ_m:

α) The projection $\mathrm{pr}_1: \Sigma_m \to \mathbb{P}_1$ onto the first component is a proper holomorphic submersion with fibers $\mathbb{P}_1$.

β) $\Sigma_m \cong \{([x_0, x_1, x_2], [y_0, y_1]) \in \mathbb{P}_2 \times \mathbb{P}_1; x_1 y_0^m = x_2 y_1^m\} \hookrightarrow \mathbb{P}_2 \times \mathbb{P}_1$.
(Hint: construct $\psi: \Sigma_m \to \mathbb{P}_2 \times \mathbb{P}_1$ by means of the assignment

$$(z, \zeta) \mapsto \begin{cases} ([\zeta_0 z_1^m, \zeta_1 z_1^m, \zeta_1 z_0^m], z) & \text{on } X \setminus A \\ ([\zeta_0 z_0^m, \zeta_1 z_1^m, \zeta_1 z_0^m], z) & \text{on } U' \end{cases} \quad .)$$

γ) Interpret $\Sigma_1 \hookrightarrow \mathbb{P}_2 \times \mathbb{P}_1 \xrightarrow{\mathrm{pr}_1} \mathbb{P}_2$ as a quadratic transformation (see 32 B.10iv)) of $\mathbb{P}_2$ at $[1, 0, 0]$.

32 B.10. iv) *Quadratic transformation* of a complex surface X: A nonsingular point a of X is to be replaced by the set of all directions of curves through a (that set forms a copy of $\mathbb{P}_1$!): without loss of generality, set $A = \{0\} \hookrightarrow U \Subset \mathbb{C}^2 = X$. Now, by 32.8,

$$U' = \{(z_1, z_2; [\zeta_1, \zeta_2]) \in U \times \mathbb{P}_1; z_1 \zeta_2 = z_2 \zeta_1\}$$

is a submanifold of $U \times \mathbb{P}_1$. Letting $\pi: U' \to U$ denote the restriction of the canonical projection, we see that $\pi^{-1}(0) = 0 \times \mathbb{P}_1 =: A'$; moreover, $\pi: U' \setminus A' \to U \setminus A$ is biholomorphic, due to the fact that

$$\pi^{-1}(z_1, z_2) = (z_1, z_2; [z_1, z_2])$$

if $z_1 \neq 0$ or $z_2 \neq 0$. One says that $X' = X \setminus \{0\} \cup_{\pi^{-1}} U'$ arises from X through a *quadratic transformation* at 0; one also refers to the "*blowing up*" of the point 0 in X.

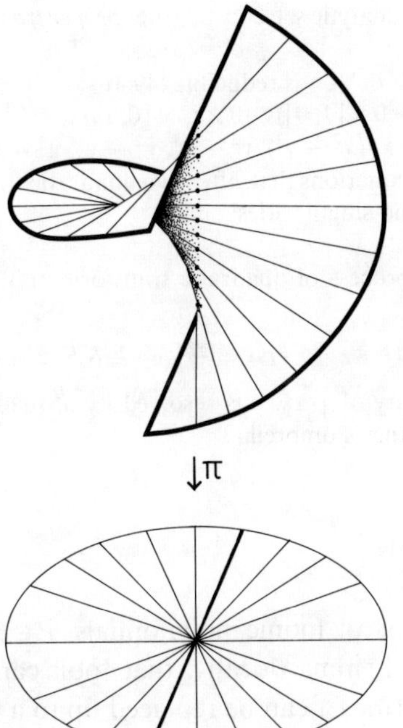

That operation is of fundamental importance for the resolution of singularities:

E. 32 Bh. Prove the following statements for a quadratic transformation:

i) The projection $\pi : U' \to U$ is proper.

ii) *Representation in local coordinates*: For $j = 1, 2$, let the mappings $\psi_j : \mathbb{C}^2 \to \mathbb{C}^2$ be defined by setting

$$\psi_1(x_1, y_1) := (x_1, x_1 y_1) \quad \text{and} \quad \psi_2(x_2, y_2) := (x_2 y_2, y_2),$$

and put $W_j := \psi_j^{-1}(U)$. Then the following statements hold for $U_j := \{(z_1, z_2) \in \mathbb{C}^2; z_j \neq 0\}$:

α) The restrictions $\psi_j : \psi_j^{-1}(U_j) \to U_j$ are biholomorphic, and the coordinate transformation $\psi_2^{-1}\psi_1 : \psi_1^{-1}(U_1) \to \psi_2^{-1}(U_2)$ is given by the assignment $(x_1, y_1) \mapsto (1/y_1, x_1 y_1)$.

β) The mappings $\varphi_j : W_j \to U'$ defined by

$$\varphi_1(x_1, y_1) := (\psi_1(x_1, y_1), [1, y_1]) \text{ and } \varphi_2(x_2, y_2) := (\psi_2(x_2, y_2), [x_2, 1])$$

determine a commutative diagram

$$\begin{array}{ccc} W_1 \cup_{\psi_2^{-1}\psi_1} W_2 & \xrightarrow{\varphi} & U' \\ & \psi \searrow \quad \swarrow \pi & \\ & U & \end{array}$$

in which φ is biholomorphic.

iii) The graph of the canonical projection $\tau|_{U^*}: U^* \to \mathbb{P}_1$ is $\pi^{-1}(U^*)$; that set is open and dense in U' (see also E. 32 g).

iv) The projection $\mathrm{pr}_2: U' \to \mathbb{P}_1$ is holomorphic; determine its fibers $\mathrm{pr}_2^{-1}[\zeta_1, \zeta_2]$.

v) Let $C = N(U; f)$ be a "curve" with $0 \in C$.

α) There is precisely one analytic set $\hat{C} \subset U'$ (the "*proper transform*" of C) such that $\pi: \hat{C} \to C$ is finite.

β) The "*total transform*" $\pi^{-1}(C)$ is reducible (49.4).

γ) For $f = xy$, $\hat{C} = (C \times 0 \times [1, 0]) \cup (0 \times C \times [0, 1]) \subset U'$.

δ) Determine $\hat{C}$ for $f = xy(x^2 - y^2)$, $y^2 - x^3$, $y^2 - x^5$, $y^3 - x^5$.

ε) Show for the last two equations that, after two quadratic transformations at the singularity, the proper transform has no singularities.

E. 32 Bi. i) Generalize the process of quadratic transformation for n-dimensional X with the help of

$$U' = \{(z, \zeta) \in \boldsymbol{P}^n(1) \times \mathbb{P}_{n-1};\ z_j \zeta_k = z_k \zeta_j,\ 1 \leqq j,\ k \leqq n\}.$$

ii) Show that the singularity of $\sqrt{z_1 z_2}$ is resolved by a quadratic transformation at $0 \in \mathbb{C}^3$. Does that also hold for Whitney's umbrella?

§ 33 Zeros of Polynomials

The behavior of the zeros of monic polynomials $P \in \mathcal{O}(G)[T]$ was characterized *algebraically* in Hensel's Lemma (locally, that took care of the general case, since every nonconstant $f \in \mathcal{O}(G \times \mathbb{C})$ can be replaced up to a unit factor by a Weierstrass polynomial in appropriate local coordinates by the Preparation Theorem). Now we want to describe the *geometry* of $N(P)$. To that end, we generalize the classical *Theorem on the Continuity of Roots* (simple zeros of a polynomial depend continuously on its coefficients) in 33.1, presenting the behavior at multiple zeros at the same time. The projection π appearing there is the local prototype of a finite holomorphic mapping (§ 33 B).

33.1 Theorem. *For $G \Subset \mathbb{C}^n$ and a nonconstant monic polynomial $P \in \mathcal{O}(G)[T]$, provide $X := N(G \times \mathbb{C}; P)$ with the reduced complex structure. Then the canonical projection $\pi: X \to G$ is a branched covering, i.e.,*

i) it is a surjective finite open holomorphic mapping, and

ii) there exists a proper analytic subset A of G such that $\pi: X \backslash \pi^{-1}(A) \to G \backslash A$ is a finite (unbranched holomorphic) covering, $X \backslash \pi^{-1}(A)$ is dense in X, and $G \backslash A$ is dense in G.

Proof. i) π is *surjective* and *finite*: fix

$$P(z, T) = \sum_{j=0}^{b-1} f_j(z) T^j + T^b;$$

for $a \in G$, by the Fundamental Theorem of Algebra, the set $\pi^{-1}(a) = \{(a, t); P(a, t) = 0\}$ contains at least one and at most b elements, so π is discrete and surjective. For each

$(a, t) \in X$ we see (E. 33a) that

$$|t| \leqq \max\{1, \sum_{j=0}^{b-1} |f_j(a)|\} =: h(a).$$

For a compactum $K \subset G$, we have that $\|h\|_K < \infty$. Since X is closed in $G \times \mathbb{C}$, it follows that $\pi^{-1}(K)$, as a closed subset of the compact set $K \times \boldsymbol{P}^1(\|h\|_K)$, is compact; hence, $\pi: X \to G$ is proper. In addition, π is open: it suffices to show that π *is open at each point* $a \in X$, i.e., that π maps each neighborhood of a in X onto a neighborhood of $\pi(a)$ in G; without loss of generality, suppose that $a = (0, 0)$. We apply the factorization from the Preparation Theorem, $P = e\omega$, which holds near $(0, 0)$; there, $N(P) = N(\omega)$, so we may assume that P is a Weierstrass polynomial. Thus, we have that $\pi^{-1}(0) = (0, 0)$. By 33 B. 1, since π is proper, the open sets $\pi^{-1}(V)$ with $0 \in V \subset\subset G$ form a fundamental system of neighborhoods of $(0, 0)$ in X; since π is surjective, $\pi(\pi^{-1}(V)) = V$ for each such V, so the images are neighborhoods of 0 in G.

ii) Without loss of generality, let G be connected. Then $\mathcal{O}(G)[T]_{\text{mon}}$ is factorial, by 23.9; thus, P has a factorization $P = P_1^{\nu_1} \ldots P_m^{\nu_m}$ into powers of distinct irreducible monic polynomials. Since we are considering the reduced structure on $X = N(P) = N(P_1 \ldots P_m) = \bigcup_{j=1}^{m} N(P_j)$, we may assume that $\nu_j = 1$ for every j. Then the following statements hold:

a) For each $j \neq k$, there exists an analytic set $A_{jk} \subsetneqq G$ with $N(P_j) \cap N(P_k) \subset \subset \pi^{-1}(A_{jk})$ (by 33.2).

b) For each j, there exists an analytic set $B_j \subsetneqq G$ such that, for each $a \in G \backslash B_j$, the polynomial $P_j(a, T) \in \mathbb{C}[T]$ has only simple zeros (by 33.3).

Now ii) follows easily: $A := (\bigcup_{j \neq k} A_{jk}) \cup (\bigcup_j B_j)$ is a proper analytic subset of G, and the analytic set $\pi^{-1}(A)$ is nowhere dense in X (by 33.4 and 7.4). For $a \in G \backslash A$, P has exactly $b := \deg P$ distinct zeros (every ν_j is 1); thus, all of those zeros are simple and $\frac{\partial P}{\partial T}(a, t) \neq 0$ for $(a, t) \in \pi^{-1}(G \backslash A)$. It follows from 8.9 that $\pi^{-1}(G \backslash A) = X \backslash \pi^{-1}(A)$ is a submanifold of $(G \backslash A) \times \mathbb{C}$; by 8.6, $\pi: X \backslash \pi^{-1}(A) \to G \backslash A$ is locally biholomorphic. Thus, $\pi: X \backslash \pi^{-1}(A) \to G \backslash A$ is an (unbranched holomorphic) covering of degree b. ∎

E. 33 a. For the zeros w of a monic polynomial $\sum_{j=0}^{b} a_j T^j \in \mathbb{C}[T]$, prove that

$$|w| \leqq \max(1, \sum_{j=0}^{b-1} |a_j|).$$

Now we prove the lemmata used above:

33.2 Lemma. *If the polynomials* $P, Q \in \mathcal{O}(G)[T]_{\text{mon}}$ *are relatively prime, then the set*

$$A := \{z \in G;\ \exists t \in \mathbb{C},\ P(z,t) = 0 = Q(z,t)\}$$

is a proper analytic subset of G.

Proof. By applying 33 A.2 with $R = \mathbb{C}$, we see that, for $a \in G$, $a \in A$ iff $\operatorname{Res}(P,Q)(a) \underset{33\text{A.1 ii)}}{=} \operatorname{Res}(P(a,T), Q(a,T)) = 0 \in \mathbb{C}$; hence, $A = N(G; \operatorname{Res}(P,Q))$. Application of 33A.3 to the normal ring $R = \mathcal{O}(G)$ (see 23.9) yields that $\operatorname{Res}(P,Q) \neq 0 \in \mathcal{O}(G)$. Thus, A is a proper analytic subset. ■

33.3 Lemma. *For an irreducible monic polynomial $P \in \mathcal{O}(G)[T]$, the set*

$$A := \{z \in G;\ P(z,T) \text{ has a multiple zero}\}$$

is a proper analytic subset of G.

Proof. By 33 A.4, $0 \neq \operatorname{Dis}(P) \in \mathcal{O}(G)$. By applying 33 A.1 i) to the evaluation ε_z for $z \in G$, we see that $A = N(G; \operatorname{Dis}(P))$. ■

33.4 Lemma. *If $\varphi: X \to Y$ is an open mapping of topological spaces, and if W is a dense subset of Y, then $\varphi^{-1}W$ is dense in X.*

Proof. If the assertion did not hold, then there would exist a $U \subset\!\subset X$ such that $U \subset X \backslash \varphi^{-1}W$. Thus the open set $\varphi(U)$ would be contained in $Y \backslash W$, so that W could not be dense in Y. ↯ ■

The algebraic factorization of a monic polynomial $P \in \mathcal{O}(G)[T]$ into powers of irreducible polynomials corresponds to a geometric partition of $X = N(P)$ into "*irreducible components*" (the closures of the connected components of $X \backslash \pi^{-1}(A)$ (see 33.1), which are investigated systematically in § 49):

33.5 Proposition. *Let $G \subset\!\subset \mathbb{C}^n$ be a domain, and let $P \in \mathcal{O}(G)[T]$ be a monic polynomial. If $P_1, \ldots, P_m$ are the prime factors of P, then $N(G \times \mathbb{C}; P) = \bigcup_{j=1}^{m} N(G \times \mathbb{C}; P_j)$ is the partition into pairwise distinct irreducible components.*

The proof is based on the following fact:

33.6 Lemma. *Consider a domain $G \subset\!\subset \mathbb{C}^n$, an analytic subset $A \subsetneqq G$,* the domain *$G' := G \backslash A$, and a submanifold Y' of $G' \times \mathbb{C}$. If*

i) the projection $\pi: Y' \to G'$ is an (unbranched holomorphic) covering of degree b, and

ii) $Y' \cap (K \times \mathbb{C})$ is bounded for every $K \subset\!\subset G$, then

a) there exists a unique polynomial $Q \in \mathcal{O}(G)[T]_{\text{mon}}$ of degree b with $N(G \times \mathbb{C}; Q) = \overline{Y'}$;

b) for every $P \in \mathcal{O}(G)[T]_{\text{mon}}$, Q divides P iff $N(G \times \mathbb{C}; Q) \subset N(G \times \mathbb{C}; P)$; and

c) Q is irreducible iff Y' is connected.

Proof of 33.5. As in the proof of 33.1, we may suppose that $P = \prod_{j=1}^{m} P_j$. For the polynomial P, we determine A according to 33.1, and thus $G' := G \backslash A$ and $Y' := N(G' \times \mathbb{C}; P)$ as well. Let X_k' be a connected component of Y', and let X_k be its closure in $N(G \times \mathbb{C}; P)$. Then $X_k' \to G'$ is a covering, since $Y' \to G'$ is; by 33.1, $X_k \to G$ is finite, so the hypotheses of 33.6 are satisfied for X_k' in place of Y'. If Q_k is the polynomial in a) determined uniquely by $N(G \times \mathbb{C}; Q_k) = X_k$, then all of the Q_k are distinct; by b) and c), they are prime factors of P, so $\prod Q_k$ divides P. To show that $P = \prod Q_k$, and thus that $P_j = Q_j$ for appropriate indices, it suffices to show that $\deg P = \deg \prod Q_k$. In the proof of 33.1 ii), it was shown that the degree of a polynomial without multiple factors is equal to the degree of the associated covering. Thus, it follows that

$$\deg P = \sum_k \deg Q_k = \deg \prod_k Q_k .$$

Therefore, $N(P) = N(Q_1 \cdot \ldots \cdot Q_n) = \bigcup_j N(Q_j) = \bigcup_j X_j$. ■

Proof of 33.6. The (unbranched holomorphic) coverings $Y' \to G'$ of degree b can be identified with the monic polynomials $Q \in \mathcal{O}(G')[T]$ of degree b without multiple zeros (i.e., for $z \in G'$, $Q(z, T) \in \mathbb{C}[T]$ has no multiple zero): Y' corresponds to the polynomial $Q(g, T) := \prod_{\substack{z \\ (g,z) \in Y'}} (T - z) = T^b - \sigma_1(g) T^{b-1} + \ldots \pm \sigma_b(g)$, $g \in G'$ in which $\sigma_j(g) \in \mathbb{C}$ is the value of the j-th elementary symmetric function of the b points of $Y' \cap (\{g\} \times \mathbb{C})$. Since $\pi: Y' \to G'$ is locally biholomorphic, every σ_j must be holomorphic, and it follows that $Q \in \mathcal{O}(G')[T]_{\text{mon}}$. Conversely, it was shown in 33.1 that each Q yields a covering of G' of degree $\deg Q$.

a) We need to see that the polynomial Q determined by the covering π has an extension in $\mathcal{O}(G)[T]_{\text{mon}}$ (as a continuous function, the extension is unique, since $G' \times \mathbb{C}$ is dense in $G \times \mathbb{C}$). Every σ_j is locally bounded at A: for $K \subset\subset G$, by ii), $Y' \cap (K \times \mathbb{C})$ is bounded; i.e., every σ_j is bounded on $K \cap G'$. By the First Removable Singularity Theorem, every σ_j can be extended holomorphically to G; hence, Q has a holomorphic extension to G as well. Finally, since $\pi: N(Q) \to G$ is open, it follows from 33.4 with $W = G'$ that $\overline{Y'} = N(G \times \mathbb{C}; Q)$.

c) *Only if.* If Y' is not connected, then Y' has a decomposition $Y' = Y_1' \cup Y_2'$ such that $\pi: Y_j' \to G'$ is a covering for $j = 1, 2$, with $Y_1' \cap Y_2' = \emptyset$; by a), each of these two coverings corresponds to a polynomial $Q_j \in \mathcal{O}(G)[T]_{\text{mon}}$. By construction, $Q = Q_1 Q_2$ in $\mathcal{O}(G')[T]$; by virtue of continuity, that equality holds in $\mathcal{O}(G)[T]$ as well.

If. For a nontrivial factorization $Q = Q_1 \cdot Q_2$ in $\mathcal{O}(G)[T]_{\text{mon}}$, and $Y_j' := N(G' \times \mathbb{C}; Q_j)$, we have that $\pi: Y_j' \to G'$ is a covering, and it follows that each Y_j' is a union of connected components of Y' (see § 33 B). Since Q has no multiple zeros in G', we see that $Y_1' \cap Y_2' = \emptyset$; hence, Y' is not connected.

b) *If.* Without loss of generality, suppose that P has no multiple prime factors and that Q is prime, i.e., Y' is connected (by c)). If $Y' \subset X' := N(G' \times \mathbb{C}; P)$, then Y' is a connected component of X', since $\pi: Y' \to G'$ is a covering. For $Z' := X' \setminus Y'$ and the associated polynomial R determined in a), $P = Q \cdot R$ over G', and thus over G as well. ■■

With those methods of proof and Remmert's Mapping Theorem 45.17, we obtain the following (see also E. 49o):

33.7 Proposition. *Let $\varphi : X \to Y$ be a holomorphic finite open mapping between connected manifolds; then*

i) φ is a branched covering, and

ii) $\mathcal{O}(X)$ is integral over $\mathcal{O}(Y)$; i.e., for each $f \in \mathcal{O}(X)$ there exists a monic polynomial $Q_f \in \mathcal{O}(Y)[T]$ of minimal degree such that $Q_f(f) = 0$. Moreover, for each $f \in \mathcal{O}(X)$, $\deg Q_f \leqq$ degree of φ.

Proof. i) First, we note that φ is surjective, for it is open and finite, and Y is connected. By E. 8f, there exists a proper analytic subset A of X such that φ is locally biholomorphic on $X' := X \setminus \varphi^{-1}(\varphi(A))$ (by 45.17, $\varphi(A)$ is analytic). By E. 33 Ba, then, $\varphi|_{X'}: X' \to Y' := Y \setminus \varphi(A)$ is a covering.

ii) As in the proof of 33.6, the assignment $Q(y, T) := \prod_{x \in \pi^{-1}(y)} (T - f(x))$ on Y' extends to a polynomial $Q \in \mathcal{O}(Y)[\mathrm{T}]_{\mathrm{mon}}$ such that $Q(f) = 0$; for the degree, see E. 33 Bc. ■

The number of irreducible components of $N(G; P)$, and, correspondingly, the number of prime factors of P in the factorial monoid $\mathcal{O}(G)[T]_{\mathrm{mon}}$ (see 23.9) depends on the domain G, as the example of Whitney's umbrella shows:

33.8 Example. The following statements hold for $G \Subset \mathbb{C}^2$ and $P := T^2 - z_1^2 z_2 \in \mathcal{O}(G)[T]$:

i) If $\sqrt{z_2}$ determines a holomorphic function on G (e.g., if G is included in $\{(z_1, z_2) \in \mathbb{C}^2;\ z_2 \notin \mathbb{R}_{\geq 0}\}$), then P is reducible:

$$P = (T + z_1\sqrt{z_2})(T - z_1\sqrt{z_2}).$$

ii) Otherwise (then $\sqrt{z_2}$ is many-valued), P is irreducible. ■

E. 33 b. i) Interpret and prove 33.8 geometrically.

ii) Use i) to show that P is irreducible in ${}_2\mathcal{O}_0[T]$.

E. 33 c. For $G \Subset \mathbb{C}^n$ and $P \in \mathcal{O}(G)[T]_{\mathrm{mon}}$, fix $X := N(G \times \mathbb{C}; P)$, and let $\pi : X \to G$ be the projection. Prove the following statements:

i) For $U \Subset \mathbb{C}$, the restriction $\pi : X \cap (G \times U) \to G$ is finite iff $\overline{X \cap (G \times U)} \cap (G \times \partial U) = \emptyset$ ("X doesn't leave $G \times U$ through $G \times \partial U$").

ii) For each $a \in X$, there exist arbitrarily small polydisks $\boldsymbol{P} = \boldsymbol{P}^n \times \boldsymbol{P}^1$ with center a such that $\pi : X \cap \boldsymbol{P} \to \boldsymbol{P}^n$ is finite (hint: Proof of 33 B. 2. ii); for 45.2).

Now we bring the section to a close with a warning: real representations of complex zero-sets lead easily to misconceptions. For example, consider the zero-sets $X_{\mathbb{C}} \hookrightarrow \mathbb{C}^2$ of the polynomials

$$\text{i) } z_1^2 + z_2 - z_2^3, \quad \text{ii) } z_1 z_2, \quad \text{and iii) } z_1 - z_2^2.$$

They are connected; moreover, i) and iii) are irreducible (see 33.5 and 23.7). However, the real portions $X_{\mathbb{R}} := X_{\mathbb{C}} \cap \mathbb{R}^2$ have the following forms:

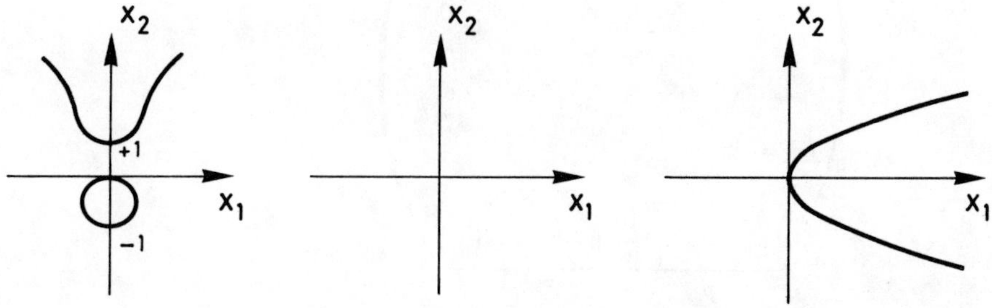

Note that

in i), $X_{\mathbb{R}}$ is disconnected, but $X_{\mathbb{C}}$ is connected;

in ii), $X_{\mathbb{R}} \backslash 0$ has four connected components, while $X_{\mathbb{C}} \backslash 0$ has only two;

in iii), there are no points of $X_{\mathbb{R}}$ lying over $x < 0$, although there are two points of $X_{\mathbb{C}}$ lying over each such x; hence, one might be inclined to indicate the projection $\mathrm{pr}_1 : X_{\mathbb{C}} \to \mathbb{C}$ as follows:

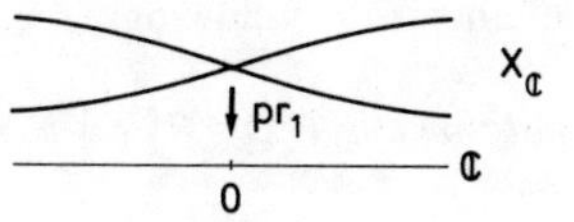

However, that suggests that $X_{\mathbb{C}} \backslash \mathrm{pr}_1^{-1}(0)$ has four connected components, although it is connected.

§ 33A Supplement: Resultants and Discriminants

The resultant of two polynomials serves to test the existence of common factors: Let

$$P = \sum_{j=0}^{p} r_j T^j \quad \text{and} \quad Q = \sum_{k=0}^{q} s_k T^k \in R[T]$$

be polynomials with degrees p and q, respectively. Then the homomorphism of R-modules

(33 A.1.1) $\varphi : R[T]_p \oplus R[T]_q \to R[T]_{p+q}, \ (f, g) \mapsto Qf + Pg,$

has the following matrix-representation with respect to the canonical basis $\{1, T, \ldots, T^{m-1}\}$ of $R[T]_m$:

$$\varphi = \left(\begin{array}{cccccccc} s_0 & 0 & \cdots & 0 & r_0 & 0 & \cdots & 0 \\ s_1 & s_0 & \ddots & \vdots & r_1 & r_0 & \ddots & \vdots \\ \vdots & & \ddots & 0 & \vdots & & \ddots & 0 \\ \vdots & & & s_0 & \vdots & & & r_0 \\ s_q & \cdots & \cdots & s_1 & r_p & \cdots & \cdots & r_1 \\ 0 & \ddots & & \vdots & 0 & \ddots & & \vdots \\ \vdots & & \ddots & \vdots & \vdots & & \ddots & \vdots \\ 0 & \cdots & 0 & s_q & 0 & \cdots & 0 & r_p \end{array}\right)$$
$$\underbrace{\qquad\qquad\qquad}_{p}\quad\underbrace{\qquad\qquad\qquad}_{q}$$

The *resultant* of P and Q is defined to be

$$\operatorname{Res}(P, Q) := \det\varphi \in R.$$

33 A. 1 Remarks. i) *If* $\vartheta: R \to S$ *is a ring-homomorphism, then*

$$\vartheta\operatorname{Res}(P, Q) = \operatorname{Res}(\vartheta P, \vartheta Q).$$

ii) $\operatorname{Res}(P, Q) = (-1)^{pq}\operatorname{Res}(Q, P)$, *and, for each* $r \in R$, $\operatorname{Res}(rP, Q) = r^q\operatorname{Res}(P, Q)$. ■

The central fact concerning resultants is the following:

33 A. 2 Proposition. *If one of the polynomials* $P, Q \in R[T]$ *is monic, then the following statements are equivalent:*

i) $\operatorname{Res}(P, Q)$ *is a unit in* R.

ii) $R[T] = R[T]\cdot P + R[T]\cdot Q$.

Proof. In the notation of (33 A.1.1), statement i) holds iff φ is a bijection. Since there exists a right inverse for each *surjective* homomorphism χ between free R-modules, $\det\chi$ must be a unit in R; consequently, i) is equivalent to the following statement:

iii) $\qquad R[T]_{p+q} = R[T]_p \cdot Q + R[T]_q \cdot P.$

Without loss of generality, let P be monic.

ii) $\Rightarrow$ iii) Every $h \in R[T]_{p+q}$ has a representation $h = fQ + gP$ with $f, g \in R[T]$ and $\deg f < p$, by the Euclidean Algorithm. From the inequality

$$\deg(gP) \leqq \max(\deg h, \deg(fQ)) < p + q,$$

it follows that $\deg g < q$, and thus that h lies in $R[T]_p \cdot Q + R[T]_q \cdot P$.

iii) $\Rightarrow$ ii) Applying the Euclidean Algorithm again, we see that each $h \in R[T]$ has a representation $h = gP + k$ with $\deg k < p \leqq p + q$. Thus, since k lies in $R[T]\cdot(P, Q)$, so does h. ■

33 A. 3 Corollary. *For a normal ring* R *of characteristic* 0 *and polynomials* $P, Q \in R[T]$, *where* P *is monic,* $\operatorname{Res}(P, Q) \neq 0$ *iff* P *and* Q *are relatively prime.*

Proof. Since a normal ring is an integral domain, $Q(R)$ is a field, and $Q(R)[T]$, a principal ideal domain. Thus, the following implications hold:

P and Q are relatively prime in $R[T]$ $\underset{23\text{A}.10}{\Leftrightarrow}$ P and Q are relatively prime in $Q(R)[T]$ $\Leftrightarrow$

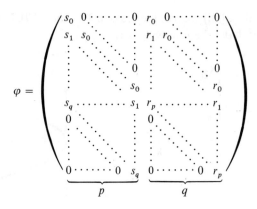

$$Q(R)[T] = Q(R)[T]\cdot(P,Q) \underset{33A.2}{\Leftrightarrow} \mathrm{Res}(P,Q) \neq 0 \text{ in } Q(R) \underset{33A.1\,i)}{\Leftrightarrow} \mathrm{Res}(P,Q)\neq 0 \text{ in } R. \quad \blacksquare$$

The *discriminant* $\mathrm{Dis}(P) := \mathrm{Res}\left(P, \dfrac{dP}{dT}\right) \in R$ is used to check for multiple factors of a monic polynomial $P \in R[T]$.

33 A. 4 Corollary. *For a normal ring R of characteristic* 0, *a monic polynomial $P \in R[T]$ has a multiple factor iff* $\mathrm{Dis}(P) = 0 \in R$.

Proof. By 33 A. 3, we need only show that P has a multiple factor iff P and $P' = \dfrac{dP}{dT}$ have a common irreducible factor H. *If.* For $P = H\cdot F$ and $P' = H\cdot K$ in $R[T]$, we see that $P' = H'F + HF'$, and thus that $H'F = H(K-F')$; hence, H is a divisor of $H'F$. Since $\deg H > \deg H'$, it follows that $H|F$; i.e., H is a multiple factor of P. *Only if.* This direction is trivial. $\blacksquare$

E. 33 Aa. Determine $\mathrm{Dis}(T^2 - z_1^2 z_2) \in \mathcal{O}(\mathbb{C}^2)$.

§ 33B Supplement: Proper Mappings and Equivalence Relations

Throughout this section, X and Y are understood to be *locally compact* spaces. In part v) of the following, we give an intuitive characterization of proper mappings:

33 B. 1 Proposition and Definition. *A continuous mapping $f: X \to Y$ is called proper if it satisfies the following equivalent conditions:*

i) Inverse images of compact sets are compact.

ii) Each fiber $f^{-1}y$ of f is compact and has a fundamental system of neighborhoods consisting of open sets of the form $f^{-1}V$, with $y \in V \Subset Y$.

iii) f is closed and every fiber of f is compact.

If X and Y have countable topologies, we have in addition the following equivalent statements:

iv) Every sequence in X whose image-sequence converges has a convergent subsequence.

v) If (x_j) is a sequence in X such that $x_j \to \partial X$, then $f(x_j) \to \partial Y$ (the notation "$x_j \to \partial X$" means that every compactum K in X contains only finitely many x_j).

Proof. The first part is proved in [Bou GT I. 10. Th. 1 and 10.3 Prop. 7]. We show all equivalences only for spaces with countable topologies, using the fact that the concepts "compact" and "sequentially compact" coincide for such spaces.

i) $\Rightarrow$ v) Suppose that $x_j \to \partial X$, and let $K \subset Y$ be compact. Then $f^{-1}K$, being compact, contains only finitely many x_j, so K contains only finitely many $f(x_j)$.

v) $\Rightarrow$ iv) If (x_j) is a sequence such that $f(x_j) \to y \in Y$, then $f(x_j) \not\to \partial Y$, and thus $x_j \not\to \partial X$; hence, there exists a compact $K \subset X$ containing infinitely many x_j and thus a convergent subsequence.

iv) $\Rightarrow$ iii) This implication is immediate.

iii) $\Rightarrow$ ii) If $U \Subset X$ is an open neighborhood of a fiber $f^{-1}y$, then so is $f^{-1}V \subset U$ for $V := Y \setminus f(X \setminus U)$, since $f(X\setminus U)$ is closed.

ii) $\Rightarrow$ i) If $K \subset Y$ is compact, then every sequence (x_j) in $f^{-1}K$ has an accumulation point: without loss of generality, suppose that $f(x_j)$ converges to $y \in K$; since Y is locally compact, the compact fiber f^{-1}y has a relatively compact neighborhood W; by assumption, we may suppose that W is of the form $f^{-1}(V)$. Since W contains all but finitely many x_j, the sequence (x_j) has an accumulation point x in $\overline{f^{-1}(V)}$, for which $f(x) = y$. $\blacksquare$

A continuous mapping is called *discrete* if its fibers are discrete topological subspaces. A proper discrete *mapping* is called *finite*.

33 B. 2 Remarks. i) *If the mapping $f: S \to T$ of topological spaces has the property that inverse images of compact sets are compact, then, for each $B \subset T$, the mapping $f|_{f^{-1}(B)}: f^{-1}(B) \to B$ has the same property.* ■

ii) If $f: X \to Y$ is discrete, then, for each $a \in X$, there exist arbitrarily small open neighborhoods U of a and V of $f(U)$ such that $f|_U: U \to V$ is finite.

iii) If X is connected and $f: X \to Y$ is holomorphic, closed, and discrete, then f is finite.

Proof. ii) If W is a compact neighborhood of a such that $W \cap f^{-1}f(a) = \{a\}$, then $f(\partial W)$ is compact, and $f(a)$ lies in $Y \setminus f(\partial W) \Subset Y$. Since W is compact, $f|_W: W \to Y$ satisfies condition 33 B. 1 i); by i), that also holds for each restriction $f: W \cap f^{-1}(V) \to V$ with $f(a) \in V \Subset Y \setminus f(\partial W)$. For such sets V, the set $U := W \cap f^{-1}(V) = \mathring{W} \cap f^{-1}(V)$ is open in X.

iii) This follows analogously; see E. 45k. ■

One important example of discrete mappings is given by the (holomorphic) coverings: a surjective holomorphic mapping $\pi: X \to Y$ of complex spaces is called a (holomorphic) *covering* if Y has an open cover consisting of connected sets U such that π maps every connected component of $\pi^{-1}(U)$ biholomorphically onto U.

If Y is connected, then clearly every fiber $\pi^{-1}(y)$ has the same number of elements, called the *degree* of the covering.

E. 33 Ba. Show that every locally biholomorphic finite mapping into a connected complex space is a covering (for 46B. 4).

Since complex spaces are locally connected (that is obviously the case for manifolds; see E. 48n), the following statement holds for connected Y and a connected subset U of X: *$\pi|_U: U \to Y$ is a covering iff U is a connected component of X* (see [Sp 2.1.14]).

In an example in § 50A, we apply the universal covering $\pi: \tilde{Y} \to Y$ for connected Y. It is determined (up to isomorphism) by the fact that the covering $\tilde{Y}$ is connected and simply connected [Sp 2.5.7]. An $f \in \mathrm{Aut}(\tilde{Y})$ is called a *covering transformation* if $\pi \circ f = \pi$. If two covering transformations coincide at a point, then they are identical [Sp 2.6.1].

An open finite holomorphic mapping $\psi: X \to Y$ of reduced complex spaces is called a *branched covering* of degree b if there exists a nowhere dense analytic set $A \hookrightarrow Y$ such that $\psi^{-1}(A)$ is nowhere dense in X and $\psi: X \setminus \psi^{-1}(A) \to Y \setminus A$ is a covering of degree b.

E. 33 Bb. Show that the mapping $\mathbb{C} \to \mathbb{C}^2$, $t \mapsto (t^3, t^2)$, determines a branched covering $\mathbb{C} \to N(\mathbb{C}^2; z_1^2 - z_2^3)$ of degree 1 (see E. 321 ii)).

E. 33 Bc. If $\psi: X \to Y$ is a branched covering of degree b and y lies in Y, then $c(y) \leq c(y') \leq b$ for y' near y, with $c(y)$ denoting the number of points in $\psi^{-1}(y)$ (for 33.7).

Now we want to carry the properties of mappings over to *equivalence relations*. The most important examples for us are the following:

33 B. 3 Examples. i) If $f: S \to T$ is a mapping, then $s_1 \sim s_2 :\Leftrightarrow f(s_1) = f(s_2)$ determines an equivalence relation R_f on S whose equivalence classes are the fibers of f.

ii) If G is a group of transformations on X, then

$$x_1 \sim x_2 :\Leftrightarrow \exists g \in G \text{ with } gx_1 = x_2$$

determines an equivalence relation R_G on X whose equivalence classes are the *orbits* $G(x) := \{gx; g \in G\}$ with respect to G. The quotient space X/R_G is called *orbit space*; for *reduced* X, we also write X/G (see E. 49 Ao v)).

An equivalence relation R on T can be described in terms of its graph $R = \{(t, u) \in T \times T; t \sim_R u\}$. For $A \subset T$, the set $R(A) := \{t \in T; \exists a \in A \text{ with } a \sim t\} = \mathrm{pr}_1(R \cap (X \times A))$ is called the *R-saturated hull* of A; moreover, A is called *R-saturated* if $A = R(A)$.

33 B. 4 Definition and Proposition. *An equivalence relation R on X is called proper if it satisfies the following equivalent conditions:*

i) If K is compact, then $R(K)$ is compact.

ii) Every $R(x)$ is compact and possesses a fundamental system of neighborhoods of saturated open sets.

iii) With the quotient topology, X/R is locally compact and $\pi: X \to X/R$ is proper.

Proof. i) $\Rightarrow$ ii) For $R(x) \subset U \sqsubset\!\subset X$[9] we have that $W := R(\bar{U}) \setminus R(R(\bar{U}) \setminus U))$ is saturated; moreover, $R(x) \subset W \subset U \subset R(\bar{U})$, so $W = U \setminus R(R(\bar{U}) \setminus U)$. Since $R(R(\bar{U}) \setminus U)$ is compact, W is open.

ii) $\Rightarrow$ iii) First, X/R is Hausdorff: disjoint compact sets $R(x_1)$ and $R(x_2)$ have disjoint saturated open neighborhoods, since X is Hausdorff. Second, X/R is locally compact: if $U \sqsubset\!\subset X$ is a saturated neighborhood of x, then $\pi U \sqsubset\!\subset X/\mathrm{R}$ is a neighborhood of πx. Finally, π is proper: for each $\bar{x}$ in a compact $K \subset X/R$, choose a saturated neighborhood $\pi^{-1}(V(\bar{x})) \sqsubset\!\subset X$ of $R(x)$; then K is covered by finitely many of such $V(\bar{x})$, so $\pi^{-1}(K) \subset \bigcup_{\text{finite}} \pi^{-1}(V(\bar{x})) \sqsubset\!\subset X$; since $\pi^{-1}(K)$ is closed, it must be compact.

iii) $\Rightarrow$ i) If K is compact, then so is $R(K) = \pi^{-1}\pi K$. ■

E. 33 Bd. For the equivalence relation $R \subset X \times X$, show that R is proper iff $\mathrm{pr}_1 : R \to X$ is proper iff $\mathrm{pr}_2 : R \to X$ is proper.

33 B. 5 Definition. *An equivalence relation R on X is called*

i) finite, if R is proper and every $R(x)$ is finite, and

ii) open, if $R(U)$ is open in X for each open U.

E. 33 Be. Prove the following statements:

i) For a continuous mapping $f: X \to Y$, if f is open, then so is R_f.

ii) If G is a transformation group on X, then the equivalence relation R_G on X is open (for 55.8).

E. 33 Bf. Show that, for each equivalence relation R on X, R is finite iff $\pi : X \to X/R$ is finite.

Even if $R(x)$ is finite for every $x \in X$, it may be that R is not a finite equivalence relation, as shown by the *shunt* (in $\mathbb{R} \cup \mathbb{R}$, identify equal strictly negative numbers).

E. 33 Bg. Let $f: S \to T$ be a finite surjective mapping of locally compact spaces, and let $\mathfrak{U} = (U_j)_{j \in J}$ be a basis for the topology of S. Prove that

$$\{\overset{\circ}{\widehat{f(U_{j_1} \cup \ldots \cup U_{j_m})}};\ j_1, \ldots, j_m \in J,\ m \in \mathbb{N}\}$$

is a basis for the topology of T; hence, *T has a countable topology if S does* (for 51 A. 3).

Chapter 4: Complex Spaces

In this chapter, we develop the fundamentals of a systematic theory of complex spaces based on local models of the form $V(G; f_1, \ldots, f_m)$. Taking further advantage of the scope granted by the concept of a ringed space, we provide the reduced com-

9) $A \sqsubset\!\subset B :\Leftrightarrow A \subset B$ and $A \sqsubset B$.

plex spaces of Chapter 3 with a richer, not necessarily reduced, structure sheaf, the essential property of which is its coherence. That this process actually generalizes the original concept of a complex space is a nontrivial result (Coherence Theorems of Oka and Cartan, 42.1 and 47.1, respectively); it provides the following chain of inclusions (of full subcategories):

$$\text{manifolds} \subset \text{reduced complex spaces} \subset \text{complex spaces} \subset \text{ringed spaces.}$$

The coherence of the structure sheaf is decisive for the primary proof-technique of this chapter, namely, deriving local statements of a geometric function-theoretic nature from the punctual statements of an algebraic nature in Chapter 2. A word about our terminology seems to be in order: we call a statement

punctual (at a) if it is applied only to the point a (i.e., only to the stalk ${}_X\mathcal{O}_a$ or to a homomorphism ${}_Y\mathcal{O}_b \to {}_X\mathcal{O}_a$),
local (at a) if it applies to all points near a,
global if it applies to all of X.

Although a strict definition of the concepts "punctual" and "local" is problematic, we find the "punctual-local" principle to be particularly helpful as a leitmotiv for investigations in connection with questions of coherence. Essential examples of propositions of the type "punctual ⇒ local" are the *Antiequivalence Theorem* (which was formulated earlier in the introduction to the Second Part), *Hilbert's Nullstellensatz* (the geometric-function-theoretically defined reduction of a complex space can be determined purely algebraically, and thus punctually: the concepts "geometrically reduced" and "algebraically reduced" are equivalent for complex spaces), and the *Representation Theorem for Prime Germs.* As an illustration of this approach, we mention §48, on dimension theory, in which we use the Antiequivalence Theorem to provide algebraic proofs for all of the geometric statements.

Among the *global statements* treated in Chapter 4, the decomposition of reduced complex spaces into irreducible components and the resulting Identity Theorem are particularly noteworthy; the theory of quotients of complex spaces, presented in a supplement to the chapter, also falls into this category.

In our construction of the local theory of complex spaces, finite mappings and the corresponding Finite Coherence Theorem play a central role: since, locally, complex spaces can be mapped finitely onto domains in complex number spaces (to interpret Noether's Normalization Theorem geometrically), the Finite Coherence Theorem provides an efficient aid for the transformation of function-theoretic statements about domains into statements about complex spaces.

§ 41 Coherent Sheaves

In the theory of modules over a ring R, the finitely-generated modules play an essential role. The same holds for modules $\mathcal{G}$ over a sheaf of rings $\mathcal{R}$; of course , in this case, one demands not only that each $\mathscr{G}_t$ be a finitely-generated $\mathscr{R}_t$-module, but also that it be possible to choose the generators locally in such a manner that they depend "continuously" on the parameter t.

Let $\mathscr{R} = {}_T\mathscr{R}$ denote a sheaf of rings on the topological space T; we admit the case that a stalk $\mathscr{R}_t$ is the zero ring (equivalently, $1_t = 0_t$) so that we can treat sheaves of rings whose supports are proper subsets of T.

41.1 Definition. *An $\mathscr{R}$-module $\mathscr{G}$ on T is called an "$\mathscr{R}$-module of finite type" or "of finite type over $\mathscr{R}$" if it is locally the homomorphic image of a finitely-generated free $\mathscr{R}$-module: for each $t \in T$, there exists on an appropriate neighborhood U of t an exact sequence $\mathscr{R}^p|_U \to \mathscr{G}|_U \to 0$.*

An equivalent condition is that every $\mathscr{G}_t$ be a finitely generated $\mathscr{R}_t$-module, with the generators depending continuously on t:

E. 41a. For an $\mathscr{R}$-module $\mathscr{G}$ over T, there exists an exact sequence $\mathscr{R}^p \to \mathscr{G} \to 0$ of $\mathscr{R}$-modules iff there exist sections $s_1, \dots, s_p \in \mathscr{G}(T)$ that generate every stalk $\mathscr{G}_t$ over $\mathscr{R}_t$ (for 41.2).

41.2 Proposition. *If $\mathscr{G}$ is an $\mathscr{R}$-module of finite type, and if $s_1, \dots, s_p \in \mathscr{G}(T)$ generate a stalk $\mathscr{G}_a$ over $\mathscr{R}_a$, then, for every y near a, $s_1, \dots, s_p$ are generators of $\mathscr{G}_y$ over $\mathscr{R}_y$.*

Proof. We may assume the existence of a surjection $\varphi: \mathscr{R}^q \to \mathscr{G}$. If $e_1, \dots, e_q \in \mathscr{R}^q(T)$ are the canonical generators of $\mathscr{R}^q$, then, by E. 41 a, $\tau_1 := \varphi(e_1), \dots, \tau_q := \varphi(e_q)$ generate each $\mathscr{G}_t$ over $\mathscr{R}_t$. In $\mathscr{G}_a$, we have by assumption that $\tau_j = \sum_{k=1}^{p} a_{jk} s_k$ for appropriate $a_{jk} \in \mathscr{R}_a$, $j = 1, \dots, q$; since those equations hold near a by 30.1 vi), the assertion follows. ■

41.3 Remarks. i) Every *free sheaf* $\mathscr{R}^p$ (with $\mathscr{R}^0 := 0$) has the *canonical generators* $e_j := (0, \dots, 0, \underset{\substack{\uparrow \\ j}}{1}, 0, \dots, 0)$ and is thus of finite type.

ii) If $\varphi: \mathscr{F} \to \mathscr{G}$ is $\mathscr{R}$-linear and $\mathscr{F}$ is of finite type, then $\mathscr{I}m\, \varphi$ is of finite type. This example is used frequently: if $\mathscr{H}$ is an $\mathscr{R}$-module and if $s_1, \dots, s_p \in \mathscr{H}(T)$ are global sections, then the $\mathscr{R}$-submodule $\mathscr{G} := \mathscr{R} \cdot (s_1, \dots, s_p) := \mathscr{R} s_1 + \dots + \mathscr{R} s_p$ generated by $s_1, \dots, s_p$ is finitely-generated, and the $\mathscr{R}$-linear mapping

$$\varphi: \mathscr{R}^p \to \mathscr{G}, \sum_{j=1}^{p} r_j e_j \mapsto \sum_{j=1}^{p} r_j s_j,$$

is surjective.

iii) Not every $\mathscr{R}$-module of finite type is generated by *global* sections: fix $T = \mathbb{P}_1 = \mathbb{C} \cup \{\infty\}$ and $\mathscr{R} = {}_{\mathbb{P}_1}\mathscr{O}$, and let $\mathscr{I}$ be the nullstellen ideal of $0 \in \mathbb{P}_1$; then $\mathscr{I}$ is not the zero sheaf, but every global section s of $\mathscr{I}$ is 0: as a holomorphic function on the compact set $\mathbb{P}_1$, s is constant by E. 32b iii); since $s(0) = 0$, we conclude that $s = 0$.

iv) Submodules of an $\mathscr{R}$-module of finite type are not necessarily of finite type: for $T = \mathbb{C}^1$ and $\mathscr{R} = {}_1\mathcal{O}$, if $\mathscr{J}$ is the characteristic ideal of $0 \in \mathbb{C}$ (i.e., $\mathscr{J}_0 = 0$ and $\mathscr{J}_z = {}_1\mathcal{O}_z$ for $z \neq 0$), then 0 is a generator of $\mathscr{J}_0$, but not of $\mathscr{J}_z$ for $z \neq 0$; by 41.2, $\mathscr{J}$ is not of finite type.

v) The kernel of a homomorphism between $\mathscr{R}$-modules of finite type is not necessarily of finite type (for the sheaf $\mathscr{J}$ of iv), consider the canonical projection $\mathscr{R} \to \mathscr{R}/\mathscr{J}$).

E. 41b. Prove the following statements for an $\mathscr{R}$-module $\mathscr{G}$ of finite type and a section $r \in \mathscr{R}(T)$:

i) The support $\operatorname{supp}\mathscr{G} = \{t \in T;\ \mathscr{G}_t \neq 0_t\}$ is closed (hint: use $1 \in \mathscr{R}(T)$);

ii) $\operatorname{supp}(\mathscr{R}^p/r \cdot \mathscr{R}^p) = \{t \in T;\ r_t \notin \mathscr{R}_t^*\} \cap \operatorname{supp}\mathscr{R}$.

An essential step in the development of complex analysis is the selection, from the category of $\mathscr{R}$-modules of finite type, of an appropriate full subcategory in which kernels and cokernels of homomorphisms exist. In 41.19, we characterize that subcategory of "coherent sheaves of modules" for those sheaves of rings which interest us; however, in order to give a systematic presentation, we must begin more generally.

For global sections $s_1, \ldots, s_p$ in an $\mathscr{R}$-module $\mathscr{G}$ on T, the kernel $\mathscr{K}er\,\varphi$ of the homomorphism

$$(41.4.1)\quad \varphi : \mathscr{R}^p \to \mathscr{G},\ \sum_{j=1}^{p} r_j e_j \mapsto \sum_{j=1}^{p} r_j s_j,$$

characterizes the relations between $s_1, \ldots, s_p$; for that reason, $\mathscr{K}er\,\varphi$ is called the *sheaf of relations* $\mathscr{K}er(s_1, \ldots, s_p)$ *of* $s_1, \ldots, s_p$.

41.4 Definition. *An $\mathscr{R}$-module $\mathscr{G}$ of finite type is called a coherent $\mathscr{R}$-module if, for each $U \subset\subset T$ and each collection of sections $s_1, \ldots, s_p \in \mathscr{G}(U)$, the sheaf of relations $\mathscr{K}er(s_1, \ldots, s_p)$ on U is of finite type.*

41.5 Remark. *For a submodule $\mathscr{F}$ of a coherent $\mathscr{R}$-module, $\mathscr{F}$ is coherent iff $\mathscr{F}$ is of finite type.* ∎

Whether for given sections $s_1, \ldots, s_p \in \mathscr{G}(U)$, $U \subset\subset T$, the sheaf of relations $\mathscr{K}er(s_1, \ldots, s_p)$ is of finite type, depends only on the submodule $\mathscr{R}|_U \cdot (s_1, \ldots, s_p) \subset \mathscr{G}|_U$:

41.6 Lemma. *For two $\mathscr{R}$-homomorphisms $\varphi : \mathscr{R}^p \to \mathscr{G}$ and $\psi : \mathscr{R}^q \to \mathscr{G}$ with the same image, $\mathscr{K}er\,\varphi$ is of finite type iff $\mathscr{K}er\,\psi$ is of finite type.*

Proof. Without loss of generality, set $\mathscr{G} = \mathscr{I}m\,\varphi$ and assume that $\mathscr{K}er\,\varphi$ is globally finitely-generated, i.e., that there exists a surjection $\mathscr{R}^{\tilde{p}} \to \mathscr{K}er\,\varphi$. Let γ denote the composition $\mathscr{R}^{\tilde{p}} \twoheadrightarrow \mathscr{K}er\,\varphi \hookrightarrow \mathscr{R}^p$. If we can introduce $\mathscr{R}$-homomorphisms λ, μ, and δ such that the diagram

$$\begin{array}{ccccccc}
\mathscr{R}^{\tilde{p}} & \xrightarrow{\gamma} & \mathscr{R}^{p} & \xrightarrow{\varphi} & \mathscr{G} & \to & 0 \\
 & & \lambda \downarrow \uparrow \mu & & \| & & \\
\mathscr{R}^{\tilde{p}} \oplus \mathscr{R}^{q} & \xrightarrow{\delta} & \mathscr{R}^{q} & \xrightarrow{\psi} & \mathscr{G} & \to & 0
\end{array}$$

commutes, and such that $\mathscr{I}m\,\delta = \mathscr{K}er\,\psi$, then $\mathscr{K}er\,\psi$ is of finite type. Let e_i and f_j denote the canonical generators of $\mathscr{R}^p$ and $\mathscr{R}^q$, respectively; we may assume that there exist $\alpha_{ij}, \beta_{ji} \in \mathscr{R}(T)$ for $i = 1, \ldots, p$ and $j = 1, \ldots, q$ such that

$$\varphi(e_i) = \sum_{j=1}^{q} \alpha_{ij}\psi(f_j),\ \psi(f_j) = \sum_{i=1}^{p} \beta_{ji}\varphi(e_i).$$

We define the $\mathscr{R}$-homomorphisms λ and μ by linearly extending

$$\lambda(e_i) := \sum_{j=1}^{q} \alpha_{ij} f_j \quad \text{and} \quad \mu(f_j) := \sum_{i=1}^{p} \beta_{ji} e_i;$$

then the diagram commutes. Now

$$\delta(r,s) := s + \lambda(\gamma r - \mu s)$$

provides an exact sequence in the lower line, because $\mathscr{K}er\,\psi = \mathscr{I}m\,\delta$: if $\psi(s) = 0$, then $\varphi\mu(s) = 0$; hence, there exists an $r \in \mathscr{R}^{\tilde{p}}$ such that $\gamma(r) = \mu(s)$ and $\delta(r,s) = s + \lambda(\gamma r - \mu s) = s$; that $\psi\delta = 0$ follows by substitution. ■

41.7 Remarks. i) If the topological space T contains only one point, then our investigation reduces to the theory of coherent modules over a ring; hence, the subsequent propositions generalize statements about finitely-generated modules over noetherian rings.

ii) Coherence is obviously a local property; therefore, investigations of coherence may be restricted to small open sets.

iii) *A coherent $\mathscr{R}$-module is locally isomorphic to the cokernel of a homomorphism of free $\mathscr{R}$-modules*, as can be deduced easily from the definition. In 41.19, we prove the converse for modules over a coherent sheaf of rings $\mathscr{R}$.

iv) *If X is a manifold and $\mathscr{G}$ is a coherent ${}_X\mathcal{O}$-module, then* $\operatorname{supp}\mathscr{G}$ *is an analytic set in X*: by iii), $\mathscr{G}$ has local representations of the form

$${}_X\mathcal{O}^p \xrightarrow{\varphi} {}_X\mathcal{O}^q \to \mathscr{G} \to 0,$$

where φ is representable by a $(q \times p)$-matrix of holomorphic functions; hence,

$$\begin{aligned}
\mathscr{G}_x \neq 0 &\Leftrightarrow \varphi_x \text{ is not surjective} \\
&\underset{\text{E.24c}}{\Leftrightarrow} \varphi(x) \text{ is not surjective} \\
&\Leftrightarrow \text{the rank of the matrix } \varphi(x) \text{ is less than } q \\
&\Leftrightarrow p < q \text{ or } \varphi(x) \text{ is not of maximal rank.}
\end{aligned}$$

By E. 8f, then, $\operatorname{supp}\mathscr{G}$ is analytic. ■

v) The zero sheaf 0 is coherent.

Before investigating nontrivial examples of coherent sheaves, we show which algebraic operations are possible in the category of coherent $\mathcal{R}$-modules. The existence of kernels and cokernels is fundamental:

41.8 Theorem. *Let* $0 \to \mathcal{G}' \to \mathcal{G} \to \mathcal{G}'' \to 0$ *be an exact sequence of* $\mathcal{R}$*-modules on* T. *If two of the three sheaves* $\mathcal{G}', \mathcal{G}, \mathcal{G}''$ *are coherent, then the third sheaf is also coherent.*

Proof. Since the zero sheaf is coherent, the theorem follows immediately from the following fact:

41.9 Lemma. *If* $\mathcal{G}_1 \xrightarrow{\varphi_1} \mathcal{G}_2 \xrightarrow{\varphi_2} \mathcal{G} \xrightarrow{\varphi} \mathcal{G}_3 \xrightarrow{\varphi_3} \mathcal{G}_4$ *is an exact sequence of* $\mathcal{R}$*-modules, and if each* $\mathcal{G}_i$ *is coherent, then* $\mathcal{G}$ *is also coherent.*

Proof. a) First we show that $\mathcal{G}$ is of finite type. Fix $t \in T$; near t, the exact sequence of the lemma induces a diagram of the form

$$\text{(41.9.1)} \qquad \begin{array}{ccccccc} \mathcal{R}^{p_2} & & \mathcal{R}^{p} & \overset{\psi}{\twoheadrightarrow} & \mathcal{K}er\,(\varphi_3 \circ \tau) \hookrightarrow \mathcal{R}^{p_3} & & \\ \downarrow{\scriptstyle \chi} & & \vdots{\scriptstyle \omega} & & \downarrow{\scriptstyle \tau} & \searrow{\scriptstyle \varphi_3 \circ \tau} & \\ \mathcal{G}_2 & \xrightarrow{\varphi_2} & \mathcal{G} & \xrightarrow{\qquad\varphi\qquad} & \mathcal{G}_3 & \xrightarrow{\varphi_3} & \mathcal{G}_4, \end{array}$$

since $\mathcal{G}_2, \mathcal{G}_3$, and (due to the coherence of $\mathcal{G}_4$) also the sheaf of relations $\mathcal{K}er\,(\varphi_3 \circ \tau)$ are of finite type. Since $\tau\psi(\mathcal{R}^p) \subset \mathcal{K}er\,\varphi_3 = \mathcal{I}m\,\varphi$, it is possible to construct, by choosing $g_1, \ldots, g_p \in \mathcal{G}_t$ such that $\varphi(g_j) = \tau\psi(e_j)$, a sheaf-homomorphism ω such that the diagram (41.9.1) commutes. To prove that $\mathcal{G}$ is of finite type, it suffices to show that

$$\varphi_2\chi + \omega : \mathcal{R}^{p_2} \oplus \mathcal{R}^p \to \mathcal{G},\ (r, s) \mapsto \varphi_2\chi(r) + \omega(s),$$

is surjective. For $g \in \mathcal{G}$, the fact that $\varphi_3\varphi(g) = 0$ implies the existence of an $s \in \mathcal{R}^p$ such that $\tau\psi(s) = \varphi(g)$. Since $\varphi(g - \omega(s)) = 0$, there exists an $r \in \mathcal{R}^{p_2}$ such that $\varphi_2\chi(r) = g - \omega(s)$. Hence, $(\varphi_2\chi + \omega)(r, s) = g$.

b) Now we show that every sheaf of relations of $\mathcal{G}$ is of finite type. By 41.3 ii), the fact that $\mathcal{G}_1$ is of finite type implies that $\mathcal{I}m\,\varphi_1$ is of finite type; hence, we may assume that $\mathcal{G}_1 = \mathcal{K}er\,\varphi_2$. For a given homomorphism $\beta : \mathcal{R}^p \to \mathcal{G}$, we obtain near a fixed point t a commutative diagram

$$\begin{array}{ccccccc} & \mathcal{R}^{p_1} & & \mathcal{R}^{\tilde{p}} \overset{\psi}{\twoheadrightarrow} \mathcal{K}er\,(\varphi \circ \beta) \hookrightarrow & \mathcal{R}^{p} & & \\ & \downarrow{\scriptstyle \chi} & & \vdots{\scriptstyle \omega} & \downarrow{\scriptstyle \beta} & \searrow{\scriptstyle \varphi \circ \beta} & \\ \mathcal{K}er\,\varphi_2 = & \mathcal{G}_1 & \hookrightarrow & \mathcal{G}_2 \xrightarrow{\qquad\varphi_2\qquad} & \mathcal{G} & \xrightarrow{\varphi} & \mathcal{G}_3 \end{array},$$

due to the coherence of $\mathcal{G}_3$ and the fact that $\mathcal{G}_1$ is of finite type; the homomorphism

ω can be constructed as in a). Since $\mathscr{G}_2$ is coherent, the homomorphism $\chi + \omega$: $\mathscr{R}^{p_1} \oplus \mathscr{R}^{\tilde{p}} \to \mathscr{G}_2$ has a sheaf of relations $\mathscr{K}er(\chi + \omega)$ of finite type. Let π: $\mathscr{R}^{p_1} \oplus \mathscr{R}^{\tilde{p}} \to \mathscr{R}^{\tilde{p}}$ denote the canonical projection. It suffices to show that $\mathscr{K}er(\beta) = \psi\pi\,\mathscr{K}er(\chi + \omega)$, for then $\mathscr{K}er(\beta)$ is obviously of finite type. Each pair $(a_1, a_2) \in \mathscr{K}er(\chi + \omega)$ satisfies the equation $\chi(a_1) = -\omega(a_2)$, so $0 = \varphi_2 \omega a_2 = \beta\psi\pi(a_1, a_2)$; i.e., $\psi\pi(a_1, a_2) \in \mathscr{K}er(\beta)$. Conversely, if $a \in \mathscr{K}er(\beta) \subset \mathscr{K}er(\varphi \circ \beta)$, then there exists a $b \in \mathscr{R}^{\tilde{p}}$ such that $\psi(b) = a$. Hence, $0 = \beta\psi(b) = \varphi_2\omega(b)$, so there exists a $c \in \mathscr{R}^{p_1}$ such that $\chi(c) = \omega(b)$. Now $(-c, b) \in \mathscr{K}er(\chi + \omega)$ and $\psi\pi(-c, b) = \psi(b) = a$. ■

41.10 Corollary. *The direct sum of a finite family of coherent $\mathscr{R}$-modules is coherent.* ■

41.11 Corollary. *If $\varphi: \mathscr{F} \to \mathscr{G}$ is a homomorphism between coherent $\mathscr{R}$-modules, then $\mathscr{K}er\,\varphi$, $\mathscr{C}oker\,\varphi$, and $\mathscr{I}m\,\varphi$ are coherent $\mathscr{R}$-modules.*

Proof. The $\mathscr{R}$-module $\mathscr{F}$, and thus also $\mathscr{I}m\,\varphi$, is of finite type; by 41.5, $\mathscr{I}m\,\varphi$ is coherent. By 41.8, the exact sequences $0 \to \mathscr{K}er\,\varphi \to \mathscr{F} \to \mathscr{I}m\,\varphi \to 0$ and $0 \to \mathscr{I}m\,\varphi \to \mathscr{G} \to \mathscr{C}oker\,\varphi \to 0$ ensure the coherence of $\mathscr{K}er\,\varphi$ and that of $\mathscr{C}oker\,\varphi$, respectively. ■

41.12 Corollary. *If $\mathscr{F}$ and $\mathscr{G}$ are coherent $\mathscr{R}$-submodules of a coherent $\mathscr{R}$-module $\mathscr{H}$, then also $\mathscr{F} + \mathscr{G}$ and $\mathscr{F} \cap \mathscr{G}$ are coherent $\mathscr{R}$-modules.*

Proof. The addition $\varphi: \mathscr{F} \oplus \mathscr{G} \xrightarrow{+} \mathscr{H}$ is a homomorphism of coherent $\mathscr{R}$-modules; hence, by 41.11, $\mathscr{I}m\varphi = \mathscr{F} + \mathscr{G}$ is coherent. Likewise $\mathscr{F} \cap \mathscr{G}$, as the kernel of the natural homomorphism $\mathscr{F} \to \mathscr{H}/\mathscr{G}$, is coherent. ■

41.13 Corollary. *If $\mathscr{F}$ and $\mathscr{G}$ are coherent $\mathscr{R}$-modules, then so is $\mathscr{F} \otimes_{\mathscr{R}} \mathscr{G}$. If $\mathscr{I}$ is a coherent $\mathscr{R}$-ideal, then $\mathscr{I} \cdot \mathscr{G}$ is a coherent $\mathscr{R}$-module.*

Proof. We may assume the existence of an exact sequence $\mathscr{R}^p \xrightarrow{\varphi} \mathscr{R}^q \to \mathscr{F} \to 0$; with that, we can construct another exact sequence:

$$\mathscr{R}^p \otimes_{\mathscr{R}} \mathscr{G} \xrightarrow{\varphi \otimes \mathrm{id}_{\mathscr{G}}} \mathscr{R}^q \otimes_{\mathscr{R}} \mathscr{G} \to \mathscr{F} \otimes_{\mathscr{R}} \mathscr{G} \to 0.$$

By 41.10, the $\mathscr{R}$-modules $\mathscr{R}^p \otimes_{\mathscr{R}} \mathscr{G} \cong \mathscr{G}^p$ and $\mathscr{R}^q \otimes_{\mathscr{R}} \mathscr{G} \cong \mathscr{G}^q$ are coherent; hence, 41.11 implies that $\mathscr{C}oker(\varphi \otimes \mathrm{id}_{\mathscr{G}}) = \mathscr{F} \otimes_{\mathscr{R}} \mathscr{G}$ is coherent. Finally, $\mathscr{I} \cdot \mathscr{G}$ is coherent, since it is the image of a homomorphism $\mathscr{I} \otimes_{\mathscr{R}} \mathscr{G} \to \mathscr{G}$. ■

41.14 Corollary. *If $\mathscr{F}$ and $\mathscr{G}$ are coherent $\mathscr{R}$-modules, then $\mathscr{H}om_{\mathscr{R}}(\mathscr{F}, \mathscr{G})$ is a coherent $\mathscr{R}$-module.*

Proof. We may assume the existence of an exact sequence $\mathscr{R}^p \to \mathscr{R}^q \to \mathscr{F} \to 0$; that provides an exact sequence of the form

$$0 \to \mathcal{H}om_{\mathcal{R}}(\mathcal{F}, \mathcal{G}) \to \mathcal{H}om_{\mathcal{R}}(\mathcal{R}^q, \mathcal{G}) \xrightarrow{\varphi} \mathcal{H}om_{\mathcal{R}}(\mathcal{R}^p, \mathcal{G}).$$

Since $\mathcal{H}om_{\mathcal{R}}(\mathcal{R}^m, \mathcal{G}) \cong \mathcal{G}^m$, 41.10 and 41.11 imply that the $\mathcal{R}$-module $\mathcal{H}om_{\mathcal{R}}(\mathcal{F}, \mathcal{G}) = \mathcal{K}er\,\varphi$ is coherent. ■

In contrast to the situation for arbitrary $\mathcal{R}$-modules, we have the following fact in the coherent case:

41.15 Proposition. *If $\mathcal{F}$ is a coherent $\mathcal{R}$-module, then for every $t \in T$, the canonical homomorphism*

$$\mu : \mathcal{H}om_{\mathcal{R}}(\mathcal{F}, \mathcal{G})_t \to \mathrm{Hom}_{\mathcal{R}_t}(\mathcal{F}_t, \mathcal{G}_t)$$

is an isomorphism.

Proof. An exact sequence $\mathcal{R}^p \to \mathcal{R}^q \to \mathcal{F} \to 0$ near t provides in a canonical manner a commutative diagram with exact lines

$$\begin{array}{ccccccc} 0 \to & \mathcal{H}om_{\mathcal{R}}(\mathcal{F}, \mathcal{G})_t & \to & \mathcal{H}om_{\mathcal{R}}(\mathcal{R}^q, \mathcal{G})_t & \to & \mathcal{H}om_{\mathcal{R}}(\mathcal{R}^p, \mathcal{G})_t \\ & \downarrow \mu & & \downarrow \lambda & & \downarrow \nu \\ 0 \to & \mathrm{Hom}_{\mathcal{R}_t}(\mathcal{F}_t, \mathcal{G}_t) & \to & \mathrm{Hom}_{\mathcal{R}_t}(\mathcal{R}_t^q, \mathcal{G}_t) & \to & \mathrm{Hom}_{\mathcal{R}_t}(\mathcal{R}_t^p, \mathcal{G}_t). \end{array}$$

By the Five Lemma E. 41d, it suffices to show that λ and ν are isomorphisms; that follows from the fact that

$$\mathcal{H}om_{\mathcal{R}}(\mathcal{R}^q, \mathcal{G})_t \cong G_t^q \cong \mathcal{H}om_{\mathcal{R}_t}(R_t^q, \mathcal{G}_t).$$

Incidentally, the proof shows that μ is injective if $\mathcal{F}$ is of finite type (without assumption of coherence). ■

41.16 Proposition. *If two submodules $\mathcal{F}$ and $\mathcal{G}$ of an $\mathcal{R}$-module $\mathcal{H}$ are of finite type, and if $\mathcal{F}$ and $\mathcal{G}$ coincide over a point a, then they coincide near a.*

Proof. If $s_1, \dots, s_p \in \mathcal{F}_a = G_a$ are generators over $\mathcal{R}_a$, then, by 41.2, they are generators of $\mathcal{F}$ and $\mathcal{G}$ over all points in a neighborhood of a. ■

41.17 Corollary. *If the sequence $\mathcal{F} \xrightarrow{\varphi} \mathcal{G} \xrightarrow{\psi} \mathcal{H}$ of coherent $\mathcal{R}$-modules is exact at the point $a \in T$, then it is exact near a.*

Proof. By 41.11, $\mathcal{K}er\,\psi$ and $\mathcal{I}m\,\varphi$ are coherent; application of 41.16 yields the desired result. ■

E. 41c. For two coherent $\mathcal{R}$-modules $\mathcal{F}$ and $\mathcal{G}$, if a homo- (mono-, epi-)morphism $\varphi : \mathcal{F}_a \to \mathcal{G}_a$ is given at the point $a \in T$, then there exists a homo- (mono-, epi-)morphism $f : \mathcal{F} \to \mathcal{G}$ near a such that $f_a = \varphi$ (for E. 41h).

E. 41d. *Five Lemma.* Prove the following statements for the exact commutative diagram

$$\begin{array}{ccccccccc} \mathscr{F}_1 & \to & \mathscr{F}_2 & \to & \mathscr{F}_3 & \to & \mathscr{F}_4 & \to & \mathscr{F}_5 \\ \downarrow\varphi_1 & & \downarrow\varphi_2 & & \downarrow\varphi_3 & & \downarrow\varphi_4 & & \downarrow\varphi_5 \\ \mathscr{G}_1 & \to & \mathscr{G}_2 & \to & \mathscr{G}_3 & \to & \mathscr{G}_4 & \to & \mathscr{G}_5 \end{array}$$

of $\mathscr{R}$-modules:

i) If φ_1 and φ_3 are epimorphisms and φ_4 is a monomorphism, then φ_2 is an epimorphism (mnemonic: epi, *epi*, epi, mono).

ii) If φ_1 is an epimorphism, and if φ_2 and φ_4 are monomorphisms, then φ_3 is a monomorphism (epi, mono, *mono*, mono).

iii) If φ_1, φ_2, φ_4, and φ_5 are isomorphisms, then φ_3 is an isomorphism.

Now we want to consider the special case in which the $\mathscr{R}$-module $\mathscr{G}$ is just the sheaf of rings $\mathscr{R}$; note that $\mathscr{R}$ is obviously of finite type over itself.

41.18 Definition. *A sheaf of rings $\mathscr{R}$ is called a coherent sheaf (of rings), if $\mathscr{R}$ is coherent as a module over itself.*

Thus $\mathscr{R}$ is a coherent sheaf of rings iff, for every homomorphism $\varphi: \mathscr{R}^p|_U \to \mathscr{R}|_U$, the sheaf of relations $\mathscr{K}er\ \varphi$ is of finite type.

Over a coherent sheaf of rings $\mathscr{R}$, the category of coherent $\mathscr{R}$-modules is the full subcategory of $\mathscr{R}$-modules that are locally cokernels of homomorphisms between (finitely-generated) free $\mathscr{R}$-modules:

41.19 Proposition. *An $\mathscr{R}$-module $\mathscr{G}$ over a coherent sheaf of rings $\mathscr{R}$ on T is coherent iff there exists an open cover of T consisting of sets U for which there is an exact sequence*

$$\mathscr{R}^p|_U \to \mathscr{R}^q|_U \to \mathscr{G}|_U \to 0 .$$

Proof. By 41.7 iii), it suffices to derive the coherence of $\mathscr{G}$ from the local existence of such exact sequences. By 41.10, the coherence of $\mathscr{R}^p|_U$ and $\mathscr{R}^q|_U$ follows from that of $\mathscr{R}$; thus $\mathscr{G}|_U$, as a cokernel, is coherent by 41.11. ∎

The coherence of an $\mathscr{R}$-module $\mathscr{G}$ depends on $\mathscr{R}$; hence we refer to $\mathscr{R}$-coherence, whenever misunderstanding is possible.

E. 41e. Give an example of a sheaf $\mathscr{G}$ that is a module over two different sheaves of rings $\mathscr{R}$ and $\mathscr{S}$, and that is $\mathscr{R}$-coherent but not $\mathscr{S}$-coherent.

The following proposition is fundamental for the transition to complex structures on closed subspaces:

41.20 Proposition. *Let $\mathscr{R}$ be a coherent sheaf of rings, let $\mathscr{I} \subset \mathscr{R}$ be a coherent sheaf of ideals, and let $\mathscr{G}$ be an $\mathscr{R}/\mathscr{I}$-module. Then $\mathscr{G}$ is $\mathscr{R}$-coherent iff $\mathscr{G}$ is $\mathscr{R}/\mathscr{I}$-coherent. In particular, $\mathscr{R}/\mathscr{I}$ is a coherent sheaf of rings.*

Proof. Obviously, $\mathscr{G}$ is of finite type over $\mathscr{R}$ iff it is over $\bar{\mathscr{R}} := \mathscr{R}/\mathscr{I}$. Moreover, the exactness of the sequence $0 \to \mathscr{I} \to \mathscr{R} \to \bar{\mathscr{R}} \to 0$ implies by 41.9 that $\bar{\mathscr{R}}$ is a coherent $\mathscr{R}$-module.

Only if. If an $\mathscr{R}$-homomorphism $\varphi: \bar{\mathscr{R}}^p \to \mathscr{G}$ is given locally, then $\mathscr{K}er\,\varphi$ is $\mathscr{R}$-coherent by 41.11; hence, the $\bar{\mathscr{R}}$-module $\mathscr{K}er\,\varphi$ is of finite type over $\mathscr{R}$ and thus over $\bar{\mathscr{R}}$. Thus $\mathscr{G}$ is $\bar{\mathscr{R}}$-coherent; as a particular case we have that $\bar{\mathscr{R}}$ is a coherent sheaf of rings.

If. By 41.19, there exist locally exact sequences $\bar{\mathscr{R}}^p \to \bar{\mathscr{R}}^q \to \mathscr{G} \to 0$; by 41.11, the $\mathscr{R}$-coherence of $\bar{\mathscr{R}}$ implies that of $\mathscr{G}$. ■

For each subset $B \subset T$, "*restriction to* B" determines an *exact functor* from the category of $\mathscr{R}$-modules on T to the category of ${}_B\mathscr{R}$-modules (${}_B\mathscr{R} := \mathscr{R}|_B$) on B. (We used that, for example, in the definition of the structure sheaf of an analytic set.) Conversely, if $B \subseteq T$ is *locally closed* (i.e., $B = A \cap U$ for a region $U \Subset T$ and a closed set $A \subset T$), then there exists an exact functor $G \mapsto G^T$, called "*trivial extension*" from ${}_B\mathscr{R}$-modules on B to ${}_B\mathscr{R}^T$-modules on T; for fixed U and A, it is determined by the presheaf on T

$$V \mapsto \begin{cases} \mathscr{G}(V \cap A) & \text{for } V \Subset U \\ 0 & \text{otherwise} \end{cases}.$$

It is easy to see that $\operatorname{supp} \mathscr{G}^T \subset U \cap A = B$ and that $\mathscr{G}^T|_B = \mathscr{G}$; in particular, $\mathscr{G}^T$ is independent of the special choice of U and A.

41.21 Proposition. *Let $B \subset T$ be closed, $\mathscr{R}$ be a sheaf of rings on B, and $\mathscr{G}$ be an $\mathscr{R}$-module. Then the following statements hold:*

i) $\mathscr{G}$ is of finite type over $\mathscr{R}$ iff $\mathscr{G}^T$ is of finite type over $\mathscr{R}^T$.

ii) $\mathscr{G}$ is $\mathscr{R}$-coherent iff $\mathscr{G}^T$ is $\mathscr{R}^T$-coherent.

iii) $\mathscr{R}$ is a coherent sheaf of rings iff $\mathscr{R}^T$ is a coherent sheaf of rings.

Proof. For $\varphi \in \operatorname{Hom}(\mathscr{R}^p, \mathscr{G})$, let $\varphi^T \in \operatorname{Hom}_{\mathscr{R}^T}((\mathscr{R}^T)^p, \mathscr{G}^T)$ be the extension to T. Since restriction and extension are exact functors, it follows that φ is surjective iff φ^T is surjective. That implies statement i); statement ii) follows similarly, and iii) is a special case of ii). ■

E. 41f. *Transitivity of coherence.* Suppose that $\mathscr{R}$ and $\mathscr{S}$ are sheaves of rings, that $\mathscr{R}$ is coherent, and that $\mathscr{S}$ is a coherent $\mathscr{R}$-module; prove the following statements:

i) $\mathscr{S}$ is a coherent sheaf of rings.

ii) Every coherent $\mathscr{S}$-module is also a coherent $\mathscr{R}$-module (for 45.1).

E. 41g. Prove the following statements, in which $\mathscr{R}$ is a coherent sheaf of rings, $\mathscr{H}$ is an $\mathscr{R}$-module, and $\mathscr{F}, \mathscr{G} \subset \mathscr{H}$ are coherent submodules:

i) The *annihilator* of $\mathscr{F}$,

$$\mathscr{A}nn(\mathscr{F}) := \mathscr{K}er\left[\mathscr{R} \to \mathscr{H}om_{\mathscr{R}}(\mathscr{F}, \mathscr{F}), \quad r \mapsto (f \mapsto rf)\right]$$

is a coherent sheaf of ideals in $\mathscr{R}$ such that

$$\mathscr{A}nn(\mathscr{F})_t = \{r \in \mathscr{R}_t;\, r\mathscr{F}_t = 0\} \quad \text{(for 45.1)}.$$

ii) The *submodule quotient* (or *ideal quotient* if $\mathcal{H} = \mathcal{R}$)

$$\mathcal{F} : \mathcal{G} := \mathcal{A}nn((\mathcal{F} + \mathcal{G}) / \mathcal{F})$$

is a coherent sheaf of ideals in $\mathcal{R}$ with

$$(\mathcal{F} : \mathcal{G})_t = \{r \in \mathcal{R}_t;\, r\mathcal{G}_t \subset \mathcal{F}_t\} \quad \text{(for 52.17)}.$$

iii) If the stalks of $\mathcal{R}$ are integral domains, then the *torsion-sheaf*

$$\mathcal{T}or^{\mathcal{R}}(\mathcal{F}) := \mathcal{K}er[\mathcal{F} \xrightarrow{\varphi} \mathcal{H}om_{\mathcal{R}}(\mathcal{H}om_{\mathcal{R}}(\mathcal{F}, \mathcal{R}), \mathcal{R})]$$

(φ is the canonical mapping) is a coherent subsheaf of $\mathcal{F}$ such that

$$\mathcal{T}or^{\mathcal{R}}(\mathcal{F})_t = \{f \in \mathcal{F}_t;\, \exists \text{ non-zero-divisor } r \in \mathcal{R}_t \text{ such that } rf = 0\} =: T(\mathcal{F}_t)$$

(hint: use that $T(\mathcal{F}_t) = \mathrm{Ker}[\mathcal{R}_t \otimes_{\mathcal{R}_t} \mathcal{F}_t \to Q(\mathcal{R}_t) \otimes_{\mathcal{R}_t} \mathcal{F}_t]$; for 46.1; this statement holds more generally for algebraically reduced $\mathcal{R}$ [GrRe Anhang §4 Satz 7]).

E. 41h. Let $\mathcal{R}$ be a coherent sheaf of rings, fix $a \in T$, and let G be an $\mathcal{R}_a$-module that can be represented as the cokernel of a homomorphism $\mathcal{R}_a^p \to \mathcal{R}_a^q$. Prove that near a there exists a coherent $\mathcal{R}$-module $\mathcal{G}$ such that $\mathcal{G}_a \cong G$. If $\mathcal{F}$ is a coherent $\mathcal{R}$-module, so that G is a submodule of $\mathcal{F}_a$, then $\mathcal{G}$ can be chosen near a as a submodule of $\mathcal{F}$ (hint: E. 41c; for 45.9).

E. 41 . Let $\mathcal{R}$ be a coherent sheaf of rings and $\mathcal{G}$ be a coherent $\mathcal{R}$-module; define the *zero-set of an ideal* $\mathcal{I} \subset \mathcal{R}$ to be

$$N(T; \mathcal{I}) := N(\mathcal{I}) := \{t \in T;\, \mathcal{I}_t \neq \mathcal{R}_t\}.$$

Prove that $\operatorname{supp}\mathcal{G} = \operatorname{supp}(\mathcal{R} / \mathcal{A}nn\,\mathcal{G}) = N(\mathcal{A}nn\,\mathcal{G})$ (for 45.2).

E. 41j. *Topological change of basis and coherence.* Let $\varphi : S \to T$ be a continuous mapping between topological spaces, $\mathcal{R}$ be a sheaf of rings on T, and $\mathcal{G}$ be an $\mathcal{R}$-module. Prove that if $\mathcal{G}$ is of finite type (resp., coherent) over $\mathcal{R}$, then $\varphi^{-1}(\mathcal{G})$ is of finite type (resp., coherent) over $\varphi^{-1}(\mathcal{R})$; in particular, $\varphi^{-1}(\mathcal{R})$ is a coherent sheaf of rings if $\mathcal{R}$ is (hint: φ^{-1} is an exact functor; for 42.1).

§ 42 The Coherence of ${}_n\mathcal{O}$

We turn now to our first nontrivial example of a coherent sheaf of rings:

42.1 Oka's Coherence Theorem. *The structure sheaf* ${}_M\mathcal{O}$ *of each manifold* M *is coherent.*

Proof. Since coherence is a local property, it suffices to demonstrate the coherence of ${}_n\mathcal{O}$ in a neighborhood U (to be diminished appropriately in the course of the proof) of 0 in $\mathbb{C}^n$. We prove by induction on n: the kernel of a homomorphism (see 41.4.1)

$$\varphi : {}_n\mathcal{O}^m \to {}_n\mathcal{O},\ (g_1, \ldots, g_m) \mapsto \sum_{j=1}^{m} g_j f_j,$$

determined by $(f_1, \ldots, f_m) \in {}_n\mathcal{O}^m(U)$ is an ${}_n\mathcal{O}$-module of finite type.

For $n = 0$, the linear subspace $\mathscr{K}er\,\varphi$ of ${}_0\mathcal{O}_0^m \cong \mathbb{C}^m$ is certainly of finite type over $\mathbb{C}$.

"$n-1 \Rightarrow n$" First we show that we may assume that $m \geqq 2$ and that $f_1(0) = \ldots = f_m(0) = 0$: for $m = 1$, either $\varphi = 0$ or φ, being multiplication by a germ of a function, is injective; on the other hand, if $f_1(0) \neq 0$, say, then we may suppose that f_1 is a unit in ${}_n\mathcal{O}(U)$; but then the mapping

$$(42.1.1)\quad {}_n\mathcal{O}^{m-1}|_U \to \mathscr{K}er(f_1,\ldots,f_m),\ (g_2,\ldots,g_m) \mapsto \left(-\frac{1}{f_1}\sum_{j=2}^{m} g_j f_j, g_2,\ldots,g_m\right)$$

is a sheaf-isomorphism, and $\mathscr{K}er(f_1,\ldots,f_m)$ is of finite type. Second, we may assume that every f_j is a Weierstrass polynomial in ${}_{n-1}\mathcal{O}_0[z_n]$ (with an appropriate transformation of coordinates (see E. 22c), the Preparation Theorem yields factorizations $f_j = e_j \cdot \omega_j$; clearly, the sheaves $\mathscr{K}er(f_1,\ldots,f_m)$ and $\mathscr{K}er(\omega_1,\ldots,\omega_m)$ are isomorphic). Finally, set $U = U' \times U'' \Subset \mathbb{C}^{n-1} \times \mathbb{C}$ with $f_j \in {}_{n-1}\mathcal{O}(U')[z_n]$, and $r := \max \deg f_j$. By the induction hypothesis, ${}_{n-1}\mathcal{O}$ is a coherent sheaf of rings on $\mathbb{C}^{n-1}$. Hence, $\mathscr{R} := \mathrm{pr}_{\mathbb{C}^{n-1}}^{-1}({}_{n-1}\mathcal{O})$ is a coherent sheaf of rings on $\mathbb{C}^n$ (see E. 41j), and, by 41.10, $\mathscr{R}[z_n]_{2r} \subset {}_n\mathcal{O}$ is a coherent $\mathscr{R}$-module. The homomorphism φ induces a commutative diagram

$$\begin{array}{ccc} {}_n\mathcal{O}^m & \xrightarrow{\ \varphi\ } & {}_n\mathcal{O} \\ \cup & & \cup \\ (\mathscr{R}[z_n]_{2r})^m & \xrightarrow{\ \psi\ } & \mathscr{R}[z_n]_{3r}, \end{array}$$

since $\deg f_j \leqq r$. By 41.11, we may assume that there exist sections $g_1,\ldots,g_s \in (\mathscr{K}er\,\psi)(U)$ such that

$$\mathscr{K}er\,\psi = \mathscr{R}g_1 + \ldots + \mathscr{R}g_s.$$

It suffices to show the following statement:

$$(42.1.2)\quad (\mathscr{K}er\,\varphi)_u = {}_n\mathcal{O}_u \cdot (\mathscr{K}er\,\psi)_u \text{ for each } u \in U.$$

Then $\mathscr{K}er\,\varphi = {}_n\mathcal{O} \cdot \mathscr{K}er\,\psi = {}_n\mathcal{O} \cdot (\mathscr{R}g_1 + \ldots + \mathscr{R}g_s) = {}_n\mathcal{O}g_1 + \ldots + {}_n\mathcal{O}g_s$ is finitely-generated over ${}_n\mathcal{O}$.

To prove (42.1.2), choose a $u \in U$, set $R := \mathscr{R}_u \subset {}_n\mathcal{O}_u =: S$ and consider the induced diagram

$$\begin{array}{ccccc} \mathrm{Ker}\,\varphi_u & \hookrightarrow & S^m & \xrightarrow{\ \varphi_u\ } & S \\ \cup & & \cup & & \cup \\ \mathrm{Ker}\,\psi_u & \hookrightarrow & (R[z_n]_{2r})^m & \xrightarrow{\ \psi_u\ } & R[z_n]_{3r}. \end{array}$$

Suppose first that f_1 is a *Weierstrass polynomial* of degree $\varrho \leqq r$ at the point u. According to the Weierstrass Formula, for fixed $(\chi_1,\ldots,\chi_m) \in \mathrm{Ker}\,\varphi_u \subset S^m$, there exist unique representations of the form

(42.1.3) $\chi_j = f_1 \cdot a_j + b_j$ with $a_j \in S$ and $b_j \in R[z_n]_\varrho$

for all $j \geqq 2$. We see then that, with $b_1 := \chi_1 + \sum_{j=2}^{m} a_j f_j$,

$$0 = \sum_{j=1}^{m} \chi_j f_j = f_1 \chi_1 + \sum_{j=2}^{m} (f_1 a_j f_j + b_j f_j) =$$

$$= f_1(\chi_1 + \sum_{j=2}^{m} a_j f_j) + \sum_{j=2}^{m} f_j b_j = f_1 b_1 + \sum_{j=2}^{m} f_j b_j .$$

With the Weierstrass Formula, the fact that

$$f_1 b_1 = - \sum_{j=2}^{m} f_j b_j \in R[z_n]_{r+\varrho}$$

implies that $b_1 \in R[z_n]_r$, since f_1 is a Weierstrass polynomial. For $A_j := (-f_j, 0, \dots, \underset{\substack{\uparrow \\ j}}{f_1}, 0, \dots, 0) \in \operatorname{Ker} \psi_u$, (42.1.3) yields that

$$(\chi_1, \dots, \chi_m) = (b_1, \dots, b_m) + \sum_{j=2}^{m} a_j A_j .$$

Since A_j and $(\chi_1, \dots, \chi_m)$ lie in $\operatorname{Ker} \varphi_u$, it follows that $(b_1, \dots, b_m) \in \operatorname{Ker} \varphi_u \cap (\mathrm{R}[z_n]_r)^m \subset \operatorname{Ker} \psi_u$; i.e., $(\chi_1, \dots, \chi_m)$ is a linear combination of elements of $\operatorname{Ker} \psi_u$ with coefficients in S.

If f_1 is not a Weierstrass polynomial at u, then there exists a representation of the form $f_1 = e \cdot f$ with $e, f \in R[z_n]_r$, f a Weierstrass polynomial at u, and e a unit in S. The homomorphism

$$\vartheta : S^m \to S, (\chi_1, \dots, \chi_m) \mapsto \varphi_u(\chi_1 e^{-1}, \chi_2, \dots, \chi_m) = f \chi_1 + \sum_{j=2}^{m} f_j \chi_j ,$$

is such that

$$(\chi_1, \dots, \chi_m) \in \operatorname{Ker} \varphi_u \Leftrightarrow (\chi_1, \chi_2 e^{-1}, \dots, \chi_m e^{-1}) \in \operatorname{Ker} \vartheta .$$

By the previously treated special case, there exist for each $(\chi_1, \dots, \chi_m) \in \operatorname{Ker} \varphi_u$ elements $d_j \in S$ and $B_j = (B_{j1}, \dots, B_{jm}) \in (R[z_n]_r)^m \cap \operatorname{Ker} \vartheta$ such that

$$(\chi_1, \chi_2 e^{-1}, \dots, \chi_m e^{-1}) = \sum_{j=1}^{t} d_j B_j, \quad \text{and}$$

$$\tilde{B}_j := (B_{j1}, B_{j2} e, \dots, B_{jm} e) \in (R[z_n]_{2r})^m \cap \operatorname{Ker} \varphi_u \subset \operatorname{Ker} \psi_u ,$$

so $(\chi_1, \dots, \chi_m) = \sum_{j=1}^{t} d_j \tilde{B}_j$ is a representation of the desired form. ■

E. 42a. Fix $G \subset\subset \mathbb{C}^n$, and let $\mathscr{I} \subset {}_G\mathcal{O}$ be an ideal of finite type, and $\mathscr{F}$, a coherent ${}_G\mathcal{O}/\mathscr{I}$-module.

Show that the following sets are analytic in G:
i) $N(\mathscr{I})$ (for 43.1);
ii) $\operatorname{supp}\mathscr{F}$ (hint: E.41i; for E.43c).

E. 42b. For $n = 1$, derive Oka's Coherence Theorem directly from (42.1.1) (hint: Without loss of generality, suppose that $f_j(z) = z^{k_j}$ and $k_1 = \min_{j=1,\dots,m} k_j$).

§ 43 Complex Spaces

Now we want toapply the methods of construction of ringed spaces described in 31.5 ii) to the definition of complex spaces.

43.1 Definition. *A closed complex subspace of a domain $G \subset\subset \mathbb{C}^n$ is a ringed subspace $V(G; \mathscr{I}) = (A, {}_A\mathcal{O}) \hookrightarrow (G, {}_n\mathcal{O})$ defined by means of a coherent ideal $\mathscr{I} \subset {}_n\mathcal{O}|_G$.*

By E. 42a, A is an analytic set in G. Every ringed space isomorphic to such an $(A, {}_A\mathcal{O})$ will be called a *local model.*

43.2 Definition. *A ringed space $(X, \mathcal{O})$ is called a complex space with structure sheaf $\mathcal{O}$ if the following conditions are satisfied:*
i) X is Hausdorff.
ii) $(X, \mathcal{O})$ has an open cover consisting of local models.
A morphism (in the category of ringed spaces) $(\varphi, \varphi^0): (X, {}_X\mathcal{O}) \to (Y, {}_Y\mathcal{O})$ of complex spaces is called a holomorphic mapping.

We generally write just X instead of $(X, \mathcal{O})$ or $(X, {}_X\mathcal{O})$, and φ instead of (φ, φ^0); $\operatorname{Hol}(X, Y)$ is the set of all holomorphic mappings $\varphi: X \to Y$.

43.3 Proposition. *The structure sheaf ${}_X\mathcal{O}$ of a complex space X is a coherent sheaf of rings; its stalks are analytic algebras.*

The proof follows by 41.20 from Oka's Coherence Theorem. ∎

43.4 Examples of complex spaces. i) Complex manifolds are complex spaces.

ii) Let $(X, \mathcal{O})$ be a complex space.

a) For $U \subset\subset X$, the complex space $(U, \mathcal{O}|_U)$ is called an *open subspace* of X.

b) For a coherent sheaf of ideals $\mathscr{I} \subset \mathcal{O}$, we call $Y^{(j)} := V(X; \mathscr{I}^{j+1}) := (N(\mathscr{I}), [\mathcal{O}/\mathscr{I}^{j+1}]|_{N(\mathscr{I})})$ the j-th *infinitesimal neighborhood* of $Y^{(0)}$ in X. There exist canonical inclusions of *closed* (complex) *subspaces* $Y^{(0)} \hookrightarrow Y^{(1)} \hookrightarrow \dots \hookrightarrow X$ with $|Y^{(j)}| = |Y^{(0)}|$ for every j; for $j \geqq 1$, if $0 \neq \mathscr{I} \neq \mathcal{O}$, then the complex space. $Y^{(j)}$ is not reduced.

c) For coherent $\mathcal{O}$-ideals $\mathscr{I}_j$ $(j = 1, 2)$, we define *union* and *intersection* of the associated subspaces by

$$V(\mathscr{I}_1) \cup V(\mathscr{I}_2) := V(\mathscr{I}_1 \cap \mathscr{I}_2) \quad \text{and} \quad V(\mathscr{I}_1) \cap V(\mathscr{I}_2) := V(\mathscr{I}_1 + \mathscr{I}_2).$$

That is possible, since $\mathscr{I}_1 \cap \mathscr{I}_2$ and $\mathscr{I}_1 + \mathscr{I}_2$ are coherent sheaves of ideals, by 41.12.

E. 43a. i) The nullstellen ideal ${}_{\{a\}}\mathscr{I}$ of a point a in a complex space X is coherent (for 45.3).
ii) Every multiple point is a complex space.

E. 43b. Prove that i) the closed complex subspaces of a complex space X form a lattice with respect to $\cup$ and $\cap$;
ii) $A_1 \cup (A_2 \cap A_3) \hookrightarrow (A_1 \cup A_2) \cap (A_1 \cup A_3)$, $|A_1 \cap A_2| = |A_1| \cap |A_2|$, and $|A_1 \cup A_2| = |A_1| \cup |A_2|$ (for 44.7); and
iii) the subspaces $A_k := V(\mathbb{C}^2; z_k)$ for $k = 1,2$ and $A_3 := V(\mathbb{C}^2; z_1^2 - z_2)$ are reduced complex spaces, but $A_2 \cap A_3$ is not reduced; moreover, $A_1 \cup (A_2 \cap A_3) \neq (A_1 \cup A_2) \cap (A_1 \cup A_3)$ (i.e., the lattice is not distributive).

E. 43c. Let X be a complex space, and $\mathscr{F}$, a coherent ${}_X\mathcal{O}$-module. Show that $\operatorname{supp} \mathscr{F}$ *is an analytic set* (defined analogously to 32.5) *in* X (hint: reduce to E. 42a with the help of 41.21; for 45.17).

E. 43d. *Punctual distinction between complex spaces* X_j *by means of infinitesimal neighborhoods.* Consider $X_1 = \mathbb{C}$, $X_2 = \mathbb{C}^2$, $X_3 = V(\mathbb{C}^2; z_1^2 - z_2^3)$, their structure sheaves $\mathcal{O}_j$, and $A_j := (\{0\}, \mathbb{C}) \hookrightarrow X_j$. For their infinitesimal neighborhoods $A_j^{(k)}$, $k \geqq 2$, show that $A_j^{(k)} \cong A_i^{(m)}$ iff $i = j$ and $k = m$ (hint: it follows from E. 32l ii) that the maximal ideal $\mathfrak{m}$ of $\mathcal{O}_{3,0}$ is isomorphic to ${}_1\mathcal{O}_0 \cdot t^2$; thus $\mathfrak{m}/\mathfrak{m}^2 \cong (t^2)/(t^4) \cong \mathbb{C}t^2 + \mathbb{C}t^3$, and the structure sheaf of $A_3^{(1)}$ is $\mathbb{C} \oplus \mathfrak{m}/\mathfrak{m}^2$).

Simply by virtue of dimension, it is not possible in general to introduce a complex structure on $\operatorname{Hol}(X,Y)$ (e.g., $\dim \operatorname{Hol}(\mathbb{C},\mathbb{C}) \geqq \dim \mathbb{C}[z] = \infty$). The situation is different if X is compact: by *Douady's Theorem* [Do Th. 1], there is a natural complex structure on $\operatorname{Hol}(X,Y)$; in the category of reduced complex spaces, it is characterized by the following universal properties:
i) The evaluation $X \times \operatorname{Hol}(X,Y) \to Y$, $(x,\varphi) \mapsto \varphi(x)$, is holomorphic.
ii) For each holomorphic mapping $\psi: T \times X \to Y$, the mapping $T \to \operatorname{Hol}(X,Y)$, $t \mapsto [x \mapsto \psi(t,x)]$ is holomorphic.

E. 43e. i) Determine the canonical complex structure on $\operatorname{Hol}(\bullet, Y)$.
ii) Give a bijective mapping from $\operatorname{Hol}(\dot{\bullet}, Y)$ onto the "tangent space" $\dot{\bigcup}_{a \in Y} T_a Y$ of Y.

43.5 Remark. *If* $\varphi: X \to Y$ *is a holomorphic mapping and* $(B, {}_B\mathcal{O}) \hookrightarrow Y$ *is a complex subspace, then the* <u>*inverse image*</u> $\varphi^{-1}(B) \hookrightarrow X$ *is a complex subspace.*

Proof. The ${}_Y\mathcal{O}$-ideal $\mathscr{J}$ determined by B is coherent; by E. 31o, then, $\varphi^* \mathscr{J}$ is of finite type in ${}_X\mathcal{O}$, and is thus coherent by 41.5. ■

In particular, the *fibers* $\varphi^{-1}(u)$ *of a holomorphic mapping* are provided thereby with a (not necessarily reduced) complex structure (see E. 31h).

E. 43f. Consider $G \subset\subset \mathbb{C}$, $0 \neq f \in \mathcal{O}(G)$, and $a \in N(f)$; show that $\dim_{\mathbb{C}}({}_{f^{-1}(0)}\mathcal{O}_a) = \min\left\{j; \frac{d^j f}{dz^j}(a) \neq 0\right\} = o_a(f)$ (order of f at a).

We saw in 32.4 iii) that, for reduced complex spaces X, the sections in ${}_X\mathcal{O}$ can be identified canonically with the holomorphic $\mathbb{C}$-valued mappings. More generally, we have the following fact:

43.6 Theorem. *Let* $(X, \mathcal{O})$ *be a complex space. Then the canonical mapping*

$$\Psi(X): \mathrm{Hol}(X, \mathbb{C}^m) \to \mathcal{O}(X)^m, (\varphi, \varphi^0) \mapsto (\varphi^0(z_1), \dots, \varphi^0(z_m)),$$

is bijective for every $m \in \mathbb{N}$ *(where* $z_j \in {}_m\mathcal{O}(\mathbb{C}^m)$ *denotes the j-th coordinate function).*

Proof. The mapping $\Psi(X)$ is *injective*: For $\varphi, \tilde{\varphi} \in \mathrm{Hol}(X, \mathbb{C}^m)$ such that $\varphi^0(z_j) = \tilde{\varphi}^0(z_j)$ in $\mathcal{O}(X)$ for $1 \leqq j \leqq m$, we first of all have for every $a \in X$ that

$$b := \varphi(a) = \tilde{\varphi}(a),$$

since $b_j = z_j(\varphi(a)) = (\varphi^0(z_j))(a)$ for every j by E. 31j. Next we obtain that $\varphi_a^0 = \tilde{\varphi}_a^0$, for $\varphi^0(z_j) = \tilde{\varphi}^0(z_j)$ implies that φ_a^0 and $\tilde{\varphi}_a^0$ agree on the set of generators $\{z_{j,b} - b_j;\ 1 \leqq j \leqq m\}$ of the maximal ideal of ${}_m\mathcal{O}_b$ and thus, by 24.2 v), on all of ${}_m\mathcal{O}_b$. Finally, E. 31e implies $\varphi^0 = \tilde{\varphi}^0$.

$\Psi(X)$ is *surjective*: For $(f_1, \dots, f_m) \in \mathcal{O}(X)^m$ and $a \in X$, there exist an open neighborhood U that may be considered as a closed subspace $i: (U, \mathcal{O}|_U) \hookrightarrow G \subset\subset \mathbb{C}^k$, and holomorphic functions $F_j \in {}_k\mathcal{O}(G)$ such that $i^0(F_j) = f_j|_U$. The holomorphic mapping $F := (F_1, \dots, F_m): G \to \mathbb{C}^m$ induces a holomorphic mapping

$$\varphi_U := F \circ i: U \to \mathbb{C}^m \quad \text{such that} \quad \varphi_U^0(z_j) = f_j|_U.$$

As $\Psi(V)$ is injective for any $V \subset\subset X$, these mappings φ_U glue together to form a globally defined $\varphi \in \mathrm{Hol}(X, \mathbb{C}^m)$ such that $\varphi^0(z_j) = f_j$. ■

Let us note that the presheaf $(\mathrm{Hol}(U, \mathbb{C}^m))_{U \subset\subset X}$ determines a sheaf $\mathcal{H}ol(X, \mathbb{C}^m)$ on X and $(\Psi(U))_{U \subset\subset X}$, a sheaf-homomorphism

(43.6.1) $\Psi: \mathcal{H}ol(X, \mathbb{C}^m) \to \mathcal{O}^m.$

Then 43.6 tells us exactly that Ψ is an isomorphism.

Categorically formulated, 43.6 says that $X \mapsto \mathcal{O}(X)^m$ is a representable functor with representing object $\mathbb{C}^m$.

E. 43g. Determine $\Psi(X)$ and $\Psi(X)^{-1}$ for a multiple point X.

In order to treat the zero-sets of polynomials from § 33 as complex spaces, we cite here the fundamental Theorem 47.1, which brings the reduced spaces into the general theory. (Of course, its consequences given here are not used in the proof of 47.1.)

43.8 Theorem. *For a complex space* $(X, \mathcal{O})$, *the sheaf* $\mathcal{N} = \sqrt{0}$ *of nilpotent elements in* $\mathcal{O}$ *is coherent, and the canonical homomorphism* $\mathcal{O}/\mathcal{N} \to \mathrm{Red}\,\mathcal{O}$ *is an isomorphism.*

43.7 Theorem. *For* $G \subset\subset \mathbb{C}^n$, *if* $\mathcal{I}$ *is an* ${}_G\mathcal{O}$*-ideal of finite type and* $A = N(G; \mathcal{I})$, *then the nullstellen ideal* ${}_A\mathcal{I}$ *of* A *is coherent, and* ${}_A\mathcal{I} = \sqrt{\mathcal{I}}$.

A statement equivalent to that (see 47.2), whose importance we discussed in connection with algebraic reduction, runs as follows:

43.9 Corollary. *Reduced complex spaces are complex spaces; in particular,* Red X *is a complex space if* X *is.*

Proof. For an analytic set A in $G \subset\subset \mathbb{C}^n$, the sheaf ${}_A\mathscr{I}$ is coherent by 43.7, so the reduced structure sheaf ${}_A\mathcal{O} = [{}_G\mathcal{O}/{}_A\mathscr{I}]|_A$ is coherent as well (see E. 31b). ■

43.10 Corollary. *In a complex space X, any analytic subset provided with the canonical reduced structure (see 31.6), is a closed complex subspace.* ■

43.11 Corollary. *The following statements are equivalent for a complex space $(X, \mathcal{O})$:*
i) X is reduced.
ii) ${}_X\mathscr{N} = 0$.
iii) In every stalk $\mathcal{O}_a$, the only nilpotent element is 0.

Proof. The implication "i) $\Rightarrow$ ii)" is trivial. By 43.8, statement ii) implies that $\mathcal{O} \to \operatorname{Red}\mathcal{O}$ is an isomorphism, and that in turn implies that X is reduced. ■

For 43.13 we require the following criterion (see also 23.6):

43.12 Proposition. *For $G \subset\subset \mathbb{C}^n$ and $f \in {}_n\mathcal{O}(G)$, the space $(A, {}_A\mathcal{O}) := V(G; f)$ is not reduced iff at least one germ f_a has a multiple factor in ${}_G\mathcal{O}_a$.*

Proof. *If*. If $f_a = gh^2$, say, then the residue class $\overline{gh} \neq 0 \in {}_A\mathcal{O}_a$ is nilpotent. *Only if*. If $h \in {}_A\mathcal{O}_a$ is a representative of a nilpotent class $0 \neq \bar{h} \in {}_A\mathcal{O}_a$, then $h \notin {}_G\mathcal{O}_a \cdot f_a$, but $h^j \in {}_G\mathcal{O}_a \cdot f_a$ for every large j. Thus every prime factor of f_a is a divisor of h, but at least one of them has a higher multiplicity in f_a than in h. ■

Thus we can extend the discussion of § 33 concerning the zero-sets of monic polynomials $P \in \mathcal{O}(G)[T]$ and their reduced structure as follows:

43.13 Example. *For a domain $G \subset\subset \mathbb{C}^n$ and a monic polynomial $P \in \mathcal{O}(G)[T] \subset \mathcal{O}(G \times \mathbb{C})$, consider the subspace $(A, {}_A\mathcal{O}) := V(G \times \mathbb{C}; P)$* of $G \times \mathbb{C}$; *let $\pi: A \to G$ be the canonical projection, and fix $b \in G$. Then the following statements hold:*

i) $(A, {}_A\mathcal{O})$ is reduced $\Leftrightarrow$ P has no multiple factors in $\mathcal{O}(G)[T]_{\text{mon}}$
$\Leftrightarrow 0 \neq \operatorname{Dis}(P) \in \mathcal{O}(G)$
$\Leftrightarrow$ *π is a branched covering of degree* $\deg P$.

ii) The fiber $\pi^{-1}(b)$ is reduced $\Leftrightarrow$ $P(b, T) \in \mathbb{C}[T]$ has no multiple factors
$\Leftrightarrow 0 \neq (\operatorname{Dis} P)(b) \in \mathbb{C}$
$\Leftrightarrow$ *$\pi^{-1}(b)$ contains* $\deg P$ *points.*

Proof. i) *A is reduced* $\underset{43.12}{\Leftrightarrow}$ no germ P_a has a multiple factor in ${}_{n+1}\mathcal{O}_a$ $\underset{23.7}{\Leftrightarrow}$ no germ P_a has a multiple factor in ${}_n\mathcal{O}_{\pi a}[T]$ $\underset{33\,A.4}{\Leftrightarrow}$ no $\operatorname{Dis}(P_a)$ is the zero germ in ${}_n\mathcal{O}_{\pi a}$ $\underset{6.1}{\Leftrightarrow}$ $0 \neq$ $\neq \operatorname{Dis}(P) \in \mathcal{O}(G)$ $\underset{33\,A.4}{\Leftrightarrow}$ P has no multiple factors in $\mathcal{O}(G)[T]_{\text{mon}}$. The final equivalence follows from 33.1 and the proof of 33.6.

ii) This follows from i) with the help of E. 43h i). ■

E. 43h. In the notation of 43.13, prove the following:

i) $\pi^{-1}(b) \cong V(\mathbb{C};\, {}_1\mathcal{O}\cdot P(b,T))$.

ii) $\deg P = \sum_{a\in\pi^{-1}(b)} \dim_{\mathbb{C}}({}_{\pi^{-1}(b)}\mathcal{O}_a)$.

Finally, we want to show that there are products in the category of complex spaces:

43.14 Definition. *A triple (X,π_1,π_2) with holomorphic mappings $\pi_j: X\to X_j$ is called a* product *of X_1 and X_2 if each diagram of the form*

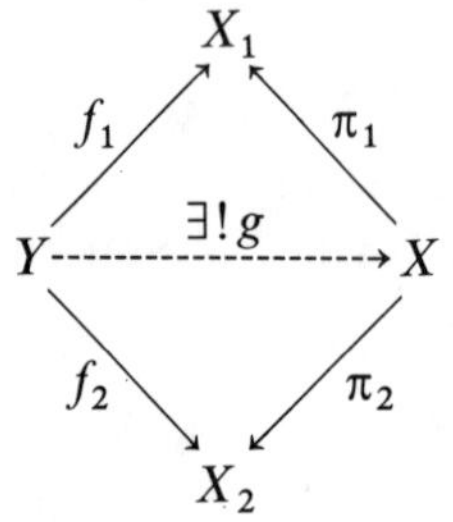

with $f_j \in \mathrm{Hol}(Y, X_j)$ can be completed commutatively by exactly one $g\in \mathrm{Hol}(Y,X)$.

E. 43i. Show, using the diagram below, that the product of complex spaces is determined uniquely "up to isomorphism" if it exists (for 43.16).

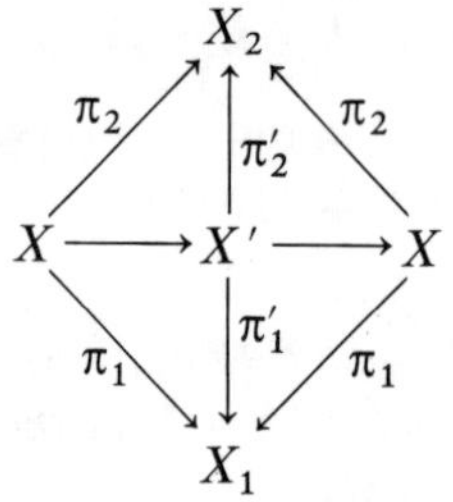

Since the product of X_1 and X_2 is unique, we will write $X_1\times X_2$ and pr_j instead of (X,π_1,π_2) and π_j.

43.15 Example. $\mathbb{C}^m\times\mathbb{C}^n = \mathbb{C}^{m+n}$.

Proof. For a complex space Y, $f=(f_1,\dots,f_m)\in\mathcal{O}(Y)^m \underset{43.6}{=} \mathrm{Hol}(Y,\mathbb{C}^m)$, and $g=(g_1,\dots,g_n)\in\mathcal{O}(Y)^n = \mathrm{Hol}(Y,\mathbb{C}^n)$, we have that $(f_1,\dots,f_m,g_1,\dots,g_n)\in\mathcal{O}(Y)^{m+n} = \mathrm{Hol}(Y,\mathbb{C}^{m+n})$. ■

43.16 Proposition. *There exist products in the category of complex spaces; they are unique up to isomorphism. Moreover,*

$$|X_1\times X_2| = |X_1|\times|X_2|.$$

Proof. Locally, we demonstrate the existence in the following fashion: For $X_j = V(\mathscr{I}_j) \hookrightarrow U_j \Subset \mathbb{C}^{n_j}$ and $\mathrm{pr}_j : U_1 \times U_2 \to U_j$, we have that

$$|X_1| \times |X_2| = |\mathrm{pr}_1^{-1} X_1| \cap |\mathrm{pr}_2^{-1} X_2| = N(\mathrm{pr}_1^* \mathscr{I}_1 + \mathrm{pr}_2^* \mathscr{I}_2).$$

The ${}_{U_1 \times U_2}\mathcal{O}$-ideal $\mathscr{I} := \mathrm{pr}_1^* \mathscr{I}_1 + \mathrm{pr}_2^* \mathscr{I}_2$ is coherent (see 41.12 and the proof of 43.5), and the complex space $V(\mathscr{I}) = (|X_1| \times |X_2|, {}_{U_1 \times U_2}\mathcal{O}/\mathscr{I})$ is a product: by the Restriction Lemma, the canonical projections $U_1 \times U_2 \to U_j$ determine mappings $\pi_j : V(\mathscr{I}) \to X_j$; since $U_1 \times U_2$ is a product, there exists for given holomorphic mappings $g_j : Y \to X_j$ a unique $g : Y \to U_1 \times U_2$ according to the diagram

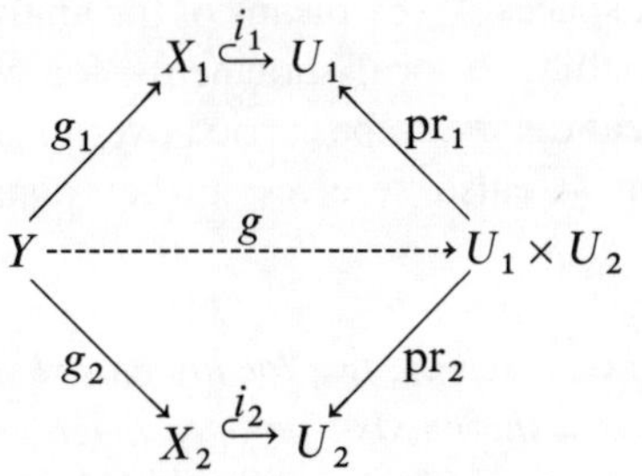

by the Restriction Lemma, g factors uniquely over $V(\mathscr{I})$, since it follows from the equality $g_j^0\, i_j^0(\mathscr{I}_j) = 0$ that $g^0(\mathscr{I}) = 0$.

The *global* existence of products of complex spaces is obtained by gluing (see E. 31m) the products of local models, for the models arising in the construction given above can be glued by virtue of their uniqueness. ■

The product of manifolds is again a manifold, since that holds locally. It can be shown that the product of reduced spaces is reduced [GrRe III. 5. Satz 17]; that is not obvious, since the analogous statement for the fiber-product is false (49 A. 6 iii)). That fact is to be used in the following exercise:

E. 43j. Show, with the help of the functor Red, that the product in the category of reduced complex spaces exists and coincides with that of 43.16.

E. 43k. Generalize the construction of products to finitely many factors.

E. 43l. Prove the following statements for the space $A := V(\mathbb{C}^2; z_1 z_2, z_2^2)$ *(line with a double point)*:

i) $A \setminus \{0\}$ is a manifold.
ii) ${}_A\mathcal{O}_0$ is not reduced.
iii) Every nonunit in ${}_A\mathcal{O}_0$ is a zero-divisor. (For E. 53Ac.)

E. 43m. Let $(X, \mathcal{O})$ be a complex space, and $\mathscr{G}$, a coherent $\mathcal{O}$-module. Then $V(X; \mathscr{A}nn\,\mathscr{G})$ is the "smallest" complex subspace $(A, {}_A\mathcal{O})$ of X on which $\mathscr{G}$ can be interpreted as a coherent ${}_A\mathcal{O}$-module (hint: E. 41g).

E. 43n. A mapping $f \in \mathrm{Hol}(X, Y)$ is biholomorphic iff for every complex space Z, f induces a bijection $f_* : \mathrm{Hol}(Z, X) \to \mathrm{Hol}(Z, Y)$.

E. 43o. If $\mathscr{G}$ is a coherent analytic sheaf on a complex space $(X, \mathcal{O})$, then for each $a \in X$ and each $j \in \mathbb{N}$, there exists near a an exact sequence with $\mathcal{O}$-homomorphisms

$$\mathcal{O}^{p_j} \to \mathcal{O}^{p_{j-1}} \to \ldots \to \mathcal{O}^{p_0} \to \mathscr{G} \to 0 .$$

(Hint: $\mathscr{K}er(\mathcal{O}^p \twoheadrightarrow \mathscr{G})$ is coherent; for 61.13.)

§ 44 Germs of Complex Spaces and Analytic Algebras

The key for translating geometric intuition into the formal language of algebra is the characterization of "germs" of complex spaces X_a by means of the analytic algebras ${}_X\mathcal{O}_a$. That permits the step from punctual to local statements, for which the concept of coherence was worked out in sheaf theory, to be carried over to geometry. Of course, the term "structure sheaf" for ${}_X\mathcal{O}$ also receives further justification. Mimicking E. 21c, we proceed as follows:

44.1 Definition. *For a point a of a complex space X, two locally closed subspaces A and B are to be considered equivalent at a if there exists an open neighborhood U of a such that $(U \cap A, {}_A\mathcal{O}|_{U \cap A}) = (U \cap B, {}_B\mathcal{O}|_{U \cap B})$. The associated equivalence class A_a is called the germ of the space A at a.*

By definition, *holomorphic mappings φ_a between germs of complex spaces* are the germs of holomorphic mappings φ between representatives; thus $\mathrm{Hol}(X_a, Y_b)$ is well-defined. The germs of complex spaces clearly form a category.

E. 44 a. For $0 \neq j \in \mathbb{N} \cup \{\infty\}$, prove the following statements for the germs of the spaces $A_j := V(\mathbb{C}^2; {}_2\mathcal{O} \cdot (z_1 z_2 (z_2 - 1/j)))$ at the point 0:
i) $A_{j,0} \cong V(\mathbb{C}^2; z_1 z_2)_0$ for $j < \infty$; ii) $A_{j,0} \not\cong A_{\infty,0}$.

For a holomorphic mapping $\varphi: X \to Y$, since the algebra-homomorphism

$$\varphi_a^0 : {}_Y\mathcal{O}_{\varphi a} \to {}_X\mathcal{O}_a$$

is determined by the direct limit of mappings

$$\varinjlim_{\varphi a \in V \subset Y} \varphi^0(V) : {}_Y\mathcal{O}(V) \to {}_X\mathcal{O}(\varphi^{-1}(V)),$$

φ_a^0 is characterized by the germ $\varphi_a \in \mathrm{Hol}(X_a, Y_{\varphi a})$. In order to prove the converse, we generalize the concept of an "immersion" introduced in § 32A for manifolds.

44.2 Definition. *A holomorphic mapping $\varphi: X \to Y$ is called a (closed) embedding if it induces an isomorphism of X onto a closed subspace of Y. The germ $\varphi_a \in \mathrm{Hol}(X_a, Y_{\varphi a})$ is called an embedding of germs if it has a representative $\varphi|_U : U \to V \subset Y$ that is an embedding (in that case, φ is called an immersion at a).*

By construction, every germ X_a can be embedded into a germ $\mathbb{C}_0^m$.
Now we generalize 21.9:

44.3 Theorem. *For germs of spaces X_a and Y_b, the canonical mapping*

$$\alpha: \mathrm{Hol}(X_a, Y_b) \to \mathrm{Hom}_{\mathrm{alg}}({}_Y\mathcal{O}_b, {}_X\mathcal{O}_a),\ \varphi_a \mapsto \varphi_a^0,$$

is bijective.

Proof. α is *injective*: Fix an embedding $i: Y_b \hookrightarrow \mathbb{C}_0^m$; it induces a commutative diagram

$$\begin{array}{ccc} \mathrm{Hol}(X_a, Y_b) & \xrightarrow[\alpha]{\varphi_a \mapsto \varphi_a^0} & \mathrm{Hom}_{\mathrm{alg}}({}_Y\mathcal{O}_b, {}_X\mathcal{O}_a) \\ \gamma \Big\downarrow \begin{array}{c}\varphi_a \\ \downarrow \\ i \circ \varphi_a\end{array} & & \Big\downarrow \begin{array}{c} f \\ \downarrow \\ f \circ i^0 \end{array} \\ \mathrm{Hol}(X_a, \mathbb{C}_0^m) & \xrightarrow[\beta]{\psi_a \mapsto \psi_a^0} & \mathrm{Hom}_{\mathrm{alg}}({}_m\mathcal{O}_0, {}_X\mathcal{O}_a)\,; \end{array}$$

hence, it suffices to show that β and γ are injective. For γ, that follows from the fact that i is a monomorphism (see E. 31n). For β, fix $\psi_a, \chi_a \in \mathrm{Hol}(X_a, \mathbb{C}_0^m)$ with $\psi_a^0 = \chi_a^0$; then $\psi_a^0(z_j) = \chi_a^0(z_j)$, $j = 1, \ldots, m$. By (43.6.1), $\psi_a = \chi_a$, so β is injective.

α is *surjective*: Let $t: {}_Y\mathcal{O}_b \to {}_X\mathcal{O}_a$ be an algebra-homomorphism, and choose representatives $X \overset{i}{\hookrightarrow} U \Subset \mathbb{C}^n$ and $Y \overset{j}{\hookrightarrow} V \Subset \mathbb{C}^m$ such that $0 = i(a) = a$ and $0 = j(b) = b$. Then, by 24.2 vii), there exists a commutative diagram of algebra-homomorphisms:

$$\begin{array}{ccc} {}_m\mathcal{O}_0 & \overset{s}{\dashrightarrow} & {}_n\mathcal{O}_0 \\ j_b^0 \Big\downarrow & & \Big\downarrow i_a^0 \\ {}_Y\mathcal{O}_b & \xrightarrow{r} & {}_X\mathcal{O}_a\,. \end{array}$$

After shrinking U and V appropriately, there exists an $F \in \mathrm{Hol}(U, V)$ such that $F_0^0 = s$ (see 21.9). For the ideals $\mathcal{I} \subset {}_U\mathcal{O}$ and $\mathcal{J} \subset {}_V\mathcal{O}$ that determine X and Y, respectively, $F_0^0(\mathcal{J}_0) \subset \mathcal{I}_0$; it follows from 41.17 that $F^0(\mathcal{J}) \subset \mathcal{I}$ near 0. Letting $\varphi: X \to Y$ denote the restriction of F, we have by construction that $\alpha(\varphi) = \varphi_0^0 = r$. ■

44.4 Corollary. *Two germs of complex spaces X_a and Y_b are isomorphic iff ${}_Y\mathcal{O}_b$ and ${}_X\mathcal{O}_a$ are isomorphic.*

Proof. The direction "*only if*" is evident. *If.* By 44.3, isomorphisms $\sigma: {}_Y\mathcal{O}_b \to {}_X\mathcal{O}_a$ and $\tau = \sigma^{-1}$ are represented by holomorphic mappings $\varphi: X \to Y$ and $\psi: Y \to X$. The equality $(\psi \circ \varphi)_a^0 = \mathrm{id}_{{}_X\mathcal{O}_a}$ yields that $\psi \circ \varphi = \mathrm{id}_X$ near a, etc. ■

E. 44 b. Let X denote the ringed subspace $(\{0\}, {}_1\mathcal{O}_0)$ of $\mathbb{C}$. Then the germs X_0 (which is similarly defined) and $\mathbb{C}_0$ are not isomorphic, although the stalks of the structure sheaves of X and $\mathbb{C}$ at 0 are.

i) Determine $\mathrm{Mor}(X_0, \mathbb{C}_0)$ and $\mathrm{Mor}(\mathbb{C}_0, X_0)$.

ii) At which point does it become impossible to carry over the proof of 44.3 (or 44.4) to $X_a = X_0$ and $Y_b = \mathbb{C}_0$?

44.5 Proposition. *For each analytic algebra R, there exists a germ of a complex space X_a such that ${}_X\mathcal{O}_a \cong R$.*

Proof. Without loss of generality, suppose that $R = {}_n\mathcal{O}_0 / (f_1, \dots, f_m)$ for appropriate $f_1, \dots, f_m \in {}_n\mathcal{O}(U)$ and $0 \in U \subset\subset \mathbb{C}^n$. Setting $X := V(U; f_1, \dots, f_m)$, we have that ${}_X\mathcal{O}_0 = R$. ∎

Results 44.3–44.5 can be summarized as follows:
The contravariant functor $X_a \mapsto {}_X\mathcal{O}_a$ from the category of isomorphy-classes of germs of complex spaces into the category of isomorphy-classes of analytic algebras is a contravariant isomorphism.

That relationship between geometry and algebra, which is based on the coherence of the structure sheaf, is fundamental for local complex analysis. For the remainder of this section, we shall present examples of the interplay between the two theories. The reader should take note of two further important applications: the characterization of finite mappings (§45), and dimension theory (§48).

44.6 Corollary. *Every holomorphic mapping between germs of complex spaces $\varphi: X_a \to Y_b$ can be represented as a restriction of a holomorphic mapping between domains of complex number spaces: for each such φ, there is a commutative diagram*

$$\begin{array}{ccc} X_a & \xrightarrow{\varphi} & Y_b \\ \downarrow & & \downarrow \\ \mathbb{C}^n_0 & \xrightarrow{\Phi} & \mathbb{C}^m_0 . \end{array}$$

Proof. It follows from the definition of analytic algebras that there exists a diagram

$$\begin{array}{ccc} {}_X\mathcal{O}_a & \xleftarrow{\varphi^0} & {}_Y\mathcal{O}_b \\ \uparrow & & \uparrow \\ {}_n\mathcal{O}_0 & \xleftarrow{\chi} & {}_m\mathcal{O}_0 \end{array}$$

that can be completed commutatively with a χ, by 24.2 vii). By 44.3, that yields the desired commutative diagram of germs. ∎

44.6 Complement. *i) n and m can be chosen as* $\operatorname{emb} X_a$ *and* $\operatorname{emb} Y_b$, *respectively (see §45).*

ii) Φ can be chosen so as to be the projection $\mathrm{pr}: \mathbb{C}^{m+n}_0 \to \mathbb{C}^m_0$.

Proof. For the graph $\Gamma(\Phi)$, E. 7d yields a commutative diagram

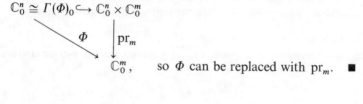

$$\begin{array}{ccc} \mathbb{C}^n_0 \cong \Gamma(\Phi)_0 & \hookrightarrow & \mathbb{C}^n_0 \times \mathbb{C}^m_0 \\ & \searrow^{\Phi} & \downarrow \mathrm{pr}_m \\ & & \mathbb{C}^m_0 , \end{array}$$

so Φ can be replaced with pr_m. ∎

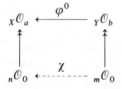

E. 44 c. Let X be a complex space, and fix $a \in X$. Show that X is a manifold near a iff ${}_X\mathcal{O}_a$ is isomorphic to a power series algebra (for 48.15).

Let X_a be the germ at a of a complex space X. We want to compare the lattices

$\mathfrak{B}$ of germs at a of analytic sub*spaces* that are closed near a,
$\mathfrak{N}$ of germs at a of analytic sub*sets* that are closed near a, and
$\mathfrak{I}$ of ideals of ${}_X\mathcal{O}_a$

with one another. The lattice-structures are provided by inclusion,

$$\begin{array}{llll} X \cap Y, & \text{and} & X \cup Y & \text{in } \mathfrak{B}, \\ |X| \cap |Y|, & \text{and} & |X| \cup |Y| & \text{in } \mathfrak{N}, \text{ and} \\ \mathfrak{a} \cap \mathfrak{b}, & \text{and} & \mathfrak{a} + \mathfrak{b} & \text{in } \mathfrak{I} \end{array}$$

($\cap$ and $\cup$ are determined by representatives). The elements of $\mathfrak{B}$ and $\mathfrak{N}$ are called *subgerms* of X_a (those of $\mathfrak{B}$ are germs of spaces, those of $\mathfrak{N}$, germs of sets).

We consider the diagram

$$(44.7.1)\qquad \begin{array}{ccc} \mathfrak{I} & \xrightarrow{\ \sqrt{\ }\ } & \mathfrak{I} \\ {\scriptstyle V}\downarrow & \searrow^{N} & \uparrow{\scriptstyle I} \\ \mathfrak{B} & \xrightarrow{\ \text{Red}\ } & \mathfrak{N} \end{array}$$

with the reduction Red, the radical $\sqrt{\ }$, and mappings V, N, and I defined by the following assignments:
If $\mathfrak{a} \in \mathfrak{I}$ is of the form $\mathfrak{a} = {}_X\mathcal{O}_a \cdot (f_{1,a}, \ldots, f_{m,a})$, where $f_1, \ldots, f_m \in {}_X\mathcal{O}(U)$ and $U \Subset X$, then (see E. 44 d)

$V(\mathfrak{a})$ is the germ of the complex space $V(U; f_1, \ldots, f_m)$ at a, and
$N(\mathfrak{a})$ is the germ of the set $N(U; f_1, \ldots, f_m)$ at a.

For $A_a \in \mathfrak{N}$, we define
$I(A_a) := \{f \in {}_X\mathcal{O}_a;\ (\text{Red} f)|_{A_a} = 0\}$

(where $\text{Red} f|_{A_a} = 0$ means: there exist $U \Subset X$ and representatives $F \in {}_X\mathcal{O}(U)$ of f and $A \hookrightarrow U$ of A_a such that $F|_A = 0$; note that $I(A_a) = {}_X\mathcal{O}_a \Leftrightarrow A_a = \emptyset_a$ (see E. 44 d)).

For $V_a \in \mathfrak{B}$, the germ $\text{Red} V_a$ is determined by the germ of the analytic set $|V|$ (see 31.6); hence, there is a well-defined mapping $\text{Red}: \mathfrak{B} \to \mathfrak{N}$.

E. 44 d. Show that $N(\mathfrak{a})$ does not depend on the choice of $f_1, \ldots, f_m$ and U, and that $I(A_a)$ does not depend on the choice of representatives of A_a.

The elementary properties of (44.7.1) can be summarized as follows:

44.7 Lemma. *The following statements hold for the diagram (44.7.1):*
i) N, V, and I are inclusion-reversing mappings, and $\sqrt{\ }$ and Red *are inclusion-preserving.*

ii) V *is bijective,* I *is injective, and* N *is surjective.*

iii) Red *is a lattice-homomorphism, and* V *and* N *are lattice-antihomomorphisms (i.e.* $N(\mathfrak{a}+\mathfrak{b}) = N(\mathfrak{a}) \cap N(\mathfrak{b})$, $N(\mathfrak{a} \cap \mathfrak{b}) = N(\mathfrak{a}) \cup N(\mathfrak{b})$, $V(\mathfrak{a}+\mathfrak{b}) = V(\mathfrak{a}) \cap V(\mathfrak{b})$, $V(\mathfrak{a} \cap \mathfrak{b}) = V(\mathfrak{a}) \cup V(\mathfrak{b})$).

iv) $\mathrm{Red} \circ V = N = N \circ \sqrt{\ }$; $N \circ I = \mathrm{id}_{\mathfrak{N}}$.

v) $I(A \cup B) = I(A) \cap I(B)$, $I(A \cap B) \supset I(A) + I(B)$, $\sqrt{\mathfrak{a}+\mathfrak{b}} \supset \sqrt{\mathfrak{a}} + \sqrt{\mathfrak{b}}$, $\sqrt{\mathfrak{a} \cap \mathfrak{b}} = \sqrt{\mathfrak{a}} \cap \sqrt{\mathfrak{b}}$.

vi) $I(N(\mathfrak{a})) \supset \sqrt{\mathfrak{a}}$.

Proof. Statement i) is clear. For iv), we restrict ourselves to showing that $N \circ I = \mathrm{id}_{\mathfrak{N}}$: as N is surjective by construction, it suffices to show that $N = N \circ I \circ N$, and that is easy to verify. Statement iii) is evident for V; it follows for Red from E. 43b, and hence for N by iv). Statement ii) follows from the definition of V and the fact that $N \circ I = \mathrm{id}_{\mathfrak{N}}$. For v), we show only that $\sqrt{\mathfrak{a}\mathfrak{b}} = \sqrt{\mathfrak{a} \cap \mathfrak{b}} = \sqrt{\mathfrak{a}} \cap \sqrt{\mathfrak{b}}$: since both $\mathfrak{a}$ and $\mathfrak{b}$ include $\mathfrak{a} \cap \mathfrak{b} \supset \mathfrak{a}\mathfrak{b}$, it follows that $\sqrt{\mathfrak{a}\mathfrak{b}} \subset \sqrt{\mathfrak{a} \cap \mathfrak{b}} \subset \sqrt{\mathfrak{a}} \cap \sqrt{\mathfrak{b}}$; for $a \in \sqrt{\mathfrak{a}} \cap \sqrt{\mathfrak{b}}$, there exist $m, n \in \mathbb{N}$ such that $a^m \in \mathfrak{a}$ and $a^n \in \mathfrak{b}$, so $a^{m+n} \in \mathfrak{a}\mathfrak{b}$ and $a \in \sqrt{\mathfrak{a}\mathfrak{b}}$. Statement vi) follows from the fact that $f^m|_A = 0 \Leftrightarrow f|_A = 0$. ∎

E. 44 e. Show that i) $\sqrt{\mathfrak{a}+\mathfrak{b}} \supsetneq \sqrt{\mathfrak{a}} + \sqrt{\mathfrak{b}}$ and ii) $I(N(\mathfrak{a}) \cap N(\mathfrak{b})) \supsetneq I(N(\mathfrak{a})) + I(N(\mathfrak{b}))$ for

$$\mathfrak{a} := {}_2\mathcal{O}_0 \cdot (z_1 - z_2^2),\ \mathfrak{b} := {}_2\mathcal{O}_0 \cdot z_1 \quad \text{and for} \quad \mathfrak{a} := {}_2\mathcal{O}_0 \cdot z_1 z_2,\ \mathfrak{b} := {}_2\mathcal{O}_0 \cdot (z_1 - z_2)$$

(the intersection of a parabola with its tangent line and the intersection of a line with the union of two other lines is not reduced).

The commutativity of the upper triangle in (44.7.1),

$$I \circ N = \sqrt{\ },$$

is the (nontrivial) *Hilbert Nullstellensatz* 47.1 i). One consequence is that $\operatorname{Im} I = \{\mathfrak{a} \in \mathfrak{I}; \mathfrak{a} = \sqrt{\mathfrak{a}}\}$; by vi), the mapping

$$N : \{\mathfrak{a} \in \mathfrak{I}; \mathfrak{a} = \sqrt{\mathfrak{a}} \to \mathfrak{N}\}$$

is thus a bijection.

Now we compare the concepts of "irreducibility" in $\mathfrak{I}$, $\mathfrak{B}$, and $\mathfrak{N}$.

44.8 Definition. *An ideal* $\mathfrak{a}$ *in a Noetherian ring* R *is called*

i) reducible, if there exist ideals $\mathfrak{a}_j \subset R$ *distinct from* $\mathfrak{a}$ *such that* $\mathfrak{a} = \bigcap_j \mathfrak{a}_j$ *(such a decomposition is called irredundant if no* $\mathfrak{a}_j$ *can be left out, i.e., if* $\bigcap_{k \neq j} \mathfrak{a}_k \not\subset \mathfrak{a}_j$ *for every* j*); otherwise* $\mathfrak{a}$ *is called irreducible;*

ii) primary, if every zero-divisor in $R/\mathfrak{a}$ *is nilpotent (*$\Leftrightarrow$ *for every* $a, b \in R$ *with* $ab \in \mathfrak{a}$*, if* $b \notin \mathfrak{a}$*, then* $a \in \sqrt{\mathfrak{a}}$*).*

44.9 Lemma. *The following statements hold for every ideal* $\mathfrak{a}$ *in a Noetherian ring* R*:*

i) $\mathfrak{a}$ *prime* $\Rightarrow$ $\mathfrak{a}$ *irreducible* $\Rightarrow$ $\mathfrak{a}$ *primary.*

ii) $\mathfrak{a}$ *is the intersection of finitely many irreducible ideals.*

iii) If $\mathfrak{a} = \sqrt{\mathfrak{a}}$*, then* $\mathfrak{a}$ *has a unique representation as an irredundant intersection of prime ideals* $\mathfrak{p}_j$*.*

iv) $\sqrt{\mathfrak{a}}$ *is irreducible* $\Leftrightarrow$ $\sqrt{\mathfrak{a}}$ *is prime.*

Proof. i) The implication "prime $\Rightarrow$ irreducible" follows from ii) and 23A. 11 ii). "irreducible $\Rightarrow$ primary": Since it is possible to move from R to $R/\mathfrak{a}$, it suffices to show that if $\mathfrak{a} = (0)$ is irreducible, then (0) is primary. Suppose that $ab = 0$ and $b \neq 0$. The ascending chain of ideals $\mathrm{Ann}(a) \subset \subset \mathrm{Ann}(a^2) \subset \ldots$ becomes stationary after finitely many steps (see E. 23d). Choose a j such that $\mathrm{Ann}(a^j) = \mathrm{Ann}(a^{j+1})$; then $(a^j) \cap (b) = 0$: for $c \in (a^j) \cap (b)$, we deduce that $ca = 0$, since $c \in (b)$, and that $c = da^j$, since $c \in (a^j)$; thus $0 = ca = da^{j+1}$ and $d \in \mathrm{Ann}(a^{j+1}) = \mathrm{Ann}(a^j)$, so $0 = da^j = c$. Since (0) is irreducible and $(b) \neq (0)$, it follows that $a^j = 0$.

ii) Let $\mathfrak{S}$ be the set of those ideals in R which are not intersections of finitely many irreducible ideals. If $\mathfrak{S}$ is not empty, then, according to E. 23d, $\mathfrak{S}$ has a maximal element $\mathfrak{a}$. Since $\mathfrak{a}$ is reducible, there exist properly larger ideals $\mathfrak{b}$ and $\mathfrak{c}$ such that $\mathfrak{a} = \mathfrak{b} \cap \mathfrak{c}$. But both $\mathfrak{b}$ and $\mathfrak{c}$ are finite intersections of irreducible ideals, so it follows that $\mathfrak{a} \notin \mathfrak{S}$. ↯

iii) By i) and ii), $\mathfrak{a}$ has a representation of the form $\mathfrak{a} = \bigcap_{j=1}^{m} \mathfrak{q}_j$ with primary ideals $\mathfrak{q}_j$. It follows from the proof of 44.7 v) that $\mathfrak{a} = \sqrt{\mathfrak{a}} = \bigcap_{j=1}^{m} \sqrt{\mathfrak{q}_j}$. By removing redundant terms, we can obtain an irredundant decomposition; radicals of primary ideals are clearly prime. Transfering to $R/\mathfrak{a}$, we deduce the uniqueness from 44.10.

iv) This follows from i) and iii). ■

E. 44 f. Prove the following statements:

i) ${}_1\mathcal{O}_0 \cdot z^2$ is irreducible, but not prime.

ii) ${}_2\mathcal{O}_0 \cdot (z_1, z_2)^2 = {}_2\mathcal{O}_0 \cdot (z_1^2, z_1 z_2, z_2^2)$ is primary, but not irreducible.

44.10 Remark. *If $\sqrt{0} = \bigcap_{j=1}^{m} \mathfrak{p}_j$ is an irredundant decomposition into prime ideals of the nilradical in a Noetherian ring R, then the $\mathfrak{p}_j$ are the minimal elements in the set of prime ideals of R.*

Proof. If $\mathfrak{p}$ is any minimal prime ideal, then $\bigcap \mathfrak{p}_j = \sqrt{(0)} \subset \mathfrak{p}$, and it follows by 23A.11 ii) that there exists a $\mathfrak{p}_j \subset \mathfrak{p}$. Consequently, $\mathfrak{p}_j = \mathfrak{p}$. Since the decomposition $\sqrt{0} = \bigcap \mathfrak{p}_j$ is irredundant, every $\mathfrak{p}_j$ must be minimal (with respect to inclusion). ■

44.11 Definition. *For an ideal $\mathfrak{a} \subset {}_X\mathcal{O}_a$, the germ $V(\mathfrak{a}) \hookrightarrow X_a$ is called prime (primary, irreducible) if $\mathfrak{a}$ is prime (primary, irreducible); the germ $N(\mathfrak{a}) \subset |X_a|$ is called irreducible (or prime) if $V(\sqrt{\mathfrak{a}})$ is irreducible.*

It is easy to see that the germ $V := V(\mathfrak{a})$ is prime (primary, irreducible) iff the ideal (0) in ${}_V\mathcal{O}_a$ is prime (primary, irreducible); hence, these properties are intrinsic, they do not depend on X_a. Note that $N(\mathfrak{a})$ may be irreducible though $V(\mathfrak{a})$ is reducible (E. 44g iii)).

For the proof of Hilbert's Nullstellensatz, we need the following fact:

44.12 Proposition. *Every space-germ X_a is a finite union of irreducible (and thus primary) subgerms.*

Proof. The decomposition $(0) = \bigcap_{j=1}^{m} \mathfrak{a}_j$ in ${}_X\mathcal{O}_0$ with irreducible ideals $\mathfrak{a}_j$ according to 44.9 ii) corresponds, by 44.7 iii), to a representation $X_a = \bigcup_{j=1}^{m} V(\mathfrak{a}_j)$. ■

For germs of analytic sets, 44.12 can be sharpened considerably. In doing that, we shall call a decomposition $X_a = \bigcup_{j=1}^{m} X_j$ *irredundant* if no X_j can be left out, i.e., if no X_j is included in $\bigcup_{k \neq j} X_k$.

44.13 Proposition. *(Punctual decomposition into irreducible components). If ${}_X\mathcal{O}_a$ is algebraically reduced, then the germ X_a has a unique representation as an irredundant union $X_a = \bigcup_{j=1}^{m} X_j$ with prime germs X_j. The X_j are called prime components of X_a; they are irreducible and algebraically reduced.*

Proof. ${}_X\mathcal{O}_a$ is algebraically reduced; i.e., $\mathfrak{n} = (0)$. By 44.9 iii) and 44.10, the minimal prime ideals $\mathfrak{p}_j$ of ${}_X\mathcal{O}_a$ provide a unique irredundant decomposition $(0) = \bigcap_{j=1}^{m} \mathfrak{p}_j$. By 44.7iii), there is a one-to-one correspondence between such decompositions and irredundant representations of the form $X_a = V(0) = \bigcup_{j=1}^{m} V(\mathfrak{p}_j)$ with prime germs $V(\mathfrak{p}_j)$. ■

44.14 Corollary. *If ${}_X\mathcal{O}_a$ is algebraically reduced, then the following statements are equivalent:*

i) The germ X_a is prime;
ii) The germ X_a is irreducible;
iii) ${}_X\mathcal{O}_a$ has no zero-divisors. ■

E. 44g. Use Hilbert's Nullstellensatz to prove these statements:

i) For an algebraically reduced subgerm Y of a germ X, Y is irreducible iff $I(Y)$ is prime.

ii) If $\mathfrak{a} \in \mathfrak{J}$ is irreducible, then $N(\mathfrak{a})$ is irreducible.

iii) For $\mathfrak{a} := {}_2\mathcal{O}_0 \cdot (z_1 z_2, z_2^2)$, the decompositions $\mathfrak{a} = (z_1, z_2^2) \cap (z_2) = (z_1, z_2)^2 \cap (z_2)$ are distinct primary decompositions; in particular, $\mathfrak{a}$ is reducible; however, $N(\mathfrak{a}) \subset \mathbb{C}_0^2$ is irreducible (for the geometry, see E. 43 l; for E. 49b).

E. 44h. A *lattice* partically ordered by inclusion $\subset$ is called *noetherian* if every "ascending chain" becomes stationary. Prove the following statements:

i) $(\mathfrak{J}, \subset)$ and $(\mathfrak{B}, \supset)$ are noetherian lattices.

ii) Every nonempty subset of a noetherian lattice contains a maximal element (geometric interpretation?).

For 74.8, we remark that a decomposition $\mathfrak{a} = \bigcap_{j=1}^{m} \mathfrak{a}_j$ with irreducible ideals induces an irredundant decomposition $\mathfrak{a} = \bigcap_{j=1}^{s} \mathfrak{q}_j$ with primary ideals $\mathfrak{q}_j$ such that the prime ideals $\sqrt{\mathfrak{q}_j}$ are pairwise distinct (since the intersection of primary ideals with the same radical $\sqrt{\mathfrak{q}}$ is primary with radical $\sqrt{\mathfrak{q}}$). Such a decomposition is called a *Lasker-Noether decomposition*. Thus we have the following fact:

44.15 Lemma. *Every minimal prime ideal* $\mathfrak{p}$ *in a noetherian ring* R *can be represented in the form* $\mathfrak{p} = (0):(s)$[1] *for an appropriate* $s \in R$.

Proof. For the Lasker-Noether decomposition $0 = \bigcap_{j=1}^{t} \mathfrak{q}_j$, we may assume, in light of 23 A. 11 ii), that $\mathfrak{q}_1 \subset \mathfrak{p}$, and, since $\mathfrak{p}$ is minimal, that $\sqrt{\mathfrak{q}_1} = \mathfrak{p}$. For an $m \in \mathbb{N}$ such that $\mathfrak{p}^m \subset \mathfrak{q}_1$, it follows that

$$0 = \mathfrak{q}_1 \cap \mathfrak{a} = \mathfrak{p}^m \cap \mathfrak{a},$$

with $\mathfrak{a} := \bigcap_{j=2}^{t} \mathfrak{q}_j$. Now choose the smallest m with that property. For every $0 \neq s \in \mathfrak{p}^{m-1} \cap \mathfrak{a}$, then, we claim that $\mathfrak{p} = (0):(s)$; in other words, for $r \in R$,

$$rs = 0 \quad \text{iff} \quad r \in \mathfrak{p}:$$

since $s \notin \mathfrak{q}_1$, the equality $rs = 0 \in \mathfrak{q}_1$ implies that $r^n \in \mathfrak{q}_1$ for some $n \in \mathbb{N}$, and thus that $r \in \sqrt{\mathfrak{q}_1} = \mathfrak{p}$; the other direction is trivial. ■

E. 44i. *Mayer-Vietoris sequence.* For $X = X_1 \cup X_2$ with $X_j \hookrightarrow X$ and $X_{12} := X_1 \cap X_2$, the sequence

$$0 \longrightarrow {}_X\mathcal{O} \longrightarrow {}_{X_1}\mathcal{O} \oplus {}_{X_2}\mathcal{O} \longrightarrow {}_{X_{12}}\mathcal{O} \longrightarrow 0$$
$$f \mapsto (f|_{X_1}, f|_{X_2})\,; \quad (g,h) \mapsto g|_{X_{12}} - h|_{X_{12}}\text{[2]}$$

is exact (hint: apply [AtMac 2.10] for each stalk). Use the example $X = N(\mathbb{C}^2; z_1 z_2(z_1 - z_2))$ to show that, for $f_j \in X_j$, the fact that f_1 and f_2 coincide as continuous functions on X_{12}, is not enough to ensure that there exists an $f \in \mathcal{O}(X)$ such that $f|_{X_j} = f_j$. (Hint: set $X_1 = N(\mathbb{C}^2; z_1 z_2)$, $X_2 = N(\mathbb{C}^2; z_1 - z_2)$, $f_1 = 0$, and $f_2 = z_1|_{X_2}$; for 53 A. 5.)

§ 45 Discrete and Finite Holomorphic Mappings

In this section, we prove a special case of the following theorem (see 45 A):

Grauert's Coherence Theorem. *The images of coherent analytic sheaves under proper holomorphic mappings are coherent.*

The proof of that theorem (see [FoKn]) lies beyond the scope of this book; we also do not use it for the foundation of the theory. Thus we restrict ourselves to indicating the importance of the theorem in its full generality by means of applications (45.17, 49 A. 18–49 A. 22), whereas we prove the special case that is necessary for our development of the theory by means of an idea of Grauert and Remmert:

45.1 Finite Coherence Theorem. *If* $f: X \to Y$ *is a finite holomorphic mapping, and if* $\mathcal{G}$ *is a coherent* ${}_X\mathcal{O}$*-module, then* $f\mathcal{G}$ *is a coherent* ${}_Y\mathcal{O}$*-module.*

---

1) $(t):(s) = \{r \in R;\, rs \in (t)\}$.

2) For $i: A \hookrightarrow X$ and $f \in \mathcal{O}(X)$, we set $f|_A := i^0(f)$.

In 45 B. 1, we shall reconstruct f from the ${}_Y\mathcal{O}$-algebra $f({}_X\mathcal{O})$ with the help of the "analytic spectrum".

We derive 45.1 from a stronger statement about discrete mappings:

45.2 Theorem. *Let $f\colon X \to Y$ be a holomorphic mapping, let $\mathcal{G}$ be a coherent ${}_X\mathcal{O}$-module, and let $a \in X$ be a point at which the mapping $f|_{\operatorname{supp}\mathcal{G}}$ is discrete. Then there exist arbitrarily small open neighborhoods U of a in X and V of $f(U)$ in Y such that $f\colon U \cap \operatorname{supp}\mathcal{G} \to V$ is finite and $(f|_U)(\mathcal{G}|_U)$ is a coherent ${}_Y\mathcal{O}|_V$-module.*

Proof of 45.1 using 45.2. We show first that $f({}_X\mathcal{O})$ is a coherent ${}_Y\mathcal{O}$*-module.* By E. 45A b, it suffices to prove that assertion near points $b \in f(X)$, since f is closed according to 33 B. 1. For $f^{-1}(b) = \{a_1, \dots, a_m\}$, determine neighborhoods V_j of b and pairwise disjoint neighborhoods U_j of a_j as provided by 45.2: for $V := \bigcap V_j$ and $U := f^{-1}(V) \cap (\bigcup_j U_j)$, it follows from 41.10 that $(f|_U)({}_X\mathcal{O}|_U) \cong \bigoplus_{j=1}^{m} (f|_{U\cap U_j})({}_X\mathcal{O}|_{U\cap U_j})$ is coherent. By E. 41f i), the sheaf $f({}_X\mathcal{O})$ is a coherent sheaf of *rings* as well. If $\mathcal{G}$ is a coherent ${}_X\mathcal{O}$-module, then it follows that $f\mathcal{G}$ is a coherent $f({}_X\mathcal{O})$-module, by E. 45A a; thus $f\mathcal{G}$ is a coherent ${}_Y\mathcal{O}$-module, by E. 41f ii). ■

Proof of 45.2. a) First we treat the special case in which f is a projection,

$$f = \operatorname{pr}_Y\colon X = Y \times \mathbf{P}^1 \to Y \mathrel{\subset\!\!\!\!\!\!\;\alpha} \mathbb{C}^n,$$

$\mathcal{G} = {}_A\mathcal{O}$, the structure sheaf of $A := V(Y \times \mathbf{P}^1; P)$, where $P \in \mathcal{O}(Y)[T]_{\text{mon}}$ is a polynomial of degree $g \geq 1$, and $a = 0$. By E. 33c ii), we may assume that $f\colon A \to Y$ is finite; thus it remains to prove that $f_A\mathcal{O}$ is a coherent ${}_Y\mathcal{O}$-module. Let $\eta \in {}_A\mathcal{O}(A)$ be the restriction to A of $\operatorname{pr}_{\mathbf{P}^1}\colon Y \times \mathbf{P}^1 \to \mathbf{P}^1$. The mapping

$$(45.2.1)\quad \vartheta\colon {}_Y\mathcal{O}[T]_g \to f({}_A\mathcal{O}),\ \sum_{j=0}^{g-1} h_j T^j \mapsto \sum_{j=0}^{g-1} h_j \eta^j,$$

is a homomorphism of ${}_Y\mathcal{O}$-modules (see 45 A. 2). It is even an isomorphism: by 45A. 1, $f({}_A\mathcal{O})_z = \bigoplus_{b \in f^{-1}(z)} {}_X\mathcal{O}_b/(P)$ for every $z \in Y$; it follows from 23.11 that ϑ_z is an isomorphism of ${}_Y\mathcal{O}_z$-modules. Thus the ${}_Y\mathcal{O}$-coherence of ${}_Y\mathcal{O}[T]_g$ implies that of $f({}_A\mathcal{O})$.

b) We replace ${}_A\mathcal{O}$ in a) with an arbitrary coherent ${}_X\mathcal{O}$-module $\mathcal{G}$, such that the mapping $f|_{\operatorname{supp}\mathcal{G}}$ is discrete at the point $a = 0$. Now we look for a $P \in \mathcal{O}(G)[T]_{\text{mon}}$ that annihilates $\mathcal{G}$. By E. 41g, we have that $\mathcal{J} := \mathcal{A}nn\,\mathcal{G} \subset {}_X\mathcal{O}$ is a coherent sheaf of ideals such that $\operatorname{supp}\mathcal{G} = N(\mathcal{J})$ (see E. 41i). Now $f|_{N(\mathcal{J})}$ is discrete at 0, so 0 is an isolated point of $N(\mathcal{J}) \cap f^{-1}(0) = N(\mathcal{J}) \cap (0 \times \mathbf{P}^1)$; consequently, for an appropriate shrinking of X, there exists an $h \in \mathcal{J}(X)$ such that $N(h) \cap f^{-1}(0) = \{0\}$. Hence, h is distinguished in the last coordinate of $Y \times \mathbf{P}^1$. By the Preparation Theorem, then, we can replace h with a Weierstrass polynomial $P \in {}_Y\mathcal{O}_0[T]$; without loss of generality, suppose that $P \in \mathcal{O}(Y)[T]_{\text{mon}}$.

We have that $P \in (\mathcal{A}nn\,\mathcal{G})(X)$; by 41.20, it follows that $\mathcal{G}$ is also a coherent ${}_A\mathcal{O}$-module for $A := V(X; P)$. The ${}_Y\mathcal{O}$-module $f({}_A\mathcal{O}) = (f|_A)({}_A\mathcal{O})$, which is coherent by part a), is a coherent sheaf of rings, according to E. 41f; by E. 45Aa, then, $f\mathcal{G}$ is a coherent $f({}_A\mathcal{O})$-module, and it follows from E. 41f that it is a coherent ${}_Y\mathcal{O}$-module.

c) Now we replace $Y \times \boldsymbol{P}^1$ in part b) with $Y \times \boldsymbol{P}^m$, and prove 45.2 by inducting on m. Part b) was the case in which $m = 1$, so we proceed to the induction step. Decompose the projection $f: Y \times \boldsymbol{P}^m \to Y$ as follows:

$$Y \times \boldsymbol{P}^m = (Y \times \boldsymbol{P}^{m-1}) \times \boldsymbol{P}^1 \xrightarrow{h} Y \times \boldsymbol{P}^{m-1} \xrightarrow{g} Y.$$

By b), we may assume that $h|_{\operatorname{supp}\mathcal{G}}$ is finite and that $h\mathcal{G}$ is coherent. Then we have that $\operatorname{supp}(h\mathcal{G}) \underset{45A.1}{=} h(\operatorname{supp}\mathcal{G})$, so the induction hypothesis is applicable to $h\mathcal{G}$ and g; from that, with the observation that $f\mathcal{G} = gh\mathcal{G}$, the assertion follows immediately.

d) In the general case, there exists, by the complement to 44.6, without loss of generality, a commutative diagram of holomorphic mappings:

$$\begin{array}{ccc} X & \xrightarrow{\quad f \quad} & Y \\ {\scriptstyle i}\big\downarrow & & \big\downarrow{\scriptstyle j} \\ \boldsymbol{P} := \boldsymbol{P}^n \times \boldsymbol{P}^m & \xrightarrow{\quad \mathrm{pr}_m \quad} & \boldsymbol{P}^m. \end{array}$$

The trivial extension $\mathcal{G}^{\boldsymbol{P}}$ of $\mathcal{G}$ to $\boldsymbol{P}$ is a coherent ${}_{\boldsymbol{P}}\mathcal{O}$-module (see 41.21); with pr_m, it satisfies the hypotheses of c). By E. 45Ae, and the fact that i and j are closed, the validity of 45.2 carries over from $\mathcal{G}^{\boldsymbol{P}}$ and pr_m to $\mathcal{G}$ and f. ■■

E. 45a. Let $f: X \to Y$ be holomorphic. Show that the set $\{x \in X;\ f \text{ is discrete at } x\}$ is open in X. Show further, with the example $\mathrm{pr}_1 : N(\mathbb{R}^3; x_1^2 - x_2^2 - x_3^2) \to \mathbb{R}$, that that statement does not hold in the real analytic case.

We now present applications of the Finite Coherence Theorem that are statements of the form "punctual ⇒ local". We begin with a characterization of isolated points of complex spaces:

45.3 Proposition. *For a point a in a complex space X, the following statements are equivalent:*

i) $\{a\} \subset\!\!\!\subset X$.

ii) $\dim_{\mathbb{C}}(\mathcal{O}_a) < \infty$.

iii) $\mathfrak{m}_a^j = 0$ *for some* $j \in \mathbb{N}$.

Proof. i) ⇒ ii) The constant mapping $\varphi: (\{a\}, \mathcal{O}_a) \to (\{0\}, \mathbb{C})$ is finite and holomorphic; thus, by the Finite Coherence Theorem, $\mathcal{O}_a$ is a finite $\mathbb{C}$-module.

ii) ⇒ iii) In the finite-dimensional complex algebra $\mathcal{O}_a$, every descending chain of ideals becomes stationary (a ring with that property is called *Artinian*), so there exists a $j \in \mathbb{N}$ such that $\mathfrak{m}_a^j = \mathfrak{m}_a^{j+1}$. By Nakayama's Lemma, $\mathfrak{m}_a^j = 0$.

iii) ⇒ i) The nullstellen ideal $\mathcal{J}$ of the analytic set $\{a\}$ is a coherent ideal, by

E. 43a. Since, by assumption, $\mathscr{J}_a^j = 0$ for some $j \in \mathbb{N}$, there exists a neighborhood $U \subset\subset X$ of a such that $\mathscr{J}^j|_U = 0$, see 41.16. By construction, we have that $1 \in \mathscr{J}_u$ for every $u \neq a$; hence, $U = \{a\}$. ■

The preceding proposition yields a characterization of discrete mappings:

45.4 Theorem (Local characterization of finite morphisms). *For a holomorphic mapping $f: X \to Y$, a point $a \in X$, and $b := f(a)$, the following statements are equivalent:*

i) f is discrete at a (i.e., $\{a\} \subset\subset f^{-1}(b)$).

ii) f is finite near the points a and b (i.e., there exist arbitrarily small open neighborhoods U of a in X and V of $f(U)$ in Y such that $f: U \to V$ is finite).

iii) $f_a^0: {}_Y\mathcal{O}_b \to {}_X\mathcal{O}_a$ is finite (i.e., ${}_X\mathcal{O}_a$ is a finitely-generated ${}_Y\mathcal{O}_b$-module).

iv) $f_a^0: {}_Y\mathcal{O}_b \to {}_X\mathcal{O}_a$ is quasifinite.

We have just used the following weakened form of the definition of a finite ring-homomorphism:

45.5 Definition. *A ring-homomorphism $\varphi: S \to R$ is called* <u>*quasifinite*</u> *if S is a local ring and $\dim_{S/\mathfrak{m}_S}[R/R\cdot\varphi(\mathfrak{m}_S)]$ is finite.*

By 31.8 iv), $R/R\cdot\varphi(\mathfrak{m}_S)$ has for $\varphi = f_a^0: {}_Y\mathcal{O}_b \to {}_X\mathcal{O}_a$ the following geometric meaning: ${}_X\mathcal{O}_a/{}_X\mathcal{O}_a\cdot f_a^0(\mathfrak{m}_b)$ is the stalk ${}_{f^{-1}(b)}\mathcal{O}_a$ of the structure sheaf of the fiber $f^{-1}(b)$ at the point a.

E. 45b. The quasifinite ring-homomorphism $\mathbb{C}\{X\} \hookrightarrow \mathbb{C}[\![X]\!]$ is not finite (hint: 24.4, 45.7).

Proof of 45.4. i) $\Rightarrow$ ii) This follows from 45.2 with $\mathscr{G} = {}_X\mathcal{O}$.

ii) $\Rightarrow$ iii) Without loss of generality, suppose that $f: X \to Y$ is finite and $\{a\} = f^{-1}(b)$. Then, by 45.2, $f({}_X\mathcal{O})$ is a coherent ${}_Y\mathcal{O}$-module; in particular, $f({}_X\mathcal{O})_b \cong {}_X\mathcal{O}_a$ is a finite ${}_Y\mathcal{O}_b$-module with respect to f_a^0.

iii) $\Rightarrow$ iv) This is immediate.

iv) $\Rightarrow$ i) The stalk ${}_{f^{-1}(b)}\mathcal{O}_a = {}_X\mathcal{O}_a/{}_X\mathcal{O}_a\cdot f_a^0(\mathfrak{m}_b)$ is a finite-dimensional vector space; hence, by 45.3, a is an isolated point of the fiber $f^{-1}(b)$. ■

Since every homomorphism of analytic algebras is induced by a holomorphic mapping, according to 44.3, we have the following:

45.6 Corollary. *A homomorphism of analytic algebras is finite iff it is quasifinite.* ■

E. 45c. Show for the finite mapping $X = \mathbb{C}^2 \to \mathbb{C}^2 = Y, (z_1, z_2) \mapsto (z_1, z_2^2)$, that the ${}_Y\mathcal{O}_0$-module ${}_X\mathcal{O}_0$ is generated by $\{1, z_2\}$.

The importance of finite mappings for the development of complex analysis can be seen in the fact that 45.6 is equivalent to the Weierstrass Theorems; especially in the French school (see [Ca exposé 18 (Houzel)]), 45.6 is used as the point of departure for the theory.

E. 45d. Using 45.6, derive the existence of "Weierstrass decompositions" (see 22.3) $Q = q\cdot P + r$ in $\mathbb{C}\{X, Y\}$ for power series $P \in \mathbb{C}\{X, Y\}$ that are distinguished in Y. (Hint: i) the composition $\mathbb{C}\{X\} \to \mathbb{C}\{X, Y\} \to \mathbb{C}\{X, Y\}/(P)$ is finite; ii) $\mathbb{C}\{Y\}/(P(0, Y)) \cong \mathbb{C}[Y]_b$; iii) apply 23. A. 4.)

If $f: X_a \to Y_b$ is a holomorphic mapping of germs, then one can determine with the help of infinitesimal neighborhoods whether f is biholomorphic:

45.7 Proposition. *A homomorphism $\varphi: S \to R$ of analytic algebras is an isomorphism iff every induced mapping $\varphi_j: S / \mathfrak{m}_S^j \to R / R \cdot \varphi(\mathfrak{m}_S^j)$ is an isomorphism.*

For the proof, we use 45.6 to show the following:

45.8 Lemma. *Let $\varphi: S \to R$ be a homomorphism of analytic algebras.*

i) The following statements are equivalent:

a) φ is surjective.

b) $\bar{\varphi}: \mathfrak{m}_S / \mathfrak{m}_S^2 \to \mathfrak{m}_R / \mathfrak{m}_R^2$ is surjective.

c) $R \cdot \varphi(\mathfrak{m}_S) = \mathfrak{m}_R$.

ii) In particular, for $S = {}_m\mathcal{O}_0$ and $f_j := \varphi(z_j)$, $j = 1, \ldots, m$, the following statements are equivalent:

d) φ is surjective.

e) The residue classes $\bar{f}_1, \ldots, \bar{f}_m$ generate $\mathfrak{m}_R / \mathfrak{m}_R^2$ over $\mathbb{C} = R / \mathfrak{m}_R$.

f) $f_1, \ldots, f_m$ generate $\mathfrak{m}_R$ over R.

Proof. The implication "a) ⇒ b)" is clear. b) ⇒ c) If $g_1, \ldots, g_k$ generate $\mathfrak{m}_S$ over S, then $\bar{\varphi}(\bar{g}_1), \ldots, \bar{\varphi}(\bar{g}_k)$ generate the vector space $\mathfrak{m}_R / \mathfrak{m}_R^2$; by 23 A.4, $\varphi(g_1), \ldots, \varphi(g_k)$ generate $\mathfrak{m}_R$ over R.

c) ⇒ a) The mapping $\mathbb{C} \cong S / \mathfrak{m}_S \to R / R \cdot \varphi(\mathfrak{m}_S) \cong \mathbb{C}$ is bijective, so φ is quasifinite; hence, by 45.6, φ is finite. According to 23 A.4, the generator 1 of $R/\mathfrak{m}_R$ over $\mathbb{C}$ also generates R over S; i.e., φ is surjective. ■

Proof of 45.7. If φ_1 is surjective, then $R \cdot \varphi(\mathfrak{m}_S) = \mathfrak{m}_R$; by 45.8, φ is surjective. If every φ_j is injective, then $\operatorname{Ker} \varphi \subset \operatorname{Ker}(\varphi_j \circ (S \to S / \mathfrak{m}_S^j)) = \mathfrak{m}_S^j$ for every j; hence, 24.2 iv) implies that $\operatorname{Ker} \varphi = 0$. ■

Here is an explicit application of 45.7:

E. 45 e. Construct isomorphisms of analytic algebras

$$
\begin{aligned}
&{}_4\mathcal{O}_0/\mathfrak{a} \cong \mathbb{C}\{x^3, x^2y, xy^2, y^3\} \subset {}_2\mathcal{O}_0, \\
&{}_5\mathcal{O}_0/\mathfrak{b} \cong \mathbb{C}\{x^4, x^3y, x^2y^2, xy^3, y^4\} \subset {}_2\mathcal{O}_0, \quad \text{and} \\
&{}_4\mathcal{O}_0/\mathfrak{p} \cong \mathbb{C}\{x^4, x^3y, xy^3, y^4\} \subset {}_2\mathcal{O}_0
\end{aligned}
$$

with

$$
\begin{aligned}
\mathfrak{a} &= {}_4\mathcal{O}_0 \cdot (z_1z_3 - z_2^2, z_1z_4 - z_2z_3, z_2z_4 - z_3^2), \\
\mathfrak{b} &= {}_5\mathcal{O}_0 \cdot \mathfrak{a} + {}_5\mathcal{O}_0 \cdot (z_1z_5 - z_3^2, z_2z_5 - z_3z_4, z_3z_5 - z_4^2), \\
\mathfrak{p} &= \sqrt{\mathfrak{q}} = \mathfrak{q} + {}_4\mathcal{O}_0 \cdot (z_1z_3^2 - z_2^2z_4), \quad \text{where} \\
\mathfrak{q} &= {}_4\mathcal{O}_0 \cdot (z_1z_4 - z_2z_3, z_1^2z_3 - z_2^3, z_2z_4^2 - z_3^3).
\end{aligned}
$$

(Hint: ${}_4\mathcal{O}_0 \to {}_2\mathcal{O}_0, z_k \mapsto x^{4-k}y^{k-1}$, induces an epimorphism $\varphi: S := {}_4\mathcal{O}_0/\mathfrak{a} \twoheadrightarrow R := \mathbb{C}\{x^3, x^2y, xy^2, y^3\}$; by 45.7, it suffices to show that all φ_j are bijective. We have that $d_j(R) := \dim_{\mathbb{C}}(\mathfrak{m}_R^j/\mathfrak{m}_R^{j+1}) = 3j+1$; show recursively that $d_{j+1}(S) \leqq d_j(S) + 3 \leqq 3j + 4$: the classes of monomials $z^\nu \in {}_4\mathcal{O}_0$

with $|\nu| = j+1$ generate $\mathfrak{m}_S^{j+1} / \mathfrak{m}_S^{j+2}$; for the "weight-function" $\gamma(z^\nu) := 3\nu_1 + 2\nu_2 + \nu_3$ motivated by $z^\nu \mapsto x^{3\nu_1+2\nu_2+\nu_3} y^{\nu_2+2\nu_3+3\nu_4}$, show that $\gamma(z^\nu) \geqq 3 \Rightarrow z^\nu \equiv z_1 z^\mu$ (modulo $\mathfrak{a}$) with $|\mu| = j$. Similarly, use

$$_5\mathcal{O}_0 \longrightarrow {_2\mathcal{O}_0}, z_k \mapsto x^{5-k}y^{k-1}, \quad \text{for} \quad {_5\mathcal{O}_0}/\mathfrak{b}, \quad \text{and}$$
$$_4\mathcal{O}_0 \longrightarrow {_2\mathcal{O}_0}, z_1 \mapsto x^4, z_2 \mapsto x^3y, z_3 \mapsto xy^3, z_4 \mapsto y^4, \quad \text{for} \quad {_4\mathcal{O}_0}/\mathfrak{p}.$$

E. 49 As gives a geometric interpretation of ${_4\mathcal{O}_0}/\mathfrak{a}$ and ${_5\mathcal{O}_0}/\mathfrak{b}$; the injection ${_4\mathcal{O}_0}/\mathfrak{p} \longrightarrow {_5\mathcal{O}_0}/\mathfrak{b}$ demonstrates the fact that the *embedding-dimension* $\dim_{\mathbb{C}} \mathfrak{m}_R/\mathfrak{m}_R^2$ of an analytic algebra R can decrease with the procedure of "*Strukturausdünnung*". For E. 48g.)

By 44.4, the *tangent space of a germ* X_a can be defined representative-wise as $TX_a := T_aX$. Thus we obtain a further application of 45.8:

45.9 Proposition. *The following conditions are equivalent for a holomorphic mapping* $f: X_a \to Y_b$ *of germs of complex spaces:*

i) f *is an embedding.*

ii) $f^0: {_Y\mathcal{O}_b} \to {_X\mathcal{O}_a}$ *is surjective.*

iii) $Tf: TX_a \to TY_b$ *is injective.*

Proof. ii) $\Leftrightarrow$ iii) This follows from 45.8 i) via duals.

i) $\Rightarrow$ ii) If $f(X_a) = V(Y_b; \mathcal{I}_b)$, then f^0 is given by ${_Y\mathcal{O}_b} \twoheadrightarrow {_Y\mathcal{O}_b}/\mathcal{I}_b \cong {_X\mathcal{O}_a}$.

ii) $\Rightarrow$ i) For representatives X and Y, there exists near b a coherent sheaf of ideals $\mathcal{I} \subset {_Y\mathcal{O}}$ with $\mathcal{I}_b = \operatorname{Ker} f^0$, by E. 41h. For $Z := V(Y; \mathcal{I})$ we have that ${_X\mathcal{O}_a} \cong {_Z\mathcal{O}_b}$; by 44.3, f determines a holomorphic mapping $X \cong Z \hookrightarrow Y$ near a. ∎

The smallest m for which an embedding $X_a \cong Z_0 \hookrightarrow \mathbb{C}_0^m$ exists is called the *embedding-dimension* $\operatorname{emb} X_a$ of X_a.

45.10 Proposition. *For a germ* X_a,

$$\operatorname{emb} X_a = \dim TX_a = \dim(\mathfrak{m}_a / \mathfrak{m}_a^2).$$

Proof. The definition of TX_a yields the second equality. By E. 32s and 44.3, for $m := \dim TX_a$, there is associated to $TX_a \cong T\mathbb{C}_0^m$ a holomorphic mapping $X_a \to \mathbb{C}_0^m$, which is an embedding, by 45.9. It follows that $m \geqq \operatorname{emb} X_a$. Conversely, the inequality $\dim TX_a \leqq \operatorname{emb} X_a$ follows by 45.9 from the fact that $X_0 \hookrightarrow \mathbb{C}_0^{\operatorname{emb} X_a}$. ∎

E. 45f. Prove the following statements:

α) For two embeddings ${_jf}: X_a \to \mathbb{C}_0^{m_j}$, these conditions hold:

i) If $m_1 = \operatorname{emb} X_a$, then there exists a commutative diagram

$$\begin{array}{ccc} X_a & \xrightarrow{\ {_2f}\ } & \mathbb{C}_0^{m_2} \\ & {_1f}\searrow \quad \nearrow g & \\ & \mathbb{C}_0^{\operatorname{emb} X_a} & \end{array}$$

with a linear embedding g. (Hint: $g^0: {_{m_2}\mathcal{O}_0} \longrightarrow {_{m_1}\mathcal{O}_0}$ can be found using 24.2 vii) and 32 A. 4_{alg}; by 45.9 and 45.10, Tg is injective).

ii) If $m_1 = m_2 = \operatorname{emb} X_a$, then the g in i) is biholomorphic.

β) A holomorphic mapping $f: X \to Y$ satisfies these conditions:

i) f is an embedding iff f is injective, closed, and an embedding at each $a \in X$ (for 52.8).

ii) If f is an embedding, then $f(X)$ is the closed subspace $V(Y; \mathscr{K}er\, f^0)$ of Y.

γ) There exists an immersion $\mathbb{C} \to \mathbb{C}$ that is not an embedding.

E. 45 g. *Extension of finite mappings to embeddings.* Let $g: X \to Y$ be a finite holomorphic mapping, and let $h_1, \ldots, h_m \in \mathcal{O}(X)$ be generators of the ${}_Y\mathcal{O}$-module $g({}_X\mathcal{O})$. Then $f := (g, h_1, \ldots, h_m)$: $: X \to Y \times \mathbb{C}^m$ is an embedding. (Hint: f is injective and proper, and every f_a^0 is surjective: $\varphi \in {}_X\mathcal{O}_a$ with $\varphi = \sum_{j=1}^{m} \psi_j h_j \in g({}_X\mathcal{O})_{g(a)}$ is the image of $\sum_{j=1}^{m} \psi_j z_j \in {}_{Y \times \mathbb{C}^m}\mathcal{O}_{f(a)}$. For E. 45j.)

Now we turn to a function-theoretic description of a topological property of holomorphic mappings:

45.11 Definition. *A continuous mapping $f: T \to V$ of topological spaces is called open at a point $a \in T$ if $f(U)$ is a neighborhood of $f(a)$ for every neighborhood U of a. In that case, the associated "mapping-germ" $f_a: T_a \to V_{f(a)}$ of germs of topological spaces is called open at a, as well.*

E. 45 h. Openness at a point is not a local property: find an example of a holomorphic mapping $N(\mathbb{C}^2; z_1 z_2) \to \mathbb{C}$ that is open at 0, but not near 0. 2

45.12 Open Lemma. *Let the holomorphic mapping $f: X \to Y$ be discrete at $a \in X$. Then f is open at a iff* $\operatorname{Red} \varphi = 0$ *for every* $\varphi \in \operatorname{Ker}(f_a^0 : {}_Y\mathcal{O}_{f(a)} \to {}_X\mathcal{O}_a)$.

Proof. Only if. By the Restriction Lemma, for $\varphi \in \operatorname{Ker}(f^0 : \mathcal{O}(Y) \to \mathcal{O}(X))$, there exists a factorization $X \to V(\varphi) \hookrightarrow Y$ of f. Now $|V(\varphi)|$ includes the neighborhood $f(X)$ of $f(a)$ in Y and $\operatorname{Red} \varphi|_{|V(\varphi)|} = 0$.

If. Since $\mathfrak{a} := \operatorname{Ker} f_a^0 \subset \operatorname{Ker} \operatorname{Red}$, the set $N(\mathfrak{a})$ is a neighborhood of $b := f(a)$ (otherwise, the finitely-generated ideal $\mathfrak{a}$ would contain a φ with values different from zero in any neighborhood of a; thus $\operatorname{Red} \varphi$ would be nonzero ↯). Hence, it suffices to show that $N(\mathfrak{a})_b \subset f(X)_b$. By 45.2, we may assume that f is a finite mapping with $f^{-1}(b) = \{a\}$. The Finite Coherence Theorem and E. 41g yield that $\mathscr{A}nn\, f({}_X\mathcal{O})$ is a coherent ${}_Y\mathcal{O}$-module with

$$f(X) \underset{33\,\mathrm{B}.1}{=} \overline{f(X)} \underset{\mathrm{E}.\,45\mathrm{A\,b}}{=} \operatorname{supp} f({}_X\mathcal{O}) \underset{\mathrm{E}.\,41\mathrm{i}}{=} N(\mathscr{A}nn\, f({}_X\mathcal{O}))\,.$$

It is easy to see that

$$\mathfrak{a} = \operatorname{Ker} f_a^0 \overset{!}{=} \operatorname{Ann}_{{}_Y\mathcal{O}_b}({}_X\mathcal{O}_a) = (\mathscr{A}nn_{{}_Y\mathcal{O}} f({}_X\mathcal{O}))_b\,;$$

it follows that $N(\mathfrak{a})_b = f(X)_b$. ∎

45.13 Corollary. *For a holomorphic mapping $f: X \to Y$ that is discrete at a,*

i) if f_a^0 is injective, then f_a is open at a;

ii) if Y is reduced and f_a is open at a, then f_a^0 is injective. ∎

With the aid of the preceding result, we can interpret *Noether's Normalization Theorem* geometrically:

45.14 Proposition. *For each germ X_a, there exists a finite holomorphic mapping $f_a: X_a \to \mathbb{C}^d_0$ that is open at a.*

Proof. By 24.3, there exists a finite injective algebra-homomorphism $\varphi: {}_d\mathcal{O}_0 \to {}_X\mathcal{O}_a$. The associated (see 44.3) holomorphic mapping $f_a: X_a \to \mathbb{C}^d_0$ is finite, by 45.4, and open at a, by 45.13. ∎

In 48.8_{geo}, the number d turns out to be the "dimension" of X_a. For prime germs, 45.14 is sharpened considerably in 46.1.

We bring the section to a close with two consequences of the Finite Coherence Theorem:

Let X be a complex space, and let $f: Y \to X$ be a (finite) holomorphic mapping. Then Y – precisely, (Y,f) – is called a *(finite) complex space over* X. We want to show that every finite complex space (Y,f) over X is characterized "up to isomorphisms" by the ${}_X\mathcal{O}$-algebra $f({}_Y\mathcal{O})$ (see 45 A. 2). The set of (finite) complex spaces over X forms a category with the sets of morphisms (for (Y,f) and (Z,g))

$$\mathrm{Hol}_X(Y,Z) := \{h \in \mathrm{Hol}(Y,Z);\, f = g \circ h\}$$

of *holomorphic mappings over* X. A contravariant functor from that category into the category of coherent ${}_X\mathcal{O}$-modules is defined by the assignment $(Y,f) \mapsto f({}_Y\mathcal{O})$ and

$$\begin{aligned} \Phi: \mathrm{Hol}_X(Y,Z) &\to \mathrm{Hom}_{\mathrm{alg}}(g({}_Z\mathcal{O}), f({}_Y\mathcal{O}))\,, \\ h &\mapsto ({}_Z\mathcal{O}(g^{-1}(U)) \xrightarrow{h^0} {}_Y\mathcal{O}(f^{-1}(U)))_{U \subset X}\,; \end{aligned} \tag{45.15.1}$$

it is easy to see that Φ is compatible with the composition of mappings over X.

45.15 Lemma. *For finite spaces Y and Z over X, Φ is bijective.*

Let us first state an immediate consequence of 45.15:

45.16 Proposition. *Two finite complex spaces (Y,f) and (Z,g) over X are biholomorphically equivalent over X iff the ${}_X\mathcal{O}$-algebras $f({}_Y\mathcal{O})$ and $g({}_Z\mathcal{O})$ are isomorphic.* ∎

The finiteness condition in 45.15 is essential:

E. 45 i. Prove the following statements for spaces (Y,f) and (Z,g) over X:

i) Even if f is biholomorphic and g is injective, Φ need not be surjective (hint: $f = \mathrm{id}_{\mathbb{C}^2}$, $g: \mathbb{C}^2 \setminus 0 \subset \mathbb{C}^2$).

ii) Even if f is biholomorphic and g is proper, Φ need not be injective (hint: $X = Y = \bullet$).

For a ringed space S, the category of *(finite) ringed spaces over* S is defined similarly; as the proof indicates, corresponding versions of 45.15 and 45.16 hold for locally compact ringed spaces.

Proof of 45.15. For $\varphi \in \mathrm{Hom}_{\mathrm{alg}}(g({}_Z\mathcal{O}), f({}_Y\mathcal{O}))$, we construct an h in $\Phi^{-1}(\varphi)$; the uniqueness of h will follow from the construction.

a) *Construction of* $|h|:|Y| \to |Z|$: for $a \in X$, $B := f^{-1}(a)$, and $C := g^{-1}(a)$, the homomorphism φ determines, according to 45 A. 1, an ${}_X\mathcal{O}_a$-algebra-homomorphism between finite direct products of ${}_X\mathcal{O}_a$-algebras:

$$\psi\colon \bigoplus_{c\in C} {}_Z\mathcal{O}_c = g({}_Z\mathcal{O})_a \xrightarrow{\varphi_a} f({}_Y\mathcal{O})_a = \bigoplus_{b\in B} {}_Y\mathcal{O}_b\,.$$

By E. 24a v), with the matrix-representation $\psi = (\psi_{bc})$, there exists for each $b \in B$ precisely one $c \in C$ with $\psi_{bc} \neq 0$ (that ψ_{bc} is therefore an algebra-homomorphism). Hence, we must set $h(b) := c$.

b) *Continuity of* h: for $U \subset X$, we see that $h^{-1}(g^{-1}(U)) = f^{-1}(U) \subset Y$. If y is a point in Y and $W \subset Z$ a neighborhood of $h(y)$ such that $W \cap g^{-1}(f(y)) = \{h(y)\}$, then there exists a $W_1 \subset Z$ that does not intersect W, with $g^{-1}(f(y)) \subset W \cup W_1$. Since g is a finite mapping of locally compact spaces, we may assume by 33 B. 2 that $W \cup W_1 = g^{-1}(U)$. Hence, it suffices to show that $h^{-1}(W)$ is open for each simultaneously open and closed subset W of $g^{-1}(U)$. We may assume that $U = X$; then the characteristic function χ of W lies in $\mathcal{O}(Z)$, and $\varphi(\chi)$, in $\mathcal{O}(Y)$. Furthermore, $V := \{y \in Y;\ \varphi(\chi)(y) = 1\}$ is open in Y, since $\varphi(\chi)$ is a continuous $\mathbb{Z}$-valued function. Finally, $h^{-1}(W) = V$: with ψ_b derived from ψ by projection onto ${}_Y\mathcal{O}_b$, we see that

$$\varphi(\chi)(b) = \psi_b\Big(\sum_{z\in g^{-1}f(b)} \chi_z\Big)(b);$$

the assertion follows from the fact that $\chi_z = 1 \in {}_Z\mathcal{O}_z$ iff $z \in W$.

c) Construction of the *comorphism* h^0: for $W \subset Z$ without loss of generality as in b), $\mathcal{O}(W)$ is a direct summand of $\mathcal{O}(Z)$; hence $h^0(W)$ is determined by

$$\mathcal{O}(W) \to \mathcal{O}(Z) \xrightarrow{\varphi} \mathcal{O}(Y) \to \mathcal{O}(h^{-1}(W)). \quad \blacksquare$$

E. 45 j. Let $h\colon (S,f) \to (T,g)$ be a *surjective* morphism of *reduced* ringed spaces that are *finite* over a *complex space* X, and suppose that there exist $g_1, \ldots, g_m \in g({}_T\mathcal{A})(X) = {}_T\mathcal{A}(T)$ with $g({}_T\mathcal{A}) = {}_X\mathcal{O}\cdot g_1 + \ldots + {}_X\mathcal{O}\cdot g_m$. If S is a complex space, then so is T. (Hint: in the commutative diagram

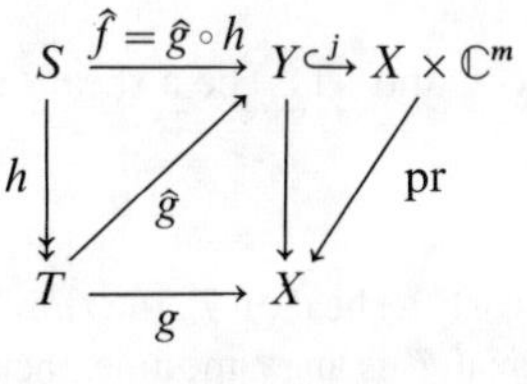

with $\hat{g} := (g, g_1, \ldots, g_m)$ and $Y := \hat{g}(T) = \hat{f}(S)$, observe that $\hat{g}$ is a homeomorphism, and $j \circ \hat{f}\colon S \to X \times \mathbb{C}^m$ is finite; by 45.17, Y is analytic; provide Y with the canonical reduced structure, and deduce as in E. 45g that $\hat{g}$ is an siomorphism of ringed spaces; E. 45j makes it possible to prove 49A. 13 without application of the analytic spectrum).

The next result is an immediate consequence of Grauert's Coherence Theorem:

45.17 Remmert's Mapping Theorem. *If $f: X \to Y$ is a proper holomorphic mapping of complex spaces, and $A \subset X$ is an analytic set, then $f(A)$ is analytic in Y.*

Proof. Since $f|_A$ is a proper holomorphic mapping as well, we may assume that $A = X$. By Grauert's Coherence Theorem, $f({}_X\mathcal{O})$ is a coherent ${}_Y\mathcal{O}$-module, and $f(X) \underset{\text{E. 45Ab}}{=} \operatorname{supp} f({}_X\mathcal{O})$ is analytic in Y, by E. 43c (for *finite* f, the presented proof uses only 45.1 and is hence complete). ∎

Remmert's Mapping Theorem holds even if f is only *quasiproper*, i.e., if there exists for each $U \Subset Y$ a compact $K \subset X$ such that $U \cap f(X) = U \cap f(K)$ (see [Kn]).

E. 45k. For a connected complex space X with a countable topology, and a *closeed* holomorphic mapping $f: X \to Y$, *if every fiber of f is nowhere dense, then f is proper*. (Hint: 51 A.2 i); for a noncompact fiber $f^{-1}(y)$, it is possible to construct a sequence $(x_j)_{j \in \mathbb{N}}$ in $X \setminus f^{-1}(y)$ with no limit point in X so that the images converge to y. Remark: If $\operatorname{Red} X$ is irreducible and f is nonconstant, then every fiber of f is nowhere dense, by 49.8).

E. 45 l. *Remmert's Mapping Theorem for closed mappings.* Use 45.17 to conclude that, if every connected component of X has a countable topology, and if $f \in \operatorname{Hol}(X, Y)$ is closed, then $f(X)$ is analytic in Y (hint: E. 45k with the proof of 49.5).

E. 45m. Give an example of an $f \in \operatorname{Hol}(X, Y)$ that sends analytic sets to analytic sets, but does not send coherent sheaves to coherent sheaves.

E. 45n. If $f: X \to Y$ is a monomorphism, then $|f|$ is injective and f^0 is surjective.

E. 45o. Prove the following statements for a proper holomorphic mapping $f: X \to Y$:

i) The Open Lemma, as stated in 45.12, does not hold for f (hint: quadratic transformation).

ii) For $a \in X$, the image of each neighborhood of $f^{-1}(f(a))$ is a neighborhood of $f(a)$ iff $\operatorname{Ker}(f_a^0: {}_Y\mathcal{O}_{f(a)} \to \mathcal{O}(f^{-1}(f(a))) \subset \operatorname{Ker}\operatorname{Red}_a$. (Hint: Grauert's Coherence Theorem).

E. 45p. Let $f: X \to \mathbb{C}^m$ be an embedding, and fix $g \in \operatorname{Hol}(X, \mathbb{C}^s)$. Then $(f, g): X \to \mathbb{C}^{m+s}$ is an embedding, as well (hint: 45.9; for 52.8).

§ 45A Supplement: Image Sheaves

Let $f: S \to T$ be a continuous mapping of topological spaces, and let $\mathcal{G}$ be a sheaf on S. Then the presheaf

$$f\mathcal{G}: U \mapsto \mathcal{G}(f^{-1}U),\ U \Subset T$$

with the restriction-homomorphisms inherited from $\mathcal{G}$, is clearly a sheaf on T, the *(direct) image-sheaf $f\mathcal{G}$ of $\mathcal{G}$ under f.* If $\mathcal{R}$ is a sheaf of rings, then so is $f\mathcal{R}$; if $\mathcal{G}$ is an $\mathcal{R}$-module, then $f\mathcal{G}$ is an $f\mathcal{R}$-module. For each homomorphism $\varphi: \mathcal{F} \to \mathcal{G}$, there is a canonical homomorphism $f\varphi: f\mathcal{F} \to f\mathcal{G}$. Hence, f can be viewed as a covariant functor from the category of $\mathcal{R}$-modules into the category of $f\mathcal{R}$-modules; it is left-exact, since the "section-functor" Γ is left-exact and the inductive limit of exact sequences is exact. In general, f is not a right-exact functor (we may use 30.3 with $S = \mathbb{C}^{2*}$, $T = \mathbb{C}$, and $f = \operatorname{pr}_2$ as an example: the homomorphism ${}_S\mathcal{O} \to {}_S\mathcal{O}/{}_A\mathcal{I}$ is surjective, but

$$f({}_2\mathcal{O})_0 = f({}_S\mathcal{O})_0 \to f({}_S\mathcal{O}/{}_A\mathcal{I})_0 = {}_A\mathcal{O}(A)$$

is not).

45 A. 1 Remark. *If $f: S \to T$ is a closed mapping of locally compact spaces, then $(f\mathcal{G})_t = \mathcal{G}(f^{-1}(t))$. If f is finite, then $(f\mathcal{G})_t = \bigoplus_{s\in f^{-1}(t)} \mathcal{G}_s$. In particular, for a finite mapping, the formation of image-sheaves is an exact functor.*

Proof. The first assertion follows from the fact that each fiber $f^{-1}(t)$ has a fundamental system of neighborhoods consisting of open sets $f^{-1}(T\setminus f(S\setminus U))$ with $f^{-1}(t) \subset U \Subset S$. For a finite mapping f, it follows that $(f\mathcal{G})_t \cong \bigoplus_{s\in f^{-1}(t)} \mathcal{G}_s$; consequently, for each surjective sheaf-homomorphism $\varphi: \mathcal{F} \to \mathcal{G}$, the associated $f\varphi$ is surjective as well. ■

For nonfinite mappings, “higher image-sheaves” $f_q\mathcal{G}$ with $f_0\mathcal{G} = f\mathcal{G}$ come under consideration as a means for replacing the missing exactness of the image-sheaf functor (§62).

E. 45 A a. Suppose that $f: S \to T$ is a finite mapping, and that $\mathcal{R}$ is a sheaf of rings on S such that $f\mathcal{R}$ is a coherent sheaf of rings. Show that if $\mathcal{G}$ is a coherent $\mathcal{R}$-module, then $f\mathcal{G}$ is a coherent $f\mathcal{R}$-module (hint: 41.19 and 45 A.1; for 45.1).

E. 45 A b. For $f \in \mathrm{Hol}(X,Y)$, show that $\overline{f(X)} = \operatorname{supp} f({}_X\mathcal{O})$ (for 45.1).

E. 45 A c. Let $f: S \to T$ be a finite surjective mapping, and $(S, \mathcal{A})$, a ringed space. Then every stalk of $f\mathcal{A}$ is a local algebra iff f is bijective.

E. 45 A d. *Canonical factorization of morphisms.* Let S and T be ringed spaces, and $\tau: |S| \to |T|$, a continuous mapping. Prove that the following statements hold:

i) ${}_S\tau^0 := (\mathrm{id}_{{}_S\mathcal{A}(\tau^{-1}V)})_{V \subset T}: \tau({}_S\mathcal{A}) \to {}_S\mathcal{A}$ is a canonical comorphism for τ.

ii) For each morphism $\varphi = (\tau, \varphi^0): S \to T$ of ringed spaces, there exists precisely one morphism ${}_T\varphi^0: {}_T\mathcal{A} \to \tau({}_S\mathcal{A})$ of algebras on $|T|$ such that for ${}_S\varphi^0$ and ${}_T\tau^0$ from E. 31 f, the diagram

$$\begin{array}{ccc} S = (|S|, {}_S\mathcal{A}) & \xrightarrow{(\mathrm{id}_S,\, {}_S\varphi^0)} & (|S|, \tau^{-1}{}_T\mathcal{A}) \\ \downarrow{\scriptstyle (\tau,\, {}_S\tau^0)} & \searrow{\scriptstyle \varphi} & \downarrow{\scriptstyle (\tau,\, {}_T\tau^0)} \\ (|T|, \tau_S\mathcal{A}) & \xrightarrow{(\mathrm{id}_T,\, {}_T\varphi^0)} & (|T|, {}_T\mathcal{A}) \end{array}$$

commutes; the assignment $\varphi \mapsto {}_T\varphi^0$ is bijective. By E. 31 f, it follows that

$$\mathrm{Hom}_{\mathrm{alg}}(\tau^{-1}{}_T\mathcal{A}, {}_S\mathcal{A}) \cong \mathrm{Comor}({}_T\mathcal{A}, {}_S\mathcal{A}) \cong \mathrm{Hom}_{\mathrm{alg}}({}_T\mathcal{A}, \tau_S\mathcal{A}).$$

E. 45 A e. Let $j: T \subset S$ denote the inclusion of a closed subset and let $\mathcal{G}$ be a sheaf on T. Show that $j\mathcal{G} = \mathcal{G}^S$ (for 45.2).

E. 45 A f. If $f: X \to Y$ is a finite holomorphic mapping, then each coherent ideal $\mathcal{I}$ in $f({}_X\mathcal{O})$ determines, in a canonical fashion, a coherent ideal $\mathcal{J}$ in ${}_X\mathcal{O}$ with $f\mathcal{J} = \mathcal{I}$ (hint: E. 24 a; for 45 B. 1).

45 A. 2 Remark. *Let $f: X \to Y$ be a holomorphic mapping, and $\mathcal{G}$, a coherent ${}_X\mathcal{O}$-module. With ${}_Yf^0: {}_Y\mathcal{O} \to f({}_X\mathcal{O})$ from E. 45 A d, $f\mathcal{G}$ is a ${}_Y\mathcal{O}$-module (for $\alpha \in {}_Y\mathcal{O}(U)$ and $\gamma \in (f\mathcal{G})(U) = \mathcal{G}(f^{-1}(U))$, we have that $\alpha \cdot \gamma = f^0(\alpha) \cdot \gamma$).* ■

A sheaf $\mathscr{G}$ on X is called *flabby* if, for every $U \subset\subset X$, the restriction-homomorphism $\mathscr{G}(X) \to \mathscr{G}(U)$ is surjective (i.e., if every section over U has a global extension); with that terminology, we obviously have the following fact:

45 A. 3 Remark. *If $f: S \to T$ is a continuous mapping of topological spaces, and if $\mathscr{G}$ is a flabby sheaf on S, then $f\mathscr{G}$ is a flabby sheaf on T.* ∎

E. 45Ag. i) For $m \in \mathbb{N}$ and $f: \mathbb{C} \to \mathbb{C}, z \mapsto z^m$, show that $f({}_1\mathcal{O})$ is a free ${}_1\mathcal{O}$-module with m generators, but that it is not a free ${}_1\mathcal{O}$-algebra for $m > 1$ (hint: (45.2.1)).

ii) Construct a finite holomorphic mapping $g: X \to Y$ such that $g({}_X\mathcal{O})$ is not a locally free ${}_Y\mathcal{O}$-module.

§ 45B Supplement: The Analytic Spectrum

Let X be a complex space. To each finite complex space (Y,f) over X, there corresponds, by means of the assignment $(Y,f) \mapsto f({}_Y\mathcal{O})$, an ${}_X\mathcal{O}$-algebra that (by the Finite Coherence Theorem) is coherent as an ${}_X\mathcal{O}$-module. Conversely, every ${}_X\mathcal{O}$-algebra that is coherent as an ${}_X\mathcal{O}$-module can be realized in that way:

45 B. 1 Theorem. *Let X be a complex space, and let $\mathscr{A}$ be an ${}_X\mathcal{O}$-algebra that is coherent as an ${}_X\mathcal{O}$-module. Then, up to isomorphism, there exists precisely one complex space (Y,f) that is finite over X with an ${}_X\mathcal{O}$-algebra-isomorphism $\mu: f({}_Y\mathcal{O}) \to \mathscr{A}$; (Y,f) is called the analytic spectrum* Specan $\mathscr{A}$ *of* $\mathscr{A}$.

The mapping f is constructed in such a way that, for each $x \in X, |f^{-1}(x)|$ is the *maximal-spectrum* (i.e., the set of maximal ideals) of the algebra $\mathscr{A}_x$; that motivates the terminology "analytic spectrum".

Proof. Since 45.16 implies that Specan $\mathscr{A}$ is *unique* if it exists, it suffices to show the *existence* locally. Thus, without loss of generality, let $X = V(G; \mathscr{I})$ be a local model in $G \subset\subset \mathbb{C}^n$. According to 41.21 ii) and 41.20, the trivial extension $\mathscr{A}^G$ of $\mathscr{A}$ is an ${}_G\mathcal{O}$-algebra that is coherent as a module; hence, we may assume that there exist a surjective homomorphism $\varphi: {}_G\mathcal{O}[T_1, \ldots, T_m] \twoheadrightarrow \mathscr{A}^G$ of ${}_G\mathcal{O}$-algebras and polynomials $P_j \in \mathcal{O}(G)[T_j]_{\text{mon}}$ such that $P_j(\varphi(T_j)) = 0,\ j = 1, \ldots, m$ (see 23 A. 7). For

$$Z := V(G \times \mathbb{C}^m; P_1, \ldots, P_m) \quad \text{and} \quad \pi: Z \to G, (g, z) \mapsto g,$$

the coherent ${}_G\mathcal{O}$-modules $\pi({}_Z\mathcal{O})$ and $\mathscr{R} := {}_G\mathcal{O}[T_1, \ldots, T_m] / (P_1, \ldots, P_m)$ are isomorphic, even as ${}_G\mathcal{O}$-algebras, by 23.11; i.e., $(Z, \pi) =$ Specan $\mathscr{R}$. We obtain Specan $\mathscr{A}$ as a closed subspace $(Y,f) \hookrightarrow (Z, \pi)$ in the following way: the homomorphism φ induces a surjective homomorphism $\bar{\varphi}: \mathscr{R} \cong \pi({}_Z\mathcal{O}) \twoheadrightarrow \mathscr{A}^G$; its kernel $\mathscr{K}$ is an ideal in $\pi({}_Z\mathcal{O})$ and a coherent ${}_G\mathcal{O}$-module. According to E. 45Af, there exists a coherent ${}_Z\mathcal{O}$-ideal $\mathscr{J}$ with $\pi\mathscr{J} = \mathscr{K}$; consequently, we have that

$$\mathscr{A}^G \cong \mathscr{R} / \mathscr{K} \cong \pi({}_Z\mathcal{O}) / \pi\mathscr{J} \underset{45\text{A}.1}{\cong} \pi({}_Z\mathcal{O}/\mathscr{J}) = \pi({}_Y\mathcal{O}),$$

with $Y := V(Z; \mathscr{J})$. It remains to show that the finite mapping $\pi|_Y: Y \to G$ can be factored by $X = V(G; \mathscr{I}) \hookrightarrow G$. By the Restriction Lemma, we need only show that $\pi^*\mathscr{I} \subset \mathscr{J}$. Since $\mathscr{A}^G$ is an ${}_G\mathcal{O}/\mathscr{I}$-module, we have that $\mathscr{I} \subset \mathscr{K} = \pi\mathscr{J}$. ∎

45 B. 2 Corollary. *Let $f: T \to X$ be a finite morphism of a locally compact ringed space T into a complex space X. If $f({}_T\mathscr{A})$ is a coherent ${}_X\mathcal{O}$-algebra, then T is a complex space.*

Proof. We have seen that 45.15 and 45.16 also hold for locally compact ringed spaces. Hence T is isomorphic to $\operatorname{Specan} f({}_T\mathscr{A})$, which is a complex space by 45 B. 1. ■

§ 46 The Representation Theorem for Prime Germs

The goal of this section is to strengthen the Noether Normalization Theorem for prime germs, i. e., germs X_a such that ${}_X\mathcal{O}_a$ is an integral domain. Aside from the algebraic Proposition 46.2, the essential tool for that is the Finite Coherence Theorem. The geometric approach used in the proof consists in comparing X_a with the germ Y_b of zeros of a monic polynomial $P \in \mathcal{O}(G)[T]$. Additional singularities may appear in Y_b, but they are of a type that we have already analyzed sufficiently in 33.1. Example 46.3 illustrates the relationship between X_a and Y_b.

46.1 Representation Theorem for Prime Germs. *Let X_a be a prime germ. Then every holomorphic mapping $f_a: X_a \to \mathbb{C}^n_0$ that is discrete and open at a has, as a representative, a finite open holomorphic mapping $f: X \to G \subset\subset \mathbb{C}^n$ (where X and G can be chosen so as to be connected and arbitrarily small) such that the following conditions hold:*

i) There exists an irreducible monic polynomial $P \in \mathcal{O}(G)[T]$ and a $\sigma \in \mathcal{O}(X)$ such that the diagram

$$\begin{array}{ccccc} X & \xrightarrow{\;F=(f,\sigma)\;} & Y := V(G \times \mathbb{C}; P) & \hookrightarrow & G \times \mathbb{C} \\ & {\scriptstyle f}\searrow & \Big\downarrow {\scriptstyle \pi = \mathrm{pr}_G|_Y} & & \\ & & G & & \end{array}$$

commutes, F is finite and surjective, and $(f^0 P)(\sigma) = 0 \in \mathcal{O}(X)$.

ii) f is a branched covering; more precisely, there exists an $\alpha \in {}_n\mathcal{O}(G)$ such that the subsets

$$G' := G \backslash N(\alpha), \quad X' := f^{-1}(G'), \quad \textit{and} \quad Y' := \pi^{-1}(G')$$

are connected and dense, and, in the induced commutative diagram

$$\begin{array}{ccc} X' & \subset\subset & X \\ \Big\downarrow {\scriptstyle F'} & & \Big\downarrow {\scriptstyle F} \\ Y' & \subset\subset & Y \\ \Big\downarrow {\scriptstyle \pi'} & & \Big\downarrow {\scriptstyle \pi} \\ G' & \subset\subset & G \end{array} \qquad (f' = \pi' \circ F',\; f = \pi \circ F),$$

F' is biholomorphic, and π' and f' are (unbranched holomorphic) coverings.

iii) For every $x \in X$ and every nonzero $\psi \in {}_G\mathcal{O}_{fx}$, $f_x^0(\psi)$ is not a zero-divisor in ${}_X\mathcal{O}_x$; in particular, $f_x^0: {}_G\mathcal{O}_{fx} \to {}_X\mathcal{O}_x$ is injective.

For the proof of 46.1, we need an extension of the Theorem on the Primitive Element [La VII. 6 Th. 14]:

46.2 Proposition. *Let $R \subset S$ be analytic algebras without zero-divisors, and suppose that S is a finitely-generated R-module. Then the following statements hold:*

α) There exist a $\sigma \in S$ and a nonzero $\varrho \in R$ such that $\varrho S \subset R[\sigma] \subset S$.

β) There exists an irreducible monic polynomial $P \in R[T]$ with $P(\sigma) = 0$ such that substitution of σ yields an isomorphism $R[T]_{\deg P} \cong R[T]/(P) \to R[\sigma]$.

Proof. For the quotient fields of R and S, we have the commutative diagram

$$\begin{array}{ccc} R & \subset & Q(R) \\ \cap & & \cap \\ S & \subset & Q(S). \end{array}$$

First we show that $Q(S) = S \cdot Q(R)$: By 23 A.6, each nonzero $s \in S$ satisfies an equation $H(s) = 0$ over R, where we may assume that $H = \sum_{j=0}^{p} a_j T^j \in R[T]_{\text{mon}}$ is of minimal degree. Hence, $u := \sum_{j=1}^{p} a_j s^{j-1}$ is nonzero, and $su = -a_0 \in R$; consequently, $1/s = -u/a_0 \in S \cdot Q(R)$.

Analytic algebras have characteristic zero; the field-extension $Q(S):Q(R)$ is finite algebraic, since S is finitely-generated. The Theorem on the Primitive Element yields the existence of a $\tau \in S \cdot Q(R)$ with $Q(S) = Q(R)[\tau]$; after multiplication by an appropriate common denominator in R, we can even have $\tau \in S$. The minimal polynomial of τ in $Q(R)[T]$ yields an irreducible polynomial $\tilde{P} = \sum_{j=0}^{m} r_j T^j \in R[T]$, $r_m \neq 0$, $P(\tau) = 0$, by means of multiplication by the common denominator. Then $P := (r_m)^{m-1} \tilde{P}(T/r_m) \in R[T]_{\text{mon}}$ is the minimal polynomial of $\sigma := r_m \tau$ in $Q(R)[T]$. That implies β), since $R[T]_{\deg P} \to R[T]/(P)$ is an isomorphism of R-modules, by the Euclidean Division Theorem; furthermore, $R[T]/(P) \to R[\sigma]$ is obviously surjective, and also injective, since it is induced by the field-isomorphism $Q(R)[T]/(P) \to Q(R)[\sigma]$.

For α), we observe that, for each σ_j in a finite system of generators of S over R, there exist a polynomial $P_j \in R[T]$ and an element $r_j \in R$ such that $\sigma_j = P_j(\sigma)/r_j$, since $S \subset Q(R)[\sigma]$; with $\varrho := \prod r_j$, we have that $\varrho \sigma_j \in R[\sigma]$ for every j. ■

Proof of 46.1. Let $f: X \to G \subset\!\!\!\!\subset \mathbb{C}^n$ be a finite holomorphic representative of f_a with $f^{-1}(0) = \{a\}$, according to 45.4. By 45.13 ii), $f_a^0 : {}_n\mathcal{O}_0 \to {}_X\mathcal{O}_a$ is injective. Let G and X be chosen so as to be so small that

a) the elements σ, ϱ, and P corresponding to $R := {}_n\mathcal{O}_0 \subset {}_X\mathcal{O}_a =: S$ in 46.2 lie in $\mathcal{O}(X)$, $\mathcal{O}(G)$ and $\mathcal{O}(G)[T]$, respectively, and P is irreducible in $\mathcal{O}(G)[T]$ as well;

b) multiplication by ϱ is an injective endomorphism of $f({}_X\mathcal{O})$, and there exist natural inclusions

$$(46.1.1)\quad \varrho\cdot f({}_X\mathcal{O}) \subset {}_G\mathcal{O}[\sigma] \subset f({}_X\mathcal{O})$$

(the "inclusions" (see 45 A. 2)

$$\varrho_0\cdot {}_X\mathcal{O}_a \cong \varrho_0\cdot f({}_X\mathcal{O})_0 \subset {}_G\mathcal{O}_0[\sigma_a] \subset f({}_X\mathcal{O})_0 \cong {}_X\mathcal{O}_a$$

extend to inclusions near a, by 41.17, as ${}_G\mathcal{O}[\sigma] = {}_G\mathcal{O}[\sigma]_{\deg P}$ is coherent);

c) $P(\sigma) = 0 \in \mathcal{O}(X)$ (as we have ${}_G\mathcal{O} \subset f({}_X\mathcal{O})$, we identify P and $f^0(P)$);

d) F induces an isomorphism of ${}_G\mathcal{O}$-modules

$$(46.1.2)\quad \varphi : \pi({}_Y\mathcal{O}) \xrightarrow{\cong} {}_G\mathcal{O}[\sigma] \subset f({}_X\mathcal{O}).$$

(By the Restriction Lemma, F maps X into Y, and thus the diagram

$$\begin{array}{ccc} X & \xrightarrow{\;F\;} & Y \hookrightarrow G\times\mathbb{C} \\ {\scriptstyle f}\searrow & & \swarrow{\scriptstyle \pi} \\ & G & \end{array}$$

makes sense, since $F^0(P) = P(\sigma) = 0$. On G, it induces a commutative diagram of coherent ${}_G\mathcal{O}$-modules

$$\begin{array}{ccc} f({}_X\mathcal{O}) & \xleftarrow{\;\varphi\;} & \pi({}_Y\mathcal{O}) \\ \cup & & \nearrow{\scriptstyle \psi} \\ {}_G\mathcal{O} & & \end{array},$$

where ψ and $\varphi = \Phi(F)$ are defined as in 45A. 2 and (45.15.1), respectively. For $\eta := \mathrm{pr}_{\mathbb{C}}|_Y \in {}_Y\mathcal{O}(Y) = \pi({}_Y\mathcal{O})(G)$, we have that $\varphi(\eta) = \sigma$; consequently, with $b := \deg P$, $\hat{f}(T) := \sigma$, and $\hat{\pi}(T) := \eta$, there is a commutative diagram of ${}_G\mathcal{O}$-homomorphisms

$$\begin{array}{ccccc} f({}_X\mathcal{O}) \supset {}_G\mathcal{O}[\sigma]_b & \xleftarrow{\;\varphi\;} & {}_G\mathcal{O}[\eta]_b & \underset{(45.2.1)}{=} & \pi({}_Y\mathcal{O}). \\ {\scriptstyle \hat{f}}\nwarrow & & \nearrow{\scriptstyle \hat{\pi}} & & \\ & {}_G\mathcal{O}[T]_b & & & \end{array}$$

It was shown in (45.2.1) that $\hat{\pi}$ is an isomorphism. By 46.2 β), $\hat{f}_0 : {}_G\mathcal{O}_0[T]_b \to {}_G\mathcal{O}_0[\sigma] = {}_G\mathcal{O}_0[\sigma]_b$ is an isomorphism; by 41.17, $\hat{f}$, and thus φ, is an isomorphism of sheaves near a).

In particular, f is open by the Open Lemma, since ${}_G\mathcal{O} \subset f({}_X\mathcal{O})$. On $G\backslash N(\varrho)$, (46.1.1) implies that

$$f({}_X\mathcal{O})|_{\varrho\neq 0} = \varrho\cdot f({}_X\mathcal{O})|_{\varrho\neq 0} \subset {}_G\mathcal{O}[\sigma]|_{\varrho\neq 0} \underset{(46.1.2)}{=} \varphi(\pi_Y\mathcal{O})|_{\varrho\neq 0} \subset f({}_X\mathcal{O})|_{\varrho\neq 0};$$

as a result,

$$\varphi : \pi({}_Y\mathcal{O})|_{\varrho \neq 0} \to f({}_X\mathcal{O})|_{\varrho \neq 0}$$

is an isomorphism. By 45.16, $F|_{\varrho \neq 0}$ is biholomorphic.

Since P is irreducible, we have that $\operatorname{Dis} P \neq 0 \in \mathcal{O}(G)$, by 33 A.4. With $\alpha := \varrho \cdot \operatorname{Dis} P$, the mapping $\pi : Y' \to G'$ is a covering of degree b (see 33.1); therefore, $f: X' \to G'$ is a covering of degree b, as well. By 33.6 c), Y', and thus X' also, is connected. By 33.4, X' is dense in X (in particular, then, X is connected). Since f is finite, so is $F = (f, \sigma)$. Moreover, F is surjective, because $F(X)$ is closed and includes Y'.

It remains to show that iii) holds for every point x near a. By E. 41g iii), that assertion is equivalent to the statement that $\mathcal{T}_{fx} = 0$ for $\mathcal{T} := \mathcal{T}or^{G\mathcal{O}} f({}_X\mathcal{O})$. For $x = a$, that follows from the fact that ${}_G\mathcal{O}_0 \subset f({}_X\mathcal{O})_0 \cong {}_X\mathcal{O}_a$, because ${}_X\mathcal{O}_a$ is an integral domain. By E. 41g, $\mathcal{T}$ is a coherent ${}_G\mathcal{O}$-module; by 41.17, then, $\mathcal{T} = 0$ near 0. ■■

46.3 Example (for 46.1 and 46.2). Let $\mathcal{H}$ be the structure sheaf on $\mathbb{C}$ defined as follows:

$$\mathcal{H}_z := \begin{cases} {}_1\mathcal{O}_z & z \neq 0 \\ \{\varphi \in {}_1\mathcal{O}_0;\ \varphi'(0) = \varphi''(0) = 0\} & z = 0\,. \end{cases}$$

The mapping $g: (\mathbb{C}, \mathcal{H}) \to \mathbb{C}^3$, $t \mapsto (t^3, t^4, t^5)$, determines an isomorphism from $(\mathbb{C}, \mathcal{H})$ onto the reduced subspace $\operatorname{Im} g \hookrightarrow \mathbb{C}^3$ (see 32.9); thus we see that $X = (\mathbb{C}, \mathcal{H})$ is a complex space, and X_0 is a prime germ. We want to verify 46.1 and 46.2 for the open discrete mapping $f: X \to \mathbb{C}$, $t \mapsto t^3$; to that end, put $R := \mathbb{C}\{t^3\} \subset \mathcal{H}_0 =: S$. Then the mapping F in 46.1 cannot be biholomorphic, since $Y \hookrightarrow \mathbb{C}^2$, but $\operatorname{emb} X_0 = 3$ (the fact that $\mathfrak{m} := \mathfrak{m}_{\mathcal{H}_0} = (t^3, t^4, t^5)$ implies that $\dim \mathfrak{m}/\mathfrak{m}^2 = 3$; see 45.10). Therefore, we need to find a nonzero $\varrho \in R$, a $\sigma \in S$, and an irreducible $P \in R[T]_{\text{mon}}$ with

$$\varrho S \subset R[\sigma] \quad \text{and} \quad P(\sigma) = 0\,.$$

With the representation $\mathcal{H}_0 = \{a_0 + a_3 t^3 + a_4 t^4 + \ldots \in {}_1\mathcal{O}_0\}$, we choose

$$\varrho := t^3,\ \sigma := t^4, \quad \text{and} \quad P := T^3 - (t^3)^4\,.$$

Then $P(\sigma) = 0$; the factorization of P into prime factors in ${}_1\mathcal{O}_0[T]_{\text{mon}}$ is

$$P = (T - t^4)(T - e^{2\pi i/3} t^4)(T - e^{-2\pi i/3} t^4)\,;$$

since the product of each pair of those factors is not in $R[T] = \mathbb{C}\{t^3\}[T]$, P is irreducible in $R[T]$. Every $\varphi \in t^3 \cdot S$ has a representation of the form

$$\varphi = a_3 t^3 + \sum_{j=6}^{\infty} a_j t^j = \sum_{j=1}^{\infty} a_{3j} t^{3j} + t^4 \sum_{j=1}^{\infty} a_{3j+4} t^{3j} + t^8 \sum_{j=0}^{\infty} a_{3j+8} t^{3j}$$
$$=: \varphi_0 + \varphi_1 t^4 + \varphi_2 t^8\,,$$

in which the φ_j converge, by 21.5. Thus $t^3 \cdot S \subset R[t^4]$. In the diagram

$$\begin{array}{ccc} X & \xrightarrow{\quad F \quad} & Y = V(P) \hookrightarrow \mathbb{C}^2 \\ & {\scriptstyle f} \searrow \quad \swarrow {\scriptstyle \pi} & \\ & \mathbb{C} & \end{array}$$

with $F(t) = (t^3, t^4)$, both f and π are branched holomorphic coverings of degree 3. Moreover, F is bijective, but not biholomorphic, as $F_0^0({}_Y\mathcal{O}_0) = \{\varphi \in \mathcal{H}_0; \varphi^{(5)}(0) = 0\}$ is a proper subalgebra of $\mathcal{H}_0$. ∎

E. 46a. For $X = (\mathbb{C}, \mathcal{H})$ as in 46.3 and $f: X \to \mathbb{C}, t \mapsto t^4$, determine ϱ, σ, and P.

E. 46b. Let $0 \in G \Subset \mathbb{C}^n$ be a domain, suppose that $P \in \mathcal{O}(G)[T]_{\text{mon}}$ is irreducible in ${}_n\mathcal{O}_0[T]$, and put $X := V(G \times \mathbb{C}; P) \hookrightarrow G \times \mathbb{C}$, $a := 0 \in X$, and $f := \text{pr}_G|_X$. Show that, in 46.1, it is possible to choose $\sigma := z_{n+1}|_X \in \mathcal{O}(X)$, $\alpha := \text{Dis}\, P \in \mathcal{O}(G)$, and P to be the polynomial originally given.

E. 46c. Show that, for $N(\mathbb{C}^3; z_1 z_3, z_2 z_3)_0$, there exists no mapping f_0 that satisfies 46.1 ii).

46.4 Maximum Principle. *If ${}_X\mathcal{O}_a$ has no nilpotent elements, and if $\varphi \in \mathcal{O}(X)$ assumes a local maximum in absolute value at a, then φ is constant near a.*

Proof. Since it suffices to prove the assertion for each component of the punctual decomposition 44.13 of X_a into prime germs, we may assume that X_a is irreducible; suppose further that $\varphi(a) = 1$. By 45.14, there exists a mapping $f: X \to G \Subset \mathbb{C}^n$ according to 46.1. On G', the assignment $h(z) := \sum_{x \in f^{-1}(z)} \varphi(x)$ determines a holomorphic function that is bounded by the degree b of the covering $f: X' \to G'$; by 7.3, then, it can be extended holomorphically at 0. Since $\lim_{x \to a} \varphi(x) = 1$, it must be that h assumes a local maximum b at 0; by 6.4, h is the constant function with value b, and it follows that $\varphi = 1$ near a. ∎

E. 46d. Show that, for a connected compact complex space X, all functions $\text{Red}\, f$ for $f \in \mathcal{O}(X)$ are constant (hint: use 43.9).

§ 46 A Supplement: Injective Holomorphic Mappings between Manifolds of the Same Dimension

Proposition 8.5 can be deduced from the Representation Theorem for Prime Germs. By applying sections 48 and 49, we obtain the following more general statement (see also 72.2):

46 A. 1 Proposition. *Let $f: X \to Z$ be an injective holomorphic mapping from a pure-dimensional reduced complex space X into a manifold Z of the same dimension. Then f is open and $f: X \to f(X)$ is biholomorphic.*

Proof. By 48.10, f is open; we may assume that f is a homeomorphism. By E. 49g, since Z is irreducible at every point, X is also; consequently, none of the algebras ${}_X\mathcal{O}_a$ has zero-divisors (44.14). In the notation of the Representation Theorem for Prime Germs, we may assume that $Z = G \Subset \mathbb{C}^n$; without loss of generality, let f be chosen as in 46.1. Since f is injective, it determines a one-sheeted covering. Thus, P is of the form $T - \gamma$; hence, $f^0 P(\sigma) = 0$ implies that $\sigma = f^0(\gamma) \in f^0(\mathcal{O}(G)) \subset \mathcal{O}(X)$. For a $\varrho \in \mathcal{O}(G)$ chosen according to 46.2, we thus have that $\varrho \cdot f({}_X\mathcal{O}) \subset {}_G\mathcal{O}$ in (46.1.1); in other words, $\varrho_{f(a)}$ is a universal denominator (46 A.2) for the inclusion $f_a^0 : {}_G\mathcal{O}_{f(a)} \hookrightarrow {}_X\mathcal{O}_a$. By 23.8, ${}_G\mathcal{O}_{f(a)}$ is normal; hence, f_a^0 is an isomorphism, according to 46 A.3 iii), and f is biholomorphic near a, by 44.4. ■

We have made use of the following algebraic concept:

46 A. 2 Definition. *For a subring R of a ring S, let u be an element of R that is not a zero-divisor in S. If $uS \subset R$, then u is called a "universal denominator for S with respect to R". If S is the integral closure $\hat{R}$ of R in $Q(R)$, then we refer to u as a universal denominator for R.*

E. 46 Aa. Show that the residue class of z_1 is a universal denominator for ${}_2\mathcal{O}_0/(z_1^2 - z_2^3)$. More generally, determine a universal denominator for $R = \mathbb{C} \oplus {}_n\mathfrak{m}^k \subset {}_n\mathcal{O}_0$.

46 A. 3 Lemma. *Let R be a noetherian subring of S, and suppose that there exists a universal denominator u for S with respect to R. Then the following statements hold:*

i) S is a finite R-module (and is thus noetherian).

ii) There exists precisely one (injective) homomorphism of R-algebras $j: S \to \hat{R}$; i.e., S is, in a canonical fashion, an R-subalgebra of $\hat{R}$.

iii) If $S = \hat{S}$ or $R = \hat{R}$, then j is an isomorphism, i.e., $S = \hat{R}$.

Proof. i) The multiplication $S \xrightarrow{\cdot u} R$ is an injective R-algebra-homomorphism whose image, as an ideal in the noetherian ring R, is finitely-generated.

ii) For j, we must have that $j(s) = j\left(\frac{u}{u} \cdot s\right) = \frac{j(us)}{j(u)} = \frac{us}{u} \in Q(S)$. By defining j in that way, we obtain an injective homomorphism of R-algebras. By i) and 23 A.6, it follows that $j(S) \subset \hat{R}$.

iii) We have that $R \subset S \subset \hat{R} \subset \hat{S}$. ■

§ 46B Supplement: Universal Denominators for Prime Germs

A second application of the existence of an appropriate universal denominator is the adaptation of the first Riemann Removable Singularity Theorem for reduced complex spaces. Since its construction depends primarily on the Representation Theorem for Prime Germs, we carry it out here, although we do not use the result until §71. For the adaptation, it is useful to reformulate the Removable Singularity Theorem with the help of the sheaf $\tilde{\mathcal{O}}$ of weakly holomorphic functions:

46 B. 1 Definition. *A closed subset A of a complex space X is called (analytically) thin if, for each $U \Subset X$, the restriction-mapping $\mathcal{O}(U) \to \mathcal{O}(U \setminus A)$ is injective.*

46 B. 2 Remark. A closed subset of a reduced space is thin iff it is nowhere dense.

E. 46 Ba. For the line with double point $X = V(\mathbb{C}; z_1 z_2, z_2^2)$, and $a \in X$, show that $\{a\}$ is thin iff $a \neq 0$.

46 B. 3 Definition. *Let U be an open subset of a reduced complex space X. A "weakly holomorphic* *function on U" is a holomorphic function $f: U\backslash A \to \mathbb{C}$ that is defined outside of a thin analytic set A in U and locally bounded at A (§7). The $\mathcal{O}(U)$-module of weakly holomorphic functions on U is denoted by $\tilde{\mathcal{O}}(U)$.*

E. 46 Bb. Use 49.1 to show that there is no loss of generality in assuming that $A = S(U)$ in 46 *B.* 3.

E. 46 Bc. Show that the presheaf $\tilde{\mathcal{O}} = (\tilde{\mathcal{O}}(U))_{U \subset X}$ is a sheaf that contains $\mathcal{O}$ in a canonical fashion; in particular, $\tilde{\mathcal{O}}$ is an $\mathcal{O}$-algebra.

It follows from 7.3 that the sheaves $\tilde{\mathcal{O}}$ and $\mathcal{O}$ coincide on manifolds.

46 B. 4 Proposition. *Let the reduced complex space $(X, \mathcal{O})$ have a prime germ at the point $a \in X$. Then there exist an open neighborhood W of a and a $u \in \mathcal{O}(W)$ that induces a non-zero-divisor in every stalk $\mathcal{O}_z$ on W, and for which*

$$u\tilde{\mathcal{O}}|_W \subset {}_W\mathcal{O}.$$

Such a u is called a *universal denominator* for W (precisely, for $(W, \mathcal{O}|_W)$).

Proof. By 45.14, we may assume that there exists a finite open mapping $f: X \to G \subset \mathbb{C}^n$ with the properties given in 46.1. In the notation of 46.1, the composition

$$u: X \xrightarrow{(f,\sigma)} Y \hookrightarrow G \times \mathbb{C} \xrightarrow{\frac{\partial P}{\partial T}} \mathbb{C}, \; x \mapsto \frac{\partial P}{\partial T}(f(x), \sigma(x)),$$

is a holomorphic function, none of whose germs is a zero-divisor: for $z \in G'$, the polynomial $P(z,T)$ has only simple zeros, by construction; thus $\frac{\partial P}{\partial T}(z,y)$ has no zeros in $Y' = \pi^{-1}(G')$, and u has no zeros in the dense subset X' of X; the assertion follows by E. 53a. It remains to show that $u\tilde{\mathcal{O}} \subset \mathcal{O}$. For $V \subset\subset X$ and $h \in \tilde{\mathcal{O}}(V)$, we prove that $u \cdot h$ can be extended holomorphically to V. Without loss of generality, suppose that $h: V \backslash S(V) \to \mathbb{C}$ is holomorphic and bounded; moreover, let $f: V \to f(V) \subset G$ be finite (see 33 B.2).

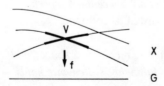

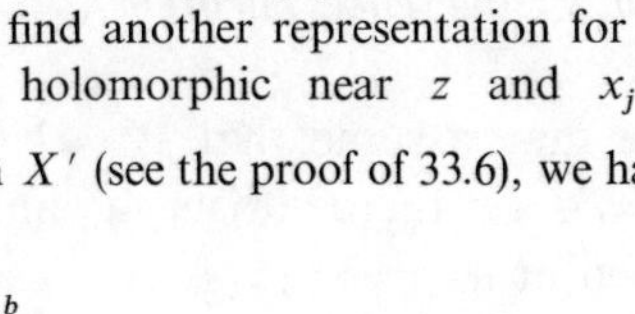

By E. 33Ba, with $V' := V \cap X'$, the mapping $f: V' \to f(V')$ is a covering with sheet-number $c \leqq b = \deg P$; hence, we can find another representation for $u \cdot h$: For $z \in f(V')$, set $f^{-1}(z) = \{x_1(z), \ldots, x_b(z)\}$ with x_j holomorphic near z and $x_j(z) \in V'$ for $1 \leqq j \leqq c$. Since $P(z,T) = \prod_{k=1}^{b} (T - \sigma(x_k(z)))$ on X' (see the proof of 33.6), we have that

(46. B. 4.1) $$\frac{P}{T}(z,T) = \sum_{i=1}^{b} \prod_{\substack{k=1 \\ k \neq i}}^{b} (T - \sigma(x_k(z)))$$

on X'. The assignment

$$(z, t) \mapsto \sum_{j=1}^{c} h(x_j(z)) \prod_{\substack{k=1 \\ k \neq j}}^{b} (t - \sigma(x_k(z)))$$

determines a holomorphic function g on $f(V') \times \mathbb{C}$; to show $u \cdot h \in \mathcal{O}(V)$, it suffices to check that

a) $u \cdot h = g \circ (f, \sigma)$ on V', and

b) g is the restriction of a holomorphic function on $f(V) \times \mathbb{C}$.

For a), we choose $x \in V', z := f(x)$, and, without loss of generality, $x = x_1(z)$. On the one hand, we have that

$$g(f(x), \sigma(x)) = \sum_{j=1}^{c} h(x_j(z)) \prod_{\substack{k=1 \\ k \neq j}}^{b} (\sigma(x) - \sigma(x_k(z))) = h(x) \prod_{k=2}^{b} (\sigma(x) - \sigma(x_k(z))),$$

since the factor $\sigma(x) - \sigma(x_1(z)) = 0$ appears in the products, except for $j = 1$. On the other hand, it follows similarly from (46 B.4.1) that

$$h(x) \cdot u(x) = h(x) \sum_{i=1}^{b} \prod_{\substack{k=1 \\ k \neq i}}^{b} (\sigma(x) - \sigma(x_k(z))) = h(x) \prod_{k=2}^{b} (\sigma(x) - \sigma(x_k(z))).$$

Statement b) follows from the first Riemann Removable Singularity Theorem, because g is bounded, since h is bounded and $\pi: Y \to G$ is proper. ■

§ 47 Hilbert's Nullstellensatz and Cartan's Coherence Theorem

We indicated the importance of the following result in § 43:

47.1 Theorem. *For $W \Subset \mathbb{C}^m$, if $\mathcal{I} \subset_W \mathcal{O}$ is a coherent sheaf of ideals and $A := N(W; \mathcal{I})$, then the following statements hold for the nullstellen-ideal ${}_A\mathcal{I}$ of A:*

i) ${}_A\mathcal{I} = \sqrt{\mathcal{I}}$ (Hilbert's Nullstellensatz).

ii) ${}_A\mathcal{I}$ is coherent (Cartan's Coherence Theorem).

That theorem can be reworded as follows:

47.2 Theorem. *Let $(X, \mathcal{O})$ be a complex space, and let $\mathcal{N} \subset \mathcal{O}$ be the sheaf of nilpotent elements. Then the following statements hold:*

iii) The canonical homomorphism $\mathcal{O}/\mathcal{N} \to \operatorname{Red} \mathcal{O}$ is an isomorphism (Hilbert's Nullstellensatz).

iv) $\mathcal{N}$ is coherent (Cartan's Coherence Theorem).

Proof of 47.1 and 47.2. We may assume that $X = V(W; \mathcal{I})$, which implies that $A = |X|$, and that, for $a \in X$, there exists a finite holomorphic mapping $f: X \to G (\subset \mathbb{C}^n)$ that is open at a, by 45.15. If X_a is a prime germ, then we may choose f according to 46.1.

iii) We need to show that, for $\varphi \in \mathcal{O}(X)$ with $\operatorname{Red} \varphi = 0$, every germ φ_x in $\mathcal{O}_x$ is nilpotent. Assume, on the contrary, that φ_a is not nilpotent for some $\alpha \in X$. We may assume that $f(a) = 0$. By the Open Lemma and 45.4, the homomorphism

$f_a^0: {}_n\mathcal{O}_0 \to \mathcal{O}_a$ is a finite injection. According to 23A.6, then, φ_a is integral over ${}_n\mathcal{O}_0$, and therefore satisfies an integral equation

$$\varphi_a^p = \sum_{j=s}^{p-1} \alpha_j \varphi_a^j \quad \text{with} \quad \alpha_j \in {}_n\mathcal{O}_0 \subset \mathcal{O}_a;$$

since $\varphi_a^p \neq 0$ by assumption, we may assume that $\alpha_s \neq 0$. It follows that

$$\varphi_a^s \cdot (\varphi_a^{p-s} - \sum_{j=s}^{p-1} \alpha_j \varphi_a^{j-s}) = 0 \in \mathcal{O}_a.$$

Suppose additionally that the ideal (0) in $\mathcal{O}_a$ is *primary*; then the second factor is nilpotent, since $\varphi_a^s \neq 0$. Hence, there exists a $q > 0$ with $(\varphi_a^{p-s} - \sum_{j=s}^{p-1} \alpha_j \varphi_a^{j-s})^q = 0$, so that

$$0 \neq \alpha_s^q = \varphi_a \cdot (\varphi_a^{q(p-s)-1} + \ldots) \in \mathcal{O}_a \cdot \varphi_a.$$

On the other hand, according to the Open Lemma, the germ $V(X;\varphi)_a \hookrightarrow X_a$ is open, since $\operatorname{Red} \varphi = 0$. Consequently, the composition $\psi: V(X;\varphi)_a \hookrightarrow X_a \xrightarrow{f_a} \mathbb{C}_0^n$ is open at a. Since $\alpha_s^q \in \mathcal{O}_a \cdot \varphi_a \subset \operatorname{Ker} \psi^0$, it follows that $0 \underset{45.13}{=} \operatorname{Red} \alpha_s^q = \alpha_s^q$ in ${}_n\mathcal{O}_0$. ↯

Now suppose that (0) is not a primary ideal. By 44.12, then, there exists a decomposition $X_a = \bigcup_{j=1}^{m} X_j$ with primary germs X_j. Thus the zero ideals in the ${}_{X_j}\mathcal{O}_a$ are primary ideals; since $\operatorname{Red} \varphi = 0$, also $\operatorname{Red} \varphi|_{X_j} = 0$; by what has already been shown, every $\varphi|_{X_j}$ is nilpotent, and it follows that φ is nilpotent.

iii) ⇔ i) Obviously, $\sqrt{\mathcal{I}} \subset {}_A\mathcal{I}$. The assertion follows from the fact that

$$\mathcal{N} = (\sqrt{\mathcal{I}}/\mathcal{I})|_A \subset ({}_A\mathcal{I}/\mathcal{I})|_A \underset{\text{E. 31b}}{=} \mathcal{K}er({}_X\mathcal{O} \to \operatorname{Red}_X \mathcal{O}).$$

ii) ⇔ iv) In the proof "iii) ⇔ i)", we saw that $\mathcal{N} = {}_A\mathcal{I}/\mathcal{I}$; hence, we have that $\mathcal{N}$ is coherent $\underset{41.8}{\Leftrightarrow}$ ${}_A\mathcal{I}$ is coherent.

iv) First let X_a be a *prime germ*; in particular, then, $\mathcal{N}_a = 0$. We wish to show that $\mathcal{N} = 0$ near a. By E. 41h, for each $x \in X$, there exists a coherent sheaf of ideals $\mathcal{J} \subset \mathcal{O}$ near x with $\mathcal{J}_x = \mathcal{N}_x$; obviously $\mathcal{J} \subset \mathcal{N}$ near x. With 46.1 ii), it follows that $\operatorname{supp} \mathcal{J} \subset \operatorname{supp} \mathcal{N} \subset N(\alpha)$. Now we apply the following lemma to $\mathcal{I} = \mathcal{O} \cdot (f^0 \alpha)$ and $\mathcal{G} = \mathcal{J}$:

47.3 Lemma. *Suppose that X is a complex space, $\mathcal{I}$ is a coherent ${}_X\mathcal{O}$-ideal, and $\mathcal{G}$ is a coherent ${}_X\mathcal{O}$-module with $\operatorname{supp} \mathcal{G} \subset N(\mathcal{I})$. Then there exists for each $a \in X$ a $k \in \mathbb{N}$ such that $\mathcal{I}^k \mathcal{G} = 0$ near a.*

Thus we obtain for x a k such that $0 = (f^0\alpha)^k \mathcal{J}_x = (f^0\alpha)^k \mathcal{N}_x$. According to 46.1 iii), $(f^0\alpha)_x$ is not a zero-divisor, and it follows that $\mathcal{N}_x = 0$.

Now let X_a be a *primary germ*. Then $\mathcal{I}_a$ is a primary ideal. Choose, according to E. 41h, a coherent sheaf of ideals $\mathcal{J} \subset {}_W\mathcal{O}$ near a with $\mathcal{J}_a = \sqrt{\mathcal{I}_a}$. We may apply the argument above to the prime germ $V(W;\mathscr{J})$ in order to obtain $\mathscr{J}=\sqrt{\mathscr{J}}$ near a. Moreover, it is easy to see that $\mathscr{I}\subset\mathscr{J}\subset\sqrt{\mathscr{I}}$ near a, and this implies that

$$\sqrt{\mathscr{I}}\subset\sqrt{\mathscr{J}}=\mathscr{J}\subset\sqrt{\mathscr{I}},$$

and thus that $\sqrt{\mathscr{I}}=\mathscr{J}$. We conclude that $\mathscr{N}=\sqrt{\mathscr{I}}/\mathscr{I}$ is coherent.

For *arbitrary* X_a, consider a decomposition $X_a=\bigcup_{j=1}^{k} X_j$ into primary germs according to 44.12, and let $\mathscr{I}_a=\bigcap_{j=1}^{k}\mathfrak{a}_j$ be the associated ideal-decomposition. By choosing coherent sheaves of ideals ${}_j\mathscr{I}$ near a with ${}_j\mathscr{I}_a=\mathfrak{a}_j$ and $\mathscr{I}=\bigcap_{j=1}^{k}{}_j\mathscr{I}$, we find that $\sqrt{\mathscr{I}}\underset{44.7.5}{=}\bigcap_{j=1}^{k}\sqrt{{}_j\mathscr{I}}$, as an intersection of coherent sheaves, is coherent (see 41.12), and thus that $\mathscr{N}$ is coherent as well. ■

Proof of 47.3. Since $\mathscr{I}^k\mathscr{G}$ is coherent, it suffices by 41.16 to find for each $a\in X$ a k such that $\mathscr{I}_a^k\mathscr{G}_a=0$. For the coherent sheaf of ideals $\mathscr{J}:=\mathscr{A}nn(\mathscr{G})$, we have that $N(\mathscr{J})\underset{\text{E. 41 i}}{=}\operatorname{supp}\mathscr{G}$. Without loss of generality, let $\mathscr{I}$ be a principal ideal ${}_X\mathscr{O}\cdot(\beta)$ near a; since $N(\mathscr{J})\subset N(\beta)$, we see that $\beta\in I(N(\mathscr{J})_a)\underset{\text{i)}}{=}\sqrt{\mathscr{J}_a}$. Hence, there exists a $k\in\mathbb{N}$ with $\beta^k\in\mathscr{J}_a$; i.e., $\mathscr{I}_a^k\mathscr{G}_a=\beta^k\mathscr{G}_a=0$. ■■

E. 47 a. Generalize 47.1 to complex spaces.

E. 47 b. Prove the equivalence of the following statements for a complex space X:

i) X is reduced.

ii) For every complex space Y, each $U\Subset X$, and all $f=(|f|,f^0)$, $g=(|g|,g^0)\in\operatorname{Hol}(U,Y)$, if $|f|=|g|$, then $f=g$.

iii) The conclusion of ii) holds for every complex number space $Y=\mathbb{C}^n, n\geqq 0$.

E. 47 c. Use 47.2 to show that *the set of points at which a complex space is not reduced is analytic;* in particular, the set of reduced points is open (for 49.1).

E. 47 d. For $\varphi,\psi\in{}_n\mathscr{O}_0$, if φ is irreducible and $\psi|_{N(\varphi)}=0$, then φ divides ψ (hint: 47.1 i); for E. 47 e).

E. 47 e. For $\varphi\in{}_n\mathscr{O}_0$, let $\varphi=\varphi_1^{m_1}\cdot\ldots\cdot\varphi_k^{m_k}$ be the prime factorization. If the reduced germ $N(\varphi)_0=\operatorname{Red}V(\varphi)_0$ is a germ of a manifold, then $k=1$ and $\frac{\partial\varphi_1}{\partial z}(0)\neq 0$. (Hint: by E. 22 c and 22.4, it follows that ${}_n\mathscr{O}_0/I\cong{}_{n-1}\mathscr{O}_0$ for the principal ideal $I(N(\varphi)_0)\underset{\text{E. 47 d}}{=}(\varphi_1\cdot\ldots\cdot\varphi_k)$; for 32.8.)

§ 48 Dimension Theory

In this section, we show that the algebraic and geometric concepts of dimension for complex spaces are compatible. In so doing, we develop dimension theory from a purely *algebraic* standpoint, and provide the equivalent *geometric* interpretation at the appropriate places. Proceeding from the intuitive idea that $\mathbb{C}^n$ and its finitely branched covering spaces should be n-dimensional, we find the following definition to be reminiscent of the Representation Theorem for Prime Germs:

48.1 Definition and Proposition. *For a germ X_a of a complex space and $R := {}_X\mathcal{O}_a$, the numbers*

$$\dim X_a := \min\{d;\ \exists \text{ a finite holomorphic mapping } f\colon X_a \to \mathbb{C}^d_0\}$$

and

$$\dim R := \min\{d;\ \exists \text{ a finite homomorphism } f\colon {}_d\mathcal{O}_0 \to R\}$$

agree. They are called the dimension *of X_a (or of X at a) and the dimension of R, respectively.*

The proof is an immediate consequence of 45.4. ■

The numbers $\dim {}_X\mathcal{O}_a$ and $\dim X_a$ are biholomorphic invariants. We have seen in the proof of 24.3 that, for $d = \dim X_a$, every finite homomorphism $f\colon {}_d\mathcal{O}_a \to R$ is injective. Theorem 45.4 provides the following additional interpretation (*Chevalley's* concept of dimension: the set of solutions of a "generic equation" in X is $(\dim X - 1)$-dimensional):

(48.1_{alg}) $\dim R = \min\{d;\ \exists f_1, \dots, f_d \in \mathfrak{m}_R,\ \dim_{\mathbb{C}} R/R\cdot(f_1, \dots, f_d) < \infty\}$;

(48.1_{geo}) $\dim X_a = \min\{d;\ \exists \text{ hypersurfaces } X_1, \dots, X_d \hookrightarrow X_a \text{ with } \{a\} \subset\!\!\!\!\subset \bigcap_{j=1}^{d} X_j\}$.

Namely, if $\varphi\colon {}_n\mathcal{O}_0 \to {}_X\mathcal{O}_a = R$ is a homomorphism, then $R\cdot\varphi({}_n\mathfrak{m}_0) = R\cdot(\varphi(z_1), \dots, \varphi(z_n))$, and we obtain the following equivalences:

$$\begin{aligned}\varphi \text{ finite} &\Leftrightarrow \varphi \text{ quasifinite}\\ &\Leftrightarrow \dim_{\mathbb{C}} R/R\cdot\varphi({}_n\mathfrak{m}_0) = \dim_{\mathbb{C}} R/R\cdot(\varphi(z_1), \dots, \varphi(z_n)) < \infty\\ &\underset{45.3}{\Leftrightarrow} \{a\} \subset\!\!\!\!\subset N(\varphi(z_1)) \cap \dots \cap N(\varphi(z_n)).\end{aligned}$$ ■

We begin by characterizing zero-dimensional algebras:

48.2_{alg} Proposition. *For an analytic algebra R, the following statements are equivalent:*

i) $\dim R = 0$.

ii) $\dim_{\mathbb{C}} R < \infty$.

iii) R is Artinian (i.e., it satisfies the descending chain condition).

iv) $\mathfrak{m}_R$ is a minimal (and thus the only) prime ideal of R.

v) $\mathfrak{m}_R = \mathfrak{n}_R$ *(nilradical).*
vi) There exists an $s \in \mathbb{N}$ *such that* $\mathfrak{m}_R^s = 0$.

Proof. The equivalence of i) and ii) follows from (48.1_{alg}), and the implication "ii) $\Rightarrow$ iii)" is trivial. To prove "iii) $\Rightarrow$ iv)", we need to see that every prime ideal $\mathfrak{p}$ in R is maximal, i.e., that $\bar{R} := R/\mathfrak{p}$ is a field. Now $\bar{R}$ is an Artinian integral domain; hence, for a nonzero $f \in \bar{R}$, the eventually stationary chain $\bar{R}\cdot f \supset \bar{R}\cdot f^2 \supset \ldots$ provides a $p \in \mathbb{N}$ and a $g \in \bar{R}$ with $f^p = gf^{p+1}$. Cancellation yields that $1 = gf$; i.e., f is a unit. The implication "iv) $\Rightarrow$ v)" follows from the fact that, by 44.10, $\mathfrak{n}_R$ is the intersection of all minimal prime ideals. To see that "v) $\Rightarrow$ vi)", choose generators $m_1, \ldots, m_s$ of $\mathfrak{m}_R$; if $m_j^p = 0$ for $j = 1, \ldots, s$, then $\mathfrak{m}_R^{sp} = 0$. Finally, it was shown in 45.3 that vi) $\Rightarrow$ ii). ∎

Propositions 45.3 and 44.7 ii) yield this translation of 48.2_{alg}:

48.2_{geo} Proposition. *For a germ X_a, the following statements are equivalent:*
i) $\dim X_a = 0$.
ii) $\{a\}$ *is open in* X.
iii) Every ascending chain of closed subgerms of X_a becomes stationary. ∎

Now we come back to arbitrary dimensions:

48.3_{alg} Remark. *The following statements hold for analytic algebras R and S:*
i) If there exists a finite homomorphism $\varphi: R \to S$, *then* $\dim S \leqq \dim R$.
ii) $f \in \mathfrak{m}_R \Rightarrow \dim(R/R\cdot f) \geqq \dim R - 1$.
iii) $0 \neq f \in {}_n\mathfrak{m} \Rightarrow \dim({}_n\mathcal{O}_0/{}_n\mathcal{O}_0\cdot f) \leqq n - 1$.

Proof. i) This follows from the transitivity of finiteness.

ii) For $\bar{R} := R/(f)$ and $f_1, \ldots, f_d \in \mathfrak{m}_R$, we have that

$$R/R\cdot(f, f_1, \ldots, f_d) \cong \bar{R}/\bar{R}\cdot(\bar{f}_1, \ldots, \bar{f}_d).$$

The assertion follows from (48.1_{alg}).

iii) For $0 \neq f \in {}_n\mathfrak{m}$, in appropriate coordinates the homomorphism ψ in the diagram

$$\begin{array}{ccc} {}_n\mathcal{O}_0 & \searrow & \\ \uparrow & & {}_n\mathcal{O}_0/{}_n\mathcal{O}_0\cdot f \underset{22.4}{\cong} {}_{n-1}\mathcal{O}_0[z_n]_b \\ {}_{n-1}\mathcal{O}_0 & \overset{\psi}{\nearrow} & \end{array}$$

is finite; hence, we have that $\dim {}_n\mathcal{O}_0/{}_n\mathcal{O}_0\cdot f \leqq \dim {}_{n-1}\mathcal{O}_0 \leqq n - 1$. ∎

By applying 45.4, we obtain this fact:

48.3_{geo} Remark. *The following statements hold for a germ X_a:*
i) If $\varphi: X_a \to Y_b$ is a finite holomorphic mapping, then $\dim X_a \leqq \dim Y_b$.
ii) If Y_a is a germ of a hypersuface in X_a, then $\dim Y_a \geqq \dim X_a - 1$. ∎

We turn now to the problem of determining when equality holds in 48.3 ii). We need this weakened version of the concept of a "*non-zero-divisor*" (i. e., an element that is not a zero-divisor) in an analytic algebra R:

48.4$_{\text{alg}}$ Definition and Proposition. *An element* $f \in \mathfrak{m}_R$ *is called* <u>*active*</u> *if it satisfies the following equivalent conditions:*

i) $\operatorname{Red} f$ *is not a zero-divisor in* $\operatorname{Red} R = R/\mathfrak{n}_R$.

ii) $f \cdot g \in \mathfrak{n}_R \Rightarrow g \in \mathfrak{n}_R$ *for each* $g \in R$.

iii) f *lies in no minimal prime ideal of* R.

Proof. The equivalence of i) and ii) is evident. For the proof that ii) and iii) are equivalent, let $\mathfrak{p}_1, \ldots, \mathfrak{p}_t$ be the minimal prime ideals; by 44.10, $\mathfrak{n}_R = \bigcap_{j=1}^{t} \mathfrak{p}_j$.

iii) $\Rightarrow$ ii) For $f \notin \bigcup \mathfrak{p}_j$, if $f \cdot g \in \mathfrak{n}_R$, then $g \in \mathfrak{n}_R$.

ii) $\Rightarrow$ iii) For $f \in \mathfrak{p}_1$, choose $g_j \in \mathfrak{p}_j \backslash \mathfrak{p}_1$, $j = 2, \ldots, t$. We have that $f \cdot g_2 \cdot \ldots \cdot g_t \in \mathfrak{n}_R$, but $g_2 \cdot \ldots \cdot g_t \notin \mathfrak{n}_R$; hence, ii) is not satisfied. ■

E. 48 a. Prove the following statements:

i) If (0) is a primary ideal in R, and $f \in \mathfrak{m}_R$, then f is active $\Leftrightarrow f$ is not nilpotent.

ii) Non-zero-divisors in $\mathfrak{m}_R$ are active (for 48.10).

E. 48 b. Show that the residue class of z_1 in ${}_2\mathcal{O}_0 \,/\, (z_1 z_2, z_2^2)$ is both active and a zero-divisor.

E. 48 c. Show that $\dim R = 0$ iff R has no active elements (hint: 48.2$_{\text{alg}}$; for 48.11$_{\text{alg}}$).

Geometrically, the statement "f is active" means that f does not vanish on any irreducible component of $\operatorname{Red} X$:

48.4$_{\text{geo}}$ Remark. *For* $f \in \mathfrak{m}({}_X\mathcal{O}_a)$, f *is active iff, for every minimal prime ideal* $\mathfrak{p}$ *in* ${}_X\mathcal{O}_a$, $f|_{N(\mathfrak{p})} \neq 0$. ■

Now we prove a proposition that is essential for proofs by induction:

48.5$_{\text{alg}}$ Active Lemma. *If* $f \in \mathfrak{m}_R$ *is active, then* $\dim R/R \cdot f = \dim R - 1$.

Proof. By 48.3, it suffices to prove $\dim R/R \cdot f \leqq \dim R - 1$. Set $n := \dim R$, and let $\varphi : {}_n\mathcal{O}_0 \to R$ be finite. By the proof of 24.3, φ is injective; hence, we may suppose that ${}_n\mathcal{O}_0 \subset R$. If there exists a nonzero $h \in R \cdot f \cap {}_n\mathcal{O}_0$, then the induced homomorphism ψ in the diagram

$$\begin{array}{ccc} {}_n\mathcal{O}_0 & \hookrightarrow & R \\ \downarrow & & \downarrow \\ {}_n\mathcal{O}_0 / {}_n\mathcal{O}_0 \cdot h & \xrightarrow{\psi} & R/R \cdot f \end{array}$$

must be finite, and the lemma follows from 48.3$_{\text{alg}}$ i) and iii). It remains to construct

h: since R is finite over ${}_n\mathcal{O}_0$, f satisfies an integral equation

$$f^p + h_1 f^{p-1} + \ldots + h_p = 0$$

with $h_j \in {}_n\mathcal{O}_0$. If $h_p \neq 0$, then h_p is the desired element h; otherwise, choose the maximal s such that $h_s \neq 0$. Then $0 = f^{p-s}(f^s + h_1 f^{s-1} + \ldots + h_s)$; since f is active, it follows from 48.4_{alg} that $f^s + h_1 f^{s-1} + \ldots + h_s$ is nilpotent. Thus, for an appropriate q, $f^{sq} + \ldots + h_s^q = 0$, and h_s^q is the desired element h. ■

E. 48 d. Find an $f \in R = {}_3\mathcal{O}_0 / (z_1 z_2, z_1 z_3)$ such that $\dim R / R \cdot f = \dim R - 1$, and such that f is not active.

48.5_{geo} Proposition. *If $f \in \mathfrak{m}({}_X\mathcal{O}_a)$ is active, then* $\dim V(f)_a = \dim X_a - 1$. ■

48.6 Corollary. $\dim {}_n\mathcal{O}_0 = n = \dim \mathbb{C}^n_0$.

Proof. Since z_n is active in ${}_n\mathcal{O}_0$ and ${}_n\mathcal{O}_0/(z_n) \cong {}_{n-1}\mathcal{O}_0$, 48.5_{alg} enables us to induct on n. ■

48.7_{alg} Proposition. *If* $(0) = \bigcap_{j=1}^{t} \mathfrak{q}_j$ *is a decomposition with primary ideals in* R, *then*

$$\dim R = \max\{\dim R/\mathfrak{q}_j;\, j = 1, \ldots, t\}.$$

Proof. "$\geqq$" This follows from 48.3_{alg} i).

"$\leqq$" Set $n := \dim R$; without loss of generality, suppose that ${}_n\mathcal{O}_0 \subset R$. The compositions $\pi_j : {}_n\mathcal{O}_0 \hookrightarrow R \twoheadrightarrow R/\mathfrak{q}_j$ are finite, and at least one π_j is injective, since $\prod_j (\mathfrak{q}_j \cap {}_n\mathcal{O}_0) \subset \bigcap_j (\mathfrak{q}_j \cap {}_n\mathcal{O}_0) = 0$ and the fact that ${}_n\mathcal{O}_0$ has no zero-divisors imply that at least one of the factors $\mathfrak{q}_j \cap {}_n\mathcal{O}_0$ is zero. Hence, it suffices to prove the following proposition for the ring $R/\mathfrak{q}_j$:

48.8_{alg} Theorem. *The following statements are equivalent:*

i) $\dim R = n$.

ii) There exists a finite injective homomorphism $\varphi : {}_n\mathcal{O}_0 \to R$.

iii) Every finite homomorphism $\varphi : {}_n\mathcal{O}_0 \to R$ *is injective, and there exists such a* φ.

Proof. The implication i) ⇒ iii) was demonstrated in 24.3; iii) ⇒ ii) is obvious.

ii) ⇒ i) First, let (0) be a *primary ideal* in R. We proceed by induction on n, beginning with the observation that the assertion holds for $n = 0$ by 48.2_{alg}. "$n-1 \Rightarrow n$" If R is a finite extension of ${}_n\mathcal{O}_0$, then z_n is not in $\mathfrak{n}_R$, and is thus active, since $\mathfrak{n}_R = \sqrt{0}$ is the only minimal prime ideal in R, by 44.10. Consequently, we can apply the induction hypothesis and the Active Lemma to the associated finite mapping

$$\psi : {}_{n-1}\mathcal{O}_0 \cong {}_n\mathcal{O}_0 / {}_n\mathcal{O}_0 \cdot z_n \to R / R \cdot z_n$$

if it is injective. Therefore, we prove the following statement:

$$\text{If}\quad g \in {}_n\mathcal{O}_0 \cap R\cdot z_n, \qquad \text{then}\quad 0 = \bar{g} \in {}_n\mathcal{O}_0 / {}_n\mathcal{O}_0 \cdot z_n.$$

With $\varphi: R \xrightarrow{g} R$ and $\mathfrak{a} = {}_n\mathcal{O}_0 \cdot z_n$, Dedekind's Lemma provides for $\varphi(1) = g$ an equation of the form $g^1 = -a_1 \in {}_n\mathcal{O}_0$; hence, we have that $\bar{g} = 0$. Thus, we have proved 48.8_{alg} for the case in which $(0) \subset R$ is a primary ideal; 48.7_{alg} follows from that.

For the *general case*, let $(0) = \bigcap \mathfrak{q}_j$ be a primary decomposition in R. If $\varphi: {}_n\mathcal{O}_0 \to R$ is finite and injective, then we may assume that the finite mapping ${}_n\mathcal{O}_0 \to R \to R/\mathfrak{q}_1$ is injective (see the proof of 48.7_{alg}). Since $(\bar{0})$ is a primary ideal in $R/\mathfrak{q}_1$, it follows that $n = \dim R/\mathfrak{q}_1 \underset{48.3_{\text{alg}}}{\leqq} \dim R \leqq n$. ■■

48.7$_{\text{geo}}$ **Proposition.** *If $X_a = X_{1a} \cup \ldots \cup X_{ta}$ is a decomposition into primary germs, then*

$$\dim X_a = \max\{\dim X_{ja};\, j = 1, \ldots, t\}. \quad ■$$

In view of 45.13, we obtain in particular that the concept of dimension 48.1 corresponds to our original geometric intuition:

48.8$_{\text{geo}}$ **Theorem.** *The following statements are equivalent:*

i) $\dim X_a = n$.

ii) There exists a finite mapping $\varphi: X_a \to \mathbb{C}^n_0$ that is open at a.

iii) Every finite holomorphic mapping $\varphi: X_a \to \mathbb{C}^n_0$ is open at a, and there exists such a mapping.

Proof. Since ${}_n\mathcal{O}_0$ is reduced, φ is open at a iff $\varphi^0: {}_n\mathcal{O}_0 \to {}_X\mathcal{O}_a$ is injective (see 45.13). The assertion follows from 45.4 and 48.8_{alg}. ■

48.9$_{\text{alg}}$ **Corollary.** *Let $\varphi: R \to S$ be a finite homomorphism. Then* $\dim S = \dim(R/\operatorname{Ker}\varphi)$, *and* $\dim R = \dim \operatorname{Red} R$; *more generally, if* $\operatorname{Ker}\varphi \subset \mathfrak{n}_R$, *then* $\dim S = \dim R$.

Proof. Since $\bar{\varphi}: R/\operatorname{Ker}\varphi \to S$ is finite and injective, it follows from 48.8_{alg} that $\dim(R/\operatorname{Ker}\varphi) = \dim S$. For ${}_n\mathcal{O}_0 \subset R$ and $\operatorname{Ker}\varphi \subset \mathfrak{n}_R$, we have that ${}_n\mathcal{O}_0 \cap \operatorname{Ker}\varphi \subset {}_n\mathcal{O}_0 \cap \mathfrak{n}_R = (0)$; hence, ${}_n\mathcal{O}_0 \to R/\operatorname{Ker}\varphi$ is injective and $\dim R \underset{48.8_{\text{alg}}}{=} \dim(R/\operatorname{Ker}\varphi) = \dim S$. ■

48.9$_{\text{geo}}$ **Corollary.** *If $\varphi: X_a \to Y_b$ is finite and open at a, then* $\dim X_a = \dim Y_b$. *In particular,* $\dim X_a = \dim \operatorname{Red} X_a$. ■

We also have a partial converse:

48.10 Proposition. *If $\varphi: X_a \to Y_b$ is a finite mapping of germs of equal dimension, and if the germ $|Y_b|$ is irreducible, then φ is open at a.*

Proof. Without loss of generality, suppose that X_a and Y_b are reduced. By the Open Lemma, we need to show that $\varphi^0: {}_Y\mathcal{O}_b \to {}_X\mathcal{O}_a$ is injective. If there were an $f \neq 0$ in $\operatorname{Ker}\varphi^0$, then f would be an active element, by E. 48a; by the Active Lemma, it would then follow that $\dim {}_X\mathcal{O}_a \underset{48.9_{\text{alg}}}{=} \dim({}_Y\mathcal{O}_b/\operatorname{Ker}\varphi^0) < \dim {}_Y\mathcal{O}_b$. ↯ ■

In contrast to the situation in algebraic geometry, the dimension of complex spaces cannot be characterized in terms of injective homomorphisms, without the assumption of finiteness. That is demonstrated by an example of Osgood (for details, see [GrRe II. § 5.2]): the mapping

$$\varphi: {}_3\mathcal{O}_0 \to {}_2\mathcal{O}_0,\ X_1 \mapsto Y_1,\ X_2 \mapsto Y_1 Y_2,\ X_3 \mapsto Y_1 Y_2 e^{Y_2},$$

determines an injective (but of course not finite) homomorphism.

The concept of the *Krull dimension* is based on the fact that, through each point $x \in X$, one can find a curve-germ, and through that, a surface-germ, etc., and obtain thereby a chain of closed subgerms that is properly ascending with respect to dimension. The Krull dimension of X is defined to be the maximal length of such a chain; it is equivalent to the earlier concept of dimension:

48.11$_{\text{geo}}$ Proposition. $\dim X_a = \max\{n;\ \exists$ *prime germs* $X_{0a} \underset{\neq}{\hookrightarrow} \dots \underset{\neq}{\hookrightarrow} X_{na}$ *in* $X_a\}$.

We prove that in its algebraic form. Note that prime germs are the subgerms determined by prime ideals; in particular, they are irreducible (see 44.9).

48.11$_{\text{alg}}$ Proposition. $\dim R = \max\{n;\ \exists$ *prime ideals* $\mathfrak{p}_n \subsetneq \mathfrak{p}_{n-1} \subsetneq \dots \subsetneq \mathfrak{p}_0$ *in* $R\}$.

Proof. By 48.2$_{\text{alg}}$, the proposition holds for $\dim R = 0$; thus, suppose that $\dim R \neq 0$. For "$\geqq$", it suffices to show that, if $\mathfrak{p} \subsetneq \mathfrak{p}'$ are prime ideals in R, then $\dim R > \dim R/\mathfrak{p}'$. In the integral domain $\bar{R} := R/\mathfrak{p}$, every nonzero $\bar{f} \in \overline{\mathfrak{p}'}$ is active; by the Active Lemma and 48.3$_{\text{alg}}$, it follows that $\dim R \geqq \dim \bar{R} > \dim \bar{R}/(\bar{f}) \geqq \dim R/\mathfrak{p}'$. For "$\leqq$" and $n := \dim R$, we construct recursively a chain $\mathfrak{p}_n \subsetneq \dots \subsetneq \mathfrak{p}_0$: By E. 48c, R contains an active f; hence, $\dim \bar{R} = n - 1$ for $\bar{R} := R/R\cdot f$. There is a chain of prime ideals in $\bar{R}$ whose inverse images provide a chain $\mathfrak{p}_{n-1} \subsetneq \dots \subsetneq \mathfrak{p}_0$ in R. There exists a prime ideal $\mathfrak{p}_n \subsetneq \mathfrak{p}_{n-1}$ since $\mathfrak{p}_{n-1}$ contains the active element f, and is thus not minimal, by 48.4$_{\text{alg}}$. ■

The definition of codimension in 7.5 yields yet another concept of dimension, which is compatible with the earlier one:

48.12 Proposition. *For* $A_0 \hookrightarrow \mathbb{C}^n_0$, $\dim A_0 = n - \operatorname{codim} A_0$.

Proof. Set $d := \dim A_0$ and $c := \operatorname{codim} A_0$. According to 48.3$_{\text{geo}}$, to verify that $d \leqq n - c$, it suffices to find a finite holomorphic mapping $\pi: A_0 \to \mathbb{C}^{n-c}_0$. In appropriate coordinates, 0 is an isolated point of $A \cap N(\mathbb{C}^n; z_1, \dots, z_{n-c})$. Hence, the restriction to A of the canonical projection $\mathbb{C}^n \to N(\mathbb{C}^n; z_{n-c+1}, \dots, z_n) = \mathbb{C}^{n-c}$ is discrete at 0; by 45.4, then, $\pi: A_0 \to \mathbb{C}^{n-c}_0$ is finite. To show that $c \geqq n - d$, we need only find a linear subspace $\Gamma \subset \mathbb{C}^n$ of dimension $n-d$ such that $A_0 \cap \Gamma_0 = \{0\}$. That is accomplished by the following:

48.13 Lemma. *For an ideal* $\mathfrak{a} \subsetneqq {}_n\mathcal{O}_0$ *with* $d := \dim {}_n\mathcal{O}_0/\mathfrak{a}$, *the following statements hold (after an appropriate linear change of coordinates):*

i) ${}_d\mathcal{O}_0 \cap \mathfrak{a} = (0)$.

ii) There exist Weierstrass polynomials $P_j \in {}_{j-1}\mathcal{O}_0[z_j] \cap \mathfrak{a}$, $j = d+1, \ldots, n$.

Put $\mathfrak{b} := {}_n\mathcal{O}_0 \cdot (P_{d+1}, \ldots, P_n)$, $X_0 := N(\mathbb{C}^n_0; \mathfrak{b})$, *and* $\Gamma := N(\mathbb{C}^n; z_1, \ldots, z_d) \cong \mathbb{C}^{n-d}$. *Then*

iii) $X_0 \cap \Gamma_0 = \{0\}$.

iv) The projection $\pi: X_0 \to \mathbb{C}^d_0$, $(z_1, \ldots, z_n) \mapsto (z_1, \ldots, z_d)$, *is finite.*

Choose an ideal $\mathfrak{a}$ such that ${}_A\mathcal{O}_0 = {}_n\mathcal{O}_0/\mathfrak{a}$. Then, the ideal $\mathfrak{b}$ of 48.13 is included in $\mathfrak{a}$; hence, $|A|_0 \subset X_0$, and, by iii), $A_0 \cap \Gamma_0 = \{0\}$. ■

Proof of 48.13. By 48.6, d $= n$ for $\mathfrak{a} = 0$, thus the assertion is evident. If $\mathfrak{a}$ is not the zero-ideal, then, in appropriate coordinates, it contains a Weierstrass polynomial P_n in z_n, by the Preparation Theorem. If there still exists a $0 \neq g \in {}_{n-1}\mathcal{O}_0 \cap \mathfrak{a}$, then we may replace it, after a linear transformation of the first $n-1$ variables, with a Weierstrass polynomial $P_{n-1} \in {}_{n-2}\mathcal{O}_0[z_{n-1}] \cap \mathfrak{a}$. By proceeding in the same manner, we finally obtain an s with ${}_s\mathcal{O}_0 \cap \mathfrak{a} = (0)$. We now show that iii) and iv) hold for s in place of d: since the P_j are Weierstrass polynomials, the fact that $z \in \Gamma$ and $P_{s+1}(z) = 0$ implies that $z_{s+1} = 0$; a simple inductive argument yields that $X_0 \cap \Gamma_0 = \{0\}$. Moreover, $X_0 \cap \Gamma_0 = \pi^{-1}(0)$; hence, 45.4 implies that $\pi: X_0 \to \mathbb{C}^s_0$ is finite. Finally, 48.8$_{\text{alg}}$ ensures that $s = d$, since ${}_s\mathcal{O}_0 \to {}_n\mathcal{O}_0/\mathfrak{a}$ is injective by i), and finite, by iv) and 45.4. ■■

E. 48e. Use 48.13 to give yet another characterization of $\dim {}_n\mathcal{O}_0/\mathfrak{a}$.

E. 48f. An analytic algebra ${}_X\mathcal{O}_0$ (and, correspondingly, the germ X_0) is called a *complete intersection* if there is a representation ${}_X\mathcal{O}_0 \cong {}_{n+m}\mathcal{O}_0/(f_1, \ldots, f_m)$ with $n = \dim_X \mathcal{O}_0$. Prove the following statements:

i) If X is a complete intersection at the point $a \in X$, then it is a complete intersection at each point near a (hint: semicontinuity of dimension 49.13).

ii) For each subgerm $Y_0 \hookrightarrow \mathbb{C}^k_0$, there exists a complete intersection Z_0 with $\dim Z_0 = \dim Y_0$ and $Y_0 \hookrightarrow Z_0 \hookrightarrow \mathbb{C}^k_0$ (hint: 48.13).

E. 48g. Prove the following statements for the analytic algebras $R = {}_4\mathcal{O}_0/\mathfrak{a}$, ${}_5\mathcal{O}_0/\mathfrak{b}$, ${}_4\mathcal{O}_0/\mathfrak{q}$ defined in E. 45e:

i) $\dim R = 2$; ii) R is not a complete intersection.

iii) Construct complete intersections X_0, Y_0, and Z_0 such that $V(\mathfrak{a}) \hookrightarrow X_0 \hookrightarrow \mathbb{C}^4_0$; $V(\mathfrak{b}) \hookrightarrow Y_0 \hookrightarrow \mathbb{C}^5_0$; $V(\mathfrak{q}) \hookrightarrow Z_0 \hookrightarrow \mathbb{C}^4_0$. (Hint for ii): E.23 Ab; for E.49i and 74.6.)

Up to now, we have defined dimension only in the punctual sense; now we introduce the global concept:

48.14 Definition. *If* X *is a complex space, then its* dimension *is* $\dim X := \sup_{a \in X} \dim X_a$.

The codimension *of a closed subspace* $Y \hookrightarrow X$ *is defined punctually to be* $\operatorname{codim}_{X_a} Y_a := \dim X_a - \dim Y_a$, *and globally as* $\operatorname{codim}_X Y := \min_{a \in Y} \operatorname{codim}_{X_a} Y_a$.

E. 48h. Show, for $A \hookrightarrow G \subset\subset \mathbb{C}^n$, that the concepts of codimension in 48.14 and 7.5 coincide.

E. 48i. Find connected complex spaces $Y \hookrightarrow X$ such that $\dim Y = \infty$, $\operatorname{codim}_X Y = 0$, and $\sup_{a \in Y} \operatorname{codim}_{X_a} Y_a = \infty$.

In 49.13, we present a further, more topological, characterization of dimension. For that, we must know that the set of singularities in a reduced space is thin; we offer the following as a preparation:

48.15 Proposition. *Each germ* X_a *satisfies the following conditions:*
i) $\operatorname{emb} X_a \geqq \dim X_a$.
ii) X_a *is a germ of a manifold iff* $\operatorname{emb} X_a = \dim X_a$.

Proof. i) This follows from 48.3_{geo} i).
ii) *If.* For $n := \dim X_a$, there is an embedding $i: X_a \hookrightarrow \mathbb{C}^n_0$, which is open at the point a by 48.10. Hence, 45.13 implies that the surjection $i^0: {}_n\mathcal{O}_0 \to {}_X\mathcal{O}_a$ is even an isomorphism. The assertion follows from E. 44c. ∎

48.16 Corollary. *A homomorphism* $\varphi: R \to {}_n\mathcal{O}_0$ *is an isomorphism iff* $\bar{\varphi}: \mathfrak{m}_R/\mathfrak{m}_R^2 \to {}_n\mathfrak{m}/{}_n\mathfrak{m}^2$ *is an isomorphism.*

Proof. If. Since $\bar{\varphi}$ is bijective, $\operatorname{emb} R = n$; by 45.8 i), φ is also surjective, and thus finite. Hence, $\dim R \geqq n$, and it follows from 48.15 that R is a power series algebra. Hence, 32.13 implies the assertion. ∎

E. 48j. Prove the following statement, and interpret it geometrically: If $\varphi: R \to S$ is a finite homomorphism of equidimensional analytic algebras, and if the zero-ideal of S is primary, then $\operatorname{Ker}\varphi$ is a primary ideal, and $\sqrt{\operatorname{Ker}\varphi}$ is a minimal prime ideal.

E. 48k. *Regularity criterion.* If f is an active element in an analytic algebra R such that $\bar{R} := R/(f)$ is regular, then so is R (i.e., it is a power series algebra. Hint: with $R_1 := \mathbb{C} \oplus R \cdot f$, use the fact that 32.11 carrries over directly to local algebras; factor $Rf/(Rf)^2 \to \mathfrak{m}_R/\mathfrak{m}_R^2$ over $Rf/\mathfrak{m}_R \cdot Rf$, show that $\dim_{\mathbb{C}}(Rf/\mathfrak{m}_R \cdot Rf) \leqq 1$, and apply 48.15; for 74.3).

E. 48 l. *Subgerms of codimension one.* Let X_0 be a germ such that the analytic algebra $R := {}_X\mathcal{O}_0$ is factorial.
i) For a nonzero ideal $\mathfrak{a}$ in R the following statements are equivalent:
α) $\mathfrak{a}$ is a principal ideal.
β) $\mathfrak{a}$ has a representation $\mathfrak{a} = \mathfrak{q}_1 \cap \ldots \cap \mathfrak{q}_n$, where each $\mathfrak{q}_j$ is a primary ideal such that $\dim R/\mathfrak{q}_j = \dim R - 1$.
(Hint for β) ⇒ α)): Set $\mathfrak{p}_j := \sqrt{\mathfrak{q}_j}$ and fix a prime element $h_j \in \mathfrak{p}_j$, then $\mathfrak{p}_j = Rh_j$; if n_j is the minimal power such that $h_j^{n_j} \in \mathfrak{q}_j$, then $\mathfrak{q}_j = Rh_j^{n_j}$. Finally, $\mathfrak{a} = Rf_1^{s_1} \cdot \ldots \cdot f_m^{s_m}$).

ii) A proper subgerm $Y_0 \hookrightarrow X_0$ is the germ of a *hypersurface* iff Y_0 admits a description $Y_{1,0} \cup \ldots \cup Y_{m,0}$, where the $Y_{j,0}$ are primary germs of codimension 1 in X_0.

iii) If Y_0 is a reduced germ in X_0, then Y_0 is the germ of a hypersurface iff all prime components of Y_0 are of codimension 1 (for 74.5).

iv) Elements f and g are relatively prime in ${}_X\mathcal{O}_a$ iff $\operatorname{codim}_{X_a} N(f) \cap N(g) \geqq 2$ (for 54 A.1).

E. 48m. i) Let X be the *Segre cone* $V(\mathbb{C}^2 \times \mathbb{C}^2; z_1 w_2 - z_2 w_1)$, and set $E := V(X; z_1, z_2)$. Show that E_0 is a 1-codimensional reduced subgerm of X_0, but not a hypersurface-germ (i.e., E. 48l ii) does not carry over; hint: if $E_0 = N(X; g)_0$ with $g \in {}_X\mathcal{O}_0$, then $g|_{N(w_1, w_2)}$ has an isolated zero. For E. 54 Aa).

ii) $Y_0 := V(\mathbb{C}_0^2; z_1 z_2, z_2^2)$ is a (non reduced) subgerm of $\mathbb{C}_0^2$ such that $|Y_0|$ is of codimension 1, which is not the germ of a hypersurface.

E. 48n. *Complex spaces are locally path-connected.* (Hint: For curves X, use 46.1 or 74.4; for reduced X with $\dim X \geqq 1$, and $0 \in X \hookrightarrow G \Subset \mathbb{C}^n$, there exists a vector space $E \subset \mathbb{C}^n$ with $\dim(E \cap S(X))_0 < \dim(E \cap X)_0 = 1$. For 49 A. 22.)

§ 49 The Set of Singular Points and Decomposition into Irreducible Components

With the decomposition of a reduced germ into irreducible components in 44.13, we have described the punctual construction of a complex space X from geometric elements whose geometric-topological structure was analyzed in the Representation Theorem for Prime Germs. Before turning to the global versions of those results, we must consider more closely the set of singularities $S(X)$, which provides a certain measure of how far removed X is from the open subset $X_{\text{reg}} := X \backslash S(X)$ of its manifold-points (or regular points):

49.1 Theorem. *The set of singular points of a complex space X is an analytic subset of X. If X is reduced, then $S(X)$ is nowhere dense.*

First we prove this stronger local statement:

49.2 Lemma. *Let the reduced complex space X be irreducible at $a \in X$. Then a has arbitrarily small connected open neighborhoods U such that:*

i) $\dim_x X = \dim_a X$ for every $x \in U$ (i.e., U is of "pure dimension").

ii) $S(U)$ is nowhere dense in U.

iii) If A is a proper analytic subset of U, then A is nowhere dense in U, and $(U \backslash S(U)) \backslash A$ is connected and dense in U.

Proof. For the prime germ X_a (see 44.14), choose a representative $f: U \to G \Subset \mathbb{C}^n$ as in 46.1. Statement i) follows from 48.9_{geo}. Since $f: U' \to G'$ is a covering by 46.1 ii), and since U' is dense in U, we conclude by 33.4 that statement ii) holds. Since U' is a connected manifold, $U' \backslash A$ must be connected and dense in U, by 7.4. But then that holds for $U_{\text{reg}} \backslash A$ as well, for otherwise, since $U' \subset U_{\text{reg}}$, $U_{\text{reg}} \backslash A$ would have a nonempty open connected component $V \Subset U$ with $V \cap (U' \backslash A) = \emptyset$, although $U' \backslash A$ is dense in U. ■

E. 49a. Find a complex space X such that $X = S(X)$.

Proof of 49.1. Since $S(X)$ is the union of the set $S(\operatorname{Red} X)$ and the analytic set of nonreduced points of X (see E. 47c), we may assume that X is reduced. We will show that the closed set $S(X)$ is locally analytic. To that end, fix $a \in S(X)$.

i) Suppose that X_a is *irreducible*. By 49.2, after an appropriate shrinking, the space $X = V(G; f_1, \ldots, f_m) \hookrightarrow G \Subset \mathbb{C}^n$ is pure-dimensional, and $S(X)$ is nowhere dense; we have that $S(X)$ is analytic as an intersection of two analytic sets (see E. 8f i)):

$$S(X) \underset{49.3}{=} X \cap \{z \in G; \operatorname{rank}\left(\frac{\partial f}{\partial z}(z)\right) < n - \dim X\}.$$

ii) If X_a is *reducible*, then there exists (after an appropriate shrinking of X) a representation of the form $X = \bigcup_{j=1}^{m} X_j$, with $X_j \hookrightarrow X$, such that the representation $X_a = \bigcup_{j=1}^{m} X_{ja}$ is precisely the decomposition into prime components presented in 44.13. By 49.2, we may assume that $S(X_j)$ and $X_j \cap X_k$ are thin analytic sets in X_j, for $j \neq k$. Then it suffices to show that

$$(49.1.1)\quad S(X) = \bigcup_{j=1}^{m} S(X_j) \cup \bigcup_{j \neq k} (X_j \cap X_k).$$

"$\subset$" This follows immediately from the fact that every X_j is closed.

"$\supset$" Suppose that some $b \in \bigcup_{j \neq k} X_j \cap X_k$ does not lie in $S(X)$; then X_b is irreducible. Since $X_b = \bigcup_{j=1}^{m} X_{jb}$, there must be a j such that $X_b = X_{jb}$; for $k \neq j$, we have that $X_{kb} \subset X_{jb}$, and it follows that $X_j \cap X_k$ is not thin in X_k. ↯. Now it is easy to see that $S(X_j) \subset S(X)$. ■

We have used the following fact:

49.3 Lemma. *For* $X = V(G; f_1, \ldots, f_m) \hookrightarrow G \Subset \mathbb{C}^n$,

$$S(X) = \{x \in X; \operatorname{rank}\left(\frac{\partial f}{\partial z}(x)\right) < n - \dim_x X\}.$$

Proof. X_x is singular $\underset{48.15}{\Leftrightarrow}$ $\operatorname{emb} X_x > \dim X_x$ $\underset{32.12}{\Leftrightarrow}$ $n - \operatorname{rank}\left(\frac{\partial f}{\partial z}(x)\right) > \dim X_x$. ■■

From 49.1, we have a representation of X as a locally finite union of locally closed submanifolds $X_{\mathrm{reg}}, S(X)_{\mathrm{reg}}, S(S(X))_{\mathrm{reg}}, \ldots$, (a "*stratification*"), which makes it possible to use techniques designed for the investigation of manifolds in the singular case as well. That most immediate stratification turns out to be too crude for some purposes; refinements with consideration given to the tangential structure can be found, for example, in $[\mathrm{Wh}]_2$.

Using 49.2, we can adapt the punctual decomposition of a *reduced* space into irreducible components (44.13) to the global case. With that in mind, we introduce the following concepts:

49.4 Definition. *An analytic subset A of a complex space X is called reducible if it has a representation $A = A_1 \cup A_2$ with proper analytic subsets A_j in A. Otherwise, of course, A is called irreducible.*

The reader should bear in mind that the concepts "reduced" and "reducible" have little to do with one another:

E. 49b. Prove that the line with a double point, $X = V(\mathbb{C}^2; z_1 z_2, z_2^2)$, is not reduced and a union of two proper closed subspaces, although $\operatorname{Red} X$ is irreducible (hint: E. 44g iii)).

First we have the following generalization of 33.5:

49.5 Proposition and Definition. *Let X be a reduced complex space. If Z' is a connected component of $X \setminus S(X)$, then the closure Z of Z' in X is an irreducible analytic subset of X; it is called an irreducible component of X. The space X is the locally finite union of its irreducible components, and every germ Z_a is a union of prime components of X_a.*

Proof. a) Every irreducible component Z of X is *analytic*: Z is closed; for $a \in Z$, it remains to show that Z is analytic near a. We may assume, for an appropriate neighborhood U of a, that $U = \bigcup_{j=1}^{m} {}_jU$, where $U_a = \bigcup_{j=1}^{m} {}_jU_a$ is the decomposition into prime components according to 44.13, and where every ${}_jU$ has the properties mentioned in 49.2. For each j, then,

$$\emptyset \neq {}_jU \setminus \bigcup_{k \neq j} {}_kU = U \setminus \bigcup_{k \neq j} {}_kU \mathrel{\subset\!\!\!\!\mid} U;$$

hence, $U_j' := {}_jU \setminus S(U) \subset U_{\mathrm{reg}}$ is nonempty, by (49.1.1), and connected by 49.2 iii) (in particular, U has m irreducible components). Now Z' is a connected component of X_{reg}, and it follows that

$$Z' \cap U = \bigcup_{j=1}^{m} (Z' \cap {}_jU) = \bigcup_{J'} {}_jU', \quad \text{with} \quad J' := \{j;\, Z' \cap {}_jU \neq \emptyset\},$$

since ${}_jU' \subset Z'$ if $Z' \cap {}_jU \neq \emptyset$. Hence, the set

$$Z \cap U = \overline{Z'} \cap U = \bigcup_{J'} \overline{{}_jU'} \cap U = \bigcup_{J'} {}_jU$$

is analytic, and $Z_a = \bigcup_{J'} {}_jU_a$.

b) Every irreducible component Z of X is *irreducible*: if $Z = A \cup B$ is a union of analytic sets A and B, then so is $Z' = (A \cap Z') \cup (B \cap Z')$. If Z' is irreducible, then we may assume that $Z' \subset A$, and thus that $Z = A$. Hence, the assertion follows from the following fact:

49.6 Remark. *If a connected reduced complex space X is irreducible at every point, then X is irreducible.*

Proof. If Y is an analytic set in X with $Y_a = X_a$ for some $a \in X$, then, by construction, $\{x \in X;\ X_x = Y_x\}$ is open, nonempty, and closed (apply 49.2 at a boundary point); thus $Y = X$. The assertion follows immediately. ■■

49.7 Corollary. *A reduced complex space X is irreducible iff X_{reg} is connected.* ■

Using the preceding, we obtain an important generalization of 7.4:

49.8 Proposition. *Let A be a closed subspace of a reduced irreducible space X. Then either $A = X$ or A is nowhere dense in X. In particular, $A = X$ if $\mathring{A} \neq \emptyset$.* ■

Observe that, in contrast to the situation in 7.4, A may separate X locally:

E. 49c. Show that the following irreducible spaces X include open connected sets U for which $U \setminus S(U)$ is not connected:
i) $X = N(\mathbb{C}^3; z_1^2 z_2 - z_3^2)$ (Whitney's umbrella; see 33.8).
ii) $X = N(\mathbb{C}^2; z_1^2 - z_2^2 - z_2^3)$ (see E. 32j).

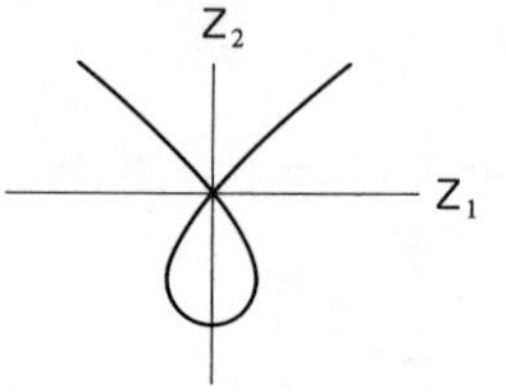

Proof of 49.8. The subset $X_{\text{reg}} \cap \mathring{A}$ of the connected manifold X_{reg} is open; by 7.4, it is also closed. If $\mathring{A}$ is nonempty, then 49.1 implies that $X_{\text{reg}} \cap \mathring{A} \neq \emptyset$; hence, $X_{\text{reg}} \subset \mathring{A}$, and $|X| = |A|$. Hence, $i: A \to X$ is an open mapping; by the open Lemma, the surjective comomorphism $i^0: {}_X\mathcal{O} \to {}_A\mathcal{O}$ is injective. Consequently, i is an isomorphism. ■

49.9 Proposition. *For analytic subsets A and B of a complex space X, if A is irreducible, and if $A_a \subset B_a$ for some $a \in A$, then $A \subset B$.*

Proof. Application of 49.8 to the closed subspace $A \cap B$ of A yields that $A \cap B = A$; i.e., $A \subset B$. ■

49.10 Corollary (Identity Theorem for irreducible analytic sets). *If A and B are irreducible analytic sets, and if $A_a = B_a$ for some $a \in A$, then $A = B$.* ■

49.11 Proposition (Identity Theorem for holomorphic mappings). *Let $f, g: X \to Y$ be holomorphic mappings, with X reduced and irreducible, and suppose that $f_a = g_a$ for some $a \in X$. Then $f = g$.*

Proof. By 31.11, it suffices to show that $\operatorname{Red} f = \operatorname{Red} g$. The set $A = \{x \in X; f(x) = g(x)\}$ is analytic, for it is closed, and – as may be seen immediately for local models (see also 49 A.2) – it is locally analytic. By hypothesis, a is an interior point of A, and it follows from 49.8 that $A = X$. ■

E. 49d. If X is an irreducible reduced space, then $\mathcal{O}(X)$ is an integral domain.

E. 49e. Show, by means of counterexamples, that 49.8–49.11 and E. 49d do not hold without the assumption of irreducibility (for nonreduced X, not even the irreducibility of $\operatorname{Red} X$ would be sufficient).

E. 49f. Prove that, for a reduced space X, the complement $X \setminus A$ of every nowhere dense analytic set A in X is connected iff X is irreducible (for E. 49h).

E. 49g. For two reduced complex structure sheaves $\mathcal{H}_1 \subset \mathcal{H}_2$ on X, show that $(X, \mathcal{H}_1)$ is irreducible (resp., irreducible at every point) iff $(X, \mathcal{H}_2)$ is (for 46 A.1).

E. 49h. Use E. 49f to show for complex germs that $\operatorname{Red} X_a$ and $\operatorname{Red} Y_b$ are irreducible iff $\operatorname{Red}(X \times Y)_{(a,b)}$ is irreducible (for 49.16).

E. 49i. For the following complex spaces X (with the notation of E. 45e), show that X and $\operatorname{Red} X$ are irreducible surfaces with $S(X) = \{0\}$:

i) $X = V(\mathbb{C}^4; \mathfrak{a} \cap \mathcal{O}(\mathbb{C}^4))$.

ii) $X = V(\mathbb{C}^5, \mathfrak{b} \cap \mathcal{O}(\mathbb{C}^5))$.

iii) $X = V(\mathbb{C}^4; \mathfrak{p} \cap \mathcal{O}(\mathbb{C}^4))$.

(Hint: investigate $\pi: X \to \mathbb{C}^2, z \mapsto (z_1, z_n)$, (with $n = 4$ or $n = 5$) on $\{z_1 z_n \neq 0\} \subset\subset \mathbb{C}^2$, and apply E. 49f; alternatively, prove the assertion for the germ X_0 (with E. 48g), and use the fact that X is a cone. For 74.6.)

Here is another characterization of the decomposition of a reduced complex space into irreducible components:

49.12 Proposition. *A reduced complex space X has exactly one representation $X = \bigcup_{j \in J} X_j$ as a locally finite union of irreducible analytic subsets X_j of X with $X_j \not\subset X_k$ for $j \neq k$, namely, the decomposition into irreducible components.*

Proof. It is sufficient to demonstrate the uniqueness. To that end, let $X = \bigcup_{j \in J} X_j = \bigcup_{i \in I} Y_i$ be such representations. For fixed j, then, $X_j = \bigcup_{i \in I} (X_j \cap Y_i)$ is a locally finite union. By choosing an $a \in (X_j)_{\text{reg}}$, we can ensure that $X_{ja} = \bigcup_{i \in I} (X_{ja} \cap Y_{ia})$ is finite;

since X_{ja} is irreducible, there exists an i with $X_j \subset Y_i$, by 49.9. Similarly, there is a k with $Y_i \subset X_k$; it follows that $j = k$, and $X_j = Y_i$. ■

E. 49j. If $f: X \to Y$ is a holomorphic mapping of reduced spaces, and if X is irreducible, then $f(X)$ is included in an irreducible component of Y. If f is finite, then $f(X)$ itself is irreducible (for 71.15).

We close the section with a further characterization of dimension: $\dim X_a = \lim\limits_{U \to a} \dim (\operatorname{Red} U)_{\text{reg}}$; precisely stated, the following holds:

49.13 Proposition (Semicontinuity of dimension). *The function $X \to \mathbb{Z}, a \mapsto \dim X_a$, is upper semicontinuous (i.e., for every x near a, $\dim X_x \leqq \dim X_a$). If a is not an isolated point of X, then, in every neighborhood U of a, there exist points $x \neq a$ of $(\operatorname{Red} X)_{\text{reg}}$ with $\dim X_x = \dim X_a$.*

Proof. By 48.9_{geo} and 48.7_{geo}, we may assume that X_a is reduced and irreducible. Then 49.2 yields the assertion. ■

E. 49k. The function $X \to \mathbb{Z}, a \mapsto \dim X_a$, is bounded on every compact subset of X.

E. 49l. If X is reduced and irreducible, then X is of pure dimension (for 54A. 1).

E. 49m. Use E.49l to prove that, for each $k \in \mathbb{N}$, $\{x \in X; \dim X_x \geqq k\}$ is an analytic subset of X (for 57.3).

E. 49n. Prove that $\dim X \times Y = \dim X + \dim Y$ (for 49.16).

49.14 Proposition (Semicontinuity of fiber dimension). *If $f: X \to Y$ is a holomorphic mapping, then the function $X \to \mathbb{Z}, x \mapsto \dim (f^{-1} f(x))_x$, is upper semicontinuous.*

The proof is based on the following fact:

49.15 Lemma. *For a holomorphic mapping $f: X \to Y$, a point $a \in X$, and $d := \dim (f^{-1} f(a))_a$, there is a neighborhood U of a, and a commutative diagram*

$$\begin{array}{ccc} U & \xrightarrow{\ \varphi\ } & Y \times \mathbb{C}^d \\ & {\scriptstyle f} \searrow \quad \swarrow {\scriptstyle \text{pr}} & \\ & Y & \end{array}$$

with φ discrete (and even finite over an appropriate neighborhood $V \Subset Y \times \mathbb{C}^d$ of $\varphi(a)$).

Proof of 49.14. For $u \in U$, the mapping φ of 49.15 induces a finite mapping $(f^{-1} f(u))_u \to (\text{pr}^{-1} f(u))_{\varphi(u)}$; it follows from 48.3_{geo} that

$$\dim (f^{-1} f(u))_u \leqq \dim (\text{pr}^{-1} f(u))_{\varphi(u)} \leqq d. \quad ■$$

Proof of 49.15. Choose $\psi : f^{-1} f(a)_a \to \mathbb{C}_0^d$ according to 48.8_{geo}. Since ${}_X\mathcal{O}_a \to {}_{f^{-1}f(a)}\mathcal{O}_a$ is surjective, ψ can be extended to a neighborhood of a in X, as in 44.6. The mapping $\varphi := (f, \psi)$ is discrete at a, since $\varphi^{-1}\varphi(a) = f^{-1}f(a) \cap \psi^{-1}\psi(a)$, and thus induces a finite holomorphic mapping near a, by 45.4. The parenthetic remark follows from 33 B.2. ■■

An example of a holomorphic mapping of complex manifolds with nonconstant fiber-dimension is provided by the quadratic transformation 32 B.10 iv).

In addition to the proper holomorphic mappings, the open holomorphic mappings also are important; they are a generalization of projections.

49.16 Proposition. *For a holomorphic mapping $f: X \to Y$ into a space Y that is irreducible at every point, f is open iff* $\dim X_a = \dim Y_{f(a)} + \dim (f^{-1}f(a))_a$, $\forall a \in X$.

Proof. It suffices to verify the assertion locally. *If.* With the notation of 49.15, let a commutative diagram

$$\begin{array}{ccc} U & \xrightarrow{\ \varphi\ } & Y \times \mathbb{C}^d \\ & {\scriptstyle f}\searrow \quad \swarrow{\scriptstyle \text{pr}} & \\ & Y & \end{array}$$

with a discrete mapping φ be given near $a \in X$. Then $\dim X_a = \dim Y_{f(a)} + d \underset{\text{E. 49n}}{=} \dim (Y \times \mathbb{C}^d)_{\varphi(a)}$. According to E. 49h, $Y \times \mathbb{C}^d$ induces an irreducible set-germ at $\varphi(a)$. By 48.10, then, φ is open at a; that fact and the openness of pr at $\varphi(a)$ imply that f is open at a, as well. That holds for every a, so f is open.

Only if. We induct on $n = \dim Y_{f(a)}$: the case in which $n = 0$ is trivial. "$n-1 \Rightarrow n$" Without loss of generality, let X_a and Y_b be reduced, with $b = f(a)$. By the Representation Theorem for Prime Germs, there exists for the prime germ Y_b an open discrete holomorphic mapping $\psi : Y \to \mathbb{C}^n$; since $f \circ \psi$ is open, we may assume that $Y = \mathbb{C}^n$ and $b = 0$. Hence, $f = (f_1, \ldots, f_n)$, and $f_n : X \to \mathbb{C}$ is open. For $N(f_n) = f^{-1}(\mathbb{C}^{n-1} \times 0)$, there is a commutative diagram

$$\begin{array}{ccc} N(f_n) & \xrightarrow{\ g = f|_{N(f_n)}\ } & \mathbb{C}^{n-1} \times 0 \\ \downarrow & & \downarrow \\ X & \xrightarrow{\quad f \quad} & \mathbb{C}^n \end{array},$$

in which g is open, since f is. Now $\dim N(f_n) = \dim X_a - 1$ (in each prime component X_{ja}, the codimension of $N(f_n)$ is 1, for otherwise, by 49.2, f_n would vanish on some X_j, and thus would not be open), so we can apply the induction hypothesis:

$$\dim X_a - 1 = \dim N(f_n)_a = n - 1 + \dim [g^{-1}g(a)]_a.$$

Since $f^{-1}f(a) = g^{-1}g(a)$, the assertion follows. ■

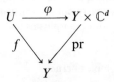

E. 49 o. Consider a reduced connected complex space X of pure dimension, a complex space Y that is irreducible at every point, and a finite holomorphic surjection $\varphi: X \to Y$. Show that, with respect to Y, φ is locally a branched covering. (Hint: find a representation $X_x \xrightarrow{\varphi} Y_{\varphi x} \xrightarrow{f} \mathbb{C}_0^n$ with branched coverings $f \circ \varphi$ and f, and use 49.16 and 48.8_{alg}; for E. 49p).

E. 49p. If $\varphi : X \to Y$ is a finite holomorphic surjection, then $\dim X = \dim Y$ (hint: E.49 o; for 49 A.17).

E. 49q. Find holomorphic mappings $f: X \to Y$ that are *counterexamples* to the following statements:

i) Also for locally reducible Y, the dimension-condition of 49.16 implies that f is open.

ii) For $Y = \mathbb{C}$ and f open at a point $a \in X$, the dimension-condition of 49.16 holds at a.

§ 49 A Supplement: Fiber Products and Quotients

In this supplement, we define and investigate kernels and cokernels of pairs of morphisms (above all we are interested in fiber products of holomorphic mappings and quotients of complex spaces with respect to equivalence relations). In the interest of clarity, we carry out the constructions in the larger category of ringed spaces; there we shall see that each pair of morphisms has a (ringed) kernel and a (ringed) cokernel.

We have these facts for pairs of holomorphic mappings:

i) The ringed kernel is a complex space; hence, it is also the complex kernel.

ii) The complex cokernel does not always exist (49 A.8 iii)).

iii) If the complex cokernel exists, it may differ from the ringed cokernel (49 A.8 ii)).

iv) If the ringed cokernel is a complex space, then it is also the complex cokernel.

Thus we give criteria for the ringed cokernel to be a complex space. In that way, many examples of complex spaces can be constructed, particularly as *quotients with respect to analytic equivalence relations.*

Every complex space is a ringed space, which rests upon a topological space, which, in turn, is built from a set. In the following, then, we shall be concerned with the categories of complex spaces, ringed spaces, topological spaces, and sets.

49 A. 1 Definition. *In a category $\mathcal{K}$, let a pair of morphisms $f, g : X_1 \to X_2$ (abbreviated $X_1 \underset{g}{\overset{f}{\rightrightarrows}} X_2$, or even $\underset{g}{\overset{f}{\rightrightarrows}}$) be given.*

i) A morphism $i : X \to X_1$ is called a kernel *(or an equalizer) in $\mathcal{K}$ of $\underset{g}{\overset{f}{\rightrightarrows}}$ if a) $f \circ i = g \circ i$, and b) for each morphism h in $\mathcal{K}$ such that $f \circ h = g \circ h$, there is a unique morphism h' in $\mathcal{K}$ making the diagram*

$$\begin{array}{ccccc} X & \xrightarrow{i} & X_1 & \underset{g}{\overset{f}{\rightrightarrows}} & X_2 \\ & \searrow^{h'} & \downarrow h & & \\ & & Z & & \end{array}$$

commute.

ii) A morphism $p : X_2 \to X$ is called a cokernel *(or a coequalizer) in $\mathcal{K}$ of $\underset{g}{\overset{f}{\rightrightarrows}}$ if a) $p \circ f = p \circ g$ and b) for each morphism h in $\mathcal{K}$ such that $h \circ f = h \circ g$, there is a unique morphism h' in $\mathcal{K}$ making the diagram*

$$\begin{array}{ccccc} X_1 & \underset{g}{\overset{f}{\rightrightarrows}} & X_2 & \xrightarrow{p} & X \\ & & \downarrow h & \swarrow_{h'} & \\ & & Z & & \end{array}$$

commute.

The next exercise provides a reformulation of Definition 49 A.1:

E. 49 A a. i) If a pair of morphisms has a (co-)kernel, then the (co-)kernel is unique up to *canonical* isomorphism.

ii) With the notation $\mathrm{Mor}_{\mathscr{K}}(X, Y)$ for the set of morphisms $f: X \to Y$ in $\mathscr{K}$, prove that $X \xrightarrow{i} X_1$ is a kernel of $X_1 \overset{f}{\underset{g}{\rightrightarrows}} X_2$ in $\mathscr{K}$ iff the mapping

$$\mathrm{Mor}_{\mathscr{K}}(Z, X) \to \{k \in \mathrm{Mor}_{\mathscr{K}}(Z, X_1);\, f \circ k = g \circ k\},\, h \mapsto i \circ h,$$

is well-defined and bijective for each object Z of $\mathscr{K}$ (for 49 A.2).

iii) Formulate and prove the analogous statement for cokernels.

Since kernels and cokernels are unique – if they exist – we use the notation $\mathrm{Ker} \overset{f}{\underset{g}{\rightrightarrows}}$ and $\mathrm{Coker} \overset{f}{\underset{g}{\rightrightarrows}}$. A *sequence* $X \xrightarrow[i]{} X_1 \overset{f}{\underset{g}{\rightrightarrows}} X_2$ in $\mathscr{K}$ is called *exact* if $i = \mathrm{Ker} \overset{f}{\underset{g}{\rightrightarrows}}$; similarly, $X_1 \overset{f}{\underset{g}{\rightrightarrows}} X_2 \xrightarrow[p]{} X$ is called *exact* if $p = \mathrm{Coker} \overset{f}{\underset{g}{\rightrightarrows}}$.

E. 49 A b. *In the category of abelian groups, kernels and cokernels always exist:* let $f: G \to H$ be a homomorphism, and let $0: G \to H$ be the zero homomorphism; show that the definitions of $\mathrm{Ker} \overset{f}{\underset{0}{\rightrightarrows}}$ and $\mathrm{Coker} \overset{f}{\underset{0}{\rightrightarrows}}$ agree with the usual definitions of $\mathrm{Ker} f$ and $\mathrm{Coker} f$, and deduce the assertion from that (hint: compare $\overset{f}{\underset{g}{\rightrightarrows}}$ and $\overset{f-g}{\underset{0}{\rightrightarrows}}$).

E. 49 A c. *In the categories of sets and topological spaces, kernels and cokernels always exist:* let $S \overset{f}{\underset{g}{\rightrightarrows}} T$ be a pair of set-mappings (continuous mappings), and define $A := \{s \in S;\, f(s) = g(s)\}$ (with the relative topology) and $\bar{T} := T/R$ (with the quotient topology), where R is the equivalence relation on T generated by "$f(s) \sim g(s)$ for every $s \in S$"; i.e., R is the intersection of all equivalence relations $Q \subset T \times T$ such that "$s \in S \Rightarrow (f(s), g(s)) \in Q$". Prove the following statements:

i) The sequences $A \xrightarrow{i} S \overset{f}{\underset{g}{\rightrightarrows}} T$ and $S \overset{f}{\underset{g}{\rightrightarrows}} T \xrightarrow{p} \bar{T}$ are exact (for 49 A.2).

ii) The topological kernel is a closed subspace if T is a Hausdorff space.

49 A. 2 Proposition. i) *In the categories of ringed and of complex spaces, every pair of morphisms has a kernel.*

ii) *If $X \xrightarrow{i} X_1 \overset{f}{\underset{g}{\rightrightarrows}} X_2$ is an exact sequence of complex spaces, then i is a (closed) embedding, and the sequence is exact for the underlying ringed spaces, topological spaces, and sets, as well.*

In particular, ii) says that the ringed kernel of a pair of holomorphic mappings is also the complex kernel.

Proof. First, let $(S, {}_S\mathscr{A}) \overset{f}{\underset{g}{\rightrightarrows}} (T, {}_T\mathscr{A})$ be a pair of morphisms of ringed spaces. By E. 49Ac, $A := \{s \in S;\, f(s) = g(s)\} \xrightarrow{i} S$ is the associated topological kernel. We construct the ringed kernel by means of the sheaf of ideals $\mathscr{I} \subset {}_S\mathscr{A}$ determined by

$$\mathscr{I}(U) = \bigcap_{s \in U \cap A,\, f(s) = g(s)} \{h \in {}_S\mathscr{A}(U);\, h_s \in {}_S\mathscr{A}_s \cdot (f_s^0 - g_s^0)({}_T\mathscr{A}_{f(s)})\}\,.$$

For $s \in A$, we have that $\mathscr{I}_s \subset {}_S\mathfrak{m}_s$, since $(f_s^0 - g_s^0)({}_T\mathscr{A}_{f(s)}) = (f_s^0 - g_s^0)({}_T\mathfrak{m}_{f(s)}) \subset {}_S\mathfrak{m}_s$. Hence, $(A, {}_A\mathscr{A})$, with ${}_A\mathscr{A} := ({}_S\mathscr{A}/\mathscr{I})|_A$, is the ringed kernel, as can be checked easily with the aid of the Restriction Lemma and E. 49 Aa ii). Note that $(A, {}_A\mathscr{A}) = V(S; \mathscr{I})$ if T is a Hausdorff space.

Now let $(X, {}_X\mathcal{O}) \overset{f}{\underset{g}{\rightrightarrows}} (Y, {}_Y\mathcal{O})$ be a pair of holomorphic mappings. We show that the ringed kernel $(A, {}_A\mathcal{O}) \hookrightarrow (X, {}_X\mathcal{O})$ is a closed complex subspace; since the kernel can be determined locally (E. 49 Ad), we may assume, after shrinking X and Y, that $Y \hookrightarrow G \Subset \mathbb{C}^n$ is a local model.

Then the composed mapping $Y \xrightarrow{j} \mathbb{C}^n$ is a monomorphism, and the two pairs of holomorphic mappings $X \underset{g}{\overset{f}{\rightrightarrows}} Y$ and $X \underset{j \circ g}{\overset{j \circ f}{\rightrightarrows}} \mathbb{C}^n$ have the same ringed kernel. Hence, the assertion follows immediately from the following fact:

49 A.3 Lemma. *For $f, g \in \mathrm{Hol}(X, \mathbb{C}^n)$, the ringed kernel of each of the pairs of holomorphic mappings $X \underset{g}{\overset{f}{\rightrightarrows}} \mathbb{C}^n$ and $X \underset{0}{\overset{f-g}{\rightrightarrows}} \mathbb{C}^n$ is the complex subspace $(f-g)^{-1}(0) \hookrightarrow X$.*

Proof. If $\mathscr{J} \subset {}_n\mathcal{O}$ is the – coherent – nullstellen ideal of $0 \in \mathbb{C}^n$, then the structure sheaf of the fiber $(f-g)^{-1}(0)$ is given by the coherent sheaf ${}_X\mathcal{O}/(f-g)^*\mathscr{J}$ (see 43.5 and 31.8 iv)). Hence, it suffices to prove stalkwise the equality of $(f-g)^*\mathscr{J}$ with the defining sheaf of ideals $\mathscr{I}$ of the ringed kernel of $\underset{g}{\overset{f}{\rightrightarrows}}$. For $x \in X$ with $f(x) \neq g(x)$, that is clear; for $f(x) = g(x) =: y$, we have that

$$\mathscr{I}_x = {}_X\mathcal{O}_x \cdot (f_x^0 - g_x^0)({}_n\mathcal{O}_y) = {}_X\mathcal{O}_x \cdot (f_x^0 - g_x^0)({}_n\mathfrak{m}_y)$$

and

$$((f-g)^*\mathscr{J})_x = {}_X\mathcal{O}_x \cdot (f-g)_x^0({}_n\mathfrak{m}_0)\,.$$

The equality of the ideals follows from the equality of all the generators:

$$(f_x^0 - g_x^0)(z_k - y_k) = f_x^0(z_k) - y_k - g_x^0(z_k) + y_k = f_x^0(z_k) - g_x^0(z_k) = (f-g)_x^0(z_k)\,. \quad \blacksquare\blacksquare$$

E. 49 Ad. *The ringed kernel (and thus also the complex kernel) can be calculated locally:* Let $(A, {}_A\mathscr{A})$ be the kernel of the pair $S \underset{g}{\overset{f}{\rightrightarrows}} T$. For $U \Subset S$ and $W \Subset T$ with $f(U) \cup g(U) \subset W$ show that $(A \cap U, {}_A\mathscr{A}|_{A \cap U})$ is the kernel of $U \underset{g}{\overset{f}{\rightrightarrows}} W$ (for 49 A.2).

49 A.4 Examples. i) $\mathrm{Ker}(\mathbb{C} \underset{-z}{\overset{z}{\rightrightarrows}} \mathbb{C}) = \bullet \hookrightarrow \mathbb{C}$, by 49 A.3.

ii) $\mathrm{Ker}(\mathbb{C} \underset{z+z^m}{\overset{z}{\rightrightarrows}} \mathbb{C})$ is the *m*-fold point, by 49 A.3.

iii) For $f \in \mathrm{Hol}(X, X)$, $\mathrm{Fix}\, f := \mathrm{Ker}(X \underset{f}{\overset{\mathrm{id}}{\rightrightarrows}} X)$ is called the *fixed point space* of *f*. In particular, by 49 A.2, $|\mathrm{Fix}\, f| = \mathrm{Fix}\, |f|$.

E. 49 Ae. Determine $\mathrm{Ker}(\mathbb{C} \underset{z^q}{\overset{z^p}{\rightrightarrows}} \mathbb{C})$ for $p, q \in \mathbb{N}$.

From our standpoint, the most important example of a kernel is given by the following:

49 A.5 Definition. *For two holomorphic mappings ${}_jf : X_j \to Y$, the complex space $X_1 \times_Y X_2 := \mathrm{Ker}(X_1 \times X_2 \underset{{}_2f \circ \mathrm{pr}_2}{\overset{{}_1f \circ \mathrm{pr}_1}{\rightrightarrows}} Y)$ is called the fiber product of X_1 and X_2 over Y.*

By 49 A.2, the fiber product exists and is unique; $X_1 \times_Y X_2$ is a closed subspace of $X_1 \times X_2$ with

$$|X_1 \times_Y X_2| = |X_1| \times_{|Y|} |X_2| = \{(x_1, x_2) \in X_1 \times X_2;\ {}_1f(x_1) = {}_2f(x_2)\}\,.$$

With the canonical holomorphic mapping

$$f := {}_1f \circ \mathrm{pr}_1 \circ i = {}_2f \circ \mathrm{pr}_2 \circ i : X_1 \times_Y X_2 \to Y,$$

the fiber product is a complex space over Y (see E. 49 Ag). For the fibers, it is easy to show that

$$f^{-1}(y) = {}_1f^{-1}(y) \times {}_2f^{-1}(y);$$

that provides motivation for the name "fiber product".

E. 49 Af. Show that a holomorphic mapping $\pi = (\pi_1, \pi_2) : F \to X_1 \times X_2$ is the fiber product of X_1 and X_2 over Y iff every commutative holomorphic diagram of the form

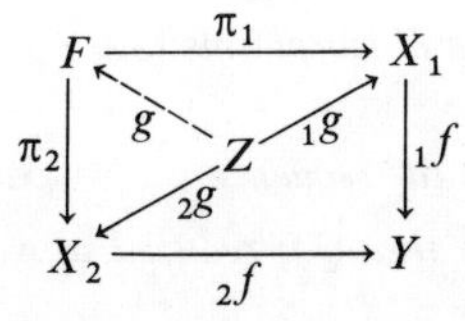

can be completed commutatively with a unique holomorphic mapping g (for 49A.6 ii)).

49 A.6 Examples. i) $X_1 \times_{\bullet} X_2 = X_1 \times X_2$, for we have that $\mathrm{Hol}(Z, X_1 \times_{\bullet} X_2) = \mathrm{Hol}(Z, X_1 \times X_2)$; thus $X_1 \times_{\bullet} X_2$ is the product of X_1 and X_2.

ii) For the *inverse image of a closed subspace* $B \hookrightarrow Y$ under a holomorphic mapping $f: X \to Y$, we have that

$$f^{-1}(B) \cong B \times_Y X,$$

by 31.9 iii) and E.49 Af, since the ringed space $f^{-1}(B)$ is a complex space, by 43.5.

iii) For $B = \{b\}$ in ii), $f^{-1}(b) \cong \{b\} \times_Y X$ is the *fiber of f at b.*

iv) *The graph of a holomorphic mapping.* The fiber product of $f: X \to Y$ and id_Y is called the graph $\Gamma(f)$ of f (see E.7d). It satisfies the following:

$$|\Gamma(f)| = \{(x, f(x)); x \in X\} = \Gamma(|f|) \subset |X| \times |Y|.$$

Letting $p: \Gamma(f) \to X$ and $q: \Gamma(f) \to Y$ denote the canonical projections, we have that p is biholomorphic and $q \circ p^{-1} = f$. For, by E.49Af, the mapping $h := (\mathrm{id}_X, f): X \to X \times Y$ determines a mapping $g: X \to \Gamma(f)$ such that the diagram

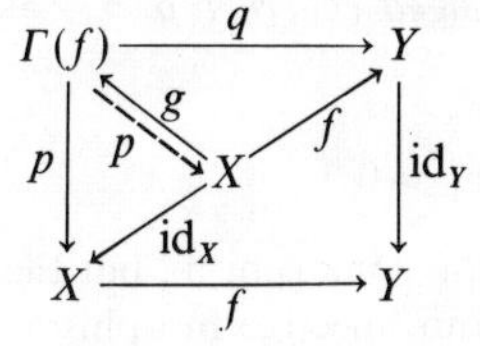

commutes. Hence, $p \circ g = \mathrm{id}_X$; with the additional entry of the arrow $\overset{p}{\dashrightarrow}$, the universal property yields immediately that $g \circ p = \mathrm{id}_{\Gamma(f)}$. Thus p is biholomorphic, and $f = g \circ p^{-1}$.

v) *The diagonal.* With $f = \mathrm{id}_X$ in iv), $\Delta_X := \Gamma(\mathrm{id}_X)$ is called the diagonal of $X \times X$; it plays an important role in the differential calculus on complex spaces. By iv), the holomorphic mapping

$$\delta: X \to \Delta_X \subset X \times X, \; x \mapsto (x, x),$$

induced by $(\mathrm{id}_X, \mathrm{id}_X)$ is biholomorphic.

vi) *Intersection of closed subspaces.* For two subspaces $j_k: A_k = V(\mathscr{I}_k) \to X$, we clearly have that $j_2^{-1}(A_1) = V(X; \mathscr{I}_1 + \mathscr{I}_2)$. Hence, it follows from ii) that $A_1 \times_X A_2 = A_1 \cap A_2$.

E. 49 Ag. *Universal property of the fiber product.* In analogy to 43.14, characterize the fiber product $X_1 \times_Y X_2$ as follows: every complex space (Z, f) over Y induces a canonical bijection

$$\mathrm{Hol}_Y(Z, X_1 \times_Y X_2) \cong \mathrm{Hol}_Y(Z, X_1) \times \mathrm{Hol}_Y(Z, X_2).$$

(The *fiber product* is thus the *product in the category of complex spaces over Y.*)

E. 49 Ah. Determine the fiber product for ${}_1f = {}_2f = \mathrm{pr}_n: \mathbb{C}^{n+m} \to \mathbb{C}^n$ and show that the canonical projections $\mathbb{C}^{n+m} \times_{\mathbb{C}^n} \mathbb{C}^{n+m} \to \mathbb{C}^{n+m}$ are submersions.

Now we come to the investigation of cokernels:

49 A. 7 Proposition. *i) In the category of ringed spaces, every pair of morphisms* $(S, {}_S\mathscr{A}) \underset{g}{\overset{f}{\rightrightarrows}} (T, {}_T\mathscr{A})$ *has a cokernel.*

ii) If $S \underset{g}{\overset{f}{\rightrightarrows}} T \xrightarrow{p} Q$ *is an exact sequence of ringed spaces, then the sequence is also exact for the associated topological spaces (respectively, sets). If T is reduced, then Q is reduced, as well.*

Proof. Let $p: T \to \bar{T} := T/R$ be the topological cokernel of $S \underset{g}{\overset{f}{\rightrightarrows}} T$ (see E.49 Ac i)). We define a structure sheaf $\bar{\mathscr{A}}$ on $\bar{T}$ as follows: for $\bar{V} \Subset \bar{T}$, $V := p^{-1}(\bar{V})$, and $U := f^{-1}(V) = g^{-1}(V) \Subset S$, set $\bar{\mathscr{A}}(\bar{V}) := \{\varphi \in {}_T\mathscr{A}(V); f^0\varphi = g^0\varphi \in {}_S\mathscr{A}(U)\}$. Clearly, $\bar{\mathscr{A}}$ is a sheaf on $\bar{T}$. To verify that $(\bar{T}, \bar{\mathscr{A}})$ is a ringed space, we apply 31.2. For a section $\varphi \in \bar{\mathscr{A}}(\bar{V}) \subset {}_T\mathscr{A}(V)$, $\operatorname{Red}\varphi \in {}_T\mathscr{C}(V)$ determines a continuous function on $\bar{V}$, which we also denote with $\operatorname{Red}\varphi$. The homomorphism $\operatorname{Red}: \bar{\mathscr{A}} \to {}_T\mathscr{C}$ induced thereby is the reduction of $\bar{\mathscr{A}}$: if $\varphi \in \bar{\mathscr{A}}(\bar{V})$ has no zeros, then φ is invertible in ${}_T\mathscr{A}(V)$. Since $f^0(\varphi^{-1}) = (f^0\varphi)^{-1} = (g^0\varphi)^{-1} = g^0(\varphi^{-1})$, we have that $\varphi^{-1} \in \bar{\mathscr{A}}(\bar{V})$; i.e., φ is invertible in $\bar{\mathscr{A}}(\bar{V})$. It follows immediately that the canonical morphism $(T, {}_T\mathscr{A}) \to (\bar{T}, \bar{\mathscr{A}})$ is the ringed cokernel of f and g. ∎

We note from the proof that

$$\bar{\mathscr{A}}(\bar{V}) = \{\varphi: \bar{V} \to \mathbb{C}; \varphi \circ p \in {}_T\mathscr{A}(p^{-1}(\bar{V}))\}$$

if $(T, {}_T\mathscr{A})$ is reduced.

49 A. 8 Examples. i) For $p, q \in \mathbb{N}_{>0}$, the pair of holomorphic mapping $\mathbb{C} \underset{\cdot e^{2\pi i/q}}{\overset{\cdot e^{2\pi i/p}}{\rightrightarrows}} \mathbb{C}$ has the mapping $\mathbb{C} \xrightarrow{z^r} \mathbb{C}$ with $r := \operatorname{lcm}(p, q) / \gcd(p, q)$ as its ringed and as its complex cokernel.

ii) *The ringed and complex cokernels may be different.* The pair of holomorphic mappings $\mathbb{P}_1 \underset{g}{\overset{f}{\rightrightarrows}} \mathbb{C} \times \mathbb{P}_1$, with $f(z) = (0, 0)$ and $g(z) = (0, z)$, has the *ringed cokernel* $p: \mathbb{C} \times \mathbb{P}_1 \to (\mathbb{C} \times \mathbb{P}_1)/\sim =: S$ with

$$x = (x_1, x_2) \sim y = (y_1, y_2) \quad \text{iff} \quad x = y \quad \text{or} \quad x_1 = y_1 = 0 .$$

The restriction $p: \mathbb{C}^* \times \mathbb{P}_1 \to p(\mathbb{C}^* \times \mathbb{P}_1) = S \backslash p(0, 0) \Subset S$ is an isomorphism, but the complex structure is not extendible to S, since, near $p(0, 0)$, there is no injective morphism of ringed spaces into a $\mathbb{C}^n$: the point $p(0, 0)$ has a fundamental system of neighborhoods in S consisting of sets $W = p(\{|\zeta| < \varepsilon\} \times \mathbb{P}_1)$, since $\{0\} \times \mathbb{P}_1$ is compact; a morphism $h: W \to \mathbb{C}^n$ is holomorphic on $W \backslash p(0, 0) \cong \{0 < |\zeta| < \varepsilon\} \times \mathbb{P}_1$, and thus constant on every compact analytic set $\{\zeta\} \times \mathbb{P}_1$, by the Maximum Principle. (Incidentally, using

$$\mathbb{C} \times \mathbb{P}_1 \to \mathbb{C} \times (\mathbb{C} \times \mathbb{R}), (z, [w_0, w_1]) \mapsto (z, |z| \cdot \varphi([w_0, w_1])),$$

with the φ of E. 32d, one can construct explicitly a homeomorphism from S onto the quadric $\{(\zeta, \eta, t) \in \mathbb{C}^2 \times \mathbb{R}; |\zeta|^2 = |\eta|^2 + t^2\}$).

The *complex cokernel* is given by the projection $\operatorname{pr}: \mathbb{C} \times \mathbb{P}_1 \to \mathbb{C}$: Clearly, $\operatorname{pr} \circ f = \operatorname{pr} \circ g$. Let $h: \mathbb{C} \times \mathbb{P}_1 \to Z$ be a holomorphic mapping with $h \circ f = h \circ g$; then h is constant on every fiber $\operatorname{pr}^{-1}(\zeta)$, since the set $A = \{\zeta \in \mathbb{C}; h|_{\operatorname{pr}^{-1}(\zeta)} = h(\zeta, 0)\}$ is closed, nonempty, and also open: for $\zeta \in A$, choose a neighborhood W of $h(\{\zeta\} \times \mathbb{P}_1)$ that is embedded in $\mathbb{C}^n$ as a local model; it follows as above that, near $\operatorname{pr}^{-1}\zeta$, h is constant on the fibers of pr.

iii) *A complex cokernel need not exist.* With the pair of holomorphic mappings in ii) restricted to $\mathbb{C} \underset{g}{\overset{f}{\rightrightarrows}} \mathbb{C}^2$, the canonical projection $p: \mathbb{C}^2 \to \mathbb{C}^2/\sim =: T$ (with $\sim$ as in ii)) is a ringed cokernel. If $q: \mathbb{C}^2 \to X$ were a complex cokernel of $\underset{g}{\overset{f}{\rightrightarrows}}$, then we would have that $\operatorname{emb} X_{q(0)} = \dim ({}_X\mathfrak{m}_{q(0)}/{}_X\mathfrak{m}^2_{q(0)}) = \infty$: let $\pi: T \to X$ be the factorization of the complex cokernel by the ringed cokernel. For each $\alpha \in \mathcal{O}(\mathbb{C}^2)$ with $\alpha \circ f = \alpha \circ g$, we obtain a commutative diagram of the form

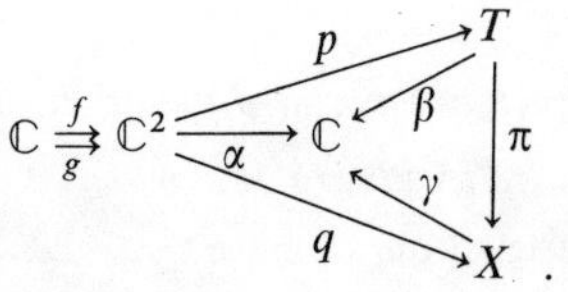

Hence, it suffices to find a sequence $(\alpha_j)_{j\in\mathbb{N}}$ in $\mathcal{O}(\mathbb{C}^2)$ with $\alpha_j \circ f = \alpha_j \circ g$ and $\alpha_j(0) = 0$ such that the induced elements $\overline{\beta_j} = \pi^0 \gamma_j$ in ${}_T\mathfrak{m}_{p(0)}/{}_T\mathfrak{m}^2_{p(0)}$ are linearly independent. Evidently, $p^0({}_T\mathcal{O}_{p(0)}) = \mathbb{C} \oplus {}_2\mathcal{O}_0 \cdot z_1$, and it follows that $p^0({}_T\mathfrak{m}_{p(0)}/{}_T\mathfrak{m}^2_{p(0)}) = {}_2\mathcal{O}_0 \cdot z_1 / {}_2\mathcal{O}_0 \cdot z_1^2$. The sequence $\alpha_j(z_1, z_2) := z_1 z_2^j$ provides the desired example.

In addition to that, T cannot carry a complex structure, since $p(0)$ does not have a countable fundamental system of neighborhoods.

iv) Comparison of ii) and iii) points out that *cokernels cannot be calculated locally* (in contrast to kernels; see E. 49 Ad).

E. 49 Ai. i) *Strukturausdünnung.* Consider the nullstellen ideal $\mathcal{J}$ of $0 \in \mathbb{C}^n$, a positive integer m, and $\varphi_m : V(\mathbb{C}^n; \mathcal{J}^m) \hookrightarrow \mathbb{C}^n$; show that

$$\operatorname{Coker}(\overset{0}{\underset{\varphi_m}{\rightrightarrows}}) = (|\mathbb{C}^n|, \mathcal{H}),$$

with $\mathcal{H}_a = {}_n\mathcal{O}_a$ for $a \neq 0$, and $\mathcal{H}_0 = \{f \in {}_n\mathcal{O}_0; D^\nu f(0) = 0 \text{ for } 0 < |\nu| < m\}$ (hint: E. 24d. For E. 51e iii)).

ii) *Intersection of structure sheaves.* Show that, if $(T, \mathcal{A})$ is a ringed space, and if ${}_1\mathcal{A}, {}_2\mathcal{A} \subset \mathcal{A}$ are subalgebras whose stalks are local algebras, then $(T, {}_1\mathcal{A} \cap {}_2\mathcal{A})$ is a ringed space as well, and the canonical sequence

$$(T, \mathcal{A}) \rightrightarrows (T, {}_1\mathcal{A}) \,\dot\cup\, (T, {}_2\mathcal{A}) \xrightarrow{p} (T, {}_1\mathcal{A} \cap {}_2\mathcal{A})$$

is exact.

iii) *The intersection of complex structure sheaves is not necessarily a complex structure sheaf.* The mapping $f : \mathbb{C}^2 \to \mathbb{C}^3, (z_1, z_2) \mapsto (z_1, z_2^2, z_2^3)$, determines on $\mathbb{C}^2$ the complex structure sheaf ${}_1\mathcal{A} = f^0({}_3\mathcal{O}) \subset {}_2\mathcal{O}$ with the stalks

$${}_1\mathcal{A}_x = \begin{cases} {}_2\mathcal{O}_x, & x_2 \neq 0 \\ \{\sum\limits_{k \neq 1} c_{jk}(z_1 - x_1)^j z_2^k \in {}_2\mathcal{O}_x\}, & x_2 = 0. \end{cases}$$

(The complex space $X_1 := (\mathbb{C}^2, {}_1\mathcal{A})$ is isomorphic to $\mathbb{C} \times$ Neil's parabola; see E. 32i and E. 32l.) From X_1, the linear isomorphism $\varphi : \mathbb{C}^2 \to \mathbb{C}^2, (z_1, z_2) \mapsto (z_1 + z_2, z_2)$, yields the complex space $X_2 := \varphi^{-1}(X_1) = (\mathbb{C}^2, {}_2\mathcal{A})$ with

$${}_2\mathcal{A}_x = \begin{cases} {}_2\mathcal{O}_x, & x_2 \neq 0 \\ \{\sum\limits_{k \neq 1} d_{jk}(z_1 - x_1 + z_2)^j z_2^k \in {}_2\mathcal{O}_x\}, & x_2 = 0. \end{cases}$$

Show that $(\mathbb{C}^2, {}_2\mathcal{O}) \rightrightarrows X_1 \,\dot\cup\, X_2$ has no complex cokernel; in particular, ${}_1\mathcal{A} \cap {}_2\mathcal{A}$ is not a complex structure sheaf (hint: the fact that $\mathfrak{m}({}_1\mathcal{A}_0 \cap {}_2\mathcal{A}_0) = {}_2\mathcal{O}_0 \cdot z_2^2$ implies that $\dim_{\mathbb{C}} \mathfrak{m}/\mathfrak{m}^2 = \infty$; see the proof of 49 A.8 iii)).

iv) Determine the ringed and complex cokernels of $\boldsymbol{P} \overset{f}{\underset{g}{\rightrightarrows}} \boldsymbol{P} = \boldsymbol{P}^1(1)$ for

α) $f = 0$ and a nonconstant $g \in \mathcal{O}(\boldsymbol{P})$; and

β) $f(z) = z^p$ and $g(z) = z^q$, for $p, q > 0$.

We now come to our most important example of a cokernel:

49 A. 9 Definition. *Let $(T, \mathscr{A})$ be a ringed space, $R \subset T \times T$, an equivalence relation on T, and*

$$\pi_j: \ (R, {}_R\mathscr{C}) \hookrightarrow (T \times T, {}_{T \times T}\mathscr{C}) \xrightarrow{\mathrm{pr}_j} (T, {}_T\mathscr{C}) \xrightarrow{\mathrm{Red}} (T, \mathscr{A})$$

the canonical projections. The ringed cokernel of $(R, {}_R\mathscr{C}) \overset{\pi_1}{\underset{\pi_2}{\rightrightarrows}} (T, \mathscr{A})$ is called the quotient space of $(T, \mathscr{A})$ with respect to the equivalence relation R; it is denoted by $(T, \mathscr{A})/R$.

Sometimes we write $(\bar{T}, \bar{\mathscr{A}})$ instead of $(T, \mathscr{A})/R$; by construction, we have that

$$\bar{\mathscr{A}}(\bar{V}) = \{f \in \mathscr{A}(p^{-1}(\bar{V})); \ \ \mathrm{Red}\, f \text{ is } R\text{-invariant}\}$$

for $\bar{V} \Subset \bar{T} = T/R$.

49 A. 10 Remark. *A morphism $h: T \to S$ of ringed spaces factors over $p: (T, \mathscr{A}) \to (T, \mathscr{A})/R$ iff the set-mapping $|h|$ is R-invariant.* ■

E. 49 Aj. For two equivalence relations R and R' on a ringed space T, show that $R \subset R'$ iff $p': T \to T/R'$ factors over $p: T \to T/R$.

In the remainder of this supplement, the central question is when the ringed quotient X/R of a complex space X with respect to an equivalence relation R is a complex space. In that case, the projection $X \to X/R$ is holomorphic, and R is thus "analytic":

49 A. 11 Definition. *An equivalence relation R on a complex space X is called analytic if R is an analytic subset of $X \times X$.*

49 A. 12 Remark. *If $h: X \to Y$ is holomorphic, then $R_h = \{(x_1, x_2); h(x_1) = h(x_2)\} = (h \times h)^{-1}(\Delta_Y)$, and R_h is thus an analytic equivalence relation.* ■

In general, a quotient X/R of a complex space is not a complex space, even if R is analytic (e.g., $\mathbb{C}^2/R_h$ for $h: \mathbb{C}^2 \to \mathbb{C}^2, (z_1, z_2) \mapsto (z_1, z_1 z_2)$; see 49 A. 8 iii), and E. 32Bh ii)). However, we do have the following result:

49 A. 13 Proposition. *Let X be a complex space, and let R be a finite analytic equivalence relation on X. Suppose that each $x \in X$ has an open R-saturated neighborhood U and a discrete R-invariant holomorphic mapping $f: U \to Y$ into a complex space Y. Then the ringed space $\bar{X} := X/R$ is a complex space.*

Proof. By 33 B. 4, $\bar{X}$ is locally compact, and thus Hausdorff. We show that $\bar{X}$ is locally a complex space. To that end, suppose without loss of generality that $U = X$, and, by 33 B. 2 ii), that $\bar{f}: \bar{X} \to \bar{Y}$ is finite; then also $f = \bar{f} \circ p$ is finite. Providing the analytic set R in $X \times X$ with the canonical reduced structure sheaf ${}_R\mathcal{O}$, we show that, in the diagram

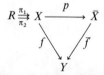

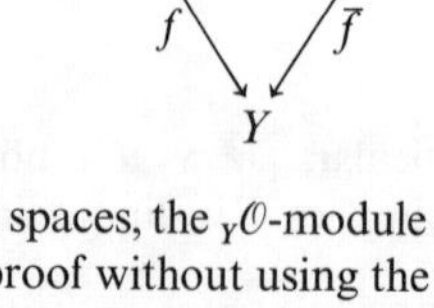

of ringed spaces, the ${}_Y\mathcal{O}$-module $\bar{f}_{\bar{X}}\mathcal{O}$ is coherent; the assertion then follows from 45 B. 2. (To carry out the proof without using the analytic spectrum, one could apply E. 45j.)

First we apply the image-sheaf functor f to the homomorphisms (see 45 A. 2) $\sigma_j: {}_X\mathcal{O} \to \pi_j({}_R\mathcal{O})$ of coherent ${}_X\mathcal{O}$-modules, and obtain a pair of morphisms

$$f({}_X\mathcal{O}) \overset{f(\sigma_1)}{\underset{f(\sigma_2)}{\rightrightarrows}} f \circ \pi_1({}_R\mathcal{O}) = f \circ \pi_2({}_R\mathcal{O})$$

between ${}_Y\mathcal{O}$-modules, which are coherent, according to the Finite Coherence Theorem. By definition, for $\bar{U} \subset\subset \bar{X}$,

$$ {}_{\bar{X}}\mathcal{O}(\bar{U}) = \{\varphi \in {}_X\mathcal{O}(U);\ \sigma_1(\varphi) = \sigma_2(\varphi)\}\ ; $$

it follows easily that $\bar{f}({}_{\bar{X}}\mathcal{O}) = \mathrm{Ker}(f(\sigma_1) - f(\sigma_2))$. By 41.11, then, $\bar{f}({}_{\bar{X}}\mathcal{O})$ is coherent. ■

The following is useful for the construction of examples of complex spaces:

49 A. 14 Corollary. *Let $A \hookrightarrow X$ be a closed subspace of the complex space X, and let R be a finite equivalence relation on A. If A/R is a complex space, then, for the trivial extension R^X of R to X (i.e., $R^X(x) = \{x\}$ for $x \in X \setminus A$), X/R^X is a complex space.*

Proof. By 49 A.12, R is analytic in $A \times A$, so $R^X = R \cup \Delta_X$ is analytic in $X \times X$; furthermore, R^X is a finite equivalence relation. By 49 A.13, it suffices to show that, for each $a \in A$, there is an open R^X-saturated neighborhood U of a in X, and a discrete R^X-invariant holomorphic mapping $\phi : U \to \mathbb{C}^n$.

Let $p : A \to A/R$ be the projection. Since A/R is a complex space, a sufficiently small neighborhood of $p(a)$ in A/R can be mapped injectively and holomorphically into some $\mathbb{C}^m$; thus there exists an open neighborhood U of $R(a) = R^X(a)$ in X with an R-invariant holomorphic mapping $(\varphi_1, \dots, \varphi_m) : U \cap A \to \mathbb{C}^m$. If U is sufficiently small, then, by 44.6, $(\varphi_1, \dots, \varphi_m)$ has a holomorphic extension $\phi : U \to \mathbb{C}^m$. Moreover, we may assume that there exist holomorphic functions $\phi_{m+1}, \dots, \phi_n : U \to \mathbb{C}$ with $U \cap A = V(U; \phi_{m+1}, \dots, \phi_n)$. The holomorphic mapping $\phi = (\phi_1, \dots, \phi_n) : U \to \mathbb{C}^n$ is R^X-invariant and discrete at a (since $\phi^{-1}(\phi(a)) = R(a)$); by 45.4, the same holds (possibly after shrinking U) on all of U. ■

E. 49 Ak. Show that 49 A.14 becomes false if

i) "finite" is replaced by "proper" (hint: 49 A.8 ii)), or ii) A is not considered with the *induced* reduced complex structure (hint: for $A = V(\mathbb{C}^2; z_1 z_2 (z_1 - z_2))$ and the equivalence relation $(t, 0) \sim (0, t) \sim (t, t)$, the ringed space $\mathbb{C}^2/\sim$ has an infinite-dimensional cotangent space $\mathfrak{m}/\mathfrak{m}^2$ at the point $(0, 0)$. Note that $N(\mathbb{C}^2; z_1 z_2 (z_1 - z_2))/\sim\ \cong \mathbb{C}$).

E. 49 Al. For a complex space X, construct a quotient space $\bar{X}$ by identifying two points $a_1 \neq a_2$ in X. Show that $\mathrm{emb}_{\bar{a}} \bar{X} = \mathrm{emb}_{a_1} X + \mathrm{emb}_{a_2} X$. (Hint: $\mathfrak{m}_{\bar{a}} = \mathfrak{m}_{a_1} \oplus \mathfrak{m}_{a_2}$ 'ring-direct"; in particular, $V(\mathbb{C}^2; z_1 z_2 (z_1 - z_2))$ and $V(\mathbb{C}^3; z_1 z_2, z_1 z_3, z_2 z_3)$ are topologically equivalent, but not biholomorphic.)

E. 49 Am. Show that, in the following cases (with the notation of 49 A.14), X/R^X is an irreducible complex space with reducible points:

i) X is an irreducible complex space, $A \subset X$ consists of two points, and A/R has only one point (for $X = \mathbb{C}$ and $A = \{1, -1\}$, the mapping $\mathbb{C} \to \mathbb{C}^2, z \mapsto (z^3 - z, z^2 - 1)$, induces a holomorphic homeomorphism $\mathbb{C}/(1 \sim -1) \to V(\mathbb{C}^2; w_1^2 - w_2^2 - w_2^3)$, which is in fact biholomorphic, by 72.2 and E. 71a).

ii) Put $A := \mathbb{C} \times \{0\} \hookrightarrow \mathbb{C}^2 =: X$, and let R be determined by $(z_1, 0) \sim (-z_1, 0)$ for every $z_1 \in \mathbb{C}$. (Show also that the set of points at which X/R^X is not irreducible is not closed in X/R^X; the mapping $\mathbb{C}^2 \to \mathbb{C}^3, (z_1, z_2) \mapsto (z_1^2, z_1 z_2, z_2)$, induces a holomorphic homeomorphism $X/R^X \to V(\mathbb{C}^3; w_1 w_3^2 - w_2^2)$ (Whitney's umbrella) that is biholomorphic, by 72.2 and E. 72e).

An important consequence of 49 A.13 is the following fact:

49 A. 15 Corollary. *Let R be a finite equivalence relation on a complex space X. The ringed space $\bar{X} := X/R$ is a complex space iff it is locally ${}_{\bar{X}}\mathcal{O}$-separable, i.e., iff every point $\bar{a} \in \bar{X}$ has an open neighborhood $\bar{U}$ such that, for all distinct $\bar{b}, \bar{c} \in \bar{U}$, there exists a $\varphi \in {}_{\bar{X}}\mathcal{O}(\bar{U})$ with $\varphi(\bar{b}) \neq \varphi(\bar{c})$ (in which case $\bar{U}$ is called ${}_{\bar{U}}\mathcal{O}$-separable).*

Proof. Since every local model is holomorphically separable, every complex space is locally $\mathcal{O}$-separable. Conversely, if $\bar{X}$ (without loss of generality) is ${}_{\bar{X}}\mathcal{O}$-separable, then $R = \bigcap_{\varphi \in \mathcal{O}(\bar{X})} R_\varphi$. For $\bar{V} \subset\subset \bar{X}$, we see that $p^{-1}(\bar{V}) =: V$ is open, R-saturated and relatively compact in X, since $p: X \to \bar{X}$ is finite. By 55.12, there exist $\varphi_1, \ldots, \varphi_n \in {}_{\bar{X}}\mathcal{O}(\bar{X})$ such that $R \cap (V \times V) = (R_{\varphi_1} \cap \ldots \cap R_{\varphi_n}) \cap (V \times V)$. It follows that $R|_V$ is a finite analytic equivalence relation on V, and $(\varphi_1, \ldots, \varphi_n): V \to \mathbb{C}^n$ is holomorphic, discrete, and R-invariant; by 49 A.13, then, $(V, \mathcal{O}|_V)/R|_V = (\bar{V}, {}_{\bar{X}}\mathcal{O}|_{\bar{V}})$ is a complex space. ■

As an application, we prove the following generalization of 32 B.8:

49 A.16 Proposition. *Let X be a complex space and $G \subset \operatorname{Aut}(X)$, a subgroup. If G is finite, or if X is reduced and G acts properly discontinuously, then the ringed space X/R_G is a complex space.*

Proof. For $x \in X$, we consider the *isotropy group* $G_x := \{g \in G; g(x) = x\}$; we can find an open neighborhood V of x with these two properties:

i) V is a local model; and ii) $g(V) \cap V \neq \emptyset$ iff $g \in G_x$.

Then $U := \bigcap_{g \in G_x} g(V)$ is an open G_x-invariant neighborhood of x.

α) Suppose that G is *finite*; then the equivalence relation R_G is open and finite. By 49 A.15, we need to show that $\bar{X} := X/R_G$ is locally ${}_{\bar{X}}\mathcal{O}$-separable. Choosing $g_1, \ldots, g_r \in G$ so that $G = \dot{\bigcup}_{1 \leqq j \leqq r} g_j G_x$, we find that

$$p^{-1}(p(U)) = G(U) = \dot{\bigcup}_{1 \leqq j \leqq r} g_j(U);$$

hence, $\bar{U} := p(U)$ is open in $\bar{X}$ (where $p: X \to \bar{X}$ is the quotient-mapping). Let us verify that $\bar{U}$ is ${}_{\bar{U}}\mathcal{O}$-separable. For $a, b \in U$ with $p(a) \neq p(b)$, there exists – since U is included in a local model – a $\varphi \in \mathcal{O}(U)$ with $\varphi(a) = 1$, $\varphi(g(a)) = 0$ for each $g \in G_x \backslash G_a$, and $\varphi|_{G_x(b)} = 0$. We define $\psi \in \mathcal{O}(p^{-1}(U))$ by setting for each j

$$\psi|_{g_j(U)} := \sum_{g \in G_x} \varphi \circ g \circ g_j^{-1}.$$

Clearly, then, Red ψ is G-invariant; hence, it induces a $\bar{\psi} \in {}_{\bar{X}}\mathcal{O}(\bar{U})$ with $\bar{\psi}(p(b)) = 0 \neq \bar{\psi}(p(a))$.

β) Suppose that X is *reduced*. The space $\bar{X}$ is Hausdorff: for $x_1, x_2 \in X$ with $p(x_1) \neq p(x_2)$, there exist disjoint neighborhoods $U_j \subset\subset X$; since $g(x_1) \neq x_2$ for every $g \in G$, and $g(U_1) \cap U_2 = \emptyset$ except for finitely many $g \in G$, we can ensure that $G(U_1)$ and $G(U_2)$ – and thus also $p(U_1)$ and $p(U_2)$ – are disjoint, by shrinking the U_j, if necessary.

All of the isotropy groups G_x are finite, since $\{x\}$ is compact. Therefore, β) will follow from α) if we can show that $p|_U: U \to p(U) =: \bar{U} \subset \bar{X}$ induces an isomorphism $\bar{p}: U/R_{G_x} \to (\bar{U}, {}_{\bar{X}}\mathcal{O}|_U)$ of ringed spaces. We see that $g(U) \cap U = \emptyset$ for every $g \in G \backslash G_x$. Since $p: X \to \bar{X}$ is open, $\bar{p}: U/R_{G_x} \to \bar{U}$ is a homeomorphism. The fact that U/R_{G_x} and $\bar{U}$ are reduced ringed spaces implies that the comorphism $\bar{p}^0$, as the lifting of functions, is injective (for $\bar{p}$ is surjective). Furthermore, $\bar{p}^0$ is surjective: if $W \subset U$ and $h \in {}_X\mathcal{O}(W)$ are both G_x-invariant, then

$$\tilde{h}|_{g^{-1}(W)} := h \circ g|_{g^{-1}(W)}$$

defines a G-invariant extension of h to $p^{-1}(p(W))$. ■

E. 49 An. Let $\mathcal{J} \subset {}_1\mathcal{O}$ be the nullstellen ideal of $\mathbb{Z} \subset \mathbb{C}$, and let $X_j := V(\mathbb{C}; \mathcal{J}^{j+1}) \hookrightarrow \mathbb{C}$ be the j-th infinitesimal neighborhood. Prove that

i) the action of $\mathbb{Z}$ on X_j by translation is properly dicontinuous; and

ii) for $j \geqq 2$, $X_j/R_{\mathbb{Z}}$ is not a complex space (however, see i) and v) in E. 49 Ao).

E. 49 A o. For a transformation group G of the complex space X, we define the sheaf ${}_{X/G}\mathcal{O}$ on the orbit space X/R_G as follows: for $\bar{W} \subset\subset |X/R_G|$ and $W := p^{-1}(\bar{W})$, set

$$ {}_{X/G}\mathcal{O}(\bar{W}) := {}_X\mathcal{O}(W)^G = \{\varphi \in {}_X\mathcal{O}(W); g^0(\varphi) = \varphi \quad \text{for all} \quad g \in G\}. $$

Prove the following statements:

i) $X/G := (|X/R_G|, {}_{X/G}\mathcal{O})$ is a ringed space, and $p: X \to X/G$ is a morphism.

ii) If the action of G is properly discontinuous, then X/G is a complex space, and, for each complex space Y,

$$ \mathrm{Hol}(X/G, Y) \cong \mathrm{Hol}(X, Y)^G := \{f \in \mathrm{Hol}(X, Y); f \circ g = f \quad \text{for all} \quad g \in G\}. $$

iii) If we add to the hypothesis of ii) the conditions that $\mathrm{Red}\, X$ be irreducible and that the mapping $G \to \mathrm{Aut}(\mathrm{Red}\, X)$, $g \mapsto \mathrm{Red}\, g$, be injective, then $p: X \to X/G$ is a (not necessarily finite) "branched covering" (hint: $\{x \in X; G_x \neq 1\} = \bigcup_{g \in G \setminus \{1\}} \{x \in X; gx = x\}$ is a proper analytic subset of X).

iv) For each normal subgroup H of G in ii), G/H acts properly discontinuously on X/H, and $(X/H)/(G/H) \cong X/G$.

v) If X is reduced, then $X/G = X/R_G$.

E. 49 A p. i) Determine $X_j/\mathbb{Z}$ from E. 49 An.

ii) Consider the action of $G := \{\pm \mathrm{id}_{\mathbb{C}}\}$ on $Y_j := V(\mathbb{C}; (z^2 - 1)^j)$; determine Y_j/G and Y_j/R_G, and compute the embedding dimensions.

iii) Let $G_k := \{\zeta \in \mathbb{C}; \zeta^k = 1\}$ act on $Z_j := V(\mathbb{C}; z^j)$ as follows: $\sum_{i=0}^{j-1} a_i z^i \mapsto \sum_{i=0}^{j-1} a_i (\zeta z)^i$. Determine Z_j/G_k and Z_j/R_{G_k}.

E. 49 A q. The group G_k of the preceding exercise acts on $\mathbb{C}$ by multiplication; prove that $\mathbb{C}/G_k \cong \mathbb{C}$. (Hint: find a G_k-invariant $\varphi \in \mathcal{O}(\mathbb{C})$ that induces $\mathbb{C}/G_k \xrightarrow{\cong} \mathbb{C}$; for $\sum a_j z^j \in {}_1\mathcal{O}_0$, show that

$$ \sum a_j z^j \quad \text{is } G_k\text{-invariant iff} \quad a_j = 0 \text{ for } \quad \text{every } j \not\equiv 0 \,(\mathrm{mod}\, k); $$

see 49 A.8 i) and 49 A.17 iii).)

E. 49 A r. Consider $G := \{\pm \mathrm{id}_{\mathbb{C}^n}\} \subset \mathrm{Aut}(\mathbb{C}^n)$. Show that $X_n := \mathbb{C}^n/G$ is a complex space whose only singularity is $\bar{0}$ (for $n = 1$, $\bar{0}$ is a regular point); the embedding dimension of X_n at $\bar{0}$ is $n(n+1)/2$. (Hint: i) for $x \neq 0$, the isotropy group G_x is trivial; ii) the projection $\mathbb{C}^n \to X_n$ determines an inclusion $\mathfrak{m} := \mathfrak{m}({}_{X_n}\mathcal{O}_0) \subset {}_n\mathfrak{m}$; iii) $\mathfrak{m} = \{\sum_{|\nu| \geq 2} a_\nu z^\nu \in {}_n\mathfrak{m}; a_\nu = 0 \text{ for odd } |\nu|\}$; iv) $\mathfrak{m}/\mathfrak{m}^2 \cong \bigoplus_{1 \leq i \leq j \leq n} \mathbb{C} z_i z_j$; v) find a G-invariant mapping $\varphi: \mathbb{C}^n \to \mathbb{C}^{n(n+1)/2}$ that induces an an embedding of X_n.)

E. 49 A s. For a fixed $k > 0$, set $G_k := \{\zeta \cdot I_2 \in GL(2, \mathbb{C}); \zeta^k = 1\}$. Show that $Y_k := \mathbb{C}^2/G_k$ is a two-dimensional complex space whose only singularity is $\bar{0}$; its embedding-dimension is $k+1$. (Hints: i) $\mathfrak{m} := {}_{Y_k}\mathfrak{m}_0 = \{\sum_{\substack{|\nu| \geq k \\ |\nu| \equiv 0 (k)}} a_\nu z^\nu \in {}_2\mathfrak{m}\}$; ii) $\mathfrak{m}/\mathfrak{m}^2 \cong \bigoplus_{j=0}^{k} \mathbb{C} z_1^j z_2^{k-j}$; for $k = 3, 4$, see E. 45e.)

E. 49 A t. For $m, n \in \mathbb{N}$ with $2 \leq n \leq m$ and $k := m - n + 1$, show that $X = Y_k \times \mathbb{C}^{n-2}$ is an n-dimensional complex space with $\mathrm{emb}(X_0) = m$. (Hint: if ${}_{Y_k}\mathfrak{m}_0/{}_{Y_k}\mathfrak{m}_0^2$ is generated by the set $\{z_1^j z_2^{k-j}; j \in J\}$ then ${}_X\mathfrak{m}_0/{}_X\mathfrak{m}_0^2$ is generated by $\{z_1^j z_2^{k-j}, w_k; j \in J, 1 \leq k \leq n-2\}$.)

49 A. 17 Examples. i) *Symmetric products.* Let X be a complex space, and fix $n \in \mathbb{N}_{>0}$. The symmetric group S_n acts on the n-fold product X^n of X by interchanging factors. The complex (by 49 A. 16) space X^n/S_n is called the n-fold symmetric product of X; its points are the "unordered n-tuples" of elements of X.

ii) In the special case in which $X = \mathbb{C}$, there exists a biholomorphic mapping $\bar{\sigma} : \mathbb{C}^n/S_n \to \mathbb{C}^n$ that is induced by the holomorphic mapping $\sigma : \mathbb{C}^n \to \mathbb{C}^n$, $z \mapsto (\sigma_j(z))$, determined by the elementary symmetric functions σ_j. To verify that $\bar{\sigma}$ is biholomorphic, it suffices, by 46 A. 1, to show that $\bar{\sigma}$ is bijective, since $\dim \mathbb{C}^n/S_n = n$ by E. 49p. To do that, we consider the commutative diagram

$$\begin{array}{ccc} \mathbb{C}^n & \xrightarrow{\ \sigma\ } & \mathbb{C}^n \\ & \searrow^{\tau} & \downarrow \alpha \\ & & V \end{array}$$

with $V := \{T^n + \sum_{k=1}^{n} a_k T^{n-k};\ a_k \in \mathbb{C}\} \subset \mathbb{C}[T]$, $\alpha(z) := T^n + \sum_{k=1}^{n} (-1)^k z_k T^{n-k}$, and $\tau(z) := \prod_{k=1}^{n} (T - z_k)$. Then $\tau^{-1}(\tau z) = S_n(z)$ for every $z \in \mathbb{C}^n$; since α is bijective, we have that $\sigma^{-1}(\sigma z) = \tau^{-1}(\tau z)$, and it follows that $\bar{\sigma}$ is injective. By the Fundamental Theorem of Algebra, τ is surjective, and we conclude that $\bar{\sigma}$ is surjective as well. ■

As a consequence, we have an analogue of the algebraic Fundamental Theorem on Symmetric Functions: *if $U \subset \mathbb{C}^n$ is an S_n-invariant subset, and if $f : U \to \mathbb{C}$ is a "symmetric" (i.e., S_n-invariant) holomorphic function, then f can be expressed uniquely as a holomorphic function of the elementary symmetric functions.*

E. 49 Au. *Projective spaces are symmetric products.* Prove that $\mathbb{P}_n = (\mathbb{P}_1)^n/S_n$ (hint: use the mapping $(\mathbb{P}_1)^n \to (\mathbb{C}[T]_{n+1} \setminus \{0\})/\mathbb{C}^*$, $([{}_1z_0, {}_1z_1], \ldots, [{}_nz_0, {}_nz_1]) \mapsto \prod_{j=1}^{n} ({}_jz_0 T - {}_jz_1)$).

49 A. 17 Examples. iii) *Weighted projective spaces.*

α) *Definition.* Consider $\mathfrak{q} = (q_0, \ldots, q_n) \in \mathbb{N}_{>0}^{n+1}$ with $\gcd(q_0, \ldots, q_n) = 1$. The group $\mathbb{C}^*$ acts on $(\mathbb{C}^{n+1})^*$ as follows: $(t, (z_0, \ldots, z_n)) \mapsto (t^{q_0} z_0, \ldots, t^{q_n} z_n)$. The orbit space $(\mathbb{C}^{n+1})^*/\mathbb{C}^*$ is called the weighted projective space $\mathbb{P}_{\mathfrak{q}}$ of type $\mathfrak{q}$.

E. 49 Av. Show that every orbit of that $\mathbb{C}^*$-action on $(\mathbb{C}^{n+1})^*$ meets the sphere S^{2n+1}. Determine the equivalence relation induced on S^{2n+1}, and show that $\mathbb{P}_{\mathfrak{q}}$ is compact (hint: 32.4 iv)).

β) *Complex structure: representation of $\mathbb{P}_{\mathfrak{q}}$ as a quotient of $\mathbb{P}_n$.* There exists a commutative diagram

$$\begin{array}{ccc} X = (\mathbb{C}^{n+1})^* & \longrightarrow\!\!\!\!\!\rightarrow & \mathbb{P}_n \\ \downarrow \varphi & & \downarrow \bar{\varphi} \\ Y = (\mathbb{C}^{n+1})^* & \longrightarrow\!\!\!\!\!\rightarrow & \mathbb{P}_{\mathfrak{q}} \end{array},$$

where $\varphi : X \to Y$, $z \mapsto (z_0^{q_0}, \ldots, z_n^{q_n})$, is equivariant with respect to the $\mathbb{C}^*$-actions $(t, z) \mapsto tz$ on X and $(t, z) \mapsto (t^{q_0} z_0, \ldots, t^{q_n} z_n)$ on Y, respectively. The mapping φ may be regarded as the quotient mapping $X \to X/G = Y$, where the group

$$G := \left\{ \begin{pmatrix} \zeta_0 & & 0 \\ & \ddots & \\ 0 & & \zeta_n \end{pmatrix} \in GL(n+1, \mathbb{C});\ \zeta_j^{q_j} = 1 \right\}$$

acts componentwise (see E. 49 Aq). Thus φ induces the mapping $\bar{\varphi}$. In particular, $\mathbb{P}_{\mathfrak{q}}$ is biholomorphic to $\mathbb{P}_n/\bar{G}$, where $\bar{G}$ denotes the image of G in $\mathbb{P}GL(\mathbb{C}^{n+1})$.

E. 49 A w. Adapt for $\mathbb{P}_q$ the construction of charts given for $\mathbb{P}_n$ in 32.4 iv): there exists an open cover by the $n+1$ sets $U_j := \mathbb{C}^n/G_j$ with the group $G_j := \{\zeta \in \mathbb{C};\ \zeta^{q_j} = 1\}$ acting as follows:

$$(z_0, \ldots, \not{z}_j, \ldots, z_n) \mapsto (\zeta^{q_0} z_0, \ldots, z_j, \ldots, \zeta^{q_n} z_n).$$

γ) *Representation of* $\mathbb{P}_n$ *as a quotient of* $\mathbb{P}_q$: Set $r_j := r/q_i$ with $r := \operatorname{lcm}(q_0, \ldots, q_n)$. Then the action of the group

$$H := \left\{ \begin{pmatrix} \xi_0 & & 0 \\ & \ddots & \\ 0 & & \xi_n \end{pmatrix} \in GL(n+1, \mathbb{C});\ \xi_j^{r_j} = 1 \right\}$$

on $(\mathbb{C}^{n+1})^*$ is compatible with the $\mathbb{C}^*$-action of α); hence, it induces an action on $\mathbb{P}_q$ (proof?). Let $\bar{H}$ denote the image of H in $\operatorname{Aut}(\mathbb{P}_q)$, then the associated orbit space $\mathbb{P}_q/\bar{H}$ can be identified with $\mathbb{P}_n$: there exists a commutative holomorphic diagram

$$\begin{array}{ccc} Y = (\mathbb{C}^{n+1})^* & \longrightarrow\!\!\!\!\!\rightarrow & \mathbb{P}_q \\ \downarrow \psi & & \downarrow \bar{\psi} \\ Z = (\mathbb{C}^{n+1})^* & \longrightarrow\!\!\!\!\!\rightarrow & \mathbb{P}_n \end{array},$$

where $\mathbb{C}^*$ acts on Z via $(t, z) \mapsto t^r z$ and $\mathbb{P}_n$ is the quotient space of Z with respect to that $\mathbb{C}^*$-action. ψ is equivariant and induces a holomorphic mapping $\bar{\psi}$. Then $Z \cong X/K$, where the group K is given by

$$K := \left\{ \begin{pmatrix} \eta_0 & & 0 \\ & \ddots & \\ 0 & & \eta_n \end{pmatrix} \in GL(n+1, \mathbb{C});\ \eta_j^r = 1 \right\};$$

hence, $Z \cong X/G/K/G = Y/H$, and $\mathbb{P}_n \cong \mathbb{P}_q/\bar{H}$.

E. 49 A x. i) Show that $[1, 0, 0]$ is the only singularity of $\mathbb{P}_{(2,1,1)}$ (hint: by E. 49 Aw, $U_0 = \mathbb{C}^2/\{\pm 1\}$; see the introduction to Chapter 3, and E. 49 Ar).

ii) The mapping $\mathbb{C}^3 \setminus \{0\} \to \mathbb{P}_3$, $(z_0, z_1, z_2) \mapsto [z_0, z_1^2, z_2^2, z_1 z_2]$, induces a holomorphic homeomorphism from $\mathbb{P}_{(2,1,1)}$ onto the projective quadric $V(\mathbb{P}_3;\ w_1 w_2 - w_3^2)$ (by 72.2 and 74.5, it is in fact biholomorphic).

iii) Show that $\mathbb{P}_{(3,1,1)}$ cannot be embedded in $\mathbb{P}_3$ (hint: E. 49 As).

Up to this point, the results on equivalence relations have been based on the Finite Coherence Theorem; as a result of that, we have had to restrict our attention to the discrete case. Application of Grauert's Coherence Theorem makes it possible to prove the following generalization:

49 A. 18 Proposition. *Let X be a complex space, and R, a proper analytic equivalence relation on X. For each $\bar{x} \in \bar{X} := X/R$, let there be an open neighborhood $\bar{U}$ in $\bar{X}$ and a discrete morphism from the ringed space $(\bar{U}, {}_{\bar{X}}\mathcal{O}|_{\bar{U}})$ into a complex space. Then $\bar{X}$ is a complex space.*

49 A. 19 Proposition. *Let R be a proper equivalence relation on a complex space X. The ringed space X/R is a complex space iff it is locally ${}_{X/R}\mathcal{O}$-separable.*

E. 49 A y. Prove 49 A. 18 and 49 A. 19, using Grauert's Coherence Theorem. ■■

For quotients with respect to group actions, we have the following result:

49 A. 20 Proposition. *Let G be a complex Lie group with at most countably many connected components, and let X be a reduced complex space on which G acts holomorphically (i.e., the mapping*

$G \times X \to X, (g, x) \mapsto g(x)$, is holomorphic). Then, in each of the following cases, the ringed space X/G is a complex space:

a) The action of G on X is proper (i.e., for compact $K \subset X$, the set $\{g \in G;\ K \cap gK \neq \emptyset\}$ is compact; equivalently, $G \times X \to X \times X, (g, x) \mapsto (g(x), x)$, is proper) [Ho Satz 19].

b) The complex structure on X is maximal (e.g., X is a manifold; see 72.1), and X/G is Hausdorff [Ho Satz 15].

For open equivalence relations, one can show the following:

49 A. 21 Proposition. *If X is a reduced complex space with maximal structure, and if R is an open analytic equivalence relation on X, then X/R is a complex space* $[\mathrm{Ka}]_{\mathrm{B}}$. ∎

E. 49 Az. Show that 49 A. 20 a) follows from Grauert's Coherence Theorem and 49 A. 21 if the complex structure of X is maximal.

If $f: X \to Y$ is a continuous mapping of locally compact spaces, then the connected component $N_f(a)$ of the fiber $f^{-1}f(a)$ that contains a is called the *level set of f through a*. With the equivalence relation N_f defined thereby, we have this result:

49 A. 22 Stein's Factorization Theorem. *If $f: X \to Y$ is a holomorphic mapping with compact level sets, then N_f is a proper analytic equivalence relation; the quotient $\pi: X \to X/N_f$ is a complex space with ${}_{X/N_f}\mathcal{O} \cong \pi({}_X\mathcal{O})$, and, in the canonical diagram*

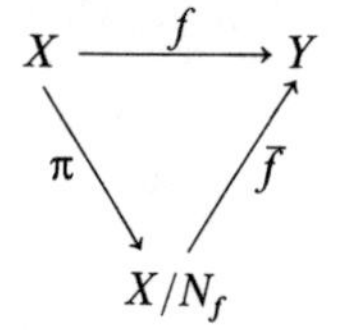

of holomorphic mappings, π is proper and $\bar{f}$ is discrete.

Proof. It suffices to prove the theorem locally with respect to Y. By 49 A. 23 and E. 48n, N_f is proper; it follows by 33 B. 4 that $\bar{X} := X/N_f$ is locally compact, and π is proper. Moreover, by 49 A. 23, we may assume that $\bar{f}$ is finite; it suffices to verify that the hypotheses of 45 B. 2 hold for $\bar{f}$. To that end, we shall show that $\bar{f}({}_{\bar{X}}\mathcal{O}) = f({}_X\mathcal{O})$, so that $\bar{f}({}_{\bar{X}}\mathcal{O})$ is a coherent ${}_Y\mathcal{O}$-module, by Grauert's Coherence Theorem: for $W \Subset \bar{X}$, we have that ${}_{\bar{X}}\mathcal{O}(W) = {}_X\mathcal{O}(\pi^{-1}W)$, since the reduction of each holomorphic function is constant on every connected compact analytic set, and thus in particular on level sets; hence, for $V \Subset Y$,

$$\bar{f}({}_{\bar{X}}\mathcal{O})(V) = {}_{\bar{X}}\mathcal{O}(\bar{f}^{-1}V) = {}_X\mathcal{O}(\pi^{-1}\bar{f}^{-1}V) = {}_X\mathcal{O}(f^{-1}V) = f({}_X\mathcal{O})(V). \quad ∎$$

In the preceding proof, we made use of this fact:

49 A. 23 Lemma. *Let S and T be locally compact topological spaces, and suppose that $f: S \to T$ is a continuous mapping with locally connected fibers and compact level sets. Then*

i) N_f is a proper equivalence relation, and

ii) for each $a \in S$, there exist an N_f-saturated open neighborhood U of a and an open neighborhood V of $f(U)$ such that $f: U \to V$ is proper.

Proof. For an arbitrary neighborhood $W \Subset S$ of a level set $N_f(a)$, we first construct an N_f-saturated neighborhood $U \Subset W$ (by 33 B. 4, then, N_f is proper): as a connected component of a locally connected set, $N_f(a)$ is open in $f^{-1}(a)$; thus we may assume that $W \cap f^{-1}f(a) = N_f(a)$. As a compact space, $\bar{W}$ is topologically normal, so one may even have that $f^{-1}(a) \cap \partial W = \emptyset$.

Then $V := T \setminus f(\partial W)$ is an open neighborhood of $f(a)$, and $U := f^{-1}(V) \cap W$ is an open neighborhood of a; moreover, U is saturated with respect to N_f: given $b \in U$, we see that $f(b) \notin f(\partial W)$ and the fact that $N_f(b)$ is connected implies that $N_f(b) \subset W$; thus $N_f(b)$ is included in U.

For ii), it suffices to show that $f|_U : U \to V$ is proper. If $K \subset V$ is compact, then $f^{-1}K$ is closed, and $f^{-1}(K) \cap \bar{W}$ is compact; since $f^{-1}(V) \cap \partial W = \emptyset$, we see that the set $(f|_U)^{-1}(K) = f^{-1}(K) \cap \bar{W} = f^{-1}(K) \cap W$ is compact. ∎

Part Three: Function Theory on Stein Spaces

Having treated both the punctual theory of complex spaces through the investigation of analytic algebras, and the transition to the local theory with the help of coherent analytic sheaves, we intend now to derive global results. Frequently, attempts to patch together local solutions to analytic problems into global solutions lead naturally to analytic or topological obstructions in the form of cohomology classes. Complex spaces with "trivial analytic cohomology" have no such analytic obstructions, so they possess a very rich global function-theoretic structure. The famous Theorem B of Cartan-Serre states that every space with "sufficiently many" global holomorphic functions (Stein space) has trivial analytic cohomology; conversely, it is easy to see that spaces with that cohomological property (which we call B-spaces) are Stein (52.6).

We present many applications of Theorem B in Chapter 5, in order to illustrate its absolutely fundamental importance, and to prepare the reader for its demonstration in Chapter 6. The proof follows ideas of Grauert and Rossi, using $\bar{\partial}$-theory. With a little extra effort, we obtain further consequences of the various steps in the proof: the solution to Levi's Problem in $\mathbb{C}^n$, the Finiteness Theorem of Cartan-Serre, and a very detailed characterization of domains of holomorphy in $\mathbb{C}^n$.

Chapter 5: Applications of Theorem B

In this chapter, we intend to demonstrate the central importance of Theorem B for the global function theory on noncompact complex spaces. For that reason, we seek to determine which conclusions can be drawn from the triviality of the analytic cohomology; at the same time, we are preparing for the proof of Theorem B in Chapter 6. The results obtained here are likewise basic for the investigation of arbitrary complex spaces: for the local theory, since every point has a fundamental system of Stein neighborhoods, and for the global theory, because analytic cohomology often is computed with the aid of Stein covers.

In § 50, we collect the concepts and results of general cohomology theory that will be essential for our purposes; we go into the proofs only as far as seems necessary for an understanding of the applications. In § 50A, using the example of additively automorphic functions, we discuss how the attempt to obtain a global solution by patching together local solutions naturally leads to an obstruction in the form of a cohomology class; we show also how one encounters such an obstruction in trying to transform a $\mathscr{C}^\infty$-solution into a holomorphic solution.

In § 51, we define Stein spaces as holomorphically separable and holomorphically convex spaces such that each connected component has a countable topology (in the supplement, we show that the third condition is superfluous).

For a complex space with the "vanishing" property of Theorem B, we introduce the terminology "B-space" in § 52. The basic properties of B-spaces are the exactness of the "analytic" section-functor (52.6); that leads to a very general existence theorem for global holomorphic functions (52.5), from which it follows that B-spaces are Stein spaces. By couching our discussion of B-spaces in terms of the more general "B-sets", we are able to subsume other important classes of geometric objects, such as the Stein compacta, without additional effort.

Sections 53 and 54 deal with the additive and multiplicative Cousin problems, the solutions to which constitute generalizations of Mittag-Leffler's Theorem and the Weierstrass Factorization Theorem, respectively. Both problems have furthered multidimensional function theory considerably. In supplements to those sections, we treat the properties of meromorphic functions on complex spaces, particularly the global representability of meromorphic functions as quotients of relatively prime holomorphic functions (Poincaré problem); we also delve briefly into the theory of vector bundles.

In the proof of Theorem B, exhaustion methods and approximation arguments à la Runge play an important role. Consequently, we introduce in § 55 the canonical topology on vector spaces of sections of coherent analytic sheaves, and state conditions in § 56 under which a space with a "B-exhaustion" is a B-space.

By means of the evaluation-homomorphisms, we define a continuous mapping from a complex space into its "spectrum". The Character Theorem in § 57 characterizes Stein spaces as those in which that mapping is a homeomorphism (in that way, it gives an indication of the analogy between Stein spaces and the affine varieties of algebraic geometry).

Finally, in § 58, we prove the holomorphic de Rham Theorem, which makes possible the determination of topological invariants of a Stein manifold using certain analytic objects, namely, global holomorphic differential forms.

§ 50 Introductory Remarks on Cohomology

As in other mathematical disciplines, cohomology plays a decisive role in complex analysis in the transition from "local" to "global". Since there are enough good presentations of cohomology theory (e.g., [Ku]), we discuss primarily results that we actually use later, without giving proofs; however, in a supplement, we use a concrete question to show how one can naturally be led in complex analysis to construct cohomology vector spaces (50 A. 1–50 A. 5). That example, although it is not used in the succeeding sections, is recommended in advance as a motivational aid, even for the reader who is familiar with cohomology theory. In most of our applications, only abstract properties of cohomology will be used. For an example of the power of cohomological methods, we refer the reader to Hartogs's Kugelsatz.

Let $\mathscr{R}$ be a sheaf of rings on a topological space T; further, let R be a subring of $\mathscr{R}(T)$, for example, $\mathbb{C} \subset \mathscr{O}(X)$.

50.1 Definition. *A cohomology theory on the topological space T with coefficients in sheaves of $\mathscr{R}$-modules is a set $H^* = \{H^q, \delta^q; q \in \mathbb{N}\}$ of functors such that*

i) $H^q: \mathscr{F} \mapsto H^q(T, \mathscr{F}) = H^q(\mathscr{F})$ is a covariant functor from the category of $\mathscr{R}$-modules on T into the category of R-modules;

ii) δ^q is a functor that assigns to each exact sequence $0 \to \mathscr{F}' \xrightarrow{\varphi} \mathscr{F} \xrightarrow{\psi} \mathscr{F}'' \to 0$ a "connecting homomorphism" $\delta^q: H^q(\mathscr{F}'') \to H^{q+1}(\mathscr{F}')$.

Moreover, the following conditions are satisfied:

a) The induced cohomology sequence

$$0 \to H^0(\mathscr{F}') \xrightarrow{H^0\varphi} H^0(\mathscr{F}) \xrightarrow{H^0\psi} H^0(\mathscr{F}'') \xrightarrow{\delta^0} H^1(\mathscr{F}') \xrightarrow{H^1\varphi} \dots$$

$$\dots \to H^q(\mathscr{F}) \xrightarrow{H^q\psi} H^q(\mathscr{F}'') \xrightarrow{\delta^q} H^{q+1}(\mathscr{F}') \xrightarrow{H^{q+1}\varphi} \dots$$

is exact.

b) For an additional exact sequence, if the diagram

$$\begin{array}{ccccccccc}
0 & \to & \mathscr{F}' & \xrightarrow{\varphi} & \mathscr{F} & \xrightarrow{\psi} & \mathscr{F}'' & \to & 0 \\
 & & \downarrow{\scriptstyle\alpha} & & \downarrow{\scriptstyle\beta} & & \downarrow{\scriptstyle\gamma} & & \\
0 & \to & \hat{\mathscr{F}}' & \xrightarrow{\hat{\varphi}} & \hat{\mathscr{F}} & \xrightarrow{\hat{\psi}} & \hat{\mathscr{F}}'' & \to & 0
\end{array}$$

commutes, then the induced cohomology ladder

$$\begin{array}{ccccccc}
\dots \to H^q(\mathscr{F}) & \xrightarrow{H^q\psi} & H^q(\mathscr{F}'') & \xrightarrow{\delta^q} & H^{q+1}(\mathscr{F}') \to \dots \\
\downarrow{\scriptstyle H^q\beta} & & \downarrow{\scriptstyle H^q\gamma} & & \downarrow{\scriptstyle H^{q+1}\alpha} \\
\dots \to H^q(\hat{\mathscr{F}}) & \xrightarrow{H^q\hat{\psi}} & H^q(\hat{\mathscr{F}}'') & \xrightarrow{\delta^q} & H^{q+1}(\hat{\mathscr{F}}') \to \dots
\end{array}$$

commutes as well.

c) $H^0 = \Gamma$; that is, $H^0(T, \mathscr{F}) = \Gamma(T, \mathscr{F}) = \mathscr{F}(T)$, and $H^0(\varphi) = \Gamma(\varphi) = \varphi(T)$ for homomorphisms $\varphi: \mathscr{F} \to \mathscr{F}'$.

d) $H^q(T, \mathscr{C}(T, \mathscr{F})) = 0$ for every $q \geqq 1$, where $\mathscr{C}(T, \mathscr{F}) = \mathscr{C}(\mathscr{F})$[1)] *is the (obviously flabby) sheaf*

$$U \mapsto \{s: U \to \mathscr{F}; \pi_{\mathscr{F}} \circ s = \mathrm{id}_U\} \cong \prod_{u \in U} \mathscr{F}_u$$

of all (not necessarily continuous) sections of the sheaf $\mathscr{F}$.

50.2 Theorem. *Up to isomorphism, there exists precisely one cohomology theory $\{H^q, \delta^q\}$ on T with coefficients in sheaves of $\mathscr{R}$-modules.*

We demonstrate the uniqueness of $H^q(\mathscr{F})$ by means of "acyclic resolutions", and only intimate the proof of its existence, using both acyclic resolutions and (for paracompact T) the Čech construction. A sheaf $\mathscr{F}$ is called *acyclic* (in the cohomology theory $\{H^q, \delta^q\}$) if $H^q(T, \mathscr{F}) = 0$ for every $q \geqq 1$. By 50.7, $\mathscr{C}(\mathscr{F})$ is an example of an acyclic sheaf. A sequence of sheaves

$$\mathscr{E}^*: \mathscr{E}^0 \xrightarrow{\vartheta^0} \mathscr{E}^1 \xrightarrow{\vartheta^1} \mathscr{E}^2 \xrightarrow{\vartheta^2} \dots$$

is called a *complex of sheaves* if $\vartheta^{j+1} \circ \vartheta^j = 0$ for every j; in particular, for every point of T one

1) That is not to be confused with $\mathscr{C}(T)$!

thus obtains a *complex of modules*. An *acyclic resolution* of $\mathscr{F}$ is an exact complex $\mathscr{F}^*$ of acyclic sheaves with an isomorphism $\mathscr{F} \cong \mathscr{K}er\, \vartheta^0$, i.e., an exact sequence

$$0 \to \mathscr{F} \to \mathscr{E}^0 \xrightarrow{\vartheta^0} \mathscr{E}^1 \xrightarrow{\vartheta^1} \mathscr{E}^2 \xrightarrow{\vartheta^2} \dots .$$

50.3 Proposition. *Every $\mathscr{R}$-module $\mathscr{F}$ has a canonical acyclic resolution $\mathscr{C}^*(\mathscr{F})$ by $\mathscr{R}$-modules of the form $\mathscr{C}(\mathscr{G})$.*

Proof. Choose $\mathscr{C}^0 = \mathscr{C}(T, \mathscr{F})$ with the canonical inclusion $\mathscr{F} \hookrightarrow \mathscr{C}(T, \mathscr{F})$. After constructing the resolution $\mathscr{C}^*$ up to $\mathscr{C}^q$, we extend it by means of the composition

$$\vartheta^q : \mathscr{C}^q \to \mathscr{C}^q / \mathscr{I}m\, \vartheta^{q-1} \hookrightarrow \mathscr{C}^{q+1} := \mathscr{C}(T, \mathscr{C}^q / \mathscr{I}m\, \vartheta^{q-1}). \quad \blacksquare$$

That canonical resolution obviously depends functorially on $\mathscr{F}$; it permits (theoretically) the determination of $H^q(\mathscr{F})$, for we have the following:

50.4 Abstract de Rham Theorem. *If $\{H^q, \vartheta^q\}$ is a cohomology theory, and if $\mathscr{E}^*$ is an acyclic resolution of $\mathscr{F}$, then there exist natural isomorphisms $H^0(T, \mathscr{F}) \cong \operatorname{Ker} \vartheta^0(T)$ and $H^q(T, \mathscr{F}) \cong$ $\cong \operatorname{Ker} \vartheta^q(T) / \operatorname{Im} \vartheta^{q-1}(T)$ for every $q \geqq 1$.*

Proof. With $\mathscr{Z}^j := \mathscr{K}er\, \vartheta^j \subset \mathscr{E}^j$, we see that $\mathscr{Z}^0 \cong \mathscr{F}$ and that, for $j \geqq 0$, the sequence

(50.4.1) $\quad 0 \to \mathscr{Z}^j \to \mathscr{E}^j \xrightarrow{\vartheta^j} \mathscr{Z}^{j+1} \to 0$

is exact. The induced cohomology sequences decompose – since every $\mathscr{E}^j$ is acyclic – into initial segments

$$0 \to H^0(\mathscr{Z}^j) \to H^0(\mathscr{E}^j) \xrightarrow{\vartheta^j(T)} H^0(\mathscr{Z}^{j+1}) \to H^1(\mathscr{Z}^j) \to H^1(\mathscr{E}^j) = 0$$

and isomorphisms $H^p(\mathscr{Z}^{j+1}) \cong H^{p+1}(\mathscr{Z}^j)$ for $p \geqq 1$; an induction argument yields that $H^q(\mathscr{F}) \cong H^1(\mathscr{Z}^{q-1})$. The initial segments yield that $H^1(\mathscr{Z}^{q-1}) \cong H^0(\mathscr{Z}^q) / \operatorname{Im} \vartheta^{q-1}(T)$. By applying the left-exact functor $H^0 = \Gamma$ to the exact sequence $0 \to \mathscr{Z}^q \to \mathscr{E}^q \to \mathscr{E}^{q+1}$, we find that $H^0(\mathscr{Z}^q) = \operatorname{Ker} \vartheta^q(T)$. $\blacksquare$

Since the canonical resolution $\mathscr{C}^*(\mathscr{F})$ is acyclic in every cohomology theory, 50.4 implies the uniqueness of the R-modules $H^q(T, \mathscr{F})$. Moreover, the proof shows that $H^q(T, \mathscr{F})$, viewed as an abelian group, is independent of $\mathscr{R}$ and R.

For a (cochain) complex of R-modules

$$E^*:\ E^0 \xrightarrow{d^0} E^1 \xrightarrow{d^1} E^2 \to \dots,$$

we define the *cohomology modules* by setting $H^q(E^*) := Z^q(E^*) / B^q(E^*)$ with $Z^q := \operatorname{Ker} d^q$ and $B^q = \operatorname{Im} d^{q-1}$. Then 50.4 states the following:

50.5 Corollary. *If $\mathscr{E}^*$ is an acyclic resolution of an $\mathscr{R}$-module $\mathscr{F}$, then $H^q(T, \mathscr{F}) \cong H^q(\mathscr{E}^*(T))$, where $\mathscr{E}^*(T)$ denotes the complex of global sections $\mathscr{E}^0(T) \xrightarrow{\vartheta^0(T)} \mathscr{E}^1(T) \xrightarrow{\vartheta^1(T)} \mathscr{E}^2(T) \to \dots$.* $\blacksquare$

To prove the *existence* in 50.2, put $H^q(T, \mathscr{F}) := H^q(\Gamma(T, \mathscr{C}^*(\mathscr{F})))$. For a homomorphism $\varphi : \mathscr{F} \to {}'\mathscr{F}$ and its canonical "extension" $\varphi^* : \mathscr{C}^*(\mathscr{F}) \to \mathscr{C}^*({}'\mathscr{F})$ to the resolutions, the homomorphisms $H^q(\varphi)$ are induced by the $\varphi^q(T)$ in a similar fashion. The connecting homomorphisms are constructed as follows: for an exact sequence $0 \to {}'\mathscr{F} \xrightarrow{\varphi} \mathscr{F} \xrightarrow{\psi} {}''\mathscr{F} \to 0$ and the exact sequence of canonical resolutions $0 \to {}'\mathscr{C}^* \xrightarrow{\varphi^*} \mathscr{C}^* \xrightarrow{\psi^*} {}''\mathscr{C}^* \to 0$ (in which all rows are exact), one obtains, with $C^q = \mathscr{C}^q(T)$, $d^q = \vartheta^q(T)$, etc., a commutative diagram with exact rows

$$\begin{array}{ccccccccc}
 & & & & C^{q-1} & \xrightarrow{\Psi^{q-1}} & {}''C^{q-1} & \longrightarrow & 0 \\
 & & & & \downarrow d^{q-1} & & \downarrow {}''d^{q-1} & & \\
0 & \longrightarrow & {}'C^{q} & \xrightarrow{\Phi^{q}} & C^{q} & \xrightarrow{\Psi^{q}} & {}''C^{q} & \longrightarrow & 0 \\
 & & \downarrow {}'d^{q} & & \downarrow d^{q} & & \downarrow {}''d^{q} & & \\
0 & \longrightarrow & {}'C^{q+1} & \xrightarrow{\Phi^{q+1}} & C^{q+1} & \xrightarrow{\Psi^{q+1}} & {}''C^{q+1} & \longrightarrow & 0 \\
 & & \downarrow {}'d^{q+1} & & \downarrow d^{q+1} & & & & \\
0 & \longrightarrow & {}'C^{q+2} & \xrightarrow{\Phi^{q+2}} & C^{q+2} & & & &
\end{array}$$

(see 50.6). The composition $(\Phi^{q+1})^{-1} \circ d^q \circ (\Psi^q)^{-1}$ obviously determines a homomorphism ${}''Z^q \to {}'Z^{q+1}/{}'B^{q+1} = H^{q+1}({}'C^*)$; that induces $\delta^q: H^q({}''\mathscr{F}) \to H^{q+1}({}'\mathscr{F})$. It is essential for that construction that Ψ^q be surjective: since sheaves $\mathscr{C}(T, \mathscr{G})$ are flabby, that follows from this fact:

50.6 Lemma. *If $0 \to {}'\mathscr{F} \to \mathscr{F} \to {}''\mathscr{F} \to 0$ is an exact sequence, and if ${}'\mathscr{F}$ is flabby, then the sequence of sections $0 \to {}'\mathscr{F}(T) \to \mathscr{F}(T) \to {}''\mathscr{F}(T) \to 0$ is exact.*

Proof. For $s'' \in {}''\mathscr{F}(T)$, there exist local inverse images $s \in \mathscr{F}(U)$; by choosing U to be maximal, we obtain that $U = T$: if there exists a $t \in T \setminus U$, then, for a neighborhood V of t, there is an inverse image $\tilde{s} \in \mathscr{F}(V)$; the section $s - \tilde{s}$ lies in ${}'\mathscr{F}(U \cap V)$, and therefore has an extension $s' \in {}'\mathscr{F}(T)$; with s and $s' + \tilde{s}$, we find an inverse image of s'' on $U \cup V$. ↯ ■

If the sheaves ${}'\mathscr{F}$ and $\mathscr{F}$ of 50.6 are flabby, then so is ${}''\mathscr{F}$ (proof?); hence, we have the following result:

50.7 Lemma. *Flabby sheaves are acyclic.*

Proof. Let $\mathscr{E}^*$ be the canonical resolution of a flabby sheaf $\mathscr{F}$. Then it follows by induction from the exact sequences (50.4.1) that the sheaves $\mathscr{Z}^j$ are flabby. By 50.6, then, the sequences

$$0 \to \mathscr{Z}^j(T) \to \mathscr{E}^j(T) \xrightarrow{\vartheta^j(T)} \mathscr{Z}^{j+1}(T) \to 0$$

are exact. We obtain that $\operatorname{Im} \vartheta^j(T) = \mathscr{Z}^{j+1}(T) = \operatorname{Ker} \vartheta^{j+1}(T)$, so $\mathscr{F}$ is acyclic, by 50.4. ■

We proceed now to the *Čech construction* for *paracompact* T (see § 51 A). Let an open cover $\mathfrak{U} = (U_j)_{j \in J}$ of T, an $\mathscr{R}$-module $\mathscr{F}$, and a subring R of $\mathscr{R}(T)$ be given.

a) A *q-cochain* with respect to $\mathfrak{U}$ with coefficients in $\mathscr{F}$ is a "tuple" $c = (c_{j_0 \ldots j_q}) \in \prod_{J^{q+1}} \mathscr{F}(U_{j_0 \ldots j_q})$; recall that $\mathscr{F}(\emptyset) = 0$. (For explicit constructions, it is frequently more economical to apply only "alternating cochains": c is called *alternating* if, for each permutation σ of the indices,

$$c_{j_{\sigma(0)} \ldots j_{\sigma(q)}} = \operatorname{sign}(\sigma)\, c_{j_0 \ldots j_q},$$

and if $c_{j_0 \ldots j_q} = 0$, whenever two indices are equal. The following Čech construction obviously can be carried out analogously with alternating cochains; by 50.2, one obtains thereby isomorphic cohomology modules $\check{H}^*_a(T, \mathscr{F}) \cong \check{H}^*(T, \mathscr{F})$, so we may leave out the subscript a.)

For each $q \geqq 0$, the set

$$C^q(\mathfrak{U}, \mathscr{F}) := \prod_{J^{q+1}} \mathscr{F}(U_{j_0 \ldots j_q})$$

of *q-cochains*, provided with componentwise operations, is an R-module.

b) "*Coboundary homomorphisms*" $d^q: C^q(\mathfrak{U}, \mathscr{F}) \to C^{q+1}(\mathfrak{U}, F)$ are defined as follows:

$$(d^q c)_{j_0 \dots j_{q+1}} := \sum_{k=0}^{q+1} (-1)^k c_{j_0 \dots \hat{j}_k \dots j_{q+1}} |_{U_{j_0 \dots j_{q+1}}}.$$

Since $d^{q+1} \circ d^q = 0$ for each $q \geqq 0$ (verify!), we obtain the *Čech complex* $\mathscr{C}^*(\mathfrak{U}, \mathscr{F})$ with respect to $\mathfrak{U}$ with coefficients in $\mathscr{F}$. The submodules $Z^q(\mathfrak{U}, \mathscr{F}) := \operatorname{Ker} d^q$ of *q-cocycles* and $B^q(\mathfrak{U}, \mathscr{F}) := \operatorname{Im} d^{q-1}$ of *q-coboundaries* lie in $C^q(\mathfrak{U}, \mathscr{F})$. The R-module $H^q(\mathfrak{U}, \mathscr{F}) := H^q(C^*(\mathfrak{U}, \mathscr{F})) = Z^q(\mathfrak{U}, \mathscr{F}) / B^q(\mathfrak{U}, \mathscr{F})$ is called the q-th *Čech cohomology module with respect to* $\mathfrak{U}$ with coefficients in $\mathscr{F}$.

If $\varphi: \mathscr{F} \to {}'\mathscr{F}$ is a homomorphism of $\mathscr{R}$-modules, then φ induces a homomorphism $\varphi^*: C^*(\mathfrak{U}, \mathscr{F}) \to C^*(\mathfrak{U}, {}'\mathscr{F})$ that is compatible with the coboundary homomorphisms. Thus $H^q(\mathfrak{U}, \cdot)$ is a covariant functor.

c) For the construction of connecting homomorphisms, we must rid ourselves of the option in the choice of $\mathfrak{U}$: for an open refinement $\mathfrak{V}$ of $\mathfrak{U}$ (§ 51 A), there exist canonical homomorphisms

$$\tau_{\mathfrak{V}}^{\mathfrak{U}}: H^q(\mathfrak{U}, \mathscr{F}) \to H^q(\mathfrak{V}, \mathscr{F}) \quad \text{with} \quad \tau_{\mathfrak{W}}^{\mathfrak{V}} \circ \tau_{\mathfrak{V}}^{\mathfrak{U}} = \tau_{\mathfrak{W}}^{\mathfrak{U}}.$$

Hence, the q-th *Čech cohomology module*

$$\check{H}^q(T, \mathscr{F}) := \varinjlim_{\mathfrak{U}} H^q(\mathfrak{U}, \mathscr{F})$$

of T with coefficients in $\mathscr{F}$ is well-defined. Clearly, $\check{H}^0(T, \mathscr{F}) = H^0(\mathfrak{U}, \mathscr{F}) = \mathscr{F}(T^0)$ for each cover $\mathfrak{U}$.

Now let $0 \to {}'\mathscr{F} \to \mathscr{F} \xrightarrow{\psi} {}''\mathscr{F} \to 0$ be an exact sequence of sheaves. Then $\delta^q: \check{H}^q(T, {}''\mathscr{F}) \to H^{q+1}(T, {}'\mathscr{F})$ is defined as follows: for the cohomology class $\bar{a} \in H^q(T, {}''\mathscr{F})$, let the *cocycle* $a \in Z^q(\mathfrak{U}, \mathscr{F}'')$ with a locally finite cover $\mathfrak{U}$ be a representative. There exist a refinement $\mathfrak{V}$ of $\mathfrak{U}$ and a *cochain* $b \in C^q(\mathfrak{V}, \mathscr{F})$ with $\overline{\psi^q(b)} = \bar{a}$ (see [Se$_2$ 25 Lemma 2]). Then $d^q b$ lies in $Z^{q+1}(\mathfrak{V}, \mathscr{F}')$ and represents $\delta^q \bar{a} \in \check{H}^{q+1}(T, {}'\mathscr{F})$.

Since that construction generally calls for the use of a refinement $\mathfrak{V}$, it is not possible to define connecting homomorphisms for $H^*(\mathfrak{U}, \cdot)$ (however, see E. 50 a).

The exact sequence $0 \to {}'\mathscr{F} \to \mathscr{F} \to {}''\mathscr{F} \to 0$ induces a cohomology sequence; it is easy to see that conditions 50.1 a) and b) are satisfied.

d) It remains to show that the sheaves $\mathscr{C}(T, \mathscr{F})$ are acyclic in the Čech theory. To that end, we generalize the concept of a "partition of unity" from functions to sheaves: let us call a sheaf $\mathscr{F}$ *fine* if, for each locally finite open cover $\mathfrak{U}$ of T, there is a partition of $\mathrm{id}_{\mathscr{F}}$ subordinate to $\mathfrak{U}$; in other words, if there exist endomorphisms $h_j: \mathscr{F} \to \mathscr{F}$ such that $\operatorname{supp} h_j := \{t \in T; h_j(\mathscr{F}_t) \neq 0\}$ is included in U_j and $\sum_{j \in J} h_j = \mathrm{id}_{\mathscr{F}}$.

50.8 Lemma. *Every fine sheaf* $\mathscr{F}$ *on a paracompact topological space* T *is acyclic.*

Proof. It suffices to show that $H^q(\mathfrak{U}, \mathscr{F}) = 0$ for every $q \geqq 1$ and every locally finite cover $\mathfrak{U}$. For a fixed partition $(h_j)_{j \in J}$ of $\mathrm{id}_{\mathscr{F}}$, and a $q \geqq 1$, we define a homomorphism $k^q: C^q(\mathfrak{U}, \mathscr{F}) \to C^{q-1}(\mathfrak{U}, \mathscr{F})$ that satisfies $d^{q-1} \circ k^q + k^{q+1} \circ d^q = \mathrm{id}_{C^q}$ by setting

$$(k^q(c))_{j_0 \dots j_{q-1}} := \sum_{j \in J} h_j(c_{j j_0 \dots j_{q-1}})$$

with $h_j(c_{j j_0 \dots j_{q-1}})$ extended to $U_{j_0 \dots j_{q-1}} \setminus U_j$ by 0. Then $Z^q(\mathfrak{U}, \mathscr{F}) = \mathrm{id}_{C^q}(Z^q(\mathfrak{U}, \mathscr{F})) \subset B^q(\mathfrak{U}, \mathscr{F})$ for every $q \geqq 1$, and it follows that $H^q(\mathfrak{U}, \mathscr{F}) = 0$. ■

50.9 Lemma. *Every sheaf* $\mathscr{C}(T, \mathscr{F})$ *on a paracompact topological space* T *is fine.*

Proof. For a locally finite cover $\mathfrak{U}$ of T, let the open cover $\mathfrak{V} = (V_j)_{j\in J}$ be a *shrinking* (i.e., $\bar{V}_j \subset U_j$ for every j [Bou GT IX §4, Prop. 4 and Th. 3]). Let $\tau: T \to J$ be a mapping with $t \in V_{\tau(t)}$ for each $t \in T$. Then

$$\tilde{h}_j: \mathscr{F} \to \mathscr{F},\ \tilde{h}_j|_{\mathscr{F}_t} := \begin{cases} \mathrm{id}_{\mathscr{F}_t}, & \text{if } j = \tau(t) \\ 0, & \text{otherwise,} \end{cases}$$

induces a partition $(h_j)_{j\in J}$ of $\mathrm{id}_{\mathscr{C}(\mathscr{F})}$, for we have that $\operatorname{supp} h_j \subset \bar{V}_j \subset U_j$ and $\sum_{j\in J} h_j = \mathrm{id}_{\mathscr{C}(\mathscr{F})}$. ■

Analogously, with alternating cochains, one can construct the R-modules C^q_a, Z^q_a, B^q_a, and H^q_a for $\mathfrak{U}$ and $\mathscr{F}$, and also $\check{H}^q_a(T, \mathscr{F})$. Hence, for paracompact spaces, $\{\check{H}^q, \delta^q\}$ and $\{\check{H}^q_a, \delta^q\}$ are cohomology theories; by 50.2, they are isomorphic to every other cohomology theory on T, so we denote them with $\{H^q, \delta^q\}$, as well. In the applications of the alternating theory we mainly use the fact that $Z^1_a = Z^1$.

We shall require the following results on cohomology modules for an arbitrary topological space T ([Ku 34.1]):

50.10 Proposition. *The homomorphism $H^1(\mathfrak{U}, \mathscr{F}) \to H^1(T, \mathscr{F})$ is always injective.* ■

For coherent analytic sheaves $\mathscr{F}$ on paracompact complex spaces X, we can do without the inductive limit in determining $H^*(X, \mathscr{F})$, as the following theorem is applicable, by 52.21:

50.11 Leray's Theorem. *For a sheaf $\mathscr{F}$ on T and a number $p \in \mathbb{N}$, let $\mathfrak{U}$ be an open cover of T such that $H^q(U_{j_0 \dots j_{p-1}}, \mathscr{F}) = 0$ for each $q = 1, \dots, p$. Then the canonical homomorphism $H^p(\mathfrak{U}, \mathscr{F}) \to H^p(T, \mathscr{F})$ is bijective.* [Ku § 34]. ■

If the cover $\mathfrak{U}$ of T satisfies the hypotheses of 50.11 for every $p \geqq 1$, then $\mathfrak{U}$ is called an (open) *Leray cover* of T with respect to $\mathscr{F}$ (see [GrRe$_2$ B § 3]).

E. 50a. Let $0 \to \mathscr{F}' \to \mathscr{F} \to \mathscr{F}'' \to 0$ be an exact sequence of sheaves, and let $\mathfrak{U}$ be an open Leray cover with respect to $\mathscr{F}'$. Construct connecting homomorphisms $\delta^q: H^q(\mathfrak{U}, \mathscr{F}'') \to H^{q+1}(\mathfrak{U}, \mathscr{F}')$ for $q \geqq 0$.

It is sometimes useful fo know that, for paracompact locally compact spaces, the cohomology with values in a constant sheaf coincides with the singular cohomology [Br III.1]. That holds for complex spaces with a countable topology (see 51 A. 2). In particular, then, open covers $\mathfrak{U}$ are Leray covers for constant sheaves if every $U_{j_0 \dots j_q}$ has contractible connected components.

E. 50b. Prove these statements for $\mathbb{P}_1$: i) The sets

$$V_0 := \{[1, z_1]; |z_1| < 1\},\ V_1 := \{[1, z_1]; z_1 \notin \mathbb{R}_{\geqq 0}\},\ V_2 := \{[1, z_1]; z_1 \notin \mathbb{R}_{\leqq 0}\}$$

and $V_3 := \{[z_0, 1]; |z_0| < 1\}$ form a Leray cover $\mathfrak{V}$ for constant sheaves.

ii) $H^q(\mathbb{P}_1, \mathbb{Z}) = \begin{cases} \mathbb{Z} & q = 0, 2 \\ 0 & \text{otherwise.} \end{cases}$

iii) Using the cohomology sequence induced by (54.3.1), one can obtain a generator $\delta^1 \xi_1 \in H^2(\mathbb{P}_1, \mathbb{Z})$, where $\xi_m \in H^1(\mathbb{P}_1, \mathcal{O}^*)$ is given by $z_1^m \in Z^1(\mathfrak{U}, \mathcal{O}^*) = \mathcal{O}^*(U_{01})$ for $m \in \mathbb{Z}$ (see 32.4 iv); for E. 54 Bc).

iv) Determine $\delta^1 \xi_m \in H^2(\mathbb{P}_1, \mathbb{Z})$.

50.12 Proposition. *If the support of a sheaf $\mathscr{F}$ is a closed discrete subset of T, then $H^q(T, \mathscr{F}) = 0$ for every $q \geqq 1$.*

Proof. If $\{t \in T; \mathscr{F}_t \neq 0\}$ is discrete and closed, then $\mathscr{F}$ is obviously flabby, and thus acyclic. ■

50.13 Proposition. *If the support of a sheaf $\mathscr{F}$ is included in a closed set $A \subset T$, then $H^*(T, \mathscr{F}) \cong H^*(A, \mathscr{F})$.*

Proof. The assertion follows from the fact that $\mathscr{C}(T, \mathscr{F}) \simeq \mathscr{C}(A, \mathscr{F})^T$. ■

50.14 Proposition. *If T is metrizable, and $B \subset T$, then, for every sheaf $\mathscr{F}$ on T,*

$$H^*(B, \mathscr{F}) = \varinjlim_{B \subset U \Subset T} H^*(U, \mathscr{F}).$$

[Go II, Th. 4.11.1]. ■

E. 50c. For a sheaf of abelian groups $\mathscr{F}$ on a paracompact complex space X, prove that $H^q(X, \mathscr{F}) = 0$ for every $q \geqq 1 + 2 \max_{x \in X} \operatorname{emb}_x(\operatorname{Red} X)$. (Hint: put $m := \max\{\operatorname{emb}_x(\operatorname{Red} X); x \in X\}$; for $X \cong \mathbb{R}^{2m}$, there exist arbitrarily fine locally finite open covers $\mathfrak{U}$ of X such that $U_{j_0 \ldots j_q} = \emptyset$ for pairwise distinct indices and $q \geqq 2m+1$; hence, $C_a^q(\mathfrak{U}, \mathscr{F}) = 0$. For general X, identify $|X|$ locally with a closed subset of $\mathbb{R}^{2m}$. For 61.12).

E. 50d. Prove that $H^*(T, \mathscr{F} \oplus \mathscr{G}) = H^*(T, \mathscr{F}) \oplus H^*(T, \mathscr{G})$ (for 61.12).

E. 50e. If $T = \bigcup_J T_j$ is the representation of a locally connected space T as the union of its connected components, then $H^q(T, \mathscr{G}) \cong \prod_J H^q(T_j, \mathscr{G})$ for every q.

E. 50f. Prove the following statements: i) If $\mathscr{F}$ is a sheaf of modules over a fine sheaf of rings $\mathscr{R}$, then $\mathscr{F}$ is fine.

ii) If M is a paracompact manifold, then ${}_M\mathscr{C}$ and ${}_M\mathscr{C}^\infty$ are fine sheaves of rings (that naturally holds analogously for differentiable manifolds. Hint: Proof of α) preceeding 50 A. 3; for 50 Ad).

E. 50g. If $\mathscr{F}$ is a sheaf on T and $A \subset T$ a closed subset such that for every subset $U \Subset T$, the restriction-mapping $\mathscr{F}(U) \to \mathscr{F}(U \setminus A)$ is bijective, then $H^1(T, \mathscr{F}) \to H^1(T \setminus A, \mathscr{F})$ is injective (hint: 50.10; for E. 74g).

Supplemental literature: [Ku], $[\mathrm{GrRe}]_2$, [Br], [Go].

§ 50 A Supplement: Automorphic Functions

We want to present an example of a problem in complex analysis for which an attempt to patch together local solutions into a global solution leads naturally to an "obstruction" in the form of a cohomology class. We choose the question of the existence of "many-valued functions" with prescribed many-valued behavior. The concept of a "many-valued function" is defined as follows:

Let X be a manifold with a countable topology, and let $\pi: \tilde{X} \to X$ be the universal covering (§ 33 B; it always exists, since the covering constructed in topology can be made into a manifold by setting ${}_{\tilde{x}}\mathcal{O} := \pi^{-1}{}_x\mathcal{O}$; then π becomes a (holomorphic) covering). We understand a *many-valued function* on X to be a function on $\tilde{X}$. Since π is surjective, we identify $\mathscr{C}^\infty(X)$ with $\pi^0(\mathscr{C}^\infty(X)) \subset \mathscr{C}^\infty(\tilde{X})$. In particular, then, functions on X are many-valued functions on X. In order to grasp the many-valued behaviour of, say, $\log z$ for $X = \mathbb{C}^*$, we use the action of the group Γ of deck transformations on $\mathscr{C}^\infty(\tilde{X})$ and $\mathcal{O}(\tilde{X})$, which is defined as follows:

$$\Gamma \times \mathscr{C}^\infty(\tilde{X}) \to \mathscr{C}^\infty(\tilde{X}), \ (\gamma, \tilde{f}) \mapsto \tilde{f} \circ \gamma^{-1} =: \gamma\tilde{f}$$

(with γ^{-1} chosen so as to ensure that $(\beta\gamma)\tilde{f} = \beta(\gamma\tilde{f})$).

50 A. 1 Definition. *Let* $a: \Gamma \to \mathcal{O}(X), \gamma \mapsto a_\gamma$, *be a group-homomorphism. An* $\tilde{f} \in \mathscr{C}^\infty(\tilde{X})$ *is called (additively) a-automorphic iff* $\gamma \tilde{f} - \tilde{f} = a_\gamma \circ \pi$ *for every* $\gamma \in \Gamma$.

50 A. 2 Example. For the complex torus $T = \mathbb{C}/\Gamma$ with $\Gamma = \mathbb{Z} + i\mathbb{Z}$, $\pi: \mathbb{C} \to T$ is the universal covering, and Γ is thus the group of deck transformations. For the homomorphism $a: \Gamma \to \mathcal{O}(T) = \mathbb{C}, p + iq \mapsto \alpha p + \beta q$, we have that $\tilde{f}(x + iy) := -(\alpha x + \beta y)$ is a-automorphic (substitute!); moreover, f is holomorphic iff $\beta = i\alpha$.

E. 50A a. Show that the assumption in 50A. 1 that a is a homomorphism is superfluous, since it $a: \Gamma \to \mathcal{O}(\mathbb{C}^*)$ such that $\mathrm{id}_{\tilde{X}}$ is a-automorphic.

For an a-automorphic function $\tilde{f}$ and a fixed $\gamma, \gamma\tilde{f} - f$ is a function on X, since its value at a point $\tilde{x}$ depends only on the point $\pi(\tilde{x})$. The set of 0-automorphic functions is precisely $\mathscr{C}^\infty(X)$. It is easy to see that, for a_j-automorphic functions $\tilde{f}_j, f_1 + f_2$ is $(a_1 + a_2)$-automorphic, and $A(a_1) := \tilde{f}_1 + \mathscr{C}^\infty(X)$ is the set of all a_1-automorphic functions.

E. 50Ab. Show that the assumption in 50A. 1 that a is a homomorphism is superfluous, since it follows from the fact that "$\gamma\tilde{f} = \tilde{f} + a_\gamma \circ \pi, \forall \gamma \in \Gamma$".

For an $a: \Gamma \to \mathcal{O}(X)$, there always exist, *locally with respect to* X, a-automorphic functions f (holomorphic ones, in fact): if Γ is provided with the discrete topology, then, for simply connected $W \Subset X, \pi^{-1}(W)$ is equivalent to $W \times \Gamma$ with the canonical action of Γ, since π is the universal covering. With

$$\tilde{f}(x, \gamma) := -a_\gamma(x)$$

on $W \times \Gamma$, it can be verified by substitution that $\delta\tilde{f} - \tilde{f} = a_\delta \circ \pi$ for every $\delta \in \Gamma$; consequently, $\tilde{f}$ is a-automorphic on $W \times \Gamma = \pi^{-1}(W)$. That leads to the following:

Problem. *For which* $a: \Gamma \to \mathcal{O}(X)$ *does there exist a global holomorphic a-automorphic function* $\tilde{f} \in \mathcal{O}(\tilde{X})$?

First Approach (using $\mathscr{C}^\infty$-differential forms; see § 58):

α) There always exists an a-automorphic $\tilde{g} \in \mathscr{C}^\infty(\tilde{X})$.

Proof. The essential tool here is the existence of a *smooth partition of unity* on X: for a locally finite cover $\mathfrak{U}$ of X, due to the countable topology of X, there exist functions $\varphi_j \in \mathscr{C}^\infty(U_j)$ such that $0 \leqq \varphi_j$, $\operatorname{supp} \varphi_j \subset U_j$, and $\sum_J \varphi_j = 1$ (see [Ns$_3$ § 1.2]). If, in addition, $\mathfrak{U}$ is chosen so that there exist a-automorphic functions $\tilde{f}_j \in \mathcal{O}(\pi^{-1}(U_j))$, then $\tilde{g} := \sum_J (\varphi_j \circ \pi) \cdot \tilde{f}_j \in \mathscr{C}^\infty(\tilde{X})$ is clearly a-automorphic.

β) Construction of the obstruction. For each a, there is a commutative diagram

$$\begin{array}{ccc}
\mathscr{C}^\infty(\tilde{X}) & \xrightarrow{\ \bar{\partial}\ } & \mathscr{E}^{0,1}(\tilde{X}) \\
\uparrow \pi^0 \quad \cup\, A(a) & & \uparrow \pi^* \\
\mathscr{C}^\infty(X) & \xrightarrow{\ \bar{\partial}\ } & \mathscr{E}^{0,1}(X).
\end{array}$$

First we show that $\bar{\partial} A(a) \subset \pi^* \mathscr{E}^{0,1}(X)$. The images of π^0 and π^*, respectively, consist precisely of the Γ-invariant elements; hence, it suffices to show that $\bar{\partial} A(a)$ is fixed pointwise under Γ. In view of the fact that $\operatorname{Ker}[\bar{\partial}: \mathscr{C}^\infty(\tilde{X}) \to \mathscr{E}^{0,1}(\tilde{X})] = \mathcal{O}(\tilde{X})$ and $(\gamma^{-1})^* \bar{\partial} = \bar{\partial} (\gamma^{-1})^0$ for holomorphic γ^{-1}, we see that, for $\tilde{g} \in A(a)$,

$$\gamma \bar{\partial} \tilde{g} = (\gamma^{-1})^* \bar{\partial} \tilde{g} = \bar{\partial}((\gamma^{-1})^0 \tilde{g}) = \bar{\partial}(\gamma \tilde{g}) = \bar{\partial} \tilde{g} + \bar{\partial}(a_\gamma \circ \pi) = \bar{\partial} \tilde{g}.$$

Since $\bar\partial \circ \bar\partial = 0$, it follows that $\bar\partial A(a) \subset \operatorname{Ker}[\bar\partial : \mathscr{E}^{0,1}(X) \to \mathscr{E}^{0,2}(X)]$. That enables us to assign to the affine space $A(a) = \tilde g + A(0)$ with $A(0) = \mathscr{C}^\infty(X)$ the equivalence class

$$\varrho_a := [\bar\partial A(a)] \in \frac{\operatorname{Ker}[\bar\partial : \mathscr{E}^{0,1}(X) \to \mathscr{E}^{0,2}(X)]}{\operatorname{Im}[\bar\partial : \mathscr{C}^\infty(X) \to \mathscr{E}^{0,1}(X)]} =: H^1_{DR}(X, \mathcal{O}).$$

γ) The kernel of the homomorphism

$$\operatorname{Hom}(\Gamma, \mathcal{O}(X)) \to H^1_{DR}(X, \mathcal{O}),\ a \mapsto \varrho_a,$$

provides a solution to the problem:

50 A. 3 Proposition. *For a homomorphism $a : \Gamma \to \mathcal{O}(X)$, there exists an a-automorphic $\tilde f \in \mathcal{O}(\tilde X)$ iff $\varrho_a = 0 \in H^1_{DR}(X, \mathcal{O})$.*

Proof. Fix a $\tilde g \in A(a)$. Then the fact that $A(a) = \tilde g + \mathscr{C}^\infty(X)$ implies that

$$\exists \tilde f \in \mathcal{O}(\tilde X) \cap A(a) \Leftrightarrow 0 \in \bar\partial A(a) \Leftrightarrow \bar\partial \tilde g \in \bar\partial \mathscr{C}^\infty(X) \Leftrightarrow \varrho_a = 0. \quad \blacksquare$$

The preceding construction leads to the *de Rham-Dolbeault cohomology*, which is determined by the fine, and thus acyclic, resolution $\mathscr{C}^\infty \xrightarrow{\bar\partial} \mathscr{E}^{0,1} \xrightarrow{\bar\partial} \mathscr{E}^{0,2} \to \dots$ of $\mathcal{O}$. Accordingly, $H^1_{DR}(X, \mathcal{O})$ is called the first de Rham cohomology vector space of X with coefficients in $\mathcal{O}$.

With 50.4, we can derive the following consequence of the central "vanishing theorem" of this chapter, namely, Theorem B (see § 52):

50 A. 4 Corollary. *If X is a Stein manifold, then, for each $a \in \operatorname{Hom}(\Gamma, \mathcal{O}(X))$, there is an a-automorphic function $\tilde f \in \mathcal{O}(\tilde X)$.* $\blacksquare$

E. 50 Ac. For the torus $X := \mathbb{C}/\Gamma$ with $\Gamma = \mathbb{Z} + \mathbb{Z}i \subset \mathbb{C}$, prove the following statements:

i) There exists an $\omega \in \mathscr{E}^{0,1}(X)$ such that $\bar\partial\omega = 0$ and $\omega \notin \bar\partial\mathscr{C}^\infty(X), \pi^*\omega = d\bar z$.

ii) $0 \neq [\omega] \in H^1_{DR}(X, \mathcal{O})$ (for E. 54 Bb).

iii) There is an $a : \Gamma \to \mathcal{O}(X) = \mathbb{C}$ with $\varrho_a = [\omega]$.

iv) There is no a-automorphic $\tilde f \in \mathcal{O}(\tilde X)$.

Second Approach. Again, let $\mathfrak{U} = (U_j)_{j \in \mathbb{N}}$ be an open cover of X with simply connected U_j. Then there is an a-automorphic $\tilde f_j \in \mathcal{O}(\tilde U_j)$ on each $\tilde U_j = \pi^{-1}(U_j)$. Every $f_{jk} := \tilde f_k - \tilde f_j$ is 0-automorphic on $\tilde U_{jk}$, and thus lies in $\mathcal{O}(U_{jk}) \subset \mathcal{O}(\tilde U_{jk})$. As a result, the "$a$-automorphic" 0-cochain $(\tilde f_j) \in C^0(\tilde{\mathfrak{U}}, \mathcal{O})$ determines a 1-chochain $(f_{jk}) \in C^1(\mathfrak{U}, \mathcal{O})$; that replaces the prescribed data on $\tilde X$ with data on X. In particular, (f_{jk}) is clearly an alternating 1-cocycle. We have that

(50 A. 5.1) $\exists \tilde f \in \mathcal{O}(\tilde X) \cap A(a)$

iff $\exists (h_j) \in C^0(\mathfrak{U}, \mathcal{O})$ with $h_k - h_j = f_{jk}$ on U_{jk}

iff $(f_{jk}) \in B^1(\mathfrak{U}, \mathcal{O})$.

Proof. The second equivalence is evident. Let us prove the first one. *Only if.* The function $h_j := \tilde f_j - \tilde f$ is clearly 0-automorphic on $\tilde U_j$, and thus lies in $\mathcal{O}(U_j)$; we see that $h_k - h_j = \tilde f_k - \tilde f - (\tilde f_j - \tilde f) = \tilde f_k - \tilde f_j = f_{jk}$. *If.* There is a unique function $f \in \mathcal{O}(\tilde X)$ such that $\tilde f|_{\tilde U_j} = \tilde f_j - h_j$; the fact that $\tilde f_j$ is a-automorphic everywhere implies that $\tilde f$ has the same property, since h_j is 0-automorphic. $\blacksquare$

Using the a-automorphic cochain $(\tilde f_j) \in C^0(\mathfrak{U}, \mathcal{O})$, we associate the cohomology class $\zeta_a = [(f_{jk})] \in H^1(\mathfrak{U}, \mathcal{O})$ to the homomorphism a; observe that ζ_a is independent of the choice of $(\tilde f_j)$: if $(\tilde g_j) \in C^0(\tilde{\mathfrak{U}}, \mathcal{O})$ is a-automorphic, then there exists an $(l_j) \in C^0(\mathfrak{U}, \mathcal{O})$ with $(\tilde f_j - \tilde g_j) = (l_j)$; hence, $f_{jk} - g_{jk} = l_k - l_j$, i.e., $(f_{jk}) - (g_{jk}) = d(l_j)$.

Thus (50 A. 5.1) states the following:

50 A. 5 Proposition. *For a homomorphism $a: \Gamma \to \mathcal{O}(X)$, there is an a-automorphic function $\tilde{f} \in \mathcal{O}(\tilde{X})$ iff $\zeta_a = 0 \in H^1(\mathfrak{U}, {}_X\mathcal{O})$.* ■

The obstruction ζ_a is also independent of the choice of the cover $\mathfrak{U}$: for a refinement $\mathfrak{V}$ of $\mathfrak{U}$, the injective (see 50.10) homomorphism $H^1(\mathfrak{U}, \mathcal{O}) \to H^1(\mathfrak{V}, \mathcal{O})$ sends $\zeta_a(\mathfrak{U})$ to $\zeta_a(\mathfrak{V})$.

Finally, we observe that, for each cocycle $(s_{jk}) \in Z^1(\mathfrak{U}, \mathcal{O})$ that represents the cohomology class ζ_a, there exists an a-automorphic $(\tilde{g}_j) \in C^0(\tilde{\mathfrak{U}}, \mathcal{O})$ with $d(\tilde{g}_j) = (s_{jk})$. For, with the originally chosen $(\tilde{f}_j)$, we have that $(s_{jk}) - (f_{jk}) \in B^1(\mathfrak{U}, \mathcal{O})$; now choose an $(h_j) \in C^0(\mathfrak{U}, \mathcal{O})$ such that $d(h_j) = (s_{jk} - f_{jk})$, and set $\tilde{g}_j := \tilde{f}_j + h_j$.

In summary, the obstruction ζ_a provides an adequate description of the problem in the context of the Čech construction.

E. 50 A d. Deduce from 50.8 that, for each $a: \Gamma \to \mathcal{O}(X)$, there exists an a-automorphic function $\tilde{f} \in \mathscr{C}^\infty(\tilde{X})$ (hint: E. 50 f ii)).

§ 51 Stein Spaces

In this section, we introduce the most important class of complex spaces with a rich global function-algebra. It constitutes a generalization of regions of holomorphy in $\mathbb{C}^n$.

51.1 Definition. *A complex space X is called a Stein space (or holomorphically complete space) if it satisfies the following conditions:*

i) Every connected component of X has a countable topology.

ii) X is holomorphically separable; i.e., for $x \neq y \in X$, there exists an $f \in \mathcal{O}(X)$ such that $f(x) \neq f(y)$.

iii) X is holomorphically convex; i.e., for $K \subset |X|$, the "holomorphically convex hull"

$$\hat{K}_{\mathcal{O}(X)} := \hat{K} := \bigcap_{f \in \mathcal{O}(X)} \{x \in X; |f(x)| \leqq \|f\|_K\}$$

of K in X is compact if K is.

51.2 Remarks. i) In the interest of clarity, we recall that $f(x)$ denotes the value of the function, and $|f(x)|$, its absolute value; moreover, $\|f\|_K := \|\mathrm{Red}\, f\|_K$.

ii) In 51 A. 3, we shall see that 51.1 i) follows from 51.1 ii). However, for singular X, the proof makes use of the Normalization Theorem 71.4, so we cannot dispense with 51.1 i) immediately.

iii) For complex spaces X with a countable topology, there is a useful criterion for holomorphic convexity. Let us consider these statements:

(51.2.1) *For each infinite discrete closed set $Z \subset X$, there exists an $f \in \mathcal{O}(X)$ such that $\sup_{z \in Z} |f(z)| = \infty$.*

(51.2.2) *For every compactum $K \subset X$, $\hat{K}_{\mathcal{O}(X)}$ is sequentially compact.*

(51.2.3) *For every compactum* $K \subset X$, $\hat{K}_{\mathcal{O}(X)}$ *is compact (i.e.,* X *is holomorphically convex).*

Then the following implications are easy to see:

(51.2.1) $\Rightarrow$ (51.2.2).

(51.2.2) and X has a countable topology $\Rightarrow$ (51.2.3).

(51.2.3) $\Rightarrow$ (51.2.2).

The implication (51.2.3) $\Rightarrow$ (51.2.1) is a consequence of E. 57g; hence, it uses Grauert's Coherence Theorem. If X is reduced and has a countable topology, then E. 55j enables us to give a proof parallel to that of 12.9, which does not use Grauert's Coherence Theorem.

51.3 Examples. i) Every complex space with only finitely many points is a Stein space.

ii) Every region in $\mathbb{C}$ is a Stein space.

iii) Regions of holomorphy in $\mathbb{C}^n$ – in particular, polydisks and hyperballs – are Stein spaces (12.8).

iv) Open Riemann surfaces are Stein spaces ([Fo 26.8]; by 71.4 and 73.1, it follows that a complex space X of pure dimension one is Stein iff $\operatorname{Red} X$ has no compact irreducible components).

51.4 Proposition. *The following statements hold for an arbitrary locally closed subspace* Y *of a complex space* X*:*

i) If X *is holomorphically separable, then* Y *is holomorphically separable.*

ii) If Y *is closed and* X *is holomorphically convex, then* Y *is holomorphically convex.*

iii) If Y *is closed and* X *is a Stein space, then* Y *is a Stein space.*

Proof. ii) If $K \subset Y$ is compact, then the closed set $\hat{K}_{\mathcal{O}(Y)}$ lies in the compact set $\hat{K}_{\mathcal{O}(X)} \cap Y$, and is therefore compact. Statement i) is trivial, and iii) follows immediately from the first two statements. ■

In particular, 51.4 holds for $\operatorname{Red} X \hookrightarrow X$; however, $\operatorname{Red} X$ being holomorphically separable or holomorphically convex does not imply that the same holds for X (see E. 51i).

E. 51a. Prove, for a compact complex space X, that

i) X is holomorphically convex;

ii) X is holomorphically separable (and thus a Stein space) iff $\dim X = 0$. In particular, then, *compact analytic subsets of holomorphically separable spaces are finite* (for 52.3 vi)).

51.5 Corollary. *The topology of each complex space* X *has a basis consisting of open Stein subspaces.*

Proof. Suppose, without loss of generality, that $X \hookrightarrow G \Subset \mathbb{C}^n$. A basis for the topology of X is given by sets $X \cap \boldsymbol{P}^n$, where $\boldsymbol{P}^n$ runs through all of the polydisks included in G. By 51.3 iii) and 51.4 iii), each $\boldsymbol{P}^n \cap X$ is a Stein space. ∎

If we call a cover consisting of open Stein subspaces a *Stein cover*, then we have the following:

51.6 Corollary. *Every complex space has arbitrarily fine Stein covers.* ∎

If X is paracompact, then we can even construct arbitrarily fine *locally finite* Stein covers $\mathfrak{W}$ of X: for each locally finite cover $\mathfrak{U}$ of X consisting of relatively compact sets, and each shrinking $\mathfrak{B}$ of $\mathfrak{U}$, every $\bar{V}_j$ can be covered by finitely many Stein open subsets of U_j.

From known Stein spaces, we obtain new ones:

51.7 Proposition. *If $X_1, X_2 \Subset X$ are Stein spaces, then $X_1 \cap X_2$ is a Stein space.*

Proof. By 51.4, we need only verify the holomorphic convexity of $X_1 \cap X_2$. For a compactum $K \subset X_1 \cap X_2$, the hull $\hat{K}_{\mathcal{O}(X_1 \cap X_2)}$, being a closed subset of the compactum $\hat{K}_{\mathcal{O}(X_1)} \cap \hat{K}_{\mathcal{O}(X_2)}$, is also compact. ∎

51.8 Proposition. *For a Stein space X, if $f \in \mathcal{O}(X)$, then $X \setminus N(f) \Subset X$ is a Stein space.*

Proof. We demonstrate the holomorphic convexity, using (51.2.1): if (x_j) is a sequence of points in $X \setminus N(f)$ with an accumulation point in $N(f)$, then $\sup_j |1/f(x_j)| = \infty$. ∎

E. 51b. *Products and fiber products of Stein spaces are Stein spaces.* To prove that, verify the following implications for complex spaces X_1 and X_2 over Y:

$X_1 \times X_2$ is holomorphically separable (spreadable (see E. 51c), convex)
$\Leftrightarrow$ X_1 and X_2 are holomorphically separable (spreadable, convex).
$\Rightarrow$ $X_1 \times_Y X_2$ is holomorphically separable (spreadable, convex).

If a complex space $f: X \to Y$ over a Stein space Y is not globally, but only locally with respect to Y, the projection of a product with typical fiber F, then X need not be a Stein space, even if F is. In [De], there is an example with $Y \Subset \mathbb{C}$ and $F = \mathbb{C}^2$ (however, see 54 B.4).

Convexity and separability can be characterized in other ways:

E. 51c. Prove the following statements for a complex space X and a set of functions $F \subset \mathcal{O}(X)$: The mapping

(E. 51c. 1) $e: X \to \mathbb{C}^F,\ x \mapsto (fx)_{f \in F},$

is continuous. If $F = \mathcal{O}(X)$, then the following statements hold:

i) If inverse images under e of compact sets are compact, then X is holomorphically convex. (Hint: $\hat{B}_{\mathcal{O}(X)} = e^{-1}(\overline{\prod_{f \in F} \boldsymbol{P}^1(\|f\|_B)})$; for spaces X with a countable topology, prove the converse by applying (52.2.1).)

ii) The mapping e is injective iff X is holomorphically separable (hint: $e^{-1}e(a) = \bigcap_{f \in F} f^{-1}f(a)$; for § 57).

iii) The mapping e is discrete iff each $a \in X$ possesses a neighborhood U such that $\{a\} = \bigcap_{f \in F} \{x \in U; f(x) = f(a)\}$ (in which case X is called *holomorphically spreadable*) iff, for each $a \in X$, there is a $\varphi \in \mathrm{Hol}(X, \mathbb{C}^{k_a})$ that is discrete at a (in which case X is called *K-complete*). (Hint: for $f \in F$, decompose $N(U; f - f(\mathrm{a}))$ near a into irreducible components A_j. If $\dim_a A_j > 0$, then there exists an $f_j \in F$ with $\dim_a N(A_j; f_j - f_j(a)) < \dim_a A_j$. For 51 A. 3.)

The mapping (E. 51c. 1) is pertinent also for other damilies F. We consider the case in which F is the vector space $B(X)$ of *bounded* holomorphic functions on a complex manifold X (the results can be genralized to reduced spaces):

E. 51d. Verify the following statements for a manifold X:

i) $B(X)$, with the norm $\|\cdot\|_X$, is a Banach space.

ii) The space $L(B(X), \mathbb{C})$ of continuous linear forms, with the norm $\|l\| := \sup\{|l(f)|; \|f\|_X \leqq 1\}$ is a normed space (a Banach space, in fact!).

iii) The mapping

$$e: X \to L(B(X), \mathbb{C}), \ x \mapsto e_x \quad \text{with} \quad e_x(f) := f(x),$$

is continuous (hint: 4.3 ii)).

iv) The mapping e is injective (discrete) iff X is separable (spreadable) with respect to bounded holomorphic functions.

v) The assignment $X \mapsto L(B(X), \mathbb{C})$ determines a covariant functor from the category of manifolds into the category of Banach spaces with the *contracting linear mappings* (i.e., $\|T(l)\| \leqq \|l\|$) as morphisms.

vi) On every manifold X that is separable with respect to $B(X)$, the inclusion $e: X \to L(B(X), \mathbb{C})$ induces a metric $d(x,y) = \|e_x - e_y\|$, called the *Carathéodory metric.* Use v) to show that d is invariant under biholomorphic mappings.

E. 51e. *Permanance of separability and convexity.* Prove the following statements for $f \in \mathrm{Hol}(X,Y)$:

i) If f is discrete and Y is holomorphically spreadable, then X is holomorphically spreadable (for 51 A. 3).

ii) If f is proper and Y is holomorphically convex, then X is holomorphically convex (for § 52).

iii) Even if f is finite and surjective, and X is holomorphically spreadable, Y need not be holomorphically spreadable. (Hint: $X = (\mathbb{C}^{2*}, {}_2\mathcal{O})$; $Y = (\mathbb{C}^{2*}, \mathscr{H})$, where $\mathscr{H}$ is derived from ${}_2\mathcal{O}$ by means of *Strukturausdünnung* at every point $(1/m, 0) \in \mathbb{C}^{2*}$ according to E. 49 Ai i).)

iv) Even if f is finite and surjective, and X is holomorphically convex, Y need not be holomorphically convex (hint: consider a space Y that has an infinity of compact irreducible components Y_j, and $X = \dot{\bigcup} Y_j$).

E. 51f. *Analytic polyhedra.* Replace the $X \subset\subset \mathbb{C}^n$ of 12.10 vi) with a complex space X, and prove the following statements for the analytic polyhedron $W = U \cap \varphi^{-1}(Z)$:

i) $f: W \to Z$ is proper, and thus W is holomorphically convex (hint: $\varphi^{-1}(K) \cap W = \varphi^{-1}(K) \cap \overline{W}$; for 63 A. 3).

ii) W is $\mathcal{O}(X)$-convex if each Z_j is simply connected (for 63.3).

E. 51g. *Stein compacta.* A compact subset K of a complex space X is called a Stein compactum if K has a fundamental system of open neighborhoods that are Stein subspaces of X. Show that

i) every compact (elementarily) convex subset of $\mathbb{C}^n$ is a Stein compactum;

ii) the closure of a Stein open set $G \subset\subset X$ need not be Stein compactum (hint: E. 12g and E. 62b).

E. 51h. For a coherent $\mathcal{O}$-module $\mathcal{G}$ on a complex space $(X, \mathcal{O})$, the coherent $\mathcal{O}$-module $\mathcal{O} \oplus \mathcal{G}$ becomes an $\mathcal{O}$-algebra when provided with the product

$$(f,g)\cdot(f_1,g_1) := (ff_1, fg_1 + f_1 g)$$

(in which case it is denoted by $\mathcal{O} \tilde{\oplus} \mathcal{G}$), and $(X, \mathcal{O} \tilde{\oplus} \mathcal{G})$ is a complex space with $\mathrm{Red}(X, \mathcal{O} \tilde{\oplus} \mathcal{G}) = (X, \mathrm{Red}\,\mathcal{O})$ (hint: apply 45 B.2 to the ringed space $(X, \mathcal{O} \tilde{\oplus} \mathcal{G})$; for E. 51i).

E. 51i. On the manifold $X = \mathbb{C} \times \mathbb{P}_1$, construct a structure sheaf $\mathcal{H}$ with $\mathrm{Red}\,\mathcal{H} = {}_X\mathcal{O}$, so that $(X, \mathcal{H})$, in contrast to $(X, \mathcal{O})$, is not holomorphically convex (according to Schuster-Horst). To that end, let $\mathcal{I}, \mathcal{J} \subset {}_X\mathcal{O}$ be the nullstellen ideals of $\mathbb{C} \times \{0\}$ and $\mathbb{C} \times \{\infty\}$, respectively, and let ∂_z denote the derivative in the direction of $\mathbb{C}$; prove that

i) the sheaf of ($\mathbb{C}$-)algebras

$$\mathcal{H} := \{(f, \sigma, \tau); \partial_z f + \sigma + \tau = 0\} \subset \mathcal{O} \tilde{\oplus} (\mathcal{I} \oplus \mathcal{J})$$

is locally isomorphic to the structure sheaf $\mathcal{O} \tilde{\oplus} \mathcal{K}$ on X with $\mathcal{K} := \mathcal{K}er(\mathcal{I} \oplus \mathcal{J} \to \mathcal{O}, (\sigma,\tau) \mapsto \sigma + \tau)$ (on $\mathbb{C} \times (\mathbb{P}_1 \backslash 0)$, by means of $(f, \sigma, \tau) \mapsto (f, \partial_z f + \sigma, \tau)$), and

ii) $\mathcal{I}(X) = 0 = \mathcal{J}(X)$ and $\mathbb{C} = \mathcal{H}(X) \subset \mathcal{O}(X) \oplus \mathcal{I}(X) \oplus \mathcal{J}(X)$.

§ 51 A Supplement: Countable Topology in Complex Spaces

In this supplement, we give conditions under which complex spaces are paracompact. That is of value for the applicability of Čech cohomology, and also for the construction of Fréchet topologies on section-modules of analytic sheaves (see § 55).

51 A.1 Remark. *Every complex space X satisfies the following conditions:*

i) X is locally compact.

ii) X is locally connected, and thus has open connected components (see also E.48n).

iii) Every relatively compact subset of X has a countable topology.

Proof. Local models, being locally closed subspaces of complex number spaces, satisfy i) and iii), and it follows that iii) holds in general. By 49.2, statement ii) holds near irreducible points; by 49.5, X is locally a finite union of irreducible components, and ii) follows by induction from 46.1 and the following simple fact: if A and B are connected subsets of X such that $A \cap B \neq \emptyset$, then $A \cup B$ is connected as well. ■

E. 51 A a. Show that, for a closed subspace A of a complex space X, every connected component of A is a closed subspace of X (for E. 63a).

A Hausdorff space T is called *paracompact* if every open cover $\mathfrak{U} = (U_j)_{j \in J}$ has a *locally finite open refinement* $\mathfrak{V} = (V_i)_{i \in I}$ (i.e., there exists a mapping $\tau: I \to J$ with $V_i \subset U_{\tau i}$, and every point has a neighborhood W that meets only finitely many V_i). *A* locally compact space T is called *countable at infinity* if it has a (countable) "relatively compact" exhaustion $T = \bigcup_{j=1}^{\infty} U_j$ with open sets $U_j \subset\subset U_{j+1}$ (so that, in the one-point compactification of T, the additional point ∞ has a countable fundamental system of neighborhoods).

The following result can be applied to connected components of complex spaces:

51 A. 2 Proposition. *Let T be a connected locally compact space in which every compact subset has a countable topology. Then the following conditions are equivalent:*

i) T is metrizable;
ii) T is paracompact;
iii) T is countable at infinity;
iv) T has a countable topology.

Proof. i) $\Rightarrow$ ii) [Bou GT IX, §45 Th. 4]. $\Rightarrow$ iii) [Bou GT I, §9.10 Th. 5]. iii) $\Rightarrow$ iv) Open relatively compact sets in T have a countable topology; thus, T, being a countable union of such sets, must also have a countable topology. iv) $\Rightarrow$ i) [Bou GT IX, Cor. to Prop. 16]. ■

E. 51 Ab. Let S and T be spaces as in 51 A.2, and suppose that $f: S \to T$ is proper. Show that S has a countable topology if T does (for E. 51 Ad).

E. 51 Ac. *A connected complex surface without a countable topology* (Calabi-Rosenlicht). Consider the disjoint union $X = \dot{\bigcup}_{s \in \mathbb{C}} \mathbb{C}^2 \times s$ as a topological sum, and introduce the following equivalence relation R on X:

$$(x, y, s) \sim (x', y', s') \Leftrightarrow \begin{cases} y = y' \\ xy + s = x'y' + s' \\ x = x', \quad \text{if} \quad s = s'. \end{cases}$$

Prove the following statements: i) R is an open equivalence relation on X. ii) $M := X/R$ is a Hausdorff space. iii) M is canonically a connected complex manifold. iv) $T := \{(0, 0, s);\ s \in \mathbb{C}\}$ is a set of pairwise nonequivalent points in X; for each $s \in \mathbb{C}$, there exists an R-saturated open neighborhood U_s in X such that $T \cap U_s = \{(0, 0, s)\}$. v) The topology of M is not countable. vi) Set $U := \mathbb{C}^2 \times 0 \subset M$. The mapping $\varphi: U \to \mathbb{C}^2, (x, y) \mapsto (xy, y)$, has a holomorphic extension $\Phi: M \to \mathbb{C}^2$ that induces an isomorphism $\Phi^0: {}_2\mathcal{O}(\mathbb{C}^2) \to {}_M\mathcal{O}(M)$. (Hint: $\sum_{j=0}^{\infty} f_j(y) x^j \in \mathcal{O}(U)$ is holomorphically extendible to M iff $o(f_j) \geq j$ at the point 0 for every j).

Now we show that every connected holomorphically separable space has a countable topology. In the singular case, we make use of the existence of a normalization, the construction of which, in § 71, relies on 61.8.

51 A. 3 Theorem. *Every connected holomorphically spreadable complex space X has a countable topology.*

Proof. By E.51e i), we may assume that X is *reduced*. We shall show that there is no loss of generality in assuming that X is irreducible, and even normal; then we prove the existence of a countable family in $\mathcal{O}(X)$ that induces a discrete mapping $X \to \mathbb{C}^{\mathbb{N}}$. The assertion follows by 51 A.4.

Step 1. If each of the irreducible components of X has a countable topology, then X has at most countably many irreducible components.

Hence, in the succeeding steps, we may assume that X is *irreducible*: let $\mathfrak{A} = \{X_j; j \in J\}$ be the set of irreducible components; then the canonical holomorphic mapping $\dot{\bigcup}_{j \in J} X_j \to X$ is finite by 49.5, and we can apply E 33 Bg.

Proof. Each $X_j \in \mathfrak{A}$ meets at most countably many X_k: by 51 A.2, X_j is a countable union of relatively compact sets, each of which meets only finitely many $X_i \in \mathfrak{A}$, by 49.5. Moreover, each pair of components can be joined with a finite chain in $\mathfrak{A}$: for a fixed $X_0 \in \mathfrak{A}$, set

$$\mathfrak{A}_k = \{X_j \in \mathfrak{A};\ \exists X_{j_0} = X_0, \ldots, X_{j_k} = X_j \ \text{ in } \mathfrak{A} \text{ with } \ X_{j_{i-1}} \cap X_{j_i} \neq \emptyset;\ \ i = 1, \ldots, k\}.$$

Every union of irreducible components is a closed (in fact, analytic) subset of X, so the connectedness of X implies that $\mathfrak{A} = \bigcup_{k=0}^{\infty} \mathfrak{A}_k$. By inducting on k, we see that each $\mathfrak{A}_k$ is at most countably infinite: for $\mathfrak{A}_0 = \{X_0\}$, that is trivial, and the induction "$k \Rightarrow k+1$" follows immediately from the fact that

$$\mathfrak{A}_{k+1} = \bigcup_{X_i \in \mathfrak{A}_k} \{X_j \in \mathfrak{A};\, X_i \cap X_j \neq \emptyset\}.$$

Step 2. A reduced complex space has a countable topology iff its normalization has a countable topology.

Since the normalization-mapping is finite and surjective, that follows from E. 33 Bg and 51 A.4.

Due to the fact that the normalization preserves the irreducibility and holomorphic spreadability of X (E. 51e), we may assume that X is *normal.* For both the next step and the completion of the proof, we use the following (purely topological) result:

51 A. 4 Lemma (Poincaré-Volterra). *Let X be a connected complex space. If there exists a continuous discrete mapping $f: X \to T$ into a Hausdorff space T with a countable topology, then the topology of X is countable, as well.*

If $\mathfrak{B}$ is a basis of the topology of T, one can see as in 33 B.2 ii) that the set of relatively compact connected components of the inverse images $f^{-1}B$ with $B \in \mathfrak{B}$ is a basis for X. It can be shown in a manner similar to Step 1 that this basis is countable if $\mathfrak{B}$ is countable (see [Fo 23.2]). ■

Step 3. If X is connected, normal, and holomorphically spreadable, then there exists a countable subset $F \subset \mathcal{O}(X)$ that is dense in the topology of compact convergence.

Proof. By the Second Riemann Removable Singularity Theorem for normal spaces, 71.12, the restriction-mapping $\mathcal{O}(X) \to \mathcal{O}(X \setminus S(X))$ is a topological isomorphism, since the set of singularities $S(X)$ has codimension at least two, by 74.3 i). Hence, by 49.7, we may assume that X is a connected *manifold.*

If $A \hookrightarrow X$ is a proper analytic subset, then, according to the First Riemann Removable Singularity Theorem 7.3, $\mathcal{O}(X)$ is a topological subspace of $\mathcal{O}(X \setminus A)$; hence, it suffices to show that $\mathcal{O}(X \setminus A)$ is metrizable and has a countable topology for an appropriate A. For then $\mathcal{O}(X)$ has those properties, and the existence of a countably dense subset follows.

Since X is holomorphically spreadable, there exists, for each point $a \in X$, a holomorphic mapping $h: X \to \mathbb{C}^k$ that is discrete at a, by E. 51c iii). According to 45.4, h is discrete near a; consequently, the semicontinuity of the rank (E. 8f) yields that the set

$$A := \left\{ b \in X;\, \operatorname{rank} \frac{\partial h}{\partial z}(b) < \dim X \right\}$$

is a proper analytic subset of X. By 8.7, h is discrete on the connected manifold $X \setminus A$, and it follows from 51 A. 4 that the topology of $X \setminus A$ is countable. Hence, in the topology of compact convergence, $\mathscr{C}(X \setminus A)$ is a Fréchet space (see § 2); it is thus metrizable, and, by [Bou GT X § 3.3 cor.], has a countable topology; finally, the same must hold for the closed subspace $\mathcal{O}(X \setminus A)$. ■

It is clearly sufficient for this step that X be normal and holomorphically spreadable at one point.

The proof of 51 A. 3 follows now from 51 A. 4 and the following:

Step 4. If X is holomorphically spreadable, then a countable dense set $F \subset \mathcal{O}(X)$ determines a continuous discrete mapping $X \to \mathbb{C}^{\mathbb{N}}$, $x \mapsto (f(x))_{f \in F}$.

Proof. We show that, for each $a \in X$, there exists a holomorphic mapping $X \to \mathbb{C}^{k_a}$ that is discrete at a, and whose component functions lie in F. By E. 51c iii), we can find a holomorphic mapping $h: X \to \mathbb{C}^{k_a}$ that is discrete at a. Now there is a relatively compact local model U at the point a in X such that $h^{-1}(h(a)) \cap \partial U = \emptyset$, and there is a sequence of holomorphic mappings ${}_jg: X \to \mathbb{C}^{k_a}$, with components in F, that approximate h on $\bar{U}$. Then some ${}_jg$ is discrete at a: otherwise, for every $j \in \mathbb{N}$, there would exist an $x_j \in {}_jg^{-1}({}_jg(a)) \cap \partial U$; an accumulation point x of the sequence (x_j) would then lie in $h^{-1}(h(a)) \cap \partial U$, in contradiction to the choice of U. ■■

E. 51 Ad. Show that *every connected holomorphically convex space has a countable topology* (hint: E. 51 Ab and 57.11; for 63 B.1).

E. 51 Ae. Prove Rado's Theorem: *every connected complex curve has a countable topology* (hint: 51.3 iv)).

§ 52 Theorem B and B-Spaces

In this section, we investigate those subsets of complex spaces whose analytic cohomology is trivial. That property has great significance in multidimensional function theory, because analytic cohomology classes often appear as obstructions in the transition from local statements to global ones. The vanishing of higher analytic cohomology is thus a fundamental property of Stein spaces:

Theorem B. *If $\mathcal{F}$ is a coherent analytic sheaf on a Stein space X, then*

$$H^q(X, \mathcal{F}) = 0 \quad \textit{for every} \quad q \geqq 1 .$$

We prove that result in Chapter 6. The consequences of the vanishing of higher analytic cohomology, which we shall discuss now, are also meaningful for subsets of complex spaces, in particular for Stein compacta. Since we shall need such results also for non-open subsets in the proof of Theorem B, we introduce the following notation for a subset L of a complex space X:

52.1 Definition. *We call L a B-set in X if, for every coherent analytic sheaf $\mathcal{F}$ defined near L, and every $q \geqq 1$, $H^q(L, \mathcal{F}) = 0$. Open B-sets in X are also called B-spaces.*

Recall that, for a subset L of a complex space, "near L" means "in an appropriately small open neighborhood of L", and that $\mathcal{F}(L) = \varinjlim_{L \subset U \Subset X} \mathcal{F}(U)$ (see 50.14) if X is a metrizable space.

52.2 Remarks. i) *B-spaces are Stein spaces (see 52.7).*

ii) If X is metrizable and $\mathcal{F}$ defined on an open neighborhood U of L, then "$H^q(L, \mathcal{F}) = 0$" means that if $\omega \in H^q(V, \mathcal{F})$ with $L \subset V \Subset U$, then there exists a W with $L \subset W \Subset V$ such that $\omega|_W = 0$ in $H^q(W, \mathcal{F})$.

52.3 Examples. i) If L is discrete and closed, then L is a B-set, by 50.12.

ii) If L has a fundamental system of neighborhoods in X consisting of open metrizable B-sets, then L is a B-set, by 50.14 (in particular, it follows from Theorem B that Stein compacta are B-sets).

iii) For B-sets L_0, L_1 in X, their union $L_0 \cup L_1$ need not be a B-set, even if $L_0 \cap L_1$ is a B-set: consider $L_j = U_j \cong \mathbb{C}$ in $\mathbb{P}_1$; by Theorem B, U_j and $U_{01} \cong \mathbb{C}^*$ are B-sets (see E. 52a).

iv) In the Exhaustion Theorem 56.2, a condition is given, under which a nested union of B-sets is a B-set.

v) For a B-set L in X, if $A \hookrightarrow X$, then $L \cap A$ is a B-set in A: if $\mathscr{G}$ is a coherent ${}_A\mathcal{O}$-module near $L \cap A$, then the trivial extension of $\mathscr{G}$ near L is a coherent ${}_X\mathcal{O}$-module (see 41.21); hence, the assertion follows from 50.13.

vi) Compact B-spaces are discrete (by 52.6 and E. 51a).

vii) $\mathbb{C}^{2*}$ is not a B-space: set $U_0 = \mathbb{C}^* \times \mathbb{C}$ and $U_1 = \mathbb{C} \times \mathbb{C}^*$, then the alternating cohomology class $s \in H^1((U_j), \mathcal{O}) \subset H^1(\mathbb{C}^{2*}, \mathcal{O})$ determined by $s_{01} := \dfrac{1}{z_0 z_1} \in \mathcal{O}(U_{01})$ is nonzero: otherwise, there would exist functions $f_j \in \mathcal{O}(U_j)$ with $f_1 - f_0 = \dfrac{1}{z_0 z_1}$ on U_{01}; it would then follow that

$$\mathcal{O}(U_0) \ni h := z_1 f_0 + \frac{1}{z_0} = z_1 \left(f_0 + \frac{1}{z_0 z_1} \right) = z_1 f_1 \in \mathcal{O}(U_1),$$

and thus that $h \in \mathcal{O}(\mathbb{C}^{2*}) = \mathcal{O}(\mathbb{C}^2)$, although $\lim_{k \to \infty} h(1/k, 0) = \infty$ ↯.

E. 52a. $\mathbb{P}_1$ *is not a* B-*space.* Prove the following statements for the ${}_{\mathbb{P}_1}\mathcal{O}$-ideal $\mathscr{I} = {}_{\{[1,0]\}}\mathscr{I}^2$ unsing the cover $\mathfrak{U} = (U_0, U_1)$ from 32.4 iv):

i) $\mathscr{I}(U_{01}) = \mathcal{O}(U_{01}) = \left\{ \sum_{j=-\infty}^{\infty} a_j z^j \text{ convergent for } z \in \mathbb{C}^* \right\} = Z_a^1(\mathfrak{U}, \mathcal{O}) = Z_a^1(\mathfrak{U}, \mathscr{I})$.

ii) Since $z_0 = 1/z_1$ on U_{01}, we have that $B_a^1(\mathfrak{U}, \mathcal{O}) = \mathcal{O}(U_{01})$ and

$$B_a^1(\mathfrak{U}, \mathscr{I}) = \left\{ \sum_{j=-\infty}^{\infty} a_j z^j \in \mathcal{O}(U_{01});\ a_1 = 0 \right\}.$$

iii) $H^p(\mathfrak{U}, {}_{\mathbb{P}_1}\mathcal{O}) = 0$ for all $p \geqq 1$, and $H^p(\mathfrak{U}, \mathscr{I}) = \begin{cases} \mathbb{C}, & p = 1 \\ 0, & p \neq 1. \end{cases}$

Remark: By Theorem B, $\mathfrak{U}$ is a Leray cover of $\mathbb{P}_1$, so that $H^*(\mathfrak{U}, \mathscr{F}) = H^*(\mathbb{P}_1, \mathscr{F})$ for coherent modules $\mathscr{F}$ on $\mathbb{P}_1$. One can show in a similar fashion that $H^q(\mathbb{P}_n, \mathcal{O}) = 0$ for every $q \geqq 1$ (see E. 74f).

Now we show that a B-set L has so many global holomorphic functions that it is holomorphically separable and holomorphically convex, and that there exists, for each point in L, local coordinates by global holomorphic functions. The essential concept in the proof of that is the "exactness of the section-functor" on B-set, and the most important result is the Existence Theorem 52.5.

52.4 Lemma. *Let L be a B-set in a complex space X, and A, a closed subspace of an* *open neighborhood U of L. Then, for each holomorphic function* $g \in {}_A\mathcal{O}(A \cap L)$, *there exists an* $f \in \mathcal{O}(L)$ *with* $f|_{A \cap L} = g$[2].

Proof. We may assume that $A = V(X; \mathcal{I})$. By 50.13, $H^0(L, {}_X\mathcal{O}/\mathcal{I}) = H^0(A \cap L, {}_X\mathcal{O}/\mathcal{I}) = {}_A\mathcal{O}(A \cap L)$; the assertion follows immediately from the exact sequence $H^0(L, \mathcal{O}) \to \to H^0(L, \mathcal{O}/\mathcal{I}) \to H^1(L, \mathcal{I}) = 0$. ∎

In particular, if A is the disjoint union of a closed subspace A_1 and a discrete subspace A_2 that consists of k_j-fold points a_j, then a holomorphic function g on A_2 is given by means of a family of "Taylor polynomials" $g_j \in {}_X\mathcal{O}_{a_j}/\mathfrak{m}_{a_j}^{k_j+1}$; and $f|_{A_2} = g$ iff $f = g_j$ in ${}_X\mathcal{O}_{a_j}/\mathfrak{m}_{a_j}^{k_j+1}$ for every j (approximation of the order k_j in the Krull topology). Hence, the following result generalizes the Weierstrass Factorization Theorem:

52.5 Existence Theorem for Global Holomorphic Functions. *If A is a closed subspace of a B-space X, then, for each* $g \in {}_A\mathcal{O}(A)$, *there exists an* $f \in {}_X\mathcal{O}(X)$ *with* $f|_A = g$. *If* A' *is a closed discrete subset of X such that* $A' \cap A = \emptyset$, *then arbitrary values can be prescribed for f at every point of* A'. ∎

E. 52b. For a B-space X, if $A \hookrightarrow X$ is an analytic subset of dimension $n \geqq 1$, then there exists an $f \in \mathcal{O}(X)$ such that $\dim N(A; f-c) \leqq n-1$ for every $c \in \mathbb{C}$ (hint: in every irreducible component of A, choose two different points. For 57.1).

E. 52c. Show that 52.4 does not hold for $L = \mathbb{P}_1$.

After 52.5, the next most important global statement about B-sets is the following:

52.6 Proposition. *For a subset* $L \subset X$, *the condition*

i) L is a B-set

implies the validity of the conditions

ii) $H^1(L, \mathcal{I}) = 0$ *for every coherent sheaf of ideals* $\mathcal{I}$ *defined near L such that* $N(\mathcal{I})$ *is closed and discrete near L;*

iii) for every exact sequence

$$0 \to \mathcal{G}' \to \mathcal{G} \to \mathcal{G}'' \to 0$$

of coherent analytic sheaves defined near L, the sequence of sections

$$0 \to \mathcal{G}'(L) \to \mathcal{G}(L) \to \mathcal{G}''(L) \to 0$$

is exact (exactness of the "analytic" section-functor).

Any one of the conditions i), ii), iii) implies that

iv) L is holomorphically (i.e., $\mathcal{O}(L)$-*)separable, and L is holomorphically (i.e.,* $\mathcal{O}(L)$-*) convex if L is metrizable.*

Theorem B yields this consequence:

[2] Recall that $f|_A = i^0(f)$ with the inclusion $i: A \hookrightarrow U$.

52.7 Remark. *If L is open in X and metrizable*[3], *then conditions i) – iv) of 52.6 are equivalent; in particular, metrizable*[3] *B-spaces are Stein spaces.* ■

Proof of 52.6. The implication "i) ⇒ ii)" is trivial, and the induced exact cohomology sequence yields that i) ⇒ iii).

ii) or iii) ⇒ iv). For a two-point set $\{a, b\} \subset L$, let $\mathscr{I} \subset \mathcal{O}$ be the nullstellen ideal. By ii) or iii), the exact cohomology sequence implies that

$$\mathcal{O}(L) \to (\mathcal{O}/\mathscr{I})(L)$$

is surjective; that, coupled with the fact that $(\mathcal{O}/\mathscr{I})(L) \cong \mathbb{C} \oplus \mathbb{C}$, implies the separability.

Suppose now that the $\mathcal{O}(L)$-convex hull $\hat{K}$ is not a compact subset of L for some compactum $K \subset L$; then there exists a closed discrete sequence of points $A = \{a_j; j \in \mathbb{N}\}$ in $\hat{K}$. Moreover, A is closed in an open neighborhood U of L, for it has no accumulation point in the closed subset $\hat{K}$ of L, and thus has no accumulation point in L. As a result, we may assume that $U = X \setminus (\bar{A} \setminus A)$. Then A is analytic in U, and, for the nullstellen ideal $\mathscr{I}$ of A, the homomorphism $\mathcal{O}(L) \to (\mathcal{O}/\mathscr{I})(L)$ is surjective, by assumption. Since $(\mathcal{O}/\mathscr{I})(L) = \mathbb{C}^{\mathbb{N}}$, there exists an $f \in \mathcal{O}(L)$ such that $f(a_j) = j \in \mathbb{N}$, in contradiction to the fact that $\|f\|_{\hat{K}} = \|f\|_K < \infty$. ■

At a regular point of a complex space X, local coordinates form a local embedding into a complex number space. More generally, for an arbitrary point $a \in X$, we call a system $f_1, \ldots, f_m \in {}_X\mathcal{O}_a$ <u>local coordinates of X at a</u> if $(f_1, \ldots, f_m): X_a \to \mathbb{C}^m_0$ is an embedding. By 45.8 ii) and 45.9, every (finite) system of generators of ${}_X\mathfrak{m}_a$ provides local coordinates.

For B-sets, we have this "quasi-global" statement:

52.8 Corollary. *If K is a compact subset of a B-set L in X, then there exists a holomorphic mapping $f: L \to \mathbb{C}^n$ that is injective near K, and that provides local coordinates (for X) at each point near K.*

Proof. Fix $a \in K$, and let $\varphi_1, \ldots, \varphi_{r_a}$ be a system of generators of ${}_X\mathfrak{m}_a$ over ${}_X\mathcal{O}_a$. By 52.4, there exist functions $g_1, \ldots, g_{r_a} \in \mathcal{O}(L)$ such that $g_j - \varphi_j \in {}_X\mathfrak{m}_a^2$, and thus such that ${}_ag = (g_1, \ldots, g_{r_a}): L \to \mathbb{C}^{r_a}$ is an embedding at a (and thus in a neighborhood U_a), by 45.8 ii). Since K is compact, it can be covered with a finite collection $U_{a_1}, \ldots, U_{a_s}$ of such neighborhoods. According to E. 45p, then, $g = ({}_{a_1}g, \ldots, {}_{a_s}g)$ is an embedding on each U_{a_j}, and therefore separates every pair (x, y) in $V := \bigcup_{j=1}^{s} U_{a_j} \times U_{a_j}$ with $x \neq y$. We need to find an extension of g that separates pairs in $K \times K \setminus V$, as well. By 52.6, L is holomorphically separable; equivalently, $L \times L \setminus \Delta_L =$

---

[3] The assumption of metrizability is superfluous: by 51 A.3 and 51 A.2, condition iv) implies that L is metrizable.

$\bigcup_{h \in \mathcal{O}(L)} (h \times h)^{-1}(\mathbb{C} \times \mathbb{C} \setminus \Delta_{\mathbb{C}})$. The compact set $K \times K \setminus V$ can be covered with finitely many open sets of the form $(h_j \times h_j)^{-1}(\mathbb{C} \times \mathbb{C} \setminus \Delta_{\mathbb{C}})$. The associated mapping $f = (g, h_1, \ldots, h_t)$ is injective near K, so it follows from E. 45p and E. 45f β) that f is an embedding. ■

Since every system of generators of ${}_X\mathfrak{m}_a$ includes one of minimal length $\operatorname{emb}_a X$ (see 23 A. 4), the preceding proof has the following additional consequence:

52.9 Remark. *If $f_1, \ldots, f_m \in \mathcal{O}(L)$ yield local coordinates at $a \in L$, then it is possible to select from them a system of local coordinates of length* $\operatorname{emb}_a X$. ■

The next result provides a bridge from punctual to global ideal theory:

52.10 Corollary. *Let $L \subset X$ be a B-set. For each coherent analytic sheaf $\mathcal{G}$ defined near L, if the sections $s_1, \ldots, s_m \in \mathcal{G}(L)$ generate every stalk $\mathcal{G}_x$, $x \in L$ (i.e., if $\mathcal{G}_x = \sum_{j=1}^{m} \mathcal{O}_x \cdot s_{jx}$), then they generate $\mathcal{G}(L)$, as well (i.e., $\mathcal{G}(L) = \sum_{j=1}^{m} \mathcal{O}(L) \cdot s_j$).*

Proof. By assumption, the homomorphism $\varphi: \mathcal{O}^m \to \mathcal{G}$, $(f_1, \ldots, f_m) \mapsto \sum_{j=1}^{m} f_j \cdot s_j$, is surjective on L; by 41.17, it is then surjective near L, and we conclude from 52.6 that $\varphi(L)$ is surjective. ■

52.11 Corollary. *If $L \subset X$ is a B-set, and if the functions $f_1, \ldots, f_m \in \mathcal{O}(L)$ have no common zeros in L, then $\mathcal{O}(L) = \sum_{j=1}^{m} \mathcal{O}(L) \cdot f_j$; in particular, there exist functions $g_1, \ldots, g_m \in \mathcal{O}(L)$ such that $1 = \sum_{j=1}^{m} f_j g_j$.*

Proof. If $f_j(a) \neq 0$, then f_j is a unit in $\mathcal{O}_a$; hence, 52.10 is applicable, with $s_j = f_j$ and $\mathcal{G} = \mathcal{O}$. ■

The preceding result does not hold for arbitrary L:

52.12 Example. The following conditions are equivalent for $L \Subset \mathbb{C}^n$:

i) L is holomorphically convex.
ii) L is a B-set.
iii) If $f_1, \ldots, f_m \in \mathcal{O}(L)$ generate every stalk $\mathcal{O}_x$, then they generate $\mathcal{O}(L)$.

Proof. i) $\Rightarrow$ ii) This follows from Theorem B, since L is holomorphically separable.
ii) $\Rightarrow$ iii) Apply 52.10.

iii) $\Rightarrow$ i) If L is not holomorphically convex, then, by 12.8, there exists a point $a \in \partial L$ at which every $f \in \mathcal{O}(L)$ can be extended holomorphically; without loss of generality, suppose that $a = 0$. The coordinate functions $z_1, \ldots, z_n$ generate every stalk $\mathcal{O}_x$, so there exist functions $h_1, \ldots, h_n \in \mathcal{O}(L)$ such that $1 = \sum_{j=1}^{m} z_j h_j$ on L, and thus also at 0; but that equation cannot hold at 0! ↯ ∎

In the historical development of complex analysis, the following property of Stein spaces has played a mayor role:

52.13 Theorem A. *Fix a point a in a B-set $L \subset X$, and let $\mathcal{G}$ be a coherent analytic sheaf defined near L. Then the $\mathcal{O}_a$-module $\mathcal{G}_a$ is generated by $m = \operatorname{corank}_{\mathcal{O}_a} \mathcal{G}_a$ global sections $s_1, \ldots, s_m \in \mathcal{G}(L)$.*

Proof. The nullstellen ideal $\mathcal{I}$ of the point a induces an exact sequence

$$0 \to \mathcal{I} \cdot \mathcal{G} \to \mathcal{G} \to \mathcal{G}/\mathcal{I} \cdot \mathcal{G} \to 0 .$$

By 41.13, $\mathcal{I} \cdot \mathcal{G}$ is coherent; moreover, $\operatorname{supp}(\mathcal{G}/\mathcal{I} \cdot \mathcal{G}) \subset \{a\}$, and it follows that $(\mathcal{G}/\mathcal{I} \cdot \mathcal{G})(L) = \mathcal{G}_a / \mathfrak{m}_a \cdot \mathcal{G}_a$. By 23 A.4, $s_1, \ldots, s_m \in \mathcal{G}_a$ generate the finitely-generated $\mathcal{O}_a$-module $\mathcal{G}_a$ iff $\bar{s}_1, \ldots, \bar{s}_m$ generate the $\mathcal{O}_a/\mathfrak{m}_a = \mathbb{C}$-vectorspace $\mathcal{G}_a/\mathfrak{m}_a \cdot \mathcal{G}_a$; by E. 23A b, we may choose $m = \operatorname{corank}_{\mathcal{O}_a} \mathcal{G}_a$. Hence, the assertion follows from the surjectivity of $\mathcal{G}(L) \to (\mathcal{G}/\mathcal{I})(L)$. ∎

52.14 Corollary. *Let L be a B-set in X and $V(X; \mathcal{I})$, a closed subspace. Then $L \cap V(X; \mathcal{I})$ can be described by means of global holomorphic functions on L:*

i) Every stalk $\mathcal{I}_x$, $x \in L$, is generated over $\mathcal{O}_x$ by $\mathcal{I}(L)$.

ii) $N(L; \mathcal{I}) = \bigcap_{f \in \mathcal{I}(L)} N(L; f)$.

Proof. Statement i) is a particular case of Theorem *A*.

ii) The inclusion "$\subset$" is trivial. Conversely, if $a \in L \backslash N(L; \mathcal{I})$, then $1_a \in \mathcal{I}_a$, and i) implies the existence of sections $f_1, \ldots, f_m \in \mathcal{I}(L)$ and functions $h_1, \ldots, h_m \in \mathcal{O}_a$ such that $1 = \sum_{j=1}^{m} h_j f_j$; hence, $f_j(a) \neq 0$ for at least one j, and $a \notin N(L; f_j)$. ∎

52.15 Remark. In a Stein space X, every analytic set has a representation of the form $N(X; f_1, \ldots, f_{\dim X})$ (see E. 56e).

E. 52d. Without the condition that L be a B-set, 52.13 and 52.14 would be false.

By definition, every coherent analytic sheaf can be represented *locally* as the cokernel of a homomorphism of free sheaves. On B-sets, the following "quasi-global" version holds:

52.16 Corollary. *Let $L \subset X$ be a B-set, and suppose that $\mathcal{G}$ is a coherent analytic sheaf defined near L. Then, for each $K \subset\subset L$, and each $m \in \mathbb{N}$, there is a sequence of sheaves*

$$\mathcal{O}^{p_m} \to \mathcal{O}^{p_{m-1}} \to \ldots \to \mathcal{O}^{p_0} \to \mathcal{G} \to 0$$

near L that is exact near K.

Proof. Without loss of generality, suppose that K is compact. For an induction on m, it suffices to find a homomorphism $\varphi: \mathcal{O}^p \to \mathcal{G}$ near L that is surjective near K (since the coherence of $\mathcal{G}$ implies that of $\mathcal{K}er\,\varphi$). By Theorem A, for each $a \in K$, there exist generators $s_1, \ldots, s_r \in \mathcal{G}(L)$ of the $\mathcal{O}_a$-module $\mathcal{G}_a$; they also generate $\mathcal{G}$ over $\mathcal{O}$ near a. In that manner, we construct a finite cover of the compact set K and thus obtain finitely many sections $t_1, \ldots, t_p \in \mathcal{G}(L)$ that generate $\mathcal{G}$ over $\mathcal{O}$ near K. We conclude that the sheaf-homomorphism

$$\varphi: \mathcal{O}^p \to \mathcal{G},\ (f_1, \ldots, f_p) \mapsto \sum_{j=1}^{p} f_j t_j,$$

which is defined near L, is surjective near K. ■

52.17 Corollary. *A holomorphic function f on a B-space X is a non-zero-divisor in $\mathcal{O}(X)$ iff f_x is a non-zero-divisor in $\mathcal{O}_x$ for every $x \in X$.*

Proof. The sheaf of ideals $\mathcal{I} = (0):(f)$ is a coherent $\mathcal{O}$-module, by E. 41g. Furthermore, $\mathcal{I}_x = 0$ iff f_x is not a zero-divisor in $\mathcal{O}_x$, and $\mathcal{I}(X) = 0$ iff f is not a zero-divisor in $\mathcal{O}(X)$. It follows from Theorem A that $\mathcal{I}(X) = 0$ iff $\mathcal{I} = 0$. ■

E. 52e. For the „*Achsenkreuz*“ X of $\mathbb{C} \times \mathbb{P}_1$, find a non-zero-divisor $f \in {}_X\mathcal{O}(X)$ such that f_0 is a zero-divisor in ${}_X\mathcal{O}_0$.

By E. 51e, a complex space that is finite over a Stein space is holomorphically convex and holomorphically spreadable. With Theorem B, the following proposition shows that such a space is in fact a Stein space:

52.18 Proposition. *Let $\varphi: X \to Y$ be a finite holomorphic mapping. If Y is a B-space, then X is a B-space, as well.*

Proof. For a coherent analytic sheaf $\mathcal{G}$ on X, we need to show that $H^q(X, \mathcal{G}) = 0$ for every $q \geqq 1$. According to the Finite Coherence Theorem, $\varphi\mathcal{G}$ is coherent, so it suffices to verify that $H^q(X, \mathcal{G}) = H^q(Y, \varphi\mathcal{G})$. For the canonical acyclic resolution $\mathcal{C}^*$ of $\mathcal{G}$, 45 A.3 and 50.7 imply that $\varphi\mathcal{C}^*$ is an acyclic resolution of φ, since the image sheaf functor φ is exact, by 45 A. 1. It follows from 50.4 that

$$H^q(X, \mathcal{G}) = H^q(\mathcal{C}^*(X)) = H^q((\varphi\mathcal{C}^*)(Y)) = H^q(Y, \varphi\mathcal{G}). \quad ■$$

E. 52f. Generalize 52.18 to B-sets in Y.

Conversely, for a finite holomorphic mapping $\varphi: X \to Y$, if X is a Stein space and φ, surjective, then Y is also a Stein space (73.1). At this point, we derive a special case from Theorem B:

52.19 Proposition. *A complex space X is Stein iff* Red X *is Stein.*

Proof. Only if. This implication follows from 51.4 and the fact that Red $X \hookrightarrow X$.

If. Let $\mathcal{N} \subset \mathcal{O}$ be the sheaf of nilpotent elements. It suffices to show that the natural homomorphism

$$\varrho_X : \mathcal{O}(X) \to (\mathcal{O}/\mathcal{N})(X) \underset{47.2}{=} {}_{(\mathrm{Red}\,X)}\mathcal{O}(X)$$

is surjective; then the fact that Red X is holomorphically separable and holomorphically convex implies that X has the same properties.

The infinitesimal neighborhoods $V(X;\mathcal{N}^j)$ of $V(X;\mathcal{N})$ enable us to construct a projective limit $\varprojlim \mathcal{O}/\mathcal{N}^j$; it coincides with $\mathcal{O}$, since every connected component of X is countable at infinity, by 51 A.2, and since $\mathcal{N}^j = 0$ for sufficiently large j on each compact subset K (that holds for each stalk $\mathcal{N}_a$, and thus near K by 41.16). In particular, $\mathcal{O}(X) = [\varprojlim \mathcal{O}/\mathcal{N}^j](X) = \varprojlim [(\mathcal{O}/\mathcal{N}^j)(X)]$, since the section-functor commutes with projective limits. If we can demonstrate that every homomorphism

$$(\mathcal{O}/\mathcal{N}^{j+1})(X) \to (\mathcal{O}/\mathcal{N}^j)(X)$$

is surjective, then so is ϱ_X. We have exact sequences

$$0 \to \mathcal{N}^j/\mathcal{N}^{j+1} \to \mathcal{O}/\mathcal{N}^{j+1} \to \mathcal{O}/\mathcal{N}^j \to 0,$$

so that it suffices to prove that $H^1(X, \mathcal{N}^j/\mathcal{N}^{j+1}) = 0$. By 41.20, the fact that $\mathcal{N}\cdot(\mathcal{N}^j/\mathcal{N}^{j+1}) = 0$ implies that the coherent $\mathcal{O}$-module $\mathcal{N}^j/\mathcal{N}^{j+1}$ is a coherent ${}_{(\mathrm{Red}\,X)}\mathcal{O}$-module, so the assertion follows from an application of Theorem B to Red X. ∎

We close the section with two further applications of Theorem B. The first is a generalization of the Kugelsatz 7.8:

52.20 Hartogs's Kugelsatz. *Fix $n \geqq 2$, and let K be a compact subset of a region $X \subset\subset \mathbb{C}^n$ such that $X \backslash K$ is connected. Then the inclusion $i: X\backslash K \subset\subset X$ induces an isomorphism of topological algebras $i^0 : \mathcal{O}(X) \to \mathcal{O}(X\backslash K)$.*

Proof. The family $\mathfrak{U} := (\mathbb{C}^n\backslash K, X)$ forms an open cover of $\mathbb{C}^n$; a fixed $f \in \mathcal{O}(X\backslash K)$ determines a cocycle in $Z^1(\mathfrak{U}, \mathcal{O})$ that is a coboundary, by Theorem B and the fact that $H^1(\mathfrak{U}, {}_n\mathcal{O}) \hookrightarrow H^1(\mathbb{C}^n, {}_n\mathcal{O}) = 0$ (50.10).

Hence, there exist an $h \in \mathcal{O}(X)$ and a $g \in \mathcal{O}(\mathbb{C}^n\backslash K)$ such that $f = (g-h)|_{X\backslash K}$. By 7.8, g can be extended holomorphically to $\mathbb{C}^n$, since $\mathbb{C}^n\backslash K$ is connected, and we may apply the Identity Theorem to conclude that $F := g - h$ is a holomorphic extension of f to X. Obviously, X must be a domain, and it follows that i^0 is an isomorphism with respect to the algebraic structure; by 3.4, i^0 is continuous, and, by 55.8 ii), it is open. ∎

E. 52 . Use 52.20 to show that, for $n \geqq 2$, the sphere $S^{2n-1} \subset \mathbb{C}^n$ is not a Stein compactum.

52.21 Proposition. *On paracompact complex spaces, there exist arbitrarily fine locally finite covers that are Leray covers for every coherent analytic sheaf.*

Proof. By 51.6, there exist arbitrarily fine locally finite Stein covers; the proposition follows from 51.7 and Theorem B. ■

E. 52h. *Example of a holomorphically spreadable space that is not holomorphically separable* (after C. Horst). Construct a space $(X, \mathscr{H})$ with $|X| \subset |\mathbb{C}^2|$ and $\mathscr{H}(X) = \mathbb{C}\{z_1, z_2^2\} \cap \mathcal{O}(\mathbb{C}^2)$, and show that (z_1, z_2) and $(z_1, -z_2)$ cannot be separated by an $f \in \mathscr{H}(X)$, although a global "spreading" $X \to \mathbb{C}^2, (z_1, z_2) \mapsto (z_1, z_2^2)$, exists. (Hints: i) the mapping $\varphi_j: \mathbb{C}^2 \to \mathbb{C}^3, (z_1, z_2) \mapsto (z_1, z_2^2, z_2^{2j+1})$, induces a structure sheaf $\mathscr{H}_j = \varphi_j^0({}_3\mathcal{O}) \subset {}_2\mathcal{O}$ on $\mathbb{C}^2$ with $\mathscr{H}_j|_{z_2 \neq 0} = {}_2\mathcal{O}|_{z_2 \neq 0}$ and $\mathscr{H}_j(\mathbb{C}^2) = \mathbb{C}\{z_1, z_2^2, z_2^{2j+1}\} \cap \mathcal{O}(\mathbb{C}^2)$;

ii) on $X := \mathbb{C}^2 \setminus \{(z_1, 0);\ |z_1| \in \mathbb{N}\}$,

$$\mathscr{H}_z := \begin{cases} {}_2\mathcal{O}_z & \text{for } z_2 \neq 0 \\ \mathscr{H}_{jz} & \text{for } z_2 = 0 \text{ and } j < |z_1| < j+1 \end{cases}$$

determines a structure sheaf;

iii) every $f \in \mathscr{H}(X)$ is holomorphically extendible to $\mathbb{C}^2$ (52.20), and $\mathscr{H}(X) = \bigcap_{j=1}^{\infty} \mathscr{H}_j(\mathbb{C}^2)$.)

§ 53 The Additive Cousin Problem

We come now to the first of the two central existence theorems for holomorphic functions of a single variable with prescribed zero- or pole-behavior, namely, Mittag-Leffler's Theorem: if a "principal part" $H_j = \sum_{k=1}^{<\infty} \alpha_{jk}(z - a_j)^{-k}$ is prescribed at each point a_j of a discrete closed subset $\{a_j\}$ of a one-dimensional region X, then there exists a meromorphic function m on X whose Laurent expansion has principal part H_j at a_j. It took fifty years for a satisfactory generalization of that theorem to higher dimensions to be proven; it was cohomology theory that made a – surprisingly simple – solution possible.

We begin by translating the necessary concepts into the general theory. Meromorphic functions on manifolds can be defined locally as quotients of holomorphic functions. Since nontrivial zero-divisors may appear in the function algebras of arbitrary complex spaces, we proceed as follows:

For each point a of the complex space X, set $\mathscr{S}_a := \{f \in \mathcal{O}_a;\ f$ is a non-zero-divisor$\}$. By E. 53a, $\mathscr{S} := \bigcup_{a \in X} \mathscr{S}_a$ is a subsheaf of $\mathcal{O}$.

E. 53a. Prove the following without applying Theorem A: If $f \in \mathcal{O}(X)$ induces a non-zero-divisor f_a in a stalk $\mathcal{O}_a$, then f_x is a non-zero-divisor in $\mathcal{O}_x$ for each x near a. The following equivalences hold if X is reduced: In every stalk of the structure sheaf, f induces a non-zero-divisor iff $N(f)$ includes no irreducible component of f iff $N(f)$ is thin in X. (Hint: 41.17 with $\psi = \cdot f$; for § 53.)

The sheaf ${}_X\mathscr{M}$ of meromorphic functions is to be constructed so that ${}_X\mathscr{M}_a = Q(\mathcal{O}_a)$ for every $a \in X$. To that end, we apply the following subrings of the total quotient ring $Q(R)$ for $R = \mathcal{O}(U)$ and $S = \mathscr{S}(U)$:

A submonoid S of the set of non-zero-divisors of a ring R is called a *multiplicative system* if it contains the unit element 1_R. Then

$$S^{-1}R := \left\{\frac{r}{s} \in Q(R);\, s \in S\right\}$$

is a subring of $Q(R)$ that includes R in a canonical manner [At Mac § 3]. For a ring-homomorphism $\varphi: R \to R'$, if $S' \subset R'$ is a multiplicative system such that $\varphi S \subset S'$, then φ has a unique homomorphic extension $\varphi: S^{-1}R \to (S')^{-1}R'$.

The presheaf $U \mapsto \mathscr{S}(U)^{-1}\mathscr{O}(U)$ on X satisfies the first axiom for sheaves (§ 30); it induces a sheaf $\mathscr{M}$ of $\mathscr{O}$-algebras called the *sheaf of meromorphic functions on X*. Its sections are called *meromorphic functions*, despite the fact that they may have poles, or even "indeterminate points" (as is the case with $1/z \in \mathscr{M}(\mathbb{C})$ and $z_1/z_2 \in \mathscr{M}(\mathbb{C}^2)$).

E. 53b. Show that there are meromorphic functions on $\mathbb{P}_1$ that cannot be represented globally as quotients of holomorphic functions.

The following fact is fundamental for the study of meromorphic functions (by a result of Thimm [Fh XIII 3.4], 53.2 and thus 53.1 hold also for nonreduced spaces):

53.1 Proposition. *On a reduced B-space X, every global meromorphic function can be represented as a quotient of global holomorphic functions in which the denominator induces a non-zero-divisor in every $\mathscr{O}_a$.*

Proof. For $m \in \mathscr{M}(X)$, $\mathscr{O}\cdot m$ is a coherent subsheaf of $\mathscr{M}$; the $\mathscr{O}$-ideal

$$\mathscr{D}_m := \mathscr{O} : (\mathscr{O}\cdot m),$$

which is coherent, by E. 41g, is called the *sheaf of denominators* of m, since $\mathscr{D}_{m,x} = \{g \in \mathscr{O}_x;\, gm \in \mathscr{O}_x\}$. According to the construction of $\mathscr{M}$, every $\mathscr{D}_{m,x}$ contains a non-zero-divisor of $\mathscr{O}_x$; we want to prove the existence of a $g \in \mathscr{D}_m(X)$ that induces a non-zero-divisor in every $\mathscr{O}_x$, for then gm is a holomorphic function and $m = gm/g$.

53.2 Lemma. *For a coherent sheaf of ideals $\mathscr{I}$ on a reduced B-space X, if every stalk of $\mathscr{I}$ contains a non-zero-divisor of ${}_X\mathscr{O}$, then there exists a global section $g \in \mathscr{I}(X)$ such that no germ g_a is a zero-divisor in $\mathscr{O}_a$.*

Proof. According to 49.1, the regular points are dense in X, since X is reduced. By E. 53a, then, it suffices to find a $g \in \mathscr{I}(X)$ that induces a non-zero-divisor at every point $a \in X \backslash S(X)$; by the Identity Theorem, that will happen iff, for every connected component V_j of $X \backslash S(X)$, g is not the zero-section on V_j. Since every stalk of $\mathscr{I}$ contains a non-zero-divisor, no V_j lies entirely in $N(X; \mathscr{I})$. For each j, choose an $a_j \in V_j \backslash N(\mathscr{I})$; then $\{a_j\}$ is a discrete closed subset of X, and $\{a_j\} \cap N(\mathscr{I}) = \emptyset$. Thus, by 52.5, there exists a $g \in \mathscr{O}(X)$ such that $g|_{V(\mathscr{I})} = 0 \in {}_{V(\mathscr{I})}\mathscr{O}(V(\mathscr{I}))$ and $g(a_j) = 1$ for every j; it follows that $g \in \mathscr{I}(X)$. ■■

Having introduced meromorphic functions, we can adapt Mittag-Leffler's Theorem to nonisolated poles (see 7.10). In particular, we want to prove a statement of the form "local ⇒ global". To that end, let $\mathfrak{U} = (U_j)_{j \in J}$ be an open cover of the complex space X, and suppose that we are given a familiy $(m_j \in \mathcal{M}(U_j))_{j \in J}$ such that

(53.3.1) $m_i - m_j \in \mathcal{O}(U_{ij})$ for every $i, j \in J$.

That "compatibility condition" means that the family $(m_j)_{j \in J}$ determines a section in the quotient sheaf $\mathcal{M}/\mathcal{O}$.

53.3 Definition. *The sheaf $\mathcal{M}/\mathcal{O}$ on the complex space X is called the sheaf of additive Cousin distributions; its global sections are called additive Cousin distributions, or Cousin I distributions.*

Every additive Cousin distribution can be represented with an appropriately chosen cover $\mathfrak{U}$ in the form (53.3.1); two such representations determine the same section iff their union satisfies (53.3.1).

E. 53c. For $X \subset\subset \mathbb{C}$, show that there is a canonical one-to-one correspondence between additive Cousin distributions on X and sets $\{(a_j, H_j); j \in J\}$, where the H_j are principal parts and $\{a_j; j \in J\}$ is a discrete closed subset of X.

In the new terminology, Mittag-Leffler's Theorem states that, for each additive Cousin distribution g on a region $X \subset\subset \mathbb{C}^1$, there exists a global meromorphic function $m \in \mathcal{M}(X)$ whose residue class in $(\mathcal{M}/\mathcal{O})(X)$ is precisely g. That leads to the following:

Additive Cousin Problem. *For a complex space X, determine the image of* $\mathcal{M}(X) \to (\mathcal{M}/\mathcal{O})(X)$ (the elements of the image are called *"solvable" additive Cousin distributions*).

There exist domains with insolvable additive Cousin distributins:

53.4 Example. On $X = \mathbb{C}^{2*}$, the additive Cousin distribution $(m_0, m_1) := (0, 1/z_0 z_1)$ associated with the cover $(\mathbb{C}^* \times \mathbb{C}, \mathbb{C} \times \mathbb{C}^*)$ is insolvable, since $d^0((m_0, m_1))$ determines the nonzero class s in $H^1(X, \mathcal{O})$ of 52.3 vii); see the proof of 53.5.

E. 53d. Prove 53.4 without using cohomology theory (hint: for a solution $m \in \mathcal{M}(X), z_0 m \in \mathcal{O}(X)$ is not holomorphically extendible to $\mathbb{C}^2$).

53.5 Theorem. *For a complex space X, if $H^1(X, \mathcal{O}) = 0$, then every additive Cousin distribution on X is solvable.*

Proof. The exact sequence of sheaves $0 \to \mathcal{O} \to \mathcal{M} \to \mathcal{M}/\mathcal{O} \to 0$ yields a surjection $\mathcal{M}(X) \to (\mathcal{M}/\mathcal{O})(X) \to H^1(X, \mathcal{O}) = 0$. ■

53.6 Corollary. *On a B-space, every additive Cousin distribution is solvable.* ■

53.7 Examples. i) A curve with no compact irreducible components is a B-space (51.3 iv)), so 53.6 is applicable.
ii) On $\mathbb{P}_n$, 53.5 is applicable (see E. 52a).

E. 53e. For a closed subspace A of a reduced B-space X, every meromorphic function on A is the restriction of a meromorphic function on X. Why does $i\colon A \hookrightarrow X$ not induce a restriction-homomorphism ${}_X\mathcal{M}(X) \to {}_A\mathcal{M}(A)$?

E. 53f. For the meromorphic function $m \in \mathcal{M}(X)$, define a "pole-variety" $P(m)$ and a "zero-variety" $N(m)$ in X (hint: consider $\mathcal{D}_m$ and $\mathcal{D}_{1/m}$ if $1/m \in \mathcal{M}(X)$).

E. 53g. Extend 53.6 to B-sets.

E. 53h. For a reduced B-space X, and $A \hookrightarrow X$, let Y be the union of the irreducible components of X not included in A. Show that there exists an $f \in \mathcal{O}(X)$ with $f|_A = 0$ that induces a non-zero-divisor in each ${}_Y\mathcal{O}_y$, $y \in Y$ (hint: 52.5, E. 53a; for 53A.5).

§ 53A Supplement: Meromorphic Functions on Reduced Complex Spaces

In this supplement, X always denotes a reduced complex space.

53 A. 1 Remark. *For each meromorphic function $m \in \mathcal{M}(X)$, there exists a maximal open dense subset $D_m \subset X$ on which m is holomorphic.*

Proof. It clearly suffices to prove that, locally, m is holomorphic on an open dense subset; without loss of generality, then, suppose that $m = f/g$ with only non-zero-divisors g_x. Then $N(g)$ is not dense in any irreducible component of X (49.8); it follows that $X \setminus N(g)$ is dense, and m is holomorphic there. ■

E. 53Aa. For $m \in \mathcal{M}(X)$, show that $D_m = X \setminus N(\mathcal{D}_m)$ (for 53 A. 2).

53 A. 2 Identity Theorem for Meromorphic Functions. *If X is irreducible, and if $m, m' \in \mathcal{M}(X)$ coincide on a nonempty set $W \subset X$, then $m = m'$.*

Proof. On $D = D_m \cap D_{m'} \underset{\text{E. 53 Aa}}{=} X \setminus (N(\mathcal{D}_m) \cup N(\mathcal{D}_{m'}))$, the meromorphic functions m and m' are holomorphic; by 49.11, they coincider there, since they coincide on $W \cap D$, and since D is irreducible, by E. 49f. For $x \in \partial(D_m \cap D_{m'})$, and local representations $m = f/g$, $m' = f'/g'$ near x, the holomorphic equation $fg' = f'g$ extends to a neighborhood of x; it follows that $m = m'$ on $\partial(D_m \cap D_{m'}) = X \setminus (D_m \cap D_{m'})$. ■

53 A. 3 Remark. *If X is irreducible and $0 \neq m \in \mathcal{M}(X)$, then m can be represented on each open B-set $U \subset X$ as a quotient of holomorphic functions that induce non-zero-divisors in each $\mathcal{O}_u$.*

Proof. Let $m = f/g$ be a representation on U according to 53.1. If $f_a \in \mathcal{O}_a$ is a zero-divisor, then there exists a $0 \neq h_a \in \mathcal{O}_a$ with $f_a h_a = 0$. By choosing a b lying in both U_{reg} and the domain of definition of h so that $g(b) \neq 0$, $h_b \neq 0$, and $f_b h_b = 0$, we obtain that $f_b = 0$; hence, $m = 0$ near b. It follows from 53 A. 2 that $m = 0$. ■

53 A. 4 Corollary. *If X is irreducible, then $\mathcal{M}(X)$ is a field.*

Proof. Since $\mathcal{M}$ is a sheaf, we need only find a local inverse for each nonzero $m \in \mathcal{M}(X)$ (in a commutative monoid, inverses are unique!). With $m = f/g$ near a as in 53 A. 3, we have that $g/f = 1/m$ is meromorphic at a. ■

With the help of Theorem B, we can generalize 53 A. 4:

53 A. 5 Remark. *For the decomposition $X = \bigcup_{j \in J} X_j$ of X into irreducible components, the mapping*

$$\varphi : \mathcal{M}(X) \to \prod_J \mathcal{M}(X_j), \quad m \mapsto (m|_{X_j})_{j \in J},$$

is an algebra-isomorphism onto a (ring-)direct product of fields.

Proof. For each $m \in \mathcal{M}(X)$, $X_j \cap D_m$ is an open dense subset of X_j; hence, φ is well-defined and injective by 53 A. 2. For the surjectivity, it suffices to show that every $m_j \in \mathcal{M}(X_j)$ has exactly one extension $m_j^X \in \mathcal{M}(X)$ such that $m_j^X|_{X \setminus X_j} = 0$ (for then $\varphi(\sum_j m_j^X) = (m_j)_{j \in J}$; the sum makes sense, since it is locally finite). We need only construct the extension m^X of $m \in \mathcal{M}(X_j)$ locally near $X_j \cap Y$ for fixed j and $Y := \bigcup_{k \neq j} X_k$. To that end, we may assume that $m = f/g$ with $f, g \in \mathcal{O}(X_j)$, and that X is a B-space (by 52.3 v), then, X_j and Y are B-spaces as well). By 52.5, there exists a non-zero-divisor $h \in \mathcal{O}(X_j)$ such that $h|_{X_j \cap Y} = 0$ (observe that $X_j \cap Y$ need not be reduced). Since, if necessary, we can multiply both f and g by such an h, there is no loss of generality in assuming that $f|_{X_j \cap Y} = g|_{X_j \cap Y} = 0$. By E. 53h, there exists a $G \in \mathcal{O}(Y)$ with $G|_{X_j \cap Y} = 0$ that induces a non-zero-divisor at every point. By E. 44i, g and G patch together to a holomorphic function on X, and f can be extended to X by defining it to be zero outside X_j. It follows that $m^X = f^X/G$. ■

E. 53A b. Derive this fact from 53 A. 5: the characteristic function χ_j of X_j determines a meromorphic function on X (for 53 A. 6).

53 A. 6 Corollary. *An (algebraically) reduced analytic algebra with $R = \hat{R}$ is normal.*

Proof. We only need to see that R has no zero-divisors. For $R = {}_X\mathcal{O}_a$, that holds iff X_a is irreducible, according to 44.14. By E. 53A b, the characteristic function of a fixed irreducible component of X_a is meromorphic; as a zero of the polynomial $T^2 - T \in {}_X\mathcal{O}_a[T]$, it is integral over ${}_X\mathcal{O}_a$, and thus holomorphic. In particular, it is continuous, so X_a has no other irreducible components. ■

The assumption "reduced" is essential:

E. 53A c. Prove the following statements for $R = {}_2\mathcal{O}_0 / (z_1 z_2, z_2^2)$:
i) R is not normal. ii) $R = \hat{R} = Q(R)$. (Hint: E. 43 l.)

A meromorphic function $m \in \mathcal{M}(X)$ can be interpreted as a true function on D_m. The behavior of m at a point $a \in X_{\text{reg}} \setminus D_m$ can be characterized as follows:

53 A. 7 Definition. *Consider a manifold M, a point $a \in M$, and a meromorphic function $m \in \mathcal{M}(M)$. If there exists an open neighborhood U of a such that $m|_{U \cap D_m}$ is bounded, then a is called a "removable singularity" of m. Otherwise, a is called a "pole" of m; moreover, it is called a*

α) "pole in the strict sense" if $\lim |m(x_j)| = \infty$ for every sequence of points $(x_j)_{j \in \mathbb{N}}$ in D_m converging to a.

β) "indeterminate point" if there exists a sequence of points $(x_j)_{j \in \mathbb{N}}$ in D_m converging to a such that $\lim |m(x_j)| \neq \infty$.

E. 53Ad. Discuss the singularities of $m \in \mathcal{M}(\mathbb{C}^2)$ for
i) $m = z_1/z_2$, ii) $m = (\sin z_1)/(\sin z_1 z_2)$.
The terminology "removable singularity of a meromorphic function m" at a manifold point $a \in M$ is justified by the fact that m can be extended holomorphically at a: if $m = f/g$ with $f, g \in \mathcal{O}_a$, then clearly $M \setminus N(g) \subset D_m$; by the first Riemann Removable Singularity Theorem, m is holomorphically extendible at a.
The situation is different at singular points of X:

E. 53Ae. Use E. 32 l b) to prove the following for Neil's parabola: t is a meromorphic function that is "bounded at 0", but it has no holomorphic extension to 0.
For indeterminate points, we have a statement à la Casorati-Weierstrass:

53 A. 8 Proposition. *For an indeterminate point of a meromorphic function m on the manifold M, the values of m approximate every $\lambda \in \mathbb{C} \cup \{\infty\}$ in each neighborhood of a.*

Proof. Otherwise, there exist an open neigborhood U of a and a $V \Subset \mathbb{C}$ with $m(U \cap D_m) \subset \mathbb{C} \setminus V$. For $\lambda \in V$, then, $g := \dfrac{1}{m - \lambda}$ is meromorphic and bounded on U, so a is a removable singularity of g; i.e., $g \in \mathcal{O}(U)$ and $m = \dfrac{1}{g} + \lambda$. If $g(a) = 0$, then m has a pole at a in the strict sense; if $g(a) \neq 0$, then m even has a removable singularity! ↯ ■
For the description of meromorphic functions at singular points of X, one can introduce a "meromorphic graph" [Fi 4.5–4.8]; that yields the following result:

53 A. 9 Kontinuitätssatz for Meromorphic Functions. *For an analytic set of codimension at least two in an irreducible complex space X, the restriction-mapping*

$$ {}_X\mathcal{M}(X) \to {}_X\mathcal{M}(X \setminus A) $$

is an algebra-isomorphism. ■

E. 53Af. *Meromorphic functions on a projective space.* Deduce from 53 A. 9 that $\mathcal{M}(\mathbb{P}_n) = Q(\mathbb{C}[z_1, \dots, z_n]) = \mathbb{C}(z_1, \dots, z_n)$ (field of rational functions). (Hint: realize m as f/g with $f, g \in \mathcal{O}(\mathbb{C}^{n+1})$ and apply the fact that $f(z)/g(z) = f(\lambda z)/g(\lambda z)$ for every $\lambda \in \mathbb{C}^*$, and the representation of holomorphic functions with homogeneous polynomials; see § 21. For E. 54 Bd.)

E. 53Ag. *Holomorphic mappings between projective spaces.* Prove that homogeneous polynomials $P_0, \dots, P_m \in \mathbb{C}[z_0, \dots, z_n]$ of the same degree with no common zeros determine a holomorphic mapping $[P_0, \dots, P_m]: \mathbb{P}_n \to \mathbb{P}_m$. Conversely, for a holomorphic mapping $\varphi: \mathbb{P}_n \to \mathbb{P}_m$ whose image is not included in any hyperplane, there exist homogeneous polynomials of the same degree, $P_0, \dots, P_m \in \mathbb{C}[z_0, \dots, z_n]$, that describe φ (hint: for $j \geqq 0$, z_j/z_0 is meromorphic on $\mathbb{P}_m$; with $(z_j/z_0) \circ \varphi = f_j/g_j$ according to E. 53 Af, choose $P_j = f_j \prod_{k \neq j} g_k$).

E. 53Ah. Use E. 53 Ag to prove that $\operatorname{Aut}(\mathbb{P}_n) = \mathbb{P}GL(\mathbb{C}^{n+1})$. More generally, one can show for Grassmann manifolds that (see [Ch])

$$ \operatorname{Aut} G_k(n) = \mathbb{P}GL(\mathbb{C}^n) \quad \text{for} \quad k \neq n/2 \quad \text{or} \quad n = 2. $$

For $n \geqq 3$, $\operatorname{Aut} G_n(2n)$ includes another connected component aside from $\mathbb{P}GL(\mathbb{C}^{2n})$; it is given by the correlation (see E. 32 Bb ii)).

E. 53Ai. Show that the ${}_n\mathcal{O}$-module ${}_n\mathcal{M}$ is not of finite type.
Since we hardly discuss the special methods for the investigation of compact complex spaces,

we want to mention at least a few of their important invariants. For an irreducible compact complex space X, the transcendence degree of the field extension $\mathcal{M}(X): \mathbb{C}$ is called the *algebraic dimension* $a(X)$ of X. By the *Weierstrass-Siegel-Thimm Theorem*, $a(X) \leqq \dim X$ [Fi 4.10]. If X is projective-algebraic, then every meromorphic function on X is *rational*, i.e., representable as a quotient of two polynomials (*Hurwitz's Theorem* [Fi 4.7]; for $X = \mathbb{P}_n$, see E. 53 Af), and it follows that $a(X) = \dim(X)$ [Sh I §§ 6.1,2]. More generally, a complex space is called a *Moishezon space* if $a(X) = \dim(X)$. For curves and for nonsingular surfaces, the classes of Moishezon spaces and of projective algebraic spaces coincide (curves are always projective-algebraic [Ht III. 5.8]; [Ko_2 I Thm. 3.1]); however, there exist normal surfaces and three-dimensional manifolds that are Moishezon spaces, but not projective-algebraic ([Gr § 4.8. d)] and [Ht App. B, 3.4.2]). Whereas curves always satisfy the equality $a(X) = 1 = \dim(X)$, the additional cases $a(X) = 0$ and $a(X) = 1$ appear even in dimension two: for $X = \mathbb{C}^{2*}/((z_1, z_2) \sim (2^j z_1, 3^j z_2))$ we have that $a(X) = 0$ [Ko_2 II § 10].

E. 53Aj. For the Hopf surface $X = \mathbb{C}^{2*}/(z \sim 2^j z)$, show that $\mathcal{M}(X) \cong \mathcal{M}(\mathbb{P}_1)$, and thus that $a(X) = 1$ (hint: if $m \in \mathcal{M}(\mathbb{C}^{2*}) \underset{53\text{A}.9}{=} \mathcal{M}(\mathbb{C}^2)$ is invariant under $z \sim 2^j z$, then it is invariant under $z \sim \lambda z$ for every $\lambda \in \mathbb{C}^*$).

The influence of the algebraic dimension on the structure of a nonsingular surface X is shown by the following results:

$a(X) = 1$. Up to a biholomorphic transformation, there is precisely one holomorphic mapping $f: X \to Y$ onto a curve Y with no singularities that induces an isomorphism $f^0: \mathcal{M}(Y) \to \mathcal{M}(X)$. The space X is called a (nonalgebraic) elliptic surface, since the fibers of f are elliptic curves (i.e., tori) with at most finitely many exceptions [Ko_2 I § 4].

$a(X) = 0$. Every meromorphic function on X is constant. Equivalently, only finitely many reduced curves exist on X [Ko_2 I Thm. 5.1]. A more detailed classification can be obtained using the Betti numbers $b_1(X)$, which can have only the values 0, 1, and 4 [Ko_3 I Th. 11]: if $b_1(X) = 0$, then, up to quadratic transformations (which does not change the field of meromorphic functions), X is diffeomorphic to $V(\mathbb{P}_3; z_0^4 + z_1^4 + z_2^4 + z_3^4)$ (K 3-surfaces); for $b_1(X) = 4$, X is a torus, up to quadratic transformations; for $b_1(X) = 1$, see [BoHu part 6], [Ue 20.18], and the literature cited there.

An overview of bimeromorphic classification in higher dimensions is given in [Ue].

Further literature: [AnSt].

§ 54 The Multiplicative Cousin Problem

Now we generalize the second central existence theorem for holomorphic functions of a single complex variable, namely, the *Weierstrass Factorization Theorem:*

For each assignment of orders $p_j \in \mathbb{Z}$ to the points a_j of a closed discrete subset $\{a_j\}$ of a region $X \subset \mathbb{C}$, there is an $m \in \mathcal{M}(X)$ that has a zero of order (precisely) p_j at each a_j, and that has no other zeros in $X \backslash \{a_j\}$ (poles are included as zeros of negative order). As in § 53, we want to find a formulation that makes sense even for nonisolated singularities. Once again, we suppose that we have a set that is given locally by the "zeros" of meromorphic functions: let X be a complex space, and let ${}_X\mathcal{O}^*$ and ${}_X\mathcal{M}^*$ be the subsheaves (of abelian groups) of all multiplicatively invertible elements of ${}_X\mathcal{O}$ and ${}_X\mathcal{M}$, respectively (neither sheaf is analytic). For an open cover $\mathfrak{U} = (U_j)_{j \in J}$ of X, suppose that we are given a family $(m_j \in \mathcal{M}^*(U_j))_{j \in J}$ such that

(54.1.1) $m_j/m_k \in \mathcal{O}^*(U_{jk})$ for every $j, k \in J$.

On the one hand, that stipulation of compatibility ensures that, at every point $a \in U_{jk}$, the zero-sets $N(m_j)$ and pole-sets $P(m_j)$ are independent of j; on the other hand, it guarantees that the family $(m_j \in \mathcal{M}^*(U_j))$ defines a global section in the sheaf

$$\mathcal{D} := \mathcal{M}^* / \mathcal{O}^* .$$

54.1 Definition. *The sheaf $\mathcal{D}$ is called the sheaf of (Cartier) divisors, or multiplicative Cousin distributions, or Cousin II distributions; correspondingly, the global sections are called divisors, etc.*

E. 54a. Let X be a Riemann surface. Prove that every divisor on X determines invertibly a discrete closed set $\{a_j; j \in J\}$ and a set of "orders" $\{p_j \in \mathbb{Z}; j \in J\}$.

For every divisor, there exists a cover $\mathfrak{U}$ and a representation satisfying (54.1.1); two such representations determine the same section in $\mathcal{D}$ iff their union satisfies (54.1.1).

The Weierstrass Factorization Theorem can now be reformulated as follows: "for each multiplicative Cousin distribution D on a region $X \subset \mathbb{C}$, there exists a global meromorphic function $m \in \mathcal{M}^*(X)$ whose equivalence class $\bar{m} \in (\mathcal{M}^* / \mathcal{O}^*)(X)$ is precisely D". That brings us to the following problem:

Multiplicative Cousin Problem. *For a complex space X, determine the image of* $\tau_X : \mathcal{M}^*(X) \to \mathcal{D}(X) = (\mathcal{M}^* / \mathcal{O}^*)(X)$ (the images $\tau_X(m) =: (m)$ are called *principal divisors* or *solvable multiplicative Cousin distributions*).

A divisor given by $(m_j \in \mathcal{M}^*(U_j))_{j \in J}$ is thus principal iff there exists a meromorphic function $m \in \mathcal{M}^*(X)$ such that $m / m_j \in \mathcal{O}^*(U_j)$ for every $j \in J$. There is an example of a nonprincipal divisor in E. 54c.

54.2 Proposition. *If $H^1(X, \mathcal{O}^*) = 0$, then every multiplicative Cousin distribution on X is solvable.*

Proof. The assertion follows immediately from the cohomology sequence induced by the exact sequence of sheaves

(54.2.1) $1 \to \mathcal{O}^* \to \mathcal{M}^* \xrightarrow{\tau} \mathcal{D} \to 0$. ∎

We thus have the condition "$H^1(X, \mathcal{O}^*) = 0$" to analyze. It is more difficult than the analogous condition "$H^1(X, \mathcal{O}) = 0$" for the additive Cousin problem, due to the fact that $\mathcal{O}^*$ is not an analytic sheaf, so that Theorem B is not applicable. Instead, we use the following fact:

54.3 Lemma. *On each complex space X, there exists an exact sequence of sheaves*

$$(54.3.1)\quad 0 \to \mathbb{Z} \xrightarrow{\cdot 2\pi i} \mathcal{O} \xrightarrow{\exp} \mathcal{O}^* \to 1$$

in which $\mathbb{Z}$ *is the constant sheaf of integers.*

Proof. By 43.6, for every $U \Subset X$, each $h \in \mathcal{O}(U)$ determines a holomorphic mapping $H: U \to \mathbb{C}$; hence, $e^h := H^0(e^z)$ lies in $\mathcal{O}(U)$ (e^z is the exponential function on $\mathbb{C}$). The mapping so defined,

$$\exp: \mathcal{O} \to \mathcal{O}^*,\ h \mapsto e^h,$$

is clearly a sheaf-homomorphism. Moreover, exp is (stalkwise) surjective, for, as above, each $h \in \mathcal{O}_a^*$ induces a $\log h \in \mathcal{O}_a$, log being a branch of the logarithmic function near $h(a) \in \mathbb{C}^*$; obviously, $e^{\log h} = h$.

It remains to show that $\mathcal{K}er(\exp) = \mathbb{Z} \cdot 2\pi i$. The inclusion "$\supset$" is trivial. "$\subset$" For $h \in \mathcal{K}er(\exp)$ (i.e., $e^h = 1$), we may assume that $h \in \mathfrak{m}_a$ (by addition of an appropriate integral multiple of $2\pi i$, we obtain that $h(a) = 0$). The power series expansion of e^z yields a $g \in \mathcal{O}_a$ such that $e^h = 1 + h - gh^2$. Since $e^h = 1$, it follows by induction that

$$h = h^2 g = \ldots = h^{j+1} g^j \in \bigcap_{j=1}^{\infty} \mathfrak{m}_a^j \underset{24.2}{=} (0);$$

in other words, $h = 0$. Therefore, $\mathcal{K}er(\exp) = \mathbb{Z} \cdot 2\pi i$. ∎

By combining the connecting homomorphisms associated with the sequences (54.2.1) and (54.3.1), we can assign to each divisor D an invariant $c(D)$,

$$\begin{array}{ccccc} H^0(X, \mathcal{D}) & \xrightarrow{\delta^0} & H^1(X, \mathcal{O}^*) & \xrightarrow{\delta^1} & H^2(X, \mathbb{Z}) \\ D & & \longmapsto & & c(D), \end{array}$$

called the "*Chern class of D*".

54.4 Theorem. *In order for a multiplicative Cousin distribution D on X to be solvable, it is necessary that its Chern class* $c(D) \in H^2(X, \mathbb{Z})$ *vanish. If* $H^1(X, \mathcal{O}) = 0$, *then that condition is also sufficient.*

Proof. The cohomology sequence

$$\mathcal{M}^*(X) \xrightarrow{\tau} \mathcal{D}(X) \xrightarrow{\delta^0} H^1(X, \mathcal{O}^*)$$

induced by (54.2.1) yields that

$$D \in \operatorname{Im} \tau \Leftrightarrow \delta^0(D) = 0 \Rightarrow c(D) = \delta^1 \delta^0(D) = 0.$$

If $H^1(X, \mathcal{O}) = 0$, then δ^1 is injective (look at the cohomology sequence induced by (54.3.1)), so the converse holds also. ∎

54.5 On a curve X with no compact irreducible components, the Weierstrass Factorization Theorem holds: by 51.3 iv), X is a Stein space; moreover, $H^2(X,\mathbb{Z}) = 0$ [AnFr].

A *divisor* is called *holomorphic*, or *positive*, if it is given by a system $(m_j \in \mathcal{O}(U_j) \cap \cap \mathcal{M}^*(U_j))_{j\in J}$ (in which case it has no "zeros of negative order", and its solutions are holomorphic). With that terminology, we have the following:

54.6 Proposition. *On a reduced B-space X, for every $\alpha \in H^2(X,\mathbb{Z})$ there is a holomorphic divisor $D = D(X)$ with Chern class $c(D) = \alpha$.*

Proof. The cohomology sequence (54.3.1) yields that $H^1(X,\mathcal{O}^*) \cong H^2(X,\mathbb{Z})$; hence, it suffices to find for each $\xi \in H^1(X,\mathcal{O}^*)$ a holomorphic divisor $D \in \mathcal{D}(X)$ with $\delta^0(D) = \xi$. To that end, let $\mathfrak{U}$ be an open cover of X, and let $\zeta \in Z^1(\mathfrak{U},\mathcal{O}^*)$ be a representative of ξ^{-1} (with 51 A. 3, it can be shown that the B-space X is paracompact). We need this:

54.7 Remark. *For an open cover $\mathfrak{U}$ of a complex spaces X, each cocycle $\zeta \in Z^1(\mathfrak{U},\mathcal{O}^*)$ determines a locally free sheaf $\mathcal{O}(\zeta)$ in a canonical fashion.*

Proof. Put $\mathfrak{U} = (U_j)_{j\in J}$ and $\zeta = (\zeta_{jk})$. With $\mathcal{G}_j := \mathcal{O}|_{U_j}$, isomorphisms are defined on every U_{jk} by

$$\varrho_{jk}: \mathcal{G}_k \to \mathcal{G}_j,\ f \mapsto \zeta_{jk} f.$$

Since ζ is a cocycle, $\varrho_{ij} \circ \varrho_{jk} = \varrho_{ik}$ holds for all i, j, and k. In analogy to E. 31m, gluing the sheaves $\mathcal{G}_j$ yields the desired sheaf $\mathcal{O}(\zeta)$ on X. ■

Let $\mathcal{O}(\zeta)$ be the sheaf belonging to $\zeta \in Z^1(\mathfrak{U},\mathcal{O}^*)$, as above; we show first that there exists a section $g \in \mathcal{O}(\zeta)(X)$ for which $\{x \in X; g(x) = 0 \in \mathcal{O}(\zeta)_x\}$ is empty: let A be a discrete subset of X that contains a regular point from each irreducible component of X. For the nullstellen ideal ${}_A\mathcal{I}$, ${}_A\mathcal{I}\cdot\mathcal{O}(\zeta)$ is a coherent analytic sheaf, so $\mathcal{O}(\zeta)(X) \to [\mathcal{O}(\zeta)/{}_A\mathcal{I}\cdot\mathcal{O}(\zeta)](X)$ is surjective. In particular, there exists a $g \in \mathcal{O}(\zeta)(X)$ that represents the zero germ at *no* point of A. Since the Identity Theorem 49.11 clearly holds for sections in $\mathcal{O}(\zeta)$, there is no point at which g induces the zero germ. By E. 31m iii), there is exactly one cochain $(g_j) \in C^0(\mathfrak{U},\mathcal{O})$ corresponding to $g \in \mathcal{O}(\zeta)(X)$ such that $g_j = \zeta_{jk} g_k$ on U_{jk}; it lies in $C^0(\mathfrak{U},\mathcal{M}^*)$, since our choice of g ensures that every set $N(U_j; g_j)$ is thin in U_j. Then (g_j) determines a holomorphic divisor $D \in \mathcal{D}(X)$, and the fact that $g_k/g_j = \zeta_{jk}^{-1}$ implies that $\delta^0(D) = \xi \in H^1(X,\mathcal{O}^*)$. ■■

The following results deal with connections between topological and holomorphic data of complex spaces:

54.8 Corollary. *For a reduced B-space X, there exist canonical isomorphisms of abelian groups $H^1(X,\mathbb{Z}) \cong \mathcal{O}^*(X)/\exp\mathcal{O}(X)$; $H^2(X,\mathbb{Z}) \cong \mathcal{D}(X)/\tau_X\mathcal{M}^*(X)$.*

Proof. The cohomology sequence

$$\mathcal{O}(X) \xrightarrow{\exp} \mathcal{O}^*(X) \to H^1(X,\mathbb{Z}) \to H^1(X,\mathcal{O}) = 0$$

induced by (54.3.1) yields the first isomorphism. The homomorphism $c:\mathcal{D}(X) \to \to H^2(X,\mathbb{Z})$ is surjective, by 54.6, and its kernel is $\tau_X(\mathcal{M}^*(X))$ (see the proof of 54.4). ■

On a simply connected region in $\mathbb{C}$, a *global logarithm* can be defined through the choice of a branch. Elementary methods of algebraic topology yield that $H^1(X,\mathbb{Z}) = 0$ for simply connected complex spaces (for $X \Subset \mathbb{C}$, the converse holds as well!). Here, we obtain the following:

54.9 Proposition. *If X is a complex space with $H^1(X,\mathbb{Z}) = 0$, then, for each holomorphic function f with no zeros, it is possible to define a holomorphic logarithm* $\log f$; *i.e.*, $\exp(\log f) = f$.

Proof. The assertion follows from the cohomology sequence $\mathcal{O}(X) \xrightarrow{\exp} \mathcal{O}^*(X) \to \to H^1(X,\mathbb{Z}) = 0$ induced by (54.3.1). ■

The importance of the logarithm for the explicit calculation of the Chern class of a divisor is shown by the following:

E. 54b. Let $D \in H^0(X,\mathcal{D})$ be given by a family $(m_j \in \mathcal{M}^*(U_j))_{j\in J}$ with respect to an open cover $\mathfrak{U} = (U_j)_{j\in J}$ such that $H^1(U_{jk},\mathbb{Z}) = 0$ for all $j,k \in J$. For $(\delta D)_{jk} = m_k / m_j \in \mathcal{O}^*(U_{jk})$ with $\delta: H^0(X,\mathcal{D}) \to H^1(X,\mathcal{O}^*)$, it is possible to choose an $l_{jk} := (\log(m_k/m_j)) / 2\pi i$ in $\mathcal{O}(U_{jk})$. The coboundary of $l = (l_{jk}) \in C^1(\mathfrak{U},\mathcal{O})$ lies in $Z^2(\mathfrak{U},\mathbb{Z})$ and represents $c(D) \in H^2(X,\mathbb{Z})$. (For E. 54c).

54.10 Example. On the Stein space $\mathbb{C}^* \times \mathbb{C}^*$ (see E. 51b), there exist unsolvable multiplicative Cousin distributions. For, with elementary methods of algebraic topology, one can show that $H^2(\mathbb{C}^* \times \mathbb{C}^*,\mathbb{Z}) = H^2(S^1 \times S^1,\mathbb{Z}) = \mathbb{Z}$; then the assertion follows from 54.8.

E. 54c. Show that the divisor $(z_2 - iz_1)$ on $X = V(\mathbb{C}^3; z_1^2 + z_2^2 + z_3^2 - 1)$ is a sum $D_1 + D_{-1}$ of two divisors that are not principal divisors. To that end, prove that

i) For the cover $U_0 = \{\operatorname{re} z_3 > 0\}$, $U_1 = \{z_2 - iz_1 \notin \mathbb{R}_{\leqq 0}\}$, $U_2 = \{z_2 - iz_1 \notin \mathbb{R}_{\geqq 0}\}$, $U_3 = \{\operatorname{re} z_3 < 0\}$, every $U_{j_0\cdots j_q}$ has contractible connected components (so $H^*(\mathfrak{U},\mathbb{Z}) \cong H^*(X,\mathbb{Z})$).

ii) If D_1 is given by $f_0 = z_2 - iz_1$, $f_1 = f_2 = f_3 = 1$, then $c(D_1)$ generates $H^2(\mathfrak{U},\mathbb{Z}) = \mathbb{Z}$.

(Hints: i) For U_0, and for U_3 with appropriate changes, use a deformation retract $\varrho(z_3,t)$ of $\{z_3 \in \mathbb{C}; \operatorname{re} z_3 > 0\}$ to $\{1\}$ with $\varrho(z_3,t)^2 = (1-t)^2 z_3^2 + t^2$; for U_1 and U_2, put $\zeta = z_2 - iz_1$, and show that $X \cap \{\zeta \neq 0\} \to \mathbb{C}^* \times \mathbb{C}, (z_1,z_2,z_3) \mapsto (\zeta, z_3)$, is biholomorphic.

ii) The characteristic functions of the connected components of $U_{j_0\cdots j_q}$ form a $\mathbb{Z}$-basis for $C_a^q(\mathfrak{U},\mathbb{Z})$; show that $H^2(\mathfrak{U},\mathbb{Z}) = C_a^2 / dC_a^1 \cong \mathbb{Z}$ is generated by the class $c(D_1)$ (which is determined according to E. 54b); also see E. 50b. Note that X is simply connected, since it is homotopically equivalent to $S^2 = X \cap \mathbb{R}^3$ [Hi-Ma § 12]. The assignment $(\lambda, z) \mapsto \sqrt{\lambda + 1} \cdot z$ maps $\boldsymbol{P}^1(1) \times X$ biholomorphically onto the simply connected domain $G = \{z \in \mathbb{C}^3; |z_1^2 + z_2^2 + z_3^2 - 1| < 1)$; consequently, there are divisors on G that are not principal divisors.)

E. 54d. For $\zeta \in H^1(X,\mathcal{O}^*)$ to lie in the image of $\mathcal{D}(X)$, it is necessary and sufficient that $\mathcal{O}(\zeta)$ possess a nowhere trivial meromorphic section (to be defined in the obvious manner; hint: proof of 54.6).

E. 54e. Prove the following for a reduced B-space X: if a multiplicative Cousin distribution $(m_j \in \mathcal{M}^*(U_j))_{j\in J}$ has a "continuous solution" $f \in \mathscr{C}(X)$ (i.e., $f_j := m_j / (f|_{U_j}) \in \mathscr{C}^*(U_j)$), then it has a holomorphic solution (hint: the inclusion $\mathcal{O}^* \subset \mathscr{C}^*$ determines an isomorphism $H^1(X, \mathcal{O}^*) \cong H^1(X, \mathscr{C}^*)$).

E. 54f. Let X be a B-space with $H^2(X, \mathbb{Z}) = 0$. If $\mathscr{I} \subset \mathcal{O}$ is a coherent sheaf of ideals whose stalks are principal ideals, then there exists an $f \in \mathcal{O}(X)$ such that $\mathscr{I} = \mathcal{O}\cdot(f)$. In particular, $V(X;\mathscr{I}) = V(X;f)$ (hint: 54.4; for 54 A.3).

E. 54g. Define divisors for subsets of complex spaces, and extend the results 54.4–54.8 to B-sets in metrizable complex spaces X (hint: 50.14).

E. 54h. Let X be a reduced B-space; for $D = (f_j \in \mathcal{O}(U_j) \cap \mathcal{M}^*(U_j))_{j\in J} \in \mathscr{D}(X)$, there is an associated coherent sheaf of ideals $\mathcal{O}(-D)$ with $\mathcal{O}(-D)|_{U_j} = \mathcal{O}\cdot f_j$. Show that there exists an $F \in \mathcal{O}(-D)(X)$ such that $g_j := F / f_j \in \mathcal{O}(U_j)$ and $\operatorname{codim}_{U_j} N(f_j, g_j) \geqq 2$ for every j. If, in addition, $A \subset X \setminus |D|$ is discrete and closed with respect to X, where $|D| := \operatorname{supp} \mathcal{O}/\mathcal{O}(-D)$, then F can be chosen such that in addition every set $N(A; g_j)$ is empty (hint: let $B \subset X$ be a discrete subset that meets each irreducible component of $|D|$ in exactly one point; apply 52.6 to $\mathcal{O}(-D) / \mathcal{O}(-D)\cdot {}_A\mathscr{I}\cdot {}_B\mathscr{I}$; for 54 A.3).

§ 54 A Supplement: The Poincaré Problem

In this supplement, we present divisibility theory in rings of global functions with the help of divisors. Specifically, we intend to investigate the *Poincaré problem*: when can a meromorphic function $m \in \mathcal{M}(X)$ be represented as a quotient $m = f/g$ of global holomorphic functions that are *relatively prime*[4] in $\mathcal{O}(X)$? The *strong Poincaré problem* is to find such f and g that are even *relatively prime everywhere* (i.e., f_x and g_x are relatively prime in $\mathcal{O}_x$ for every $x \in X$). Clearly, everywhere relatively prime functions are relatively prime, whereas the converse does not hold:

E. 54A a. Let X be the Segre cone $V(\mathbb{C}^4; z_1 w_2 - z_2 w_1)$. For $f_j := z_j|_X \in \mathcal{O}(X)$, show that f_1 and f_2 are relatively prime in ${}_X\mathcal{O}_0$, and thus also in $\mathcal{O}(X)$, but not in ${}_X\mathcal{O}_a$ for every point $a \neq 0$ (hint: E. 48m i)).

54 A. 1 Lemma. *Let $\mathcal{O}_x$ be factorial for each $x \in X$. Then two holomorphic functions that are relatively prime at a point $a \in X$ are relatively prime at every point near a.*

Proof. Without loss of generality, suppose that X is connected. Since X is irreducible at every point by 44.14, it follows from E. 49 l that X is of pure dimension. Then E. 48 l implies that the following conditions are equivalent:

α) f and g are relatively prime in $\mathcal{O}_a$;
β) $\operatorname{codim}_{X_a}[N(f) \cap N(g)]_a \geqq 2$;
γ) $\dim[N(f) \cap N(g)]_a \leqq \dim X_a - 2$.

The semicontinuity of dimension (see 49.13) implies that the last inequality holds near a. ∎

54 A. 2 Corollary. *Let $\mathcal{O}_x$ be factorial for every $x \in X$. Then, for each divisor $D \in \mathscr{D}(X)$, there*

---

[4] Elements r and s of an integral domain R are called *relatively prime* if every common divisor of r and s in R is a unit in R.

exist holomorphic divisors D_+, D_- *such that* $D = D_+ - D_-$[5]*; specifically, there exists an open cover* $(U_j)_{j\in J}$ *of* X *such that* $D = (m_j \in \mathcal{M}^*(U_j))_{j\in J}$, $D_+ = (f_j^+ \in \mathcal{O}(U_j)\cap\mathcal{M}^*(U_j))_{j\in J}$, *and* $D_- = (f_j^- \in \mathcal{O}(U_j)\cap\mathcal{M}^*(U_j))_{j\in J}$ *with* $m_j = f_j^+/f_j^-$.

Proof. By 54 A. 1, there is an open cover $\mathfrak{U}$ of X such that there exists an everywhere relatively prime representation $m_j = f_j^+/f_j^-$ on each U_j. Then D_+ and D_- are holomorphic divisors, since the equality $f_i^+ f_j^- = h_{ji} f_j^+ f_i^-$ with $h_{ji} = m_i/m_j \in \mathcal{O}^*(U_{ij})$ implies that f_i^+/f_j^+ and f_j^+/f_i^+ are in $\mathcal{O}(U_{ij})$, etc., due to the fact that f_j^+ and f_j^- were assumed to be relatively prime. ■

Now we discuss the relationship between the strong Poincaré problem and the multiplicative Cousin problem:

54 A. 3 Proposition. *Let* X *be a B-space, and suppose that* $\mathcal{O}_x$ *is factorial for every* $x \in X$. *Then the following conditions about* X *are equivalent:*

i) Every $m \in \mathcal{M}(X)$ *has a representation* $m = f/g$ *with everywhere relatively prime* $f, g \in \mathcal{O}(X)$.

ii) Every divisor on X *is a principal divisor.*

iii) $H^2(X,\mathbb{Z}) = 0$.

iv) If $\mathfrak{i} \subset \mathcal{O}(X)$ *is an ideal such that every stalk* $(\mathcal{O}\cdot\mathfrak{i})_x$ *is a principal ideal in* $\mathcal{O}_x$, *then* $\mathfrak{i}$ *itself is a principal ideal.*

Proof. i) ⇒ ii) By 54 A. 2, it suffices to find for each holomorphic divisor $D = (f_j \in \mathcal{O}(U_j)\cap\mathcal{M}^*(U_j))_{j\in J}$ an $f \in \mathcal{O}(X)$ with $D = (f)$. Without loss of generality, suppose that the cover $(U_j)_{j\in J}$ is locally finite. By E. 54h, there exists an $F \in \mathcal{O}(X)$ such that $g_j := F/f_j \in \mathcal{O}(U_j)$ and $\operatorname{codim}_{U_j} N(f_j, g_j) \geqq 2$ for every $j \in J$. By E. 48 l, then, f_j and g_j are relatively prime everywhere in U_j. Analogously, there exists a $G \in \mathcal{O}(X)$ such that $h_j := G/g_j \in \mathcal{O}(U_j)$ and $\operatorname{codim}_{U_j} N(g_j, h_j) \geqq 2$, and even so that $\operatorname{codim}_{U_j} N(f_j, h_j) \geqq 2$ (choose the A in E. 54h so that it meets every irreducible component of $|D|$); then f_j, g_j, and h_j are pairwise relatively prime. By hypothesis, F/G has an everywhere relatively prime representation $F/G = f/g$ in $\mathcal{O}(X)$; on U_j, then, $f/g = f_j/h_j$, and it follows that $f|f_j$ and $f_j|f$. Hence, $D = (f)$.

ii) ⇒ i) Without loss of generality, suppose that X is connected. Represent $m \in \mathcal{M}^*(X)$ locally according to 54 A. 1 with everywhere relatively prime $f_j, g_j \in \mathcal{O}(U_j)$. By 54 A. 2, that determines holomorphic divisors D_+ and D_-, for which there must exist $f, g \in \mathcal{O}(X)$ such that $D_+ = (f)$ and $D_- = (g)$, by assumption. Then f and g are everywhere relatively prime, and we see that $(m) = (f/g)$. Hence, there exists a unit $h \in \mathcal{O}^*(X)$ such that $m = hf/g$, and hf and g are everywhere relatively prime.

The implications "ii) ⇒ iii)" and "iii) ⇒ iv)" follow from 54.8 and E. 54f, respectively.

iv) ⇒ ii) By 54 A. 2, it suffices to show that every holomorphic divisor D is a principal divisor. The stalks of the coherent sheaf of ideals $\mathcal{O}(-D)$ are principal ideals, so there exists an f with $\mathcal{O}(-D)(X) = (\mathcal{O}\cdot f)(X)$. By Theorem A, every stalk $\mathcal{O}(-D)_x$ is generated by f, so $D = (f)$. ■

54 A. 4 Remark. *Let* X *be a complex space in which every strong Poincaré problem is solvable, and suppose that* $\mathcal{O}_x$ *is factorial for every* $x \in X$. *Then the following statements hold:*

i) $f, g \in \mathcal{O}(X)$ *are relatively prime iff they are everywhere relatively prime.*

ii) If $f, g \in \mathcal{O}(X)$ *are relatively prime, and if* $f|gh$ *for some* $h \in \mathcal{O}(X)$, *then* $f|h$.

Proof. i) For $f, g \in \mathcal{O}(X)$, it remains only to show that "relatively prime ⇒ everywhere relatively prime". We may assume that X is connected and that $g \neq 0$. Hence, $f/g \in \mathcal{M}(X)$, and there exists an everywhere relatively prime representation $f/g = \tilde{f}/\tilde{g}$. Then $f\tilde{g} = \tilde{f}g$, and $\tilde{f}$ divides f in every ring $\mathcal{O}_x$ (which is factorial by hypothesis). Consequently, $h := f/\tilde{f} = g/\tilde{g}$ is holomorphic at every point. Since $f = h\tilde{f}$ and $g = h\tilde{g}$ are relatively prime by assumption, h must be a unit

[5] The group operation in $\mathcal{D}(X)$ is written additively.

in $\mathcal{O}(X)$; hence, the fact that $\tilde{f}$ and $\tilde{g}$ are everywhere relatively prime implies that f and g are likewise everywhere relatively prime.

ii) According to i), f and g are everywhere relatively prime, so $f_x | h_x$ for every x. In other words, h/f lies in $\mathcal{O}(X)$. ■

E. 54Ab. For a complex space X, the stalks of whose structure sheaf are factorial, show that every Cartier divisor $D \in \mathcal{D}(X)$ determines a "*Weil divisor*" $((A_j, m_j))_{j \in J}$, where $(A_j)_{j \in J}$ is a locally finite family of irreducible analytic subsets of codimension one with zero-orders $m_j \in \mathbb{Z} \setminus \{0\}$. Conversely, every Cartier divisor is determined by a unique Weil divisor $((A_j, m_j))_{j \in J}$. On spaces whose structure sheaf has nonfactorial stalks, the Cartier and Weil divisors can no longer be identified in general. (Hint: E. 48l, E. 48m).

E. 54Ac. Extend 54 A. 3 to B-sets.

§ 54B Supplement: Holomorphic Line Bundles

The obstructions to the solution of the multiplicative Cousin problem appeared as cohomology classes in $H^1(X, \mathcal{O}^*)$. We want to interpret such classes as systems of "*transition functions*" for holomorphic "line bundles":

54 B. 1 Definition. *A (concrete holomorphic) vector bundle of rank r over a complex space X is a complex space (E, π) over X satisfying the following conditions:*

i) For each $x \in X$, the fiber E_x[6]$:= \pi^{-1}(x)$ has the structure of an r-dimensional vector space.

ii) There exist an open cover $(U_j)_{j \in J}$ of X and biholomorphic mappings

$$\varphi_j : \pi^{-1}(U_j) \to U_j \times \mathbb{C}^r$$

over U_j (bundle charts) that induce an isomorphism $\varphi_j|_{E_x} : E_x \to \mathbb{C}^r$ of vector spaces for each $x \in U_j$.

Vector bundles of rank 1 are called *line bundles.* A *morphism* $h : (E_1, \pi_1) \to (E_2, \pi_2)$ *of vector bundles* over X is a holomorphic mapping over X that is linear on every fiber.

54 B. 2 Examples. i) Every vector bundle isomorphic to the product bundle $\mathrm{pr}_X : X \times \mathbb{C}^r \to X$ is called an *(analytically) trivial vector bundle.*

ii) For an open cover $\mathfrak{U} = (U_j)_{j \in J}$ of X, and holomorphic functions

$$\psi_{jk} : U_{jk} \to GL(r, \mathbb{C}) \quad \text{with} \quad \psi_{jj} = I_r \quad \text{and} \quad \psi_{jk} \circ \psi_{kl} = \psi_{jl},$$

the quotient space $E(\psi) := (\bigcup_j U_j \times \mathbb{C}^r)/\sim$, with the equivalence relation

$$(u, z) \sim (v, w) \ :\Leftrightarrow\ u = v \ \text{and} \ w = \psi_{jk}(u) \cdot z$$

for $(u, z) \in U_k \times \mathbb{C}^r$ and $(v, w) \in U_j \times \mathbb{C}^r$, and the canonical projection onto X, is a vector bundle of rank r over X. The family $\psi := (\psi_{jk})_{j, k \in J}$ is called the system of *transition functions* of $E(\psi)$. With the aid of the sheaf $\mathcal{GL}(r, \mathcal{O})$ of invertible holomorphic $r \times r$-matrices (for $r > 1$, $\mathcal{GL}(r, \mathcal{O})$ is a sheaf of nonabelian groups), ψ can be interpreted as an alternating 1-cocycle in $Z^1_a(\mathfrak{U}, \mathcal{GL}(r, \mathcal{O}))$; conversely, a system of bundle charts $(\varphi_j)_{j \in J}$ of a vector bundle E of rank r determines a 1-cocycle $\psi := (\varphi_j \circ \varphi_k^{-1}) \in Z^1_a(\mathfrak{U}, \mathcal{GL}(r, \mathcal{O}))$ with $E(\psi) \cong E$.

iii) By adapting constructions from (multi-)linear algebra (see [La$_3$ II, § 4]), we may use given vector bundles E, E_1, E_2 to form new ones:

---

[6] Since $x \notin E_x$, the reader will not misinterpret E_x as a germ.

$E_1 \oplus E_2$ (*direct* or Whitney *sum*)

$E_1 \otimes E_2$ (*tensor product*)

$\mathrm{Hom}(E_1, E_2)$ (*Hom bundle*)

$E' = \mathrm{Hom}(E, X \times \mathbb{C})$ (*dual bundle*)

$\Lambda^k E$ (*k-th exterior power*).

iv) If X is a *manifold* described by charts $h_j: U_j \to V_j \subset \mathbb{C}^n$, then the cocycle given by the holomorphic functional matrices $\frac{\partial}{\partial z}(h_j \circ h_k^{-1})$ determines the *tangent bundle* TX of X. Its dual is called the *cotangent bundle* $T'X$, and the exterior power $\Lambda^{\dim X} T'X =: K_X$ is the *canonical* (*line*) *bundle* of X.

E. 54 B a. Determine the transition functions for the canonical bundle (and use them to motivate the common appelation "*determinant bundle*").

Vector bundles E of rank r on X and locally free ${}_X\mathcal{O}$-modules $\mathcal{F}$ of rank r are closely related:

i) (E, π) determines the ${}_X\mathcal{O}$-module ${}^E_X\mathcal{O}$ of holomorphic sections of π; in particular, ${}^{X\times\mathbb{C}}_X\mathcal{O} = {}_X\mathcal{O}$. For a *manifold* X, that permits the construction of the sheaf of *vector fields* ${}_X\Theta := {}^{TX}_X\mathcal{O}$ on X (tangent sheaf) and the sheaf ${}_X\Omega^p := {}^{\Lambda^p T'X}_X\mathcal{O}$ of *holomorphic p-forms*, in particular, the *canonical sheaf* ${}_X\Omega^{\dim X}$.

ii) $\mathcal{F}$ determines the vector bundle $E(\psi)$, with $\psi = (\psi_{jk}) \in Z^1_a(\mathfrak{U}, \mathcal{GL}(r, \mathcal{O}))$ defined as follows: if $\mathcal{F}$ is given by isomorphisms $\chi_j: \mathcal{F}|_{U_j} \to \mathcal{O}^r|_{U_j}$, then $\psi_{jk} \in \mathcal{GL}(r, \mathcal{O})(U_{jk})$ is the matrix of the isomorphism $\chi_j \circ \chi_k^{-1}: \mathcal{O}^r|_{U_{jk}} \to \mathcal{O}^r|_{U_{jk}}$ with respect to the canonical basis.

It can be shown that those assignments yield an equivalence of categories:

54 B. 3 Proposition. *The following conditions on cocycles* $\psi \in Z^1(\mathfrak{U}, \mathcal{GL}(r, \mathcal{O}))$ *and* $\psi' \in Z^1(\mathfrak{U}', \mathcal{GL}(r, \mathcal{O}))$ *are equivalent:*

i) $\psi = \psi'$ *in* $H^1(X, \mathcal{GL}(r, \mathcal{O}))$;

ii) $E(\psi) \cong E(\psi')$;

iii) $\mathcal{F}(\psi) \cong \mathcal{F}(\psi')$.

Condition i) means that, on a common refinement $\mathfrak{V}$ of $\mathfrak{U}$ and $\mathfrak{U}'$, there is a 0-cochain $(\varphi_j) \in C^0(\mathfrak{V}, \mathcal{GL}(r, \mathcal{O}))$ with $\psi'_{jk} = \varphi_j^{-1} \circ \psi_{jk} \circ \varphi_k$ on V_{jk}; the locally free ${}_X\mathcal{O}$-module $\mathcal{F}(\psi)$ of rank r is constructed as in the proof of 54.7.

The proof of 54 B.3 is not difficult [Ku § 36]. ■

By broadening the scope from the category of locally free sheaves to that of coherent analytic sheaves (this category is closed with respect to formation of kernels and cokernels), one can expand from vector bundles to the category of linear spaces, which is antiquivalent to the category of coherent analytic sheaves [Fi 1.4ff].

Having made those general remarks, we turn to the special cases of vector bundles on Stein spaces and on $\mathbb{P}_1$. In contrast to the case of the general locally trivial fibrations with fiber $\mathbb{C}^r$ (see the remarks in connection with E. 51b), the basis X of a vector bundle E has a strong geometric influence on the structure of E:

54 B. 4 Proposition. *The following implications*[7] *hold for a vector bundle* (E, π) *over* X:
i) *X is a B-space* $\Rightarrow$ ii) *E is a Stein space* $\Rightarrow$ iii) *X is a Stein space.*

Proof. ii) $\Rightarrow$ iii) This follows from 51.4, since the "*zero-section*" $x \mapsto 0_x \in E_x$ determines an embedding $X \hookrightarrow E$.

---

[7] By Theorem B, they are in fact equivalences.

i) $\Rightarrow$ ii) Let us see first that E is *holomorphically convex.* By 51 A.3 and 52.6, we may assume that the topology of X is countable, and thus that E has a countable topology, as well. For a sequence $(y_j)_{j\in\mathbb{N}}$ in E with no accumulation point, we seek an $f\in\mathcal{O}(E)$ such that

$$\sup|f(y_j)| = \infty .$$

If $(x_j) := (\pi(y_j))$ has no accumulation point in X, then there exists such an f in $\pi^0\mathcal{O}(X)\subset\mathcal{O}(E)$ by 52.6 iv). Otherwise, we may assume that (x_j) converges to $x\in X$. Choose an open neighborhood U of x with a bundle-isomorphism

$$\varphi : E|_U \to U\times\mathbb{C}^n .$$

Since the sequence $(\mathrm{pr}_{\mathbb{C}^n}\circ\varphi(y_j))$ cannot have an accumulation point in $\mathbb{C}^n$, we obtain, with one of the coordinate functions of $\mathbb{C}^n$, a fiberwise linear $g\in\mathcal{O}(E|_U)$ such that $\sup|g(y_j)|=\infty$. Then

$$h := (\mathrm{pr}_U, g) : E|_U \to U\times\mathbb{C}$$

is a linear form of bundles, and thus a section over U in the coherent ${}_X\mathcal{O}$-module ${}^{E'}_{X}\mathcal{O}$. By Theorem A, we may assume that there exist global sections $\sigma_1,\dots,\sigma_r\in {}^{E'}_{X}\mathcal{O}(X)$ and $f_1,\dots,f_r\in{}_X\mathcal{O}_x$ such that

$$h = \sum_{k=1}^{r} f_k\sigma_k \quad\text{in}\quad {}^{E'}_{X}\mathcal{O}(U).$$

Since $g=\mathrm{pr}_{\mathbb{C}}\circ h$, there exists a k such that

$$\sup|(\mathrm{pr}_{\mathbb{C}}\circ\sigma_k)(y_j)| = \infty .$$

It can be shown in a completely analogous fashion that E is *holomorphically separable.* ■

Now we want to present an example of *Oka's Principle*, according to which, on *reduced Stein spaces*, holomorphic problems that can be formulated in the language of cohomology, and that are solvable topologically, have holomorphic solutions, as well (see E. 54e for an other example): on a manifold X, by defining a continuous (resp., $\mathscr{C}^\infty$-)vector bundle, and, *mutatis mutandis*, replacing "holomorphic" in 54 B.1 with "continuous" (resp., $\mathscr{C}^\infty$), we obtain the following result:

54 B. 5 Proposition. *If X is a paracompact manifold with $H^1(X,\mathcal{O})=0$, and (E,π) is a line bundle on X, then E is determined by the Chern class $c(E)\in H^2(X,\mathbb{Z})$, and the following statements are equivalent:*

i) *E is analytically trivial.*

ii) *E is $\mathscr{C}^\infty$-trivial.*

iii) *E is topologically trivial.*

iv) *$c(E)=0$.*

Proof. The sheaves $\mathscr{C}$ and $\mathscr{C}^\infty$ on X are fine (E. 50f), and thus acyclic (50.8); hence, the exponential sequences (in analogy to (54.3.1)) yield a commutative diagram

$$\begin{array}{ccccc}
0 = H^1(X,\mathcal{O}) & \to & H^1(X,\mathcal{O}^*) & \xrightarrow{c} & H^2(X,\mathbb{Z}) \\
\downarrow & & \downarrow\alpha & & \| \\
0 = H^1(X,\mathscr{C}^\infty) & \to & H^1(X,\mathscr{C}^{\infty *}) & \xrightarrow{c} & H^2(X,\mathbb{Z}) \to 0 \\
\downarrow & & \downarrow\beta & & \| \\
0 = H^1(X,\mathscr{C}) & \to & H^1(X,\mathscr{C}^*) & \xrightarrow{c} & H^2(X,\mathbb{Z}) \to 0
\end{array}$$

with exact rows. It follows that α and β are injective; in each case, the trivial bundle is determined by the neutral element $1\in H^1(X,\mathcal{O}^*)$, etc. ■

B-manifolds and $\mathbb{P}_n$ satisfy the hypotheses of 54 B.5 (see E.52a and E.74f).

54 B.6 Example. *A line bundle that is topologically, but not analytically, trivial.* The space $X = \mathbb{C}^{2*}$ is homeomorphic to $S^3 \times \mathbb{R}_{>0}$; with elementary methods of algebraic topology, it is possible to show that $H^1(X,\mathbb{Z}) = 0 = H^2(X,\mathbb{Z})$, and it follows from the cohomology sequences induced by (54.3.1) that

$$H^1(X,\mathscr{C}^*) \to H^2(X,\mathbb{Z}) = 0 \quad \text{and} \quad H^1(X,\mathcal{O}) \xrightarrow{\exp^*} H^1(X,\mathcal{O}^*), [(s_{01})] \mapsto [(\exp 2\pi i s_{01})],$$

are isomorphisms (with the cover consisting of $U_0 = \mathbb{C}^* \times \mathbb{C}$ and $U_1 = \mathbb{C} \times \mathbb{C}^*$). Hence, the line bundle given by $s_{01} = 1/z_0 z_1 \in \mathcal{O}(U_{01})$ (see 52.3 vii)), determines the neutral element in $H^1(X,\mathscr{C}^*)$, but not in $H^1(X,\mathcal{O}^*)$.

E. 54 Bb. Prove that, on every one-dimensional torus T, there exist uncountably many non-isomorphic line bundles that are topologically, but not analytically, trivial (hint: apply $H^2(T,\mathbb{Z}) = \mathbb{Z}, H^1(T,\mathbb{Z}) = H^1(S^1 \times S^1,\mathbb{Z}) = \mathbb{Z} \oplus \mathbb{Z}$, and E. 50Ac ii)).

The remainder of this supplement is devoted to the investigation of vector bundles on $\mathbb{P}_1$. By the *Splitting Theorem of Grothendieck*, every vector bundle on $\mathbb{P}_1$ splits into a direct sum of *line bundles* (that is unique up to order of summation) [GrRe$_2$ VII.8.5], so it suffices to investigate just those:

E. 54 Bc. *Holomorphic line bundles on* $\mathbb{P}_1$. Prove the following statements:

i) Each line bundle on $\mathbb{P}_1$ is isomorphic to exactly one bundle $E(m)$ defined for $m \in \mathbb{Z}$ by means of the identification $U_0^* \times \mathbb{C} \to U_1^* \times \mathbb{C}$, $([1,z],\lambda) \mapsto ([1/z,1],\lambda/z^m)$ (hint: E.50b, 54 B.5).

ii) $E(0)$ is the trivial line bundle.

iii) $E(-1)$ is the "*tautological*" *line bundle* ("*Hopf bundle*")

$$\mathrm{pr}_1 : \{([z_0,z_1],\lambda z_0,\lambda z_1) \in \mathbb{P}_1 \times \mathbb{C}^2; [z_0,z_1] \in \mathbb{P}_1, \lambda \in \mathbb{C}\} \to \mathbb{P}_1$$

with the bundle charts $\varphi_j([z_0,z_1],\zeta_0,\zeta_1) \mapsto ([z_0,z_1],\zeta_j)$ (the projection $\mathrm{pr}_2 : E(-1) \to \mathbb{C}^2$ is the quadratic transformation at $0 \in \mathbb{C}^2$, see E.32 Bg γ)).

iv) $E(1)$ is given by the projection $[z_0,z_1,z_2] \mapsto [z_0,z_1]$ of $\mathbb{P}_2 \setminus \{[0,0,1]\}$ onto the hyperplane $\{z_2 = 0\} \cong \mathbb{P}_1$; the bundle charts are $\varphi_j : [z_0,z_1,z_2] \mapsto ([z_0,z_1],z_2/z_j)$. $E(1)$ is called the *normal bundle* of $\mathbb{P}_1 \hookrightarrow \mathbb{P}_2$ *(hyperplane bundle)*.

v) $E(2)$ is isomorphic to the tangent bundle $T\mathbb{P}_1$ (with $\varphi_0^{-1} : U_0 \times \mathbb{C} \to T\mathbb{P}_1|_{U_0}$, $([1,z],\lambda) \mapsto ([1,z],\lambda \cdot \partial/\partial z)$, and the corresponding φ_1^{-1}).

vi) $E(-2)$ is isomorphic to the cotangent bundle $T'\mathbb{P}_1$ (with $\varphi_0^{-1} : ([1,z],\lambda) \mapsto ([1,z],\lambda dz)$ and φ_1^{-1}; since $\dim \mathbb{P}_1 = 1$, $T'\mathbb{P}_1$ is also the canonical bundle $K_{\mathbb{P}_1}$). The mapping

$$E(-2) \to N(\mathbb{C}^3; z_1 z_2 - z_3^2), ([1,z],\lambda) \mapsto (\lambda, \lambda z^2, \lambda z), \text{ and } ([w,1],\mu) \mapsto (\mu w^2, \mu, \mu w),$$

is a resolution of the singularity of the surface $\sqrt{z_1 z_2}$ (see 32 B. 10 ii γ)).

vii) $E(m) \otimes E(m') \cong E(m+m')$.

E. 54 Bd. *Global holomorphic and meromorphic sections in* $E(m)$.

i) For the sheaf $\mathcal{O}(m) := {}^{E(m)}_{\mathbb{P}_1}\mathcal{O}$ on $\mathbb{P}_1$, show that

$$\mathcal{O}(m)(\mathbb{P}_1) \cong \begin{cases} \{P \in \mathbb{C}[z_0,z_1];\ P \text{ homogeneous of degree } m\} & \text{for } m \geq 0 \\ 0 & \text{for } m < 0. \end{cases}$$

ii) For $m, m' \geq 0$, with respect to $\mathcal{O}(m) \otimes \mathcal{O}(m') \cong \mathcal{O}(m+m')$, the tensor product of global sections corresponds to the product of homogeneous polynomials.

iii) The sheaf $\mathcal{M}(m) := \mathcal{M} \otimes_{\mathcal{O}} \mathcal{O}(m)$ is a locally free $\mathcal{M}$-module, called "the sheaf of *meromorphic sections* in $E(m)$". We have that

$$\mathcal{M}(m)(\mathbb{P}_1) \cong \{f = P/Q \in \mathbb{C}(z_0, z_1);\, f(\lambda z) = \lambda^m \cdot f(z)\};$$

in particular, every global meromorphic section $s \neq 0$ has *total order* (sum of orders of zeros minus sum of orders of poles) $o(s) = m$ (hint: every meromorphic function on $\mathbb{C}$ without an essential singularity at ∞ is meromorphic on $\mathbb{P}_1$, and thus rational, by E.53 Af).

iv) Adapt ii) for meromorphic sections.

E. 54Be. *Global holomorphic functions on* $E(m)$. Prove the following statements (where S_0 denotes the zero-section):

i)

$$_{E(m)}\mathcal{O}(S_0) \cong \begin{cases} \mathbb{C} & \text{for } m>0 \\ _{\mathbb{P}_1\times\mathbb{C}}\mathcal{O}(\mathbb{P}_1\times\{0\}) \cong {}_1\mathcal{O}_0 & \text{for } m=0 \\ \{\sum\limits_{k=0}^{\infty}\sum\limits_{j=0}^{k|m|} a_{jk} z^j \lambda^k \in {}_{U_0\times\mathbb{C}}\mathcal{O}(U_0\times\{0\})\} \cong & \\ \mathbb{C}\{\lambda, z\lambda, \dots, z^{|m|}\lambda\} \subset {}_2\mathcal{O}_0 & \text{for } m<0; \end{cases}$$

$$\mathcal{O}(E(m)) \cong \begin{cases} \mathbb{C} & \text{for } m>0 \\ \mathcal{O}(\mathbb{P}_1\times\mathbb{C}) \cong {}_1\mathcal{O}(\mathbb{C}) & \text{for } m=0 \\ \{\sum\limits_{k=0}^{\infty}\sum\limits_{j=0}^{k|m|} a_{jk} z^j \lambda^k \in \mathcal{O}(U_0\times\mathbb{C})\} & \text{for } m<0. \end{cases}$$

ii) $E(m)$ is holomorphically convex iff $m \leqq 0$.

iii) The ringed quotient $E(m)/S_0$ is separable with respect to global sections in its structure sheaf iff $m<0$.

(Hint: Prove i) and iii) with "monomials" and use the compacta

$$K_j = \{|z_1| \leqq 1, |\lambda_1| \leqq j\} \cup \{|z_0| \leqq 1, |\lambda_0| \leqq j\} \subset (U_0\times\mathbb{C}) \cup (U_1\times\mathbb{C})$$

for ii). For E. 57h.)

E. 54Bf. The Hirzebruch surfaces Σ_m are locally trivial holomorphic fiber spaces over $\mathbb{P}_1$ with fiber $\mathbb{P}_1$ that admit a "zero-section" S_0 and a "section at infinity" S_∞. Show that $\Sigma_m \setminus S_\infty$ is isomorphic to $E(m)$, and $\Sigma_m \setminus S_0$ is isomorphic to $E(-m)$. Conclude that $\Sigma_m \to \mathbb{P}_1$ has other holomorphic sections. (Σ_m is an example of a "holomorphic $\mathbb{P}_1$-bundle" over $\mathbb{P}_1$.)

Further literature: [Gu], [OkSnSp].

§ 55 Coherent Analytic Sheaves as Fréchet Sheaves

In the proof of Theorem B, we use an approximation for sections in coherent analytic sheaves (56.5). In preparation for that, we now introduce in a canonical manner the structure of a Fréchet space on the space $\mathcal{G}(X)$ of global sections of a coherent $_X\mathcal{O}$-module $\mathcal{G}$.

For simplicity, we assume that *every connected component of X has a countable topology* (the general case can be treated using projective limits; we do not delve into that generalization except for a very particular statement (see E. 55i).

Denote as usual the maximal ideal of $\mathcal{O}_x$ with $\mathfrak{m}_x$; for every $j \in \mathbb{N}$, the vector space

$\mathscr{G}_x/\mathfrak{m}_x^j\cdot\mathscr{G}_x$ is finite-dimensional and thus has a *unique* topology that makes it a Hausdorff topological vector space. The following discussion is based on these facts:

(55.1.1) *Each homomorphism* $\pi_x:\mathscr{G}_x \to \prod_{j=1}^{\infty} \mathscr{G}_x/\mathfrak{m}_x^j\cdot\mathscr{G}_x$ *is injective (see 23 A.5).*

(55.1.2) *The homomorphism* $\mathscr{G}(X) \to \prod_{x\in X} \mathscr{G}_x$ *is injective.*

(55.1.3) *If V is a Hausdorff topological vector space and $\varphi:\mathscr{G}(X)\to V$, an injective linear mapping, then there exists at most one Fréchet topology on $\mathscr{G}(X)$ that makes φ continuous (see 55.8 i)).*

55.1 Lemma and Definition. *Let $(X, \mathcal{O})$ be a complex space and $\mathscr{G}$, a coherent $\mathcal{O}$-module. Then there is at most one Fréchet topology on $\mathscr{G}(X)$ (called the canonical topology) such that one of the two following equivalent conditions holds:*

i) $\pi_{x,j}:\mathscr{G}(X)\to\mathscr{G}_x/\mathfrak{m}_x^j\cdot\mathscr{G}_x$ is continuous for every $x\in X$, $j\in\mathbb{N}$.

ii) The injective homomorphism of vector spaces

$$(\pi_{x,j})_{x,j}:\mathscr{G}(X) \to \prod_{x,j} \mathscr{G}_x/\mathfrak{m}_x^j\cdot\mathscr{G}_x$$

is continuous.

Proof. By (55.1.1) and (55.1.2), the mapping $(\pi_{x,j})_{x,j}$ is injective. Since $\prod_{x,j}\mathscr{G}_x/\mathfrak{m}_x^j\cdot\mathscr{G}_x$ is a Hausdorff space, the assertion follows from (55.1.3). ■

E. 55a. Let $\mathscr{G}(X)$ be endowed with the canonical topology and let $\varphi: F\to\mathscr{G}(X)$ be a linear mapping between Fréchet spaces. Then φ is continuous iff each composition

$$F \xrightarrow{\varphi} \mathscr{G}(X) \xrightarrow{\pi_{x,j}} \mathscr{G}_x/\mathfrak{m}_x^j\cdot\mathscr{G}_x$$

is continuous (hint: 55.8; for 55.5).

55.2 Proposition. *For $X \Subset \mathbb{C}^n$, the topology of compact convergence on $\mathcal{O}(X)$ is the canonical topology.*

Proof. By 4.3, each homomorphism

$$\pi_{x,j}:\mathcal{O}(X) \to \mathcal{O}_x/\mathfrak{m}_x^j,\ f\mapsto \sum_{|\nu|<j} \frac{D^\nu f(x)}{\nu!}(z-x)^\nu,$$

is continuous. ■

If, for a complex space X, the canonical topology on $\mathcal{O}(X)$ exists, then we see that the product topology is the canonical topology on $\mathcal{O}^p(X)$.

In order to prove the *existence* of the canonical topology, we need the following:

55.3 Lemma. *For a coherent $\mathcal{O}$-module $\mathscr{G}$ on a region $X\Subset\mathbb{C}^n$, let $\varphi:\mathcal{O}^p\to\mathscr{G}$ be an $\mathcal{O}$-homomorphism. Then* $\operatorname{Ker}(\varphi(X):\mathcal{O}^p(X)\to\mathscr{G}(X))$ *is a closed submodule of* $\mathcal{O}^p(X)$.

Proof. The commutative diagram

$$\begin{array}{ccc} \mathcal{O}^p(X) & \xrightarrow{\varphi(X)} & \mathcal{G}(X) \\ \cap\downarrow (\pi_{x,j})_{x,j} & & \cap\downarrow \\ \prod\limits_{x,j} \mathcal{O}_x^p/\mathfrak{m}_x^j\cdot\mathcal{O}_x^p & \xrightarrow{\Pi\varphi_{x,j}} & \prod\limits_{x,j} \mathcal{G}_x/\mathfrak{m}_x^j\cdot\mathcal{G}_x \end{array}$$

yields that $\operatorname{Ker}\varphi(X) = \operatorname{Ker}(\Pi\varphi_{x,j}\circ\pi_{x,j}) = \bigcap\limits_{x,j}\pi_{x,j}^{-1}(\operatorname{Ker}\varphi_{x,j})$. The assertion follows from the fact that the homomorphisms $\varphi_{x,j}$ and $\pi_{x,j}$ are continuous. ∎

E. 55b. Let $\mathcal{G}$ be a coherent analytic sheaf on a complex space X.

i) For $U, V \Subset X$, define $\beta: \mathcal{G}(U)\times\mathcal{G}(V)\to\mathcal{G}(U\cap V)$ by setting $\beta(f,g) := f|_{U\cap V} - g|_{U\cap V}$. Prove that $\operatorname{Ker}\beta$ is closed in $\mathcal{G}(U)\times\mathcal{G}(V)$ if the canonical topology on $\mathcal{G}(U)$ and $\mathcal{G}(V)$ exists (hint: $\operatorname{Ker}\beta = \bigcap\limits_{x\in U\cap V,\, j\in\mathbb{N}} \operatorname{Ker}(\mathcal{G}(U)\times\mathcal{G}(V)\to\mathcal{G}_x/\mathfrak{m}_x^j\cdot\mathcal{G}_x)$; for 55.5).

ii) For a submodule $N\subset\mathcal{G}_x$, prove that $F := \{f\in\mathcal{G}(X);\, f_x\in N\}$ is a closed submodule of $\mathcal{G}(X)$ (hint: $F = \bigcap\limits_{j\in\mathbb{N}} \pi_{x,j}^{-1}(N/(\mathfrak{m}_x^j\cdot N))$; for 55.9).

55.4 Definition. *Let $\mathcal{G}$ be a sheaf of vector spaces on a topological space T with a countable topology. Then $\mathcal{G}$ is called a* <u>*Fréchet sheaf*</u> *if*

i) $\mathcal{G}(U)$ is a Fréchet space for every $U\Subset T$, and

ii) for every $V\Subset U\Subset T$, the restriction-homomorphism $\mathcal{G}(U)\to\mathcal{G}(V)$ is continuous.

Now, we can prove the main result of this section:

55.5 Theorem. *Let $(X,\mathcal{O})$ be a complex space and $\mathcal{G}$, a coherent $\mathcal{O}$-module. Then, for every $U\Subset X$, the canonical topology on $\mathcal{G}(U)$ exists; in particular, $\mathcal{G}$ is a Fréchet sheaf.*[8]

Proof. a) Suppose that X is a region in $\mathbb{C}^n$ such that there exist exact sequences

$$\mathcal{O}^p \xrightarrow{\psi} \mathcal{G}\to 0 \quad\text{and}\quad \mathcal{O}^p(X)\xrightarrow{\psi(X)}\mathcal{G}(X)\to 0.$$

By 55.3, $\operatorname{Ker}\psi(X)$ is closed in $\mathcal{O}^p(X)$; hence, $\mathcal{G}(X)$ inherits a Fréchet structure from $\mathcal{O}^p(X)$ such that $\psi(X)$ is an open mapping (see 55.8 ii)). That structure is the canonical one: in the commutative diagram

$$\begin{array}{ccc} \mathcal{O}^p(X) & \xrightarrow{\psi(X)} & \mathcal{G}(X) \\ \downarrow \tau_{x,j} & & \downarrow \pi_{x,j} \\ \mathcal{O}_x^p/\mathfrak{m}_x^j\cdot\mathcal{O}_x^p & \xrightarrow{\psi_{x,j}} & \mathcal{G}_x/\mathfrak{m}_x^j\cdot\mathcal{G}_x, \end{array}$$

---

[8] In the proof of 55.5, we use 61.13. We want to point out that the proof of 61.13 does not depend on the validity of 55.5: according to § 2, the sheaves ${}_n\mathcal{O}$ are Fréchet sheaves; that fact justifies the application of 56.3 and 56.4 with $\mathcal{G} = {}_n\mathcal{O}$ in the proof of 61.8; finally, 61.13 is a formal consequence of 61.8.

$\pi_{x,j}$ is continuous, since $\psi(X)$ is open and surjective, and since $\tau_{x,j}$ and $\psi_{x,j}$ are continuous.

b) Now let $(X, \mathcal{O})$ be a complex space. Consider a local model $U \subset\subset X$ isomorphic to an analytic subspace $i:(A, {}_A\mathcal{O}) \hookrightarrow \boldsymbol{P} \subset\subset \mathbb{C}^n$. Then, in a canonical manner, $i(\mathcal{G}|_U)$ is a coherent ${}_P\mathcal{O}$-module. If $\boldsymbol{P}$ (and hence U) is sufficiently small, then 61.13 ensures that there exists an exact sequence ${}_P\mathcal{O}^p \to i(\mathcal{G}|_U) \to 0$ that induces an exact sequence ${}_P\mathcal{O}^p(\boldsymbol{P}) \to \mathcal{G}(U) \to 0$.
By a), then, the canonical topology on $\mathcal{G}$(U) exists.

c) Let $\mathfrak{U} = (U_j)_{j \in J}$ denote a countable open cover of the complex space X with local models U_j for which the canonical topology on $\mathcal{G}(U_j)$ exists. Then the exactness of the sequence (E. 30a. 1)

$$0 \to \mathcal{G}(X) \xrightarrow{\alpha} \prod_{j \in J} \mathcal{G}(U_j) \xrightarrow{\beta} \prod_{i,j} \mathcal{G}(U_{ij})$$

ensures that $\operatorname{Im} \alpha = \operatorname{Ker} \beta$ is a closed subspace of the Fréchet space $\prod \mathcal{G}(U_j)$ (see E. 55b). It follows that the induced topology on $\mathcal{G}(X)$ is the canonical topology, since every restriction $\mathcal{G}(X) \to \mathcal{G}(U_j)$ is continuous.

We still have to demonstrate that, for $U \subset\subset X$ without loss of generality, the restriction $\varrho: \mathcal{G}(X) \to \mathcal{G}(U)$ is continuous. By E. 55a, we need only show that each composition

$$\mathcal{G}(X) \xrightarrow{\varrho} \mathcal{G}(U) \xrightarrow{\pi_{x,j}} \mathcal{G}_x / \mathfrak{m}_x^j \cdot \mathcal{G}_x$$

is continuous; that is obviously true, since $\mathcal{G}(X)$ has the canonical topology. ∎

55.6 Proposition. *i) Let $\varphi: \mathcal{F} \to \mathcal{G}$ be a homomorphism of coherent $\mathcal{O}$-modules on the complex space $(X, \mathcal{O})$. Then $\varphi(X): \mathcal{F}(X) \to \mathcal{G}(X)$ is continuous.*

ii) Let $f: X \to Y$ be a holomorphic mapping between complex spaces. Then $f^0: {}_Y\mathcal{O}(Y) \to {}_X\mathcal{O}(X)$ is continuous.

Proof. i) By E. 55a, it suffices to prove that every composition $\pi_{x,j} \circ \varphi(X)$: $\mathcal{F}(X) \to \mathcal{G}_x / \mathfrak{m}_x^j \cdot \mathcal{G}_x$ is continuous. That follows immediately from the commutative diagrams

$$\begin{array}{ccc} \mathcal{F}(X) & \xrightarrow{\varphi(X)} & \mathcal{G}(X) \\ \downarrow \tau_{x,j} & & \downarrow \pi_{x,j} \\ \mathcal{F}_x / \mathfrak{m}_x^j \cdot \mathcal{F}_x & \xrightarrow{\varphi_{x,j}} & \mathcal{G}_x / \mathfrak{m}_x^j \cdot \mathcal{G}_x . \end{array}$$

ii) In the commutative diagrams

$$\begin{array}{ccc} \mathcal{O}(Y) & \xrightarrow{f^0} & \mathcal{O}(X) \\ \downarrow \tau_{x,j} & & \downarrow \pi_{x,j} \\ {}_Y\mathcal{O}_{f(x)} / \mathfrak{m}_{f(x)}^j & \xrightarrow{f^0_{x,j}} & {}_X\mathcal{O}_x / \mathfrak{m}_x^j , \end{array}$$

the homomorphisms $\pi_{x,j} \circ f^0 = f^0_{x,j} \circ \tau_{x,j}$ are continuous: hence, f^0 is continuous, by E. 55a. ■

We have used the following results on Fréchet spaces (see [Bou EVT]):

55.7 Theorem. *Let V be a Fréchet space.*

i) Closed Graph Theorem. If $u: V \to W$ is a linear mapping between Fréchet spaces, then u is continuous iff the graph $\Gamma(u)$ is closed in $V \times W$.

ii) A countable product of Fréchet spaces is a Fréchet space.

iii) If W is a closed linear subspace of V, then the quotient space V/W is a Fréchet space. ■

55.8 Corollary. *i) Let V be a vector space and τ_H, a Hausdorff topology on V. Then there exists at most one Fréchet topology τ_F on V that is finer than τ_H.*

ii) Open Mapping Theorem. Every continuous linear surjection between Fréchet spaces is open.

iii) For Fréchet spaces V and W and a Hausdorff space T, let $i: W \to T$ be a continuous injection. Then a linear mapping $f: V \to W$ is continuous iff $i \circ f: V \to T$ is continuous.

Proof. i) We consider the graph $\Gamma(\mathrm{id}_V)$, which obviously is the diagonal Δ in $V \times V$. Since $\mathrm{id}_V: (V, \tau_H) \to (V, \tau_H)$ is continuous, Δ is closed in $(V, \tau_H) \times (V, \tau_H)$; for Fréchet topologies τ_F and $\tilde{\tau}_F$ on V that are finer than τ_H, then, Δ is closed in $(V, \tau_F) \times (V, \tilde{\tau}_F)$ and in $(V, \tilde{\tau}_F) \times (V, \tau_F)$. By the Closed Graph Theorem, id_V: $(V, \tau_F) \to (V, \tilde{\tau}_F)$ is a homeomorphism.

ii) If $u: V \to W$ is a continuous linear surjection between Fréchet spaces, then $\operatorname{Ker} u$ is a closed linear subspace of V; by 55.7 iii), $V/\operatorname{Ker} u$ is a Fréchet space, and, by E. 33Be, the projection $V \to V/\operatorname{Ker} u$ is open. Hence, we may assume that u is bijective, and the assertion follows from i).

iii) *If.* Since $i \circ f$ is continuous, $\Gamma(i \circ f)$ is closed in $V \times T$, so $\Gamma(f)$ is closed in $V \times W$. The Closed Graph Theorem implies that f is continuous. ■

E. 55c. If $f: X \to Y$ is a holomorphic mapping and $\mathscr{F} = f^*\mathscr{G}$ is the analytic inverse image of a coherent ${}_Y\mathcal{O}$-module $\mathscr{G}$, then the induced linear mapping $\mathscr{G}(Y) \to \mathscr{F}(X)$ is continuous.

E. 55d. Let $(X, \mathcal{O})$ be a complex space and $\mathscr{G}$, a coherent $\mathcal{O}$-module.

i) $\mathscr{G}(X)$ is a topological $\mathcal{O}(X)$-module (i.e., the mappings

$$\mathscr{G}(X) \times \mathscr{G}(X) \to \mathscr{G}(X), (g, h) \mapsto g + h, \quad \text{and}$$
$$\mathcal{O}(X) \times \mathscr{G}(X) \to \mathscr{G}(X), (f, g) \mapsto f \cdot g,$$

are continous).

ii) If $X = \bigcup_{j \in J} X_j$ is the decomposition of X into connected components, and J is at most countably infinite, then the assignment $f \mapsto (f|_{X_j})_{j \in J}$ determines an isomorphism

$$\mathscr{G}(X) \to \prod_{j \in J} \mathscr{G}(X_j)$$

of Fréchet spaces.

E. 55e. Let $0 \to {}'\mathscr{F} \to \mathscr{F} \to {}''\mathscr{F} \to 0$ be an exact sequence of coherent analytic sheaves on a complex space X. Prove the following for the canonical topologies:

i) ${}'\mathscr{F}(X) \hookrightarrow \mathscr{F}(X)$ has the induced topology.

ii) If $\mathscr{F}(X) \to {}''\mathscr{F}(X)$ is surjective, then ${}''\mathscr{F}(X)$ has the quotient topology.

E. 55f. Let T be a topological space with an at most countably infinite basis $\mathfrak{U}$ of the topology. Further, let $\mathscr{G}$ be a sheaf of vector spaces on T such that, for each pair $U, V \in \mathfrak{U}$ with $V \Subset U$, the restriction-homomorphism $\mathscr{G}(U) \to \mathscr{G}(V)$ is a continuous mapping between Fréchet spaces.

i) Prove that, for $\mathfrak{W} \subset \mathfrak{U}$ and $W := \bigcup_{U \in \mathfrak{W}} U$, the projective limit topology on

$$\mathscr{G}(W) = \varprojlim_{U \in \mathfrak{W}} \mathscr{G}(U)$$

is a Fréchet topology that depends on W but not on $\mathfrak{W}$, and that each restriction-homomorphism $\mathscr{G}(W) \to \mathscr{G}(U)$, $U \in \mathfrak{W}$, is continuous (for 62.6).

ii) Use i) to define the structure of a Fréchet sheaf on $\mathscr{G}$.

We now use the notion of a Fréchet sheaf in order to investigate subsheaves of an $\mathcal{O}$-module that are generated by an infinity of global sections. Let $\mathscr{G}$ be a coherent analytic sheaf on a complex space $(X, \mathcal{O})$, and consider a subset $M \subset \mathscr{G}(X)$. Then the presheaf

$$U \mapsto \mathcal{O}(U) \cdot M \subset \mathscr{G}(U),\ U \Subset X,$$

determines a submodule $\mathcal{O} \cdot M$ of $\mathscr{G}$. Obviously, $(\mathcal{O} \cdot M)_x = \mathcal{O}_x \cdot M \subset \mathscr{G}_x$ for each x. Fix $a \in X$; since $\mathscr{G}_a$ is noetherian, there exist $f_1, \ldots, f_r \in M$ such that $(\mathcal{O} \cdot M)_a = \sum_{j=1}^{r} \mathcal{O}_a \cdot f_j$.

55.9 Lemma. *There exists a neighborhood U of a in X such that $f_1|_U, \ldots, f_r|_U$ generate $(\mathcal{O} \cdot M)(U)$ as an $\mathcal{O}(U)$-module (U is called a privileged neighborhood).*

Proof. Since the problem is local, we may assume, by 41.21 ii), that $X = \boldsymbol{P}^n(0; 1) =: \boldsymbol{P}$, and that $a = 0 \in \boldsymbol{P}$. For $x \in \boldsymbol{P}$, then, $F_x := \{g \in \mathscr{G}(\boldsymbol{P});\ g_x \in \sum \mathcal{O}_x \cdot f_{j,x} \subset \mathscr{G}_x\}$ is a closed linear subspace of $\mathscr{G}(\boldsymbol{P})$, by E. 55b ii), and hence a Fréchet space. Put

$$U_k := \boldsymbol{P}^n(0; 1/k) \text{ and } F_k := \bigcap_{x \in U_k} F_x \subset \mathscr{G}(\boldsymbol{P}) \text{ for } k \geqq 1.$$

Then the F_k form an ascending chain of Fréchet subspaces of $F_a =: F$. We see that $\bigcup_{k \geqq 1} F_k = F$, since every $g \in F$ is of the form $g = \sum g_j f_j$ near $a = 0$. By Baire's Theorem [Bou GT IX § 5.3], $F = F_k$ for some k. On U_k, then, the homomorphism

$$\varphi : \mathcal{O}^r \to \mathcal{O} \cdot M,\ (h_1, \ldots, h_r) \mapsto \sum h_j f_j,$$

is surjective, since, for $x \in U_k$, we have that

$$\sum \mathcal{O}_x \cdot f_j \subset (\mathcal{O} \cdot M)_x \subset (\mathcal{O} \cdot F)_x = (\mathcal{O} \cdot F_k)_x \subset \sum \mathcal{O}_x \cdot f_j.$$

In particular, $\mathcal{O} \cdot M|_{U_k}$ is of finite type, and thus coherent (see 41.5); that yields the result 55.10, stated below. By 61.13, for large k, the induced homomorphism

$\mathcal{O}^r(U_k) \to (\mathcal{O}\cdot M)(U_k)$ is surjective; that is, $f_1, \ldots, f_r$ generate $(\mathcal{O}\cdot M)(U_k)$ as an $\mathcal{O}(U_k)$-module. ■

55.10 Proposition. *If M is a set of global sections in a coherent $\mathcal{O}$-module $\mathcal{G}$ on a complex space X, then $\mathcal{O}\cdot M$ is a coherent submodule of $\mathcal{G}$.* ■

For a coherent ${}_X\mathcal{O}$-module $\mathcal{G}$, the set of the coherent submodules of $\mathcal{G}$ forms a lattice with respect to finite intersections and finite sums, in analogy to the lattice $\mathfrak{J}$ in § 44. This lattice is "locally noetherian", and thus admits infinite unions (see also E. 23d):[9)]

55.11 Corollary. *Let $\mathcal{G}$ be a coherent analytic sheaf on the complex space X, and fix a compact set $K \subset X$. Then every chain $\mathcal{G}_1 \subset \mathcal{G}_2 \subset \ldots$ of coherent submodules of $\mathcal{G}$ becomes stationary on K.*

Proof. Given such a chain, we need only show that each point $a \in X$ has an open neighborhood on which the chain eventually is stationary. By 41.21 and 41.20, we may assume that X is a polydisk $\boldsymbol{P}$, and that $a = 0$. Moreover, since $\mathcal{G}_0$ is noetherian, we may suppose that $\mathcal{G}_{j,0} = \mathcal{G}_{1,0}$, and (after replacing $\mathcal{G}$ with the coherent sheaf $\mathcal{G}/\mathcal{G}_1$) that $\mathcal{G}_{j,0} = 0$ for every j. Hence, we need to find a neighborhood U of 0 such that $\mathcal{G}_j|_U = 0$ for each j. To that end, put $M := \bigcup \mathcal{G}_j(\boldsymbol{P})$; by 55.10, $\mathcal{F} := \mathcal{O}\cdot M$ is a coherent subsheaf of $\mathcal{G}$. Since $\mathcal{F}_0 = 0$, 41.16 implies that there is a U such that $\mathcal{F}|_U = 0$; hence, every $g \in \mathcal{G}_j(\boldsymbol{P}) \subset \mathcal{F}(\boldsymbol{P})$ induces the zero section on U. By 62.2, $\boldsymbol{P}$ is a B-space, so it follows from Theorem A that each $\mathcal{G}_j$ is generated by its global sections; ie., $\mathcal{G}_j|_U = 0$. ■

By definition, analytic sets are determined locally by a *finite* number of holomorphic functions. At this point, we can drop that finiteness condition:

55.12 Corollary. *Let $(X, \mathcal{O})$ be a complex space and $M \subset \mathcal{O}(X)$, a set of global functions. Then, on each compactum $K \subset X$, the coherent ideal $\mathcal{O}\cdot M \subset \mathcal{O}$ is generated by finitely many elements of M.*

Proof. By 55.10, $\mathcal{O}\cdot M$ is coherent. By 55.9, on privileged neighborhoods, $\mathcal{O}\cdot M$ is generated by finitely many $f_j \in M$. ■

E. 55g. In a complex space X, let $U \subset V \cap W$ be open subsets such that, for each $f \in \mathcal{O}(W)$, there exists exactly one $\check{f} \in \mathcal{O}(V)$ with $\check{f}|_U = f|_U$. Prove that $\varphi: \mathcal{O}(W) \to \mathcal{O}(V), f \mapsto \check{f}$, is continuous (hint: 55.7 i); for 63.7).

E. 55h. *Canonical topology on cohomology modules.* Consider a complex space X (with a countable topology), a countable cover $\mathfrak{U}$ of X with open subsets, and a coherent ${}_X\mathcal{O}$-module $\mathcal{G}$. Prove the following for each $p \geqq 0$:

---

9) The proof makes use of 62.2.

i) In a canonical fashion, $C^p(\mathfrak{U}, \mathcal{G})$ admits the structure of a Fréchet space; that topology induces the structure of a topological vector space on $H^p(\mathfrak{U}, \mathcal{G})$.

ii) The coboundary operator $d^p : C^p(\mathfrak{U}, \mathcal{G}) \to C^{p+1}(\mathfrak{U}, \mathcal{G})$ is continuous; hence, $Z^p(\mathfrak{U}, \mathcal{G})$ is a Fréchet space (for 62.4).

iii) If $f: \mathcal{G}_1 \to \mathcal{G}_2$ is a homomorphism of coherent ${}_X\mathcal{O}$-modules, then the induced homomorphisms

$$C^p(\mathfrak{U}, \mathcal{G}_1) \to C^p(\mathfrak{U}, \mathcal{G}_2) \quad \text{and} \quad H^p(\mathfrak{U}, \mathcal{G}_1) \to H^p(\mathfrak{U}, \mathcal{G}_2)$$

are continuous.

iv) For every $V \Subset X$, the restriction-homomorphisms $H^p(\mathfrak{U}, \mathcal{G}) \to H^p(\mathfrak{U} \cap V, \mathcal{G})$ are continuous.

v) Let $\mathfrak{V}$ be a countable refinement of $\mathfrak{U}$ and suppose that the corresponding homomorphism $\varrho^p : H^p(\mathfrak{U}, \mathcal{G}) \to H^p(\mathfrak{V}, \mathcal{G})$ is surjective; then ϱ^p is continuous and open (hint: in the diagram

$$\begin{array}{ccc}
Z^p(\mathfrak{U}, \mathcal{G}) \times C^{p-1}(\mathfrak{V}, \mathcal{G}) & \xrightarrow{\varrho^p + d^{p-1}} & Z^p(\mathfrak{V}, \mathcal{G}) \\
\downarrow \text{pr} & & \\
Z^p(\mathfrak{U}, \mathcal{G}) & & \Big\downarrow \\
\downarrow & & \\
H^p(\mathfrak{U}, \mathcal{G}) & \xrightarrow{\quad \varrho^p \quad} & H^p(\mathfrak{V}, \mathcal{G}),
\end{array}$$

the mapping $\varrho^p + d^{p-1}$ is surjective; apply 55.8 ii)).

vi) $H^p(X, \mathcal{G}) = \varinjlim_{\mathfrak{U}} H^p(\mathfrak{U}, \mathcal{G})$ has a canonical topology (hint: 52.21).

vii) If $H^p(X, \mathcal{G})$ is finite-dimensional, then $H^p(X, \mathcal{G})$ is a Hausdorff space (hint: 50.11; if $\varphi : F \to G$ is a continuous linear mapping between Fréchet spaces such that $\dim(G/\varphi F) < \infty$, then φF is closed in G [Se$_3$, Lemma 2]).

Remark: If X is holomorphically convex, then $H^p(X, \mathcal{G})$ is Hausdorff and thus a Fréchet space [Pr II.1].

viii) Set $U := \{z \in \mathbb{C}^2; |z_2| < 1\}$ and $V := \{z \in \mathbb{C}^2; |z_1| > 1\}$. Then $H^1(U \cup V, \mathcal{O})$ is not Hausdorff (hint: $f_{U \cap V} := \dfrac{1}{z_1(1-z_2)} = (\sum_{j=0}^{\infty} z_2^j) \cdot \dfrac{1}{z_1}$ determines a cocycle $f \in Z^1((U, V), \mathcal{O})$; $f \in \overline{B^1}$, but $f \notin B^1$, otherwise, by 11.6, there would exist a $g \in \mathcal{O}(\mathbb{C}^2)$ and an $h \in \mathcal{O}(U)$ such that $1/(1-z_2) = g - z_1 \cdot h$).

E. 55i. Let Y be a complex space (whose topology need not be countable). Let $\mathfrak{U}$ be the family of open subsets of Y that have a countable topology. Then the inclusion $\mathcal{O}(Y) \to \prod_{U \in \mathfrak{U}} \mathcal{O}(U)$ induces a topology on $\mathcal{O}(Y)$ (in fact, $\mathcal{O}(Y) = \varprojlim_{U \in \mathfrak{U}} \mathcal{O}(U)$). Prove that $\|\cdot\|_K : \mathcal{O}(Y) \to \mathbb{R}$ is continuous for each compact set $K \subset Y$. (Hint: for a local model $i : A \hookrightarrow \boldsymbol{P}$ of Y, the homomorphism $i^0 : \mathcal{O}(\boldsymbol{P}) \to \mathcal{O}(A)$ is surjective for small $\boldsymbol{P}$ (see 61.13) and hence open (see 55.8 ii)); for § 57.)

E. 55j Let $(X, \mathcal{O})$ be a reduced complex space (with a countable topology); on $\mathcal{O}(X)$, then, the canonical topology τ and the topology τ_c of compact convergence coincide. (Hint: let $(X, \tilde{\mathcal{O}})$ be the maximalization of $(X, \mathcal{O})$ (see § 72). Then $\tilde{\mathcal{O}}(X)$ is a closed subalgebra of $\mathscr{C}(X)$, since $\tilde{\mathcal{O}}(X \setminus S(X)) = \mathcal{O}(X \setminus S(X))$ is a closed subalgebra of $\mathscr{C}(X \setminus S(X))$, by 2.1. By E.55b and the fact that $\tilde{\mathcal{O}}$ is a coherent $\mathcal{O}$-module, $\mathcal{O}(X) = \bigcap_{x \in X} \{f \in \tilde{\mathcal{O}}(X); f_x \in \mathcal{O}_x\}$ is closed in $\tilde{\mathcal{O}}(X)$. Hence, $(\mathcal{O}(X), \tau_c)$ is a Fréchet space. To prove that $\tau = \tau_c$, it suffices, by 55.8 i), to prove that the mapping $\mathcal{O}(X) \to \prod_{\substack{x \in X \setminus S(X) \\ j \in \mathbb{N}}} \mathcal{O}_x/\mathfrak{m}_x^j$ is injective and continuous for both τ_c and τ. For 57.7.)

§ 56 The Exhaustion Theorem

For our proof of Theorem B, we need the fact that, under certain conditions, an ascending union of B-spaces is again a B-spaces (that is not true in general; see E. 56b).

56.1 Definition. *Let X be a complex space with a countable topology and U, an open subset of X. Then (X, U) is called a* Runge pair *if the image of the restriction-mapping $\mathcal{O}(X) \to \mathcal{O}(U)$ is dense in $\mathcal{O}(U)$.*

Hence, for $U \subset \mathbb{C}^n$, the pair $(\mathbb{C}^n, U)$ is Runge iff U is a Runge region in $\mathbb{C}^n$.

56.2 Exhaustion Theorem. *The following statements about a complex space X are equivalent:*

i) X is a B-space.

ii) There exists an exhaustion $X = \bigcup_{j=1}^{\infty} X_j$ with open B-sets $X_j \subset\subset X_{j+1}$ in X such that each (X_{j+1}, X_j) is a Runge pair.

Proof. We shall demonstrate the implication "i) ⇒ ii)" only in the proof of 63.2, since we do not need it before then. The implication "ii) ⇒ i)" [10] is an immediate consequence of 56.3 and 56.5. ■

56.3 Lemma. *Let $\mathcal{G}$ be a sheaf of abelian groups on a topological space T and $T = \bigcup_{j=1}^{\infty} T_j$, an open exhaustion. If $p \in \mathbb{N}$ is such that, for $k = 0,1$ and $j \geqq 1$, each restriction-mapping*

$$H^{p+k}(T_{j+1}, \mathcal{G}) \to H^{p+k}(T_j, \mathcal{G})$$

is the zero-mapping, then $H^{p+1}(T, \mathcal{G}) = 0$.

Proof. Let $0 \to \mathcal{G} \to \mathcal{C}^0 \xrightarrow{d^0} \mathcal{C}^1 \xrightarrow{d^1} \mathcal{C}^2 \to \dots$ be the canonical resolution of the sheaf $\mathcal{G}$ as in 50.3. For each cocycle $\omega \in \mathcal{C}^{p+1}(T)$, we have to find a cochain $\gamma \in \mathcal{C}^p(T)$ such that $d\gamma = \omega$. The assumption on the restriction-homomorphisms implies that, for every cocycle $\beta \in \mathcal{C}^{p+k}(T_{j+1})$, there exists a cochain $\zeta \in \mathcal{C}^{p+k-1}(T_j)$ such that $d\zeta = \beta|_{T_j}$; as the sheaf $\mathcal{C}^{p+k-1}$ is flabby, we may even assume that $\zeta \in \mathcal{C}^{p+k-1}(T)$. Consequently, there exist cochains $\alpha_j \in \mathcal{C}^p(T)$ such that $d\alpha_j = \omega$ on T_{j+1}, and cochains $\beta_j \in \mathcal{C}^{p-1}(T)$ such that $d\beta_j = \alpha_{j+1} - \alpha_j$ on T_j. Define the cochain $\gamma \in \mathcal{C}^p(T)$ by setting

$$\gamma|_{T_j} := \alpha_j - \sum_{i<j} d\beta_i;$$

then we see that $d\gamma = \omega$. ■

---

[10] The proof uses the fact that, for every coherent ${}_X\mathcal{O}$-module $\mathcal{F}$, each restriction $\mathcal{F}|_{X_j}$ is acyclic on X_j (instead of the stronger hypothesis that X_j is a B-set).

For $p = 0$, we cannot meet the requirements of 56.3. We overcome that difficulty by means of an approximation:

56.4 Lemma. *Let $\mathcal{G}$ be a Fréchet sheaf on a topological space T and $T = \bigcup_{j=1}^{\infty} T_j$, an open exhaustion. If the image of each restriction-mapping $\mathcal{G}(T_j) \to \mathcal{G}(T_{j-1})$ is dense in $\mathcal{G}(T_{j-1})$, then the following hold:*

i) the image of the restriction-mapping $\mathcal{G}(T) \to \mathcal{G}(T_1)$ is dense in $\mathcal{G}(T_1)$, and

ii) if every restriction-mapping $H^1(T_j, \mathcal{G}) \to H^1(T_{j-1}, \mathcal{G})$ is the zero-mapping, then $H^1(T, \mathcal{G}) = 0$.

Proof. We may assume that, for every j, the topology on $G_j := \mathcal{G}(T_j)$ is induced by a translation-invariant metric δ_j such that $\delta_{j-1} \leqq \delta_j$ (if not, we replace the metric δ_j recursively with the metric defined by setting

$$\tilde{\delta}_j(f, g) := \max(\delta_j(f, g), \delta_{j-1}(\varrho f, \varrho g)),$$

where $\varrho: G_j \to G_{j-1}$ is the restriction-homomorphism; since ϱ is continuous, we easily see that δ_j and $\tilde{\delta}_j$ induce the same topology on G_j). For a $g_1 \in G_1$ and $\varepsilon > 0$, we select recursively sections $g_j \in G_j$ such that $\delta_j(g_{j+1}, g_j) \leqq 2^{-j}\varepsilon$. For each j, the Cauchy sequence $(g_k|_{T_j})_{k \geqq j}$ converges to an $f_j \in G_j$. Then the fact that $f_{j+1}|_{T_j} = f_j$ implies that there exists an $f \in \mathcal{G}(T)$ such that $f|_{T_j} = f_j$. Consequently, we see that

$$\delta_1(f, g_1) = \delta_1(f_1, g_1) \leqq \sum_{j=1}^{\infty} \delta_j(g_{j+1}, g_j) \leqq \varepsilon.$$

ii) We use the notation of the proof of 56.3; for each $\omega \in \operatorname{Ker} d^1$, we have to find a $\gamma \in \mathscr{C}^0(T)$ such that $d\gamma = \omega$. If there exists a sequence (g_j) in $\mathscr{C}^0(T)$ such that

$$\text{a)} \quad dg_j = \omega \quad \text{on} \quad T_j, \qquad \text{and} \qquad \text{b)} \quad \delta_j(0, g_{j+1} - g_j) \leqq 2^{-j},$$

then we may define γ by setting

$$\gamma|_{T_j} := g_j + \lim_{k \to \infty} (g_k - g_j):$$

by a), we see that $g_k - g_j = \sum_{i=j}^{k-1} (g_{i+1} - g_i) \in \operatorname{Ker} d\,(T_j) = G_j$; by b), the limit exists in G_j and satisfies the equality

$$d\gamma = dg_j + d(\lim_k (g_k - g_j)) = \omega + 0.$$

We now determine the g_j recursively: as in the proof of 56.3, fix $\alpha_j \in \mathscr{C}^0(T)$ such that $d\alpha_j = \omega$ on T_j. If $g_1 = \alpha_1, g_2, \ldots, g_j$ are already constructed, then $d(\alpha_{j+1} - g_j) = 0$ on T_j; it follows that $\alpha_{j+1} - g_j \in G_j$. By i), there exists an $f \in \mathcal{G}(T)$ such that $\delta_j(f, \alpha_{j+1} - g_j) \leqq 2^{-j}$. Then, a) and b) hold for $g_{j+1} := \alpha_{j+1} - f$. ∎

56.5 Lemma. *Let X be a complex space with a countable topology and $X = \bigcup_{j=1}^{\infty} X_j$, an open exhaustion such that, for each j and for every coherent ${}_X\mathcal{O}$-module $\mathcal{F}$, the following hold:*

i) (X_{j+1}, X_j) is a Runge pair,

ii) $X_j \subset\subset X_{j+1}$,

iii) $\mathcal{F}$ is acyclic on X_j.

Then the image of the restriction-homomorphism $\mathcal{F}(X) \to \mathcal{F}(X_1)$ is dense in $\mathcal{F}(X_1)$, and $H^1(X, \mathcal{F}) = 0$.

Proof. For a fixed j, it is easy to see from the proof of 52.16 that there exists an exact sequence $\mathcal{O}^{p_{n+2}} \to \ldots \to \mathcal{O}^{p_0} \xrightarrow{\varphi} \mathcal{F} \to 0$ near $\bar{X}_j$, where $n := \max\{\mathrm{emb}_a X; a \in \bar{X}_j\}$ is a finite number, since $\bar{X}_j$ is compact (note that the proof of Theorem A requires only that $H^1(X_{j+1}, {}_{\{a\}}\mathcal{I} \cdot \mathcal{F}) = 0$ for the nullstellen ideal ${}_{\{a\}}\mathcal{I} \subset {}_X\mathcal{O}$ of a point a). We may apply 61.12 to $\mathcal{K}er\,\varphi$; hence, we obtain an exact commutative diagram

$$\begin{array}{ccccc}
\mathcal{O}^{p_0}(X_j) & \longrightarrow & \mathcal{F}(X_j) & \longrightarrow & 0 \\
\varphi \big\downarrow & & \big\downarrow \psi & & \\
\mathcal{O}^{p_0}(X_{j-1}) & \to & \mathcal{F}(X_{j-1}) & \to & 0
\end{array}$$

with continuous homomorphisms. By assumption, $\mathrm{Im}\,\varphi$ is dense in $\mathcal{O}^{p_0}(X_{j-1})$; thus $\mathrm{Im}\,\psi$ is dense in $\mathcal{F}(X_{j-1})$. Then 56.4 implies the assertion. ∎

The essence of 56.2 is the fact that a "Runge exhaustion" preserves convexity:

E. 56a. If $X = \bigcup_{j=1}^{\infty} X_j$ is an exhaustion of a *holomorphically convex* complex space X with open Stein subspaces X_j, then X is Stein (hint: use E.57e).

If $X = \bigcup_{j=1}^{\infty} X_j$ is an exhaustion with open Stein subspaces X_j, then X need not be holomorphically convex (the following example is due to Fornaess):

E. 56b. Prove the following statements for $P_m(T) := \prod_{j=1}^{m} (T - 1/j) \in \mathbb{C}[T]$ and $X_m := \{(u, v, w) \in \mathbb{C}^3; uv = P_m(w)\}$:

i) Each X_m is a Stein manifold.

ii) The mapping $\varphi_m : X_{m-1} \to X_m$, $(u, v, w) \mapsto (u \cdot (w - 1/m), v, w)$, is a biholomorphic mapping onto the complement of $\{(u, 0, 1/m) \in X_m; u \in \mathbb{C}\}$.

iii) The projections $\pi_m : X_m \to \mathbb{C}^2$, $(u, v, w) \mapsto (v, w)$, determine a holomorphic mapping $\pi = \varinjlim \pi_m : X := \varinjlim X_m \to \mathbb{C}^2$.

iv) $L := \{(v, w) \in \mathbb{C}^2; |v| = 1, |w| \leqq 1\}$ and $K := \pi^{-1}(L)$ are compact.

v) $\hat{K}$ is not compact, since $\{(0, v, 1/m); |v| = 1\} \subset \pi_m^{-1}(L)$ and thus $x_m \in \hat{K}$ for $x_m = (0, 0, 1/m) \in X_m \hookrightarrow \mathbb{C}^3$ for every $m \in \mathbb{N}_{\geqq 1}$.

The implication "i) ⇒ ii)" of 56.2, which we shall prove in 63.2, yields the following stronger version of 52.10:

56.6 Proposition. *Let $\mathcal{G}$ be a coherent ${}_X\mathcal{O}$-module on a B-space X and M, a subset of $\mathcal{G}(X)$. Then $(\mathcal{O}\cdot M)(X)$ is the smallest closed $\mathcal{O}(X)$-submodule of $\mathcal{G}(X)$ that includes M.*

Proof. We have that $M \subset (\mathcal{O}\cdot M)(X)$; hence, we may assume that M is an $\mathcal{O}(X)$-module. Fix an exhaustion $X = \bigcup_{j=1}^{\infty} X_j$ according to 56.2; by 55.11, for every j, there exist finitely many $m_{jk} \in M$ that generate the coherent sheaf (see 55.10) $\mathcal{O}\cdot M$ over the relatively compact subset X_j. Hence, for $f \in (\mathcal{O}\cdot M)(X)$, there exists a representation

$$f|_{X_j} = \sum_{k=1}^{s_j} h_{jk}\, m_{jk}\,.$$

As in the proof of 56.4, for every j, fix a metric δ_j on $\mathcal{G}(X_j)$ such that $\delta_j \leqq \delta_{j+1}$. Since each (X_{j+1}, X_j) is a Runge pair, 56.4 implies the existence of sections $f_{jk} \in \mathcal{O}(X)$ such that the functions $m_j := \sum_k f_{jk}\, m_{jk} \in M$ satisfy the inequalities $\delta_j(f, m_j) \leqq 1/j$. Thus, in $\mathcal{G}(X)$, we have that $f = \lim_{j\to\infty} m_j$ and that $f \in \overline{M}$. ■

E. 56c. Show that 56.6 does not hold for $X = \mathbb{C}^{2*}$ (hint: use the kernel of the "evaluation-mapping" ε_0).

The following result is a generalization of 52.11:

56.7 Corollary. *If X is a B-space, then every closed proper ideal $\mathfrak{i}$ in $\mathcal{O}(X)$ has a nonempty zero-set.*

Proof. Otherwise, we would have that $\mathcal{O} = \mathcal{O}\cdot\mathfrak{i}$ and thus that $\mathcal{O}(X) = (\mathcal{O}\cdot\mathfrak{i})(X) \underset{56.6}{=} \mathfrak{i}$. ↯ ■

E. 56d. Let $\mathcal{O}(X)$ and $\mathcal{O}(Y)$ be Stein algebras (see § 57) and $\varphi: \mathcal{O}(Y) \to \mathcal{O}(X)$, a surjective algebra-homomorphism. For closed ideals $\mathfrak{a}, \mathfrak{b} \subset \mathcal{O}(Y)$, prove that
i) $\mathfrak{a} + \mathfrak{b}$ is closed (hint: 56.6 and the exact sequence $0 \to \mathfrak{a}\cap\mathfrak{b} \to \mathfrak{a} \oplus \mathfrak{b} \to \mathfrak{a} + \mathfrak{b} \to 0$); and
ii) $\varphi(\mathfrak{a})$ is closed (for E. 56e).

E. 56e. For a B-space X of dimension n and a closed ideal $\mathfrak{a} \subset \mathcal{O}(X)$, there exist functions $f_1, \ldots, f_n \in \mathfrak{a}$ such that $N(X;\mathfrak{a}) = N(X; f_1, \ldots, f_n)$ (hint: for $n = 1$ see 54.5; for "$n-1 \Rightarrow n$", apply the idea of E. 54h to the ideal $\mathcal{O}\cdot\mathfrak{a}$ in order to construct an $f \in \mathfrak{a}$ such that $N(f) = N(\mathfrak{a}) \cup A$, where $\dim A < n$; then use E. 56d. For E. 63e).

E. 56f. Let A be an analytic subset of a Stein space X. Prove that (in contrast to E. 56e) the nullstellen ideal ${}_A\mathcal{I}$ of A need not be globally finitely generated (hint: $A := \mathbb{N}_{\geq 1} \hookrightarrow X = (\mathbb{C}, {}_X\mathcal{O})$, where, for $j \in A$, the stalks ${}_X\mathcal{O}_j$ are given by a *Strukturausdünnung*: ${}_X\mathcal{O}_j = \mathbb{C} \oplus {}_1\mathfrak{m}_j^j \subset {}_1\mathcal{O}_j$; then $\operatorname{emb} A_j = j$).

§ 57 The Character Theorem and Holomorphic Hulls

The assignment $X \mapsto \mathcal{O}(X)$ extends to a contravariant isomorphism between the category of isomorphism classes of Stein spaces and that of isomorphism classes of Stein algebras (a topological algebra A is called a *Stein algebra* if there exists a Stein space X such that the topological algebras $\mathcal{O}(X)$ and A are isomorphic); for a complete proof, see [Fo_3 § 1]. Concerning that contravariant isomorphism, we do the following in this section: for a Stein space X, we reconstruct, via the spectrum $Sp(X)$ of X, the set $|X|$ from the algebra $\mathcal{O}(X)$ (that is, the topology of $\mathcal{O}(X)$ is not used; see 57.1 and 57.2); in 57.7, we prove that a continuous algebra-homomorphism $\sigma: \mathcal{O}(Y) \to \mathcal{O}(X)$ between reduced Stein algebras is induced by a holomorphic mapping $\tau: X \to Y$.

Let X be an arbitrary complex space. A *character* of $\mathcal{O}(X)$ is a (not necessarily continuous) homomorphism of algebras $\chi: \mathcal{O}(X) \to \mathbb{C}$. The sets $Sp(X)$ and $Sp_c(X)$ of all characters (resp., all continuous characters) of $\mathcal{O}(X)$ are called the *spectrum*, resp., the *continuous spectrum*, of X. For example, for every $a \in |X|$, the evaluation-homomorphism $\varepsilon_a: \mathcal{O}(X) \to \mathcal{O}_a / \mathfrak{m}_a \cong \mathbb{C}, f \mapsto f(a)$, is a continuous character, called a *point character*. Thus, there is a canonical mapping

$$e: X \to Sp_c(X) \subset Sp(X),\ a \mapsto \varepsilon_a,$$

and a natural injection

$$Sp(X) \to \mathbb{C}^{\mathcal{O}(X)},\ \chi \mapsto (\chi(f))_{f \in \mathcal{O}(X)}.$$

Hence, the product topology on $\mathbb{C}^{\mathcal{O}(X)}$ induces a Hausdorff topology on $Sp(X)$ and on $Sp_c(X)$, and e becomes a continuous mapping, by E. 51c. Moreover, e is injective iff X is holomorphically separable (see E. 51c ii)).

57.1 Proposition. *If X is a finite-dimensional B-space, then every character of $\mathcal{O}(X)$ is a point character; thus, $e: X \to Sp_c(X) = Sp(X)$ is a bijection.*

Proof. We may assume that X is connected. Given a character χ, we construct recursively functions $f_0 = 0, f_1, \ldots, f_m \in \operatorname{Ker} \chi$ such that $N(X; f_0, \ldots, f_m)$ contains exactly one point a: having constructed $f_0, \ldots, f_j$, we set

$$X_j := N(X; f_0, \ldots, f_j) \quad \text{and} \quad n_j := \dim X_j.$$

Then we put $f_{j+1} := g - \chi(g)$, where $g \in \mathcal{O}(X)$ is defined as follows: if $n_j \geqq 1$, then choose the g provided by E. 52b such that $\dim N(X_j; g - c) \leqq n_j - 1$ for every $c \in \mathbb{C}$; if $n_j = 0$, then X_j is discrete in X; hence, by 51 A.3 and 52.5, there exists a g that is injective when restricted to X_j (in that case, set $m := j + 1$). Note that no X_j is empty, since $f_0, \ldots, f_j \in \operatorname{Ker} \chi \neq \mathcal{O}(X)$, (see 52.11); hence, by construction, X_m consists of one point a. Since $f(a) = 0$ for every $f \in \operatorname{Ker} \chi$ by 52.11, the maximality of $\operatorname{Ker} \chi$ ensures that $\operatorname{Ker} \chi = \operatorname{Ker} \varepsilon_a$, and thus that $\chi = \varepsilon_a$. ■

E. 57a. For $X := \mathbb{P}_2 \setminus \text{point}$, prove that $e: X \to Sp_c(X) = Sp(X)$ is surjective (so surjectivity of e does not imply holomorphic convexity).

By 57.1, the set $|X|$ can be reconstructed from $\mathcal{O}(X)$ if X is a finite-dimensional Stein space. A second way to do that uses the fact that, to every character χ, there corresponds a maximal ideal $M := \operatorname{Ker} \chi$ in $\mathcal{O}(X)$: two characters agree iff the corresponding maximal ideals coincide. By 57.2, the assignment $a \mapsto \operatorname{Ker} \varepsilon_a$ determines a bijection of $|X|$ onto the set of all finitely generated maximal ideals of $\mathcal{O}(X)$.

57.2 Proposition. *For a B-space X, the following statements about a maximal ideal M in $\mathcal{O}(X)$ are equivalent:*

i) There exists a point $a \in X$ such that $M = \operatorname{Ker} \varepsilon_a$.

ii) M is finitely-generated.
iii) M is closed in $\mathcal{O}(X)$ with respect to the canonical topology.
iv) There exists a continuous character χ such that $M = \operatorname{Ker}\chi$.

Proof. i) $\Rightarrow$ ii) Denote by Y the (finite-dimensional) reduced Stein subspace of X that consists of all irreducible components of Red X passing through the point a. As in the proof of 57.1, we construct functions $g_1, \dots, g_m \in \mathcal{O}(Y)$ such that $N(Y; g_1, \dots, g_m) = \{a\}$. By 52.5, there exist functions $f_1, \dots, f_m \in \mathcal{O}(X)$ such that $f_j|_Y = g_j$, and a function $f_{m+1} \in \mathcal{O}(X)$ such that $f_{m+1}(a) = 0$ and such that $\operatorname{Red} f_{m+1}$ is the constant function 1 on every irreducible component of Red X not included in Y. Add functions $f_{m+2}, \dots, f_k \in \mathcal{O}(X)$ that generate the ${}_X\mathcal{O}_a$-module $\mathfrak{m}_a$ (their existence is guaranteed by Theorem A); then we see that the coherent ideal $\mathcal{O}\cdot(f_1, \dots, f_k)$ is the nullstellen ideal ${}_{\{a\}}\mathcal{I}$. Hence, we have that

$$M = \operatorname{Ker}\varepsilon_a = {}_{\{a\}}\mathcal{I}(X) = (\mathcal{O}\cdot(f_1, \dots, f_k))(X) \underset{56.6}{=} \mathcal{O}(X)\cdot(f_1, \dots, f_k).$$

The implication "ii) $\Rightarrow$ i)" follows from 52.11. i) $\Rightarrow$ iv) Point characters are continuous. The implication "iv) $\Rightarrow$ iii)" is obvious. iii) $\Rightarrow$ i) By 56.7, there exists an $a \in X$ such that $f(a) = 0$ for each $f \in M$; i. e., $M \subset \operatorname{Ker}\varepsilon_a$. Hence, $M = \operatorname{Ker}\varepsilon_a$, since M is maximal. ■

Not every maximal ideal M in $\mathcal{O}(X)$ is finitely-generated (at least not if X is a finite-dimensional Stein space that includes an infinite discrete closed subset A): let I denote the ideal $\{f \in \mathcal{O}(X); f(a) = 0 \text{ for almost all } a \in A\}$. Since I is a proper ideal, it is included in a maximal ideal, which has an empty zero-set as well.

We now prove the announced characterization for locally closed subspaces in $\mathbb{C}^n$ and indicate the demonstration in the general case:

57.3 Character Theorem. *A finite-dimensional complex space X is Stein iff the mapping $e: X \to Sp(X)$ is homeomorphic.*

Proof. If. By E. 51c ii), X is holomorphically separable. Thus we only need to show that, for every compact set $K \subset X$, the holomorphically convex hull $\hat{K}$ is compact. To that end, we introduce the set $Sp(K) := \{\chi \in Sp(X); |\chi(f)| \leq \|f\|_K, \forall f \in \mathcal{O}(X)\}$, which obviously includes $e(K)$. We accomplish the proof by showing that $\hat{K} = e^{-1}(Sp(K))$ and that $Sp(K)$ is compact (then, the assertion follows from the continuity of e^{-1}). We have that $\hat{K} = \{x \in X; |\varepsilon_x(f)| = |f(x)| \leq \|f\|_K, \forall f \in \mathcal{O}(X)\} = \{x; \varepsilon_x \in Sp(K)\} = e^{-1}(Sp(K))$. Secondly, $Sp(K)$ is included in the compact subset $C := \prod_{f \in \mathcal{O}(X)} \boldsymbol{P}^1(\|f\|_K)$ of $\mathbb{C}^{\mathcal{O}(X)}$ and is even closed as the zero-set of a system of quadratic polynomials:

$$Sp(K) = \{(\lambda_f) \in C; \lambda_1 = 1, \lambda_{sf+tg} = s\lambda_f + t\lambda_g, \lambda_{f\cdot g} = \lambda_f \cdot \lambda_g, \forall s, t \in \mathbb{C}, \forall f, g \in \mathcal{O}(X)\} =: D$$

(the inclusion $\subset$ is evident; on the other hand, for each $(\lambda_f) \in D$, the assignment $\chi: f \mapsto \lambda_f$ determines a continuous character of $\mathcal{O}(X)$, which satisfies $|\chi(f)| = |\lambda_f| \leq \|f\|_K$ and thus is contained in $Sp(K)$).

Only if. In this proof we use Theorem B. By E. 51c ii) and 57.2, we need only show that e is an open mapping. To that end, for a point $a \in X$, we shall construct functions $f_1, \dots, f_m \in \mathcal{O}(X)$ such that the family

$$U_j := \{x \in X; |f_k(x)| < 1/j, 1 \leq k \leq m\}, j \in \mathbb{N}$$

is a fundamental system of neighborhoods of a. Since e is surjective, we have that

$$e(U_j) = \{\varepsilon_x \in Sp_c(X); |\varepsilon_x(f_k)| = |f_k(x)| \leq 1/j, 1 \leq k \leq m\}$$

is an open subset of $Sp_c(X)$.

In the case in which $X \hookrightarrow W \Subset \mathbb{C}^m$, we may define the f_k by means of the assignment $f_k(z) := z_k - a_k$. In the general case, it suffices to find an $f \in \mathrm{Hol}(X, \mathbb{C}^m)$ such that the following hold:

i) $f^{-1}(0) = \{a\}$, and

ii) $f^{-1}(\boldsymbol{P}) \Subset X$ for an appropriate polydisk $\boldsymbol{P} \Subset \mathbb{C}^m$.

Then we may assume that $f: f^{-1}(\boldsymbol{P}) \to \boldsymbol{P}$ is finite (see 45.4), and we may apply 33 B. 2.

For the construction of f, denote with $Y \hookrightarrow X$ the union of all irreducible components of Red X passing through a, and with $Z \hookrightarrow X$ the union of the remaining components. By E. 49m, Y is the union of finitely many subspaces Y_j of pure dimension j. On each Y_j, there exists a proper mapping $g_j \in \mathrm{Hol}(Y_j, \mathbb{C}^{j+1})$ such that $g_j(a) = 0$ (see [GuRo VII. C. Th. 2 and Th. 4]; by 52.5, each g_j admits a holomorphic extension $G_j \in \mathrm{Hol}(X, \mathbb{C}^{j+1})$. Again by 52.5, there is an $h \in \mathcal{O}(X)$ such that $h|_Z = 2$ and $h(a) = 0$. That yields a holomorphic mapping $F = (h, G_1, \ldots, G_t): X \to \mathbb{C}^n$ such that $K := F^{-1}(\overline{\boldsymbol{P}^n(1)})$ is a compact neighborhood of a. According to 52.8, we can find an $H \in \mathrm{Hol}(X, \mathbb{C}^k)$ that is injective near K and such that $\|H\|_K \leq 1$. Then $f := (F, H)$ has the desired properties. ■

In the proof of 57.3, we used the fact that every finite-dimensional B-space admits a *proper* holomorphic mapping into a complex number space. That is also crucial in the proof of the following:

57.4 Remmert's Embedding Theorem. *Every connected n-dimensional Stein space X admits a holomorphic homeomorphism f onto a closed subspace of a complex number space $\mathbb{C}^m$. If X is a manifold, then f can be chosen so as to be an embedding with $m = 2n$ (except for $n = 1$, in which case $m = 3$)* [Fo]$_2$, [Wi]. ■

We now use the Character Theorem for the extension of holomorphic functions:

57.5 Definition. *In the category of those complex spaces, each of whose connected components has a countable topology, a holomorphic mapping $\varphi: X \to Y$ is called a holomorphic extension of X if $\varphi^0: \mathcal{O}(Y) \to \mathcal{O}(X)$ is an isomorphism. Such a φ is called a* <u>*holomorphic hull of X*</u> *if φ is "maximal", that is, if every holomorphic extension ψ admits a unique factorization*

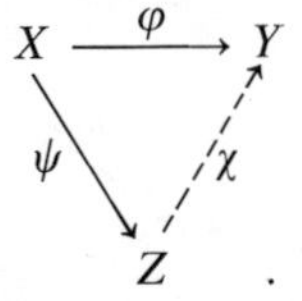

E. 57b. If the holomorphic hull of X exists, then it is unique up to isomorphism.

E. 57c. Show that the number of connected components X_j of a complex space X is determined by the algebra $\mathcal{O}(X)$. In particular, that number does not change under a holomorphic extension (hint: decompose $1 \in \mathcal{O}(X)$ using the characteristic functions χ_{X_j}).

If X and Y have a countable topology, then, by 55.6 ii) and 55.8 ii), φ^0 is an isomorphism of Fréchet algebras. In general, φ^0 is an isomorphism of products of Fréchet algebras (see E. 57c).

We now derive from Theorem B the following:

57.6 Proposition. *If $\varphi: X \to Y$ is a holomorphic extension and Y, a Stein space, then φ is the holomorphic hull of X.*

Proof. Every holomorphic extension $\psi: X \to Z$ induces an isomorphism of topological algebras $\mathcal{O}(Y) \to \mathcal{O}(Z)$. We shall see in the proof of 57.7 that there exists a unique $\chi \in \mathrm{Hol}(Z, Y)$ such that $\chi \circ \psi = \varphi$. ■

57.7 Lemma. *Let Y be a Stein space and X, a reduced*[11] *complex space such that each connected component has a countable topology. Then every continuous algebra-homomorphism $\sigma: \mathcal{O}(Y) \to \mathcal{O}(X)$ is induced by a unique holomorphic mapping $\tau = (|\tau|, \sigma): X \to Y$.*

Proof. We consider the transposed mapping

$$\sigma^t : Sp_c(X) \longrightarrow Sp_c(Y), \ \varphi \longmapsto \varphi \circ \sigma .$$

By 57.3, $e_Y : Y \to Sp_c(Y)$ is a homeomorphism; hence, $\tau := e_Y^{-1} \circ \sigma^t \circ e_X : X \to Y$ is a continuous mapping such that $(f \circ \tau)(x) = \sigma(f)(x)$ for every $f \in \mathcal{O}(Y)$ and every $x \in X$. Since X is reduced, it remains to show that $g \in {}_Y\mathcal{O}(W)$ implies that $g \circ \tau \in {}_X\mathcal{O}(\tau^{-1}(W))$ for a basis of open sets W of Y. By 63.3 i), we may assume that W is an analytic polyhedron in Y; then, by 63.6, there exists a sequence (g_j) in ${}_Y\mathcal{O}(Y)$ such that $(g_j|_W)$ converges to g. Thus the sequence of functions $g_j \circ \tau = \sigma(g_j) \in {}_X\mathcal{O}(X)$ converges in $\mathscr{C}(\tau^{-1}(W))$ to $g \circ \tau$. By E. 55j, $g \circ \tau$ lies in the Fréchet space ${}_X\mathcal{O}(\tau^{-1}(W))$. ■

Lemma 57.7 is fundamental for a further characterization of Stein spaces:

57.8 Remark. *A connected normal (see 71.1) Stein space X is determined (up to isomorphism) by the field extension $\mathscr{M}(X): \mathbb{C}$* [Is Th. II]. ■

57.9 Examples. i) Let G be a holomorphically convex domain in $\mathbb{C}^n$ for $n \geqq 2$. If K is a compact subset of G such that $G \setminus K$ is connected, then the inclusion $G \setminus K \to G$ is the holomorphic hull of $G \setminus K$ (see 52.20).

ii) If B denotes a domain in $\mathbb{R}^n$, $\hat{B}$, its (elementarily) convex hull and T_B, the tube, then $\mathcal{O}(T_B) = \mathcal{O}(T_{\hat{B}})$ (see [Hö 2.5.10]). By 14.9, $T_{\hat{B}}$ is a Stein space; thus 57.6 implies that $T_B \to T_{\hat{B}}$ is the holomorphic hull.

We call a complex space X *holomorphically precomplete* if it admits a holomorphic hull $\varphi: X \to Y$ such that Y is a Stein space.

57.10 Corollary. *The holomorphic hull determines a covariant functor from the category of reduced holomorphically precomplete spaces to the category of Stein spaces.*

Proof. Given a holomorphic mapping $f: X \to Z$ of holomorphically precomplete spaces, we have to find a holomorphic mapping between the holomorphic hulls $f': X' \to Z'$ such that the associated diagram

$$\begin{array}{ccc} X & \xrightarrow{\ f\ } & Z \\ {\scriptstyle \varphi_X}\downarrow & & \downarrow{\scriptstyle \varphi_Z} \\ X' & \xrightarrow{\ f'\ } & Z' \end{array}$$

commutes. That is possible, according to 57.7, using the continuous homomorphism $\mathcal{O}(Z') \cong$ $\cong \mathcal{O}(Z) \xrightarrow{f^\circ} \mathcal{O}(X) \cong \mathcal{O}(X')$; it is easy to see that the construction of f' is functorial. ■

By 57.6 and the following result, holomorphically convex spaces are holomorphically precomplete:

57.11 Theorem. *Let X be a holomorphically convex space. Then the equivalence relation*

$$x \underset{R}{\sim} y \quad \textit{iff} \quad f(x) = f(y) \quad \textit{for every} \quad f \in \mathcal{O}(X)$$

---

[11] The conclusion also holds for nonreduced X, see [Fo_3, § 1].

determines a Stein space X/R such that the natural projection $\pi: X \to X/R$ enjoys the following properties:

i) $\pi^0: \mathcal{O}(X/R) \to \mathcal{O}(X)$ is an isomorphism of topological algebras.

ii) π is a proper mapping with connected fibers.

iii) Every holomorphic mapping of X into a holomorphically separable space admits a unique factorization over π.

Proof. By construction, the ringed space X/R is separable by global sections. The equivalence relation R is proper: for each compactum $K \subset X$, the set $R(K) = \bigcap_{f \in \mathcal{O}(X)} f^{-1} f(K)$ is a closed subset of the compact set $\hat{K}$ and thus compact as well. Hence, π is proper, and X/R is a (holomorphically convex and holomorphically separable) complex space according to 49 A. 19; by 51 A. 3, it is a Stein space. Statement iii) is a consequence of 49 A. 10; the construction of X/R and the Open Mapping Theorem imply i).

ii) By 49 A. 22, there exists a Stein factorization

$$\begin{array}{ccc} X & \xrightarrow{\;\pi\;} & X/R \\ & {\scriptstyle\psi}\searrow \quad \nearrow{\scriptstyle\bar{\pi}} & \\ & X/N_\pi & \end{array},$$

where ψ has connected fibers, and $\bar{\pi}$ is finite. By Theorem B and 52.18, X/N_π is a Stein space; hence, by iii), $\bar{\pi}$ is bijective, and π has connected fibers. ■

E. 57d. Let X be a holomorphically convex space and Y, a Stein space. For a proper holomorphic surjection $\tau: X \to Y$, prove the equivalence of the following statements:

i) Up to an isomorphism, τ coincides with the mapping π of 57.11.

ii) $\tau^0: \mathcal{O}(Y) \to \mathcal{O}(X)$ is bijective.

iii) τ has connected fibers, and $Y = X/R_\tau$.

(Hint: for ii) $\Rightarrow$ i), use 57.7; for iii) $\Rightarrow$ i), use 57.11 iii).)

There exist manifolds that are not holomorphically precomplete: in [Gr], a holomorphically spreadable manifold X is described for which $\mathcal{O}(X)$ has closed maximal ideals that are not finitely-generated (compare 57.2).

E. 57e. From 57.11, deduce the following equivalence for a holomorphically convex space X: Every compact analytic subset of X is finite iff X is a Stein space.

E. 57f. Let X be a complex space such that every connected component of X has a countable topology. Then X is holomorphically precomplete iff $\mathcal{O}(X)$ is a Stein algebra.

E. 57g. Let X be a holomorphically convex space, Z, a discrete closed subset of X, and $r: Z \to \mathbb{R}$, a function. Prove that there exists an $f \in \mathcal{O}(X)$ such that $|f(z)| \geqq r(z)$ for every $z \in Z$ (hint: 52.5; for 51.2 iii)).

E. 57h. In the notation of E. 54Be, show that $E(m)/S_0$ is the holomorphic hull of $E(m)$ iff $m < 0$ (for 72.8).

§ 58 The Holomorphic Version of de Rham's Theorem

On a Stein space X, there is a close relation between the function theory on X and the topology of X: for instance, consider the multiplicative Cousin problem or Oka's Principle. As a further example, we shall prove that, on a Stein manifold X, the cohomology $H^*(X, \mathbb{C})$ can be calculated by means of the complex Ω^* of holomorphic differential forms on X (that corresponds to the fact that, on a paracompact $\mathscr{C}^\infty$-manifold X, $H^*(X,\mathbb{C})$ can be calculated by means of the complex $\mathscr{E}^*$ of $\mathscr{C}^\infty$-differential forms. In § 58 A we define the sheaves $\mathscr{E}^r$ using Grassmann algebras). For that purpose, and as a preparation to the Dolbeault theory (in § 61), we first introduce the sheaves Ω^r of holomorphic r-forms and the sheaves $\mathscr{E}^{p,q}$ of differentiable (p,q)-forms.

Let X be a (complex) manifold of dimension n and $z_1,\dots,z_n$, with $z_j = x_j + iy_j$, local coordinates at a point $a \in X$. Then $\{dx_1,\dots,dx_n, dy_1,\dots,dy_n\}$ and $\{dz_1,\dots,dz_n, d\bar z_1,\dots,d\bar z_n\}$ are both bases of $\mathscr{E}^1_a$. Hence, for $r \in \mathbb{N}$, the set

$$\{dz_I \wedge d\bar z_J;\ I \in \mathbb{N}\tbinom{n}{p},\ J \in \mathbb{N}\tbinom{n}{q},\ p+q=r\}$$

(where $\mathbb{N}\binom{n}{p} := \{(j_1,\dots,j_p) \in \mathbb{N}^p;\ 1 \leqq j_1 < \dots < j_p \leqq n\}$, $dz_I := dz_{i_1} \wedge \dots \wedge dz_{i_p}$ and $d\bar z_I := d\bar z_{i_1} \wedge \dots \wedge d\bar z_{i_p}$ for $I = (i_1,\dots,i_p)$, $dz_\emptyset := 1 =: d\bar z_\emptyset$) is a basis of $\mathscr{E}^r_a$.

An r-form $\omega \in \mathscr{E}^r(U)$, $U \subset X$, is called a *(p,q)-form*, if $p, q \in \mathbb{N}$, $r = p+q$ and if, in local coordinates $z_1,\dots,z_n, \omega$ can be written as $\omega = \sum_{I \in \mathbb{N}\binom{n}{p},\, J \in \mathbb{N}\binom{n}{q}} f_{IJ} dz_I \wedge d\bar z_J$ (the type (p,q) is invariant under holomorphic mappings: let $h = (h_1,\dots,h_m)$ be holomorphic; then $dh_k = \sum_{j=1}^n \frac{\partial h_k}{\partial z_j} dz_j$ is a $(1,0)$-form, and $d\bar h_k = \sum_{j=1}^n \frac{\partial \bar h_k}{\partial \bar z_j} d\bar z_j$ is a $(0,1)$-form, so $h^*\omega = \Sigma h^0(f_{IJ}) dh_I \wedge d\bar h_J$ is a (p,q)-form, as well). The presheaf $U \mapsto \mathscr{E}^{p,q}(U) := \{\omega \in \mathscr{E}^{p+q}(U);\ \omega \text{ is a } (p,q)\text{-form}\}$ obviously is a locally free $\mathscr{E}$-module $\mathscr{E}^{p,q}$ of rank $\binom{n}{p}\binom{n}{q}$, and

$$\mathscr{E}^r = \bigoplus_{p+q=r} \mathscr{E}^{p,q}, \quad \mathscr{E}^{0,0} = \mathscr{E}^0 = \mathscr{E}.$$

The exterior derivative $d: \mathscr{E}^0 \to \mathscr{E}^1 = \mathscr{E}^{1,0} \oplus \mathscr{E}^{0,1}$ splits into $d = \partial + \bar\partial$, where, in local coordinates $z_1,\dots,z_n$ (see E. 1a)

$$\partial: \mathscr{E}^{0,0} \to \mathscr{E}^{1,0},\ f \mapsto \sum_{j=1}^n \frac{\partial f}{\partial z_j} dz_j,\quad \bar\partial: \mathscr{E}^{0,0} \to \mathscr{E}^{0,1},\ f \mapsto \sum_{j=1}^n \frac{\partial f}{\partial \bar z_j} d\bar z_j.$$

That decomposition extends to all r: $d = \partial + \bar\partial: \mathscr{E}^r \to \mathscr{E}^{r+1}$, where $\partial: \mathscr{E}^{p,q} \to \mathscr{E}^{p+1,q}$ and $\bar\partial: \mathscr{E}^{p,q} \to \mathscr{E}^{p,q+1}$ are $\mathbb{C}$-linear, and, in local coordinates, are determined by

$$\partial(\Sigma f_{IJ} dz_I \wedge d\bar z_J) = \Sigma \partial f_{IJ} \wedge dz_I \wedge d\bar z_J,\ \bar\partial(\Sigma f_{IJ} dz_I \wedge d\bar z_J) = \Sigma \bar\partial f_{IJ} \wedge dz_I \wedge d\bar z_J.$$

The identity $0 = d^2 = \partial^2 + (\bar\partial\partial + \partial\bar\partial) + \bar\partial^2$ implies that

$$\partial^2 = 0,\ \bar\partial^2 = 0,\ \bar\partial\partial + \partial\bar\partial = 0.$$

Let $\varphi: X \to Y$ be a holomorphic mapping between complex manifolds. Then the comorphisms $\varphi^*: {}_Y\mathscr{E}^r \to {}_X\mathscr{E}^r$ induce comorphisms $\varphi^*: {}_Y\mathscr{E}^{p,q} \to {}_X\mathscr{E}^{p,q}$ that commute with ∂ and $\bar{\partial}$; hence, for every $p \in \mathbb{N}$, φ induces a comorphism of complexes

$$\varphi^*: ({}_Y\mathscr{E}^{p,*}, \bar{\partial}) \to ({}_X\mathscr{E}^{p,*}, \bar{\partial}).$$

A form $\omega \in \mathscr{E}^p(U)$ is called a *holomorphic p-form*, if, in local coordinates $z_1, \ldots, z_n$, there exist holomorphic functions f_I such that

$$\omega = \sum_{I \in \mathbb{N}\binom{n}{p}} f_I dz_I.$$

The presheaf $U \mapsto \Omega^p(U) := \{$holomorphic p-forms on $U\}$ is a locally free $\mathcal{O}$-module of rank$\binom{n}{p}$; moreover, $\Omega^p = \mathscr{K}er(\bar{\partial}: \mathscr{E}^{p,0} \to \mathscr{E}^{p,1})$, since, in local coordinates, we have that

$$\begin{aligned}\bar{\partial}(\Sigma f_I dz_I) = 0 \Leftrightarrow \Sigma \bar{\partial} f_I \wedge dz_I = 0 &\Leftrightarrow \bar{\partial} f_I = 0 \text{ for each } I\\ &\Leftrightarrow \text{each } f_I \text{ is holomorphic.}\end{aligned}$$

In particular, $\Omega^p \subset \mathscr{E}^{p,0}$ and $\Omega^0 = \mathcal{O}$.

Since $d|_{\Omega^p} = \partial|_{\Omega^p}$, (Ω^*, ∂) is a complex of $\mathcal{O}$-modules, and the holomorphic mapping $\varphi: X \to Y$ induces a comorphism $\varphi^*: ({}_Y\Omega^*, \partial) \to ({}_X\Omega^*, \partial)$ of complexes.

E. 58a. Find an $\omega \in \mathscr{K}er(\bar{\partial}: \mathscr{E}^{p,1} \to \mathscr{E}^{p,2})$ that is not holomorphic.

In analogy to Poincaré's Lemma in real analysis, we have the following:

58.1 Poincaré's Lemma (holomorphic version). *The complex (Ω^*, ∂) of holomorphic differential forms on an n-dimensional manifold X induces the resolution*

$$0 \to \mathbb{C} \to \mathcal{O} \xrightarrow{\partial} \Omega^1 \xrightarrow{\partial} \ldots \xrightarrow{\partial} \Omega^n \to 0$$

of the constant sheaf $\mathbb{C}$ on X.

Proof. We may assume that X is an open convex neighborhood U of 0 in $\mathbb{C}^n$. Then it suffices to show that the sequence

$$0 \to \mathbb{C} \to \mathcal{O}(U) \xrightarrow{\partial} \Omega^1(U) \to \ldots \to \Omega^n(U) \to 0$$

is exact. Since that obviously holds for $0 \to \mathbb{C} \to \mathcal{O}(U) \to \Omega^1(U)$, we only have to show for $p \geqq 1$ that, for each $\omega \in \Omega^p(U)$ such that $\partial\omega = 0$, there exists a $\sigma \in \Omega^{p-1}(U)$ with $\partial\sigma = \omega$. To that end, put $W := \{(\lambda, u) \in \mathbb{C} \times U; \lambda u \in U\} \Subset \mathbb{C}^{n+1}$ and $h: W \to U$, $(\lambda, u) \mapsto \lambda u$; then $h^*\omega \in \Omega^p(W)$ admits a decomposition

$$h^*\omega = \omega_0(\lambda) + d\lambda \wedge \omega_1(\lambda),$$

where neither ω_0 nor ω_1 contains $d\lambda$. We denote with ∂_U the derivative with respect to the variables of U. Then we see that

$$0 = h^* \partial\omega = \partial h^* \omega = \partial_U \omega_0(\lambda) + d\lambda \wedge \frac{\partial \omega_0(\lambda)}{\partial \lambda} - d\lambda \wedge \partial_U \omega_1(\lambda),$$

and thus that

$$\frac{\partial \omega_0(\lambda)}{\partial \lambda} = \partial_U \omega_1(\lambda).$$

If ω has a decomposition $\omega = \sum_{J \in \mathbb{N}\binom{n}{p}} f_J dz_J$, then $\omega_0 = \sum_J f_J(\lambda z)\lambda^p dz_J$; hence, $\omega_0(0) = 0$ and $\omega_0(1) = \omega$. For

$$\sigma := \int_0^1 \omega_1(\lambda) d\lambda \in \Omega^{p-1}(U),$$

we finally obtain that

$$\partial_U \sigma = \partial_U \int_0^1 \omega_1(\lambda) d\lambda = \int_0^1 ({}_U\omega_1(\lambda)) d\lambda = \int_0^1 \frac{\partial \omega_0(\lambda)}{\partial \lambda} d\lambda = \omega_0(1) - \omega_0(0) = \omega.$$

■

58.2 De Rham's Theorem (holomorphic version). *For every B-manifold X and every p, there is an isomorphism of vector spaces*

$$H^p(X, \mathbb{C}) \cong [\operatorname{Ker} \partial^p : \Omega^p(X) \to \Omega^{p+1}(X)] / \partial^{p-1}\Omega^{p-1}(X) (= H^p(\Omega^*(X))).$$

Proof. The sheaves Ω^p are locally free $\mathcal{O}$-modules and thus coherent; hence, the resolution $0 \to \mathbb{C} \to \Omega^*$ of $\mathbb{C}$ in 58.1 is acyclic, and 58.2 follows from 50.4. ■

E. 58b. i) Prove that $H^1(\mathbb{C}^*, \mathbb{C}) \cong \mathbb{C} \cdot \frac{dz}{z}$ (hint: Laurent series of $f \in \mathcal{O}(\mathbb{C}^*)$).

ii) Find an $f \in \mathscr{C}^\infty(\mathbb{C}^*)$ such that $\partial f = \frac{dz}{z} \in \Omega^1(\mathbb{C}^*)$ and $\bar\partial f = \frac{d\bar z}{\bar z} \in \mathscr{E}^{0,1}(\mathbb{C}^*)$.

58.3 Corollary. *If X is a B-manifold, then* $H^{\dim X + j}(X, \mathbb{C}) = 0$ *for every* $j \geqq 1$. ■

58.4 Example. According to 52.20, for $n \geqq 2$ and $0 \leqq a < b \leqq \infty$, the "annulus" $\boldsymbol{R} := \{z \in \mathbb{C}^n; a < \|z\| < b\}$ is not holomorphically convex. Here is another proof, which uses 58.3: we may assume that $a < 1 < b$. Then S^{2n-1} is a strong deformation retract of $\boldsymbol{R}$ (use the retraction $r: \boldsymbol{R} \to S^{2n-1}, z \mapsto z/\|z\|$); hence, $H^{2n-1}(\boldsymbol{R}, \mathbb{C}) = H^{2n-1}(S^{2n-1}, \mathbb{C}) \cong \mathbb{C}$. Thus, by 58.3, $\boldsymbol{R}$ is not a Stein domain. ■

The construction of holomorphic differential forms can be generalized to complex spaces: for a local model $A = V(U; \mathscr{I}) \hookrightarrow U \subset\subset \mathbb{C}^n$ and $p > 0$, it is easy to see that the sheaf

$${}_A\Omega^p := {}_U\Omega^p / ({}_U\Omega^{p-1} \wedge d\mathscr{I} + \mathscr{I} \cdot {}_U\Omega^p)$$

is a coherent ${}_A\mathcal{O}$-module.

E. 58c. Prove the following statements:

i) ${}_{\mathbb{C}^{n+1}}\Omega^p/({}_{\mathbb{C}^{n+1}}\Omega^{p-1}\wedge dz_{n+1}+z_{n+1}\cdot{}_{\mathbb{C}^{n+1}}\Omega^p)\cong{}_{\mathbb{C}^n}\Omega^p$ (here $\mathbb{C}^n=V(\mathbb{C}^{n+1};z_{n+1})\hookrightarrow\mathbb{C}^{n+1}$).

ii) Put $A:=V(\mathbb{C}^2;z_1z_2)$. Then ${}_A\Omega_a^2\neq 0$ iff a $=0$.

iii) Put $A:=V(\mathbb{C};z^2)$. Then ${}_A\Omega_0^1\neq 0$.

iv) Determine ${}_A\Omega_0^{\mathrm{emb}A_0}$ for Neil's parabola, Whitney's umbrella, $\sqrt{z_1z_2}$, and the line with a double point.

Since ${}_A\Omega^p$ does not depend on the embedding of A [GrKe], one can define coherent analytic sheaves ${}_X\Omega^p$ on every complex space X; in particular, ${}_X\Omega^0={}_X\mathcal{O}$ and ${}_X\Omega_a^p=0$ for $p>\operatorname{emb}X_a$. Now suppose that X is reduced and *locally holomorphically contractible*, i.e., for every $a\in X$ and every sufficiently small open neighborhood U of a, there exist neighborhoods V of a in U and D of $[0,1]$ in $\mathbb{C}$ and a holomorphic mapping $\varphi: D\times V\to U$, $(\lambda,v)\mapsto\varphi_\lambda(v)$, such that $\varphi_0(V)=\{a\}$ and $\varphi_1=\mathrm{id}_V$. Then the holomorphic version of Poincaré's Lemma is valid on X [Rf p. 227]; that obviously implies the holomorphic version of de Rham's Theorem for X. But ${}_X\Omega^p$ does not necessarily vanish for $p>\dim X$ (see E. 58c); hence, the Vanishing Theorem 58.3 cannot be generalized for X by simply generalizing our proof. However, 58.3 is valid for arbitrary Stein spaces; a first proof of that fact has been given in $[\mathrm{Ka}]_{\mathrm{L}}$, using a vanishing theorem for the *integral* homology groups of Stein spaces. A much stronger result is the fact that each Stein space has the homotopy-type of a *CW*-complex of real dimension $\dim_{\mathbb{C}}X$ [Hm]; that implies that $H^p(X,\mathbb{Z})=0$ for $p>\dim_{\mathbb{C}}X$.

§ 58 A Supplement: The Grassmann Algebra and Differential Forms

At the beginning of this section, we remind the reader of some basic facts about the Grassmann algebra $A^*(M)$ of a finitely generated free R-module M of rank n (we assume that $\mathrm{char}(R)=0$; for $R=\mathbb{R}$, see [Gb, Chap. 5]):

a) For $p\geqq 1$, $A^p(M)$ is the R-module of all alternating p-fold R-linear mappings ω: $M\times\ldots\times M\to R$; in particular, $A^1(M)=M'=\mathrm{Hom}_R(M,R)$ and $A^p(M)=0$ for $p>n$. By definition, $A^0(M):=R$.

$A^*(M):=\bigoplus_{p=0}^{\infty}A^p(M)$ is endowed with the following multiplication $\wedge$ (Grassmann product):

for $\omega\in A^p(M)$, $\chi\in A^q(M)$ and $r:=p+q$, the product $\omega\wedge\chi\in A^r(M)$ is determined by

$$\omega\wedge\chi(m_1,\ldots,m_r)\ :=\ \frac{r!}{p!\,q!}\sum_{j_1,\ldots,j_r}\varepsilon_{j_1,\ldots,j_r}\,\omega(m_{j_1},\ldots,m_{j_p})\cdot\chi(m_{j_{p+1}},\ldots,m_{j_r}),$$

with $\varepsilon_{j_1,\ldots,j_r}=\prod_{1\leqq i<k\leqq r}\operatorname{sign}(j_k-j_i)$.

For $p=0$ or $q=0$, $\wedge$ is nothing but multiplication by scalars.

b) $A^*(M)$, endowed with the Grassmann product $\wedge$, is a graded associative non-commutative R-algebra with unit element 1, such that $\omega\wedge\chi=(-1)^{pq}\chi\wedge\omega$ for $\omega\in A^p(M)$, $\chi\in A^q(M)$; it is called the *Grassmann algebra of M*.

c) $A^p(M)$ is a free R-module of rank $\binom{n}{p}$: if $\{\varphi_1,\ldots,\varphi_n\}$ is a basis of $A^1(M)$, and if we put

$$\varphi_J:=\varphi_{j_1}\wedge\ldots\wedge\varphi_{j_p}\quad\text{for}\quad J\in\mathbb{N}\tbinom{n}{p}:=\{(j_1,\ldots,j_p)\in\mathbb{N}^p;\ 1\leqq j_1<\ldots<j_p\leqq n\}$$

(where $\varphi_\emptyset:=1\in A^0(M)=R$), then $\{\varphi_J;\ J\in\mathbb{N}\binom{n}{p}\}$ is a basis of $A^p(M)$.

d) Grassmann algebras are used in real analysis for the construction of $\mathscr{C}^\infty$-differential forms: let X be a $\mathscr{C}^\infty$-manifold of real dimension n, and let $\mathscr{E}^{\mathbb{R}}$ be the structure sheaf of X (i.e., $\mathscr{E}^{\mathbb{R}}$ is the sheaf of real-valued $\mathscr{C}^\infty$-functions on X); moreover, let a be a point in X and $x_1, \dots, x_n$ local coordinates at a. An $\mathbb{R}$-linear mapping $\xi : \mathscr{E}^{\mathbb{R}}_a \to \mathbb{R}$ is called a *derivation* or a *real tangent vector* at a if

$$\xi(fg) = \xi(f)\cdot g(a) + f(a)\cdot \xi(g)$$

for every $f, g \in \mathscr{E}^{\mathbb{R}}_a$. The collection $T_a = T_a X$[12] of all real tangent vectors at a is an n-dimensional real vector space with basis $\left.\frac{\partial}{\partial x_1}\right|_a, \dots, \left.\frac{\partial}{\partial x_n}\right|_a$. A *vectorfield* ξ on an open subset U of X is a mapping that associates to each $x \in U$ a tangent vector $\xi(x) \in T_x X$. Near a, ξ has a representation of the form $\xi = \sum_{j=1}^{n} f_j \frac{\partial}{\partial x_j}$. The vectorfield ξ is said to be $\mathscr{C}^\infty$ iff the functions f_j are $\mathscr{C}^\infty$; that is equivalent to the condition that, for each $f \in \mathscr{E}^{\mathbb{R}}(U)$, the function $\xi(f): U \to \mathbb{R}, x \mapsto \xi(x)(f_x)$, be $\mathscr{C}^\infty$.

Now, for $r \geqq 0$, denote with

$$\mathscr{E}^r_a := A^r(T_a X) \oplus i A^r(T_a X)$$

the (complex) vector space of alternating r-fold $\mathbb{R}$-linear mappings $\varphi: T_a \times \dots \times T_a \to \mathbb{C}$. If $dx_1, \dots, dx_n$ is the basis of $A^1(T_a) = \mathrm{Hom}_{\mathbb{R}}(T_a, \mathbb{R})$ dual to the basis $\left.\frac{\partial}{\partial x_1}\right|_a, \dots, \left.\frac{\partial}{\partial x_n}\right|_a$ of T_a, then, $dx_1, \dots, dx_n$ is also a basis of the complex vector space $\mathscr{E}^1_a$, and $\{dx_J;\ J \in \mathbb{N}\binom{n}{r}\}$ is a basis of $\mathscr{E}^r_a$, where $dx_J := dx_{j_1} \wedge \dots \wedge dx_{j_r}$ for $(j_1, \dots, j_r) \in \mathbb{N}\binom{n}{r}$ and $dx_\emptyset := 1 \in \mathbb{R}$.

An *r-form* ω on an open subset U of X is a mapping that associates to each $x \in U$ an element $\omega(x) \in \mathscr{E}^r_x$. Near a, ω has a representation of the form

$$\omega = \sum_{J \in \mathbb{N}\binom{n}{r}} f_J \, dx_J .$$

By definition, ω is $\mathscr{C}^\infty$ iff the functions f_J are $\mathscr{C}^\infty$; that is equivalent to the condition that, for arbitrary $\mathscr{C}^\infty$-vectorfields $\xi_1, \dots, \xi_r$ on U, the function $\omega(\xi_1, \dots, \xi_r): U \to \mathbb{C}, x \mapsto \omega(\xi_1(x), \dots, \xi_r(x))$, is $\mathscr{C}^\infty$.

The presheaf $U \mapsto \mathscr{E}^r(U) := \{\text{all } \mathscr{C}^\infty\ r\text{-forms on } U\}$ is a locally free sheaf $\mathscr{E}^r$ of rank $\binom{n}{r}$ over the sheaf $\mathscr{E} := \mathscr{E}^0$ of ($\mathbb{C}$-valued) $\mathscr{C}^\infty$-functions on X. The $\mathbb{C}$-linear *exterior derivatives*

$$d : \mathscr{E}^r \to \mathscr{E}^{r+1}$$

are defined with respect to local coordinates $x_1, \dots, x_n$ by

$$d\left(\sum f_J dx_J\right) := \sum df_J \wedge dx_J, \quad \text{where } df := \sum_{j=1}^{n} \frac{\partial f}{\partial x_j} dx_j \ \text{ for } f \in \mathscr{E}^0 = \mathscr{E}.$$

(That definition of d is independent of the choice of the local coordinates, since $\varphi^0 d = d\varphi^0$ for $\mathscr{C}^\infty$-mappings φ between open subsets of $\mathbb{R}^n$). The exterior derivatives satisfy

$$d \circ d = 0 .$$

---

[12] If X is even a *complex* manifold of (complex) dimension $m = n/2$, then α: $\mathrm{Der}_{\mathbb{R}}(\mathscr{E}^{\mathbb{R}}_a, \mathbb{R}) \to \mathrm{Der}_{\mathbb{C}}(\mathscr{O}_a, \mathbb{C})$, defined by $\alpha(\xi)(h) := \xi(\operatorname{re} h) + i\xi(\operatorname{im} h)$, is an isomorphism of real vector spaces, since, by the Cauchy-Riemann differential equations, for holomorphic h, $\operatorname{im} h$ is determined (up to an additive constant) by $\operatorname{re} h$. So we may denote the real tangent space (of real dimension n) and the complex tangent space (as defined in § 32) of complex dimension $n/2$ by the same symbol $T_a X$.

Furthermore, $d(\omega \wedge \eta) = d(\omega) \wedge \eta + (-1)^r \omega \wedge d\eta$ if $\omega \in \mathscr{E}^r, \eta \in \mathscr{E}^s$. A $\mathscr{C}^\infty$-mapping $\varphi : X \to Y$ of $\mathscr{C}^\infty$-manifolds induces in a canonical manner for $x \in X$ the morphisms

$$T_x\varphi : T_x X \to T_{\varphi(x)} Y, \ \xi \mapsto [g \mapsto \xi(\varphi^0(g))],$$

and $\varphi_x^* : {}_Y\mathscr{E}^r_{\varphi(x)} \to {}_X\mathscr{E}^r_x, \psi \mapsto [(\xi_1, \ldots, \xi_r) \mapsto \psi((T_x\varphi)(\xi_1), \ldots, (T_x\varphi)(\xi_r))]$, and thus comorphisms $\varphi^* : {}_Y\mathscr{E}^r \to {}_X\mathscr{E}^r$.

Since the comorphisms φ^* commute with the exterior derivatives d, φ induces a comorphism of complexes

$$\varphi^* : ({}_Y\mathscr{E}^*, d) \to ({}_X\mathscr{E}^*, d).$$

Chapter 6: Proof of Theorem B

Having discussed applications of Theorem B in the preceding chapter, we turn now to its proof. Although it is fairly involved, we have attempted not to proceed in the most direct and concise manner possible, but rather to separate the proof into various parts that are useful for other applications as well.

On polynomially convex regions $X \subset\subset \mathbb{C}^n$, the partial differential equation $\bar{\partial}\xi = \omega$ has a solution $\xi \in \mathscr{E}^{0,q}(X)$ for every $\bar{\partial}$-closed differential form $\omega \in \mathscr{E}^{0,q+1}(X)$ (Dolbeault's Lemma); hence, by the $\bar{\partial}$-version of de Rham's Theorem, the structure sheaf ${}_X\mathcal{O}$ of polynomially convex regions $X \subset\subset \mathbb{C}^n$ is acyclic.

We discuss strongly pseudoconvex domains in the second section; after presenting a "vanishing theorem" on cohomology (62.1), we use the techniques developed for its proof in order to derive the Finiteness Theorem of Cartan-Serre and to solve Levi's problem.

To prove the general version of Theorem B in § 63, all we need to know from the first two sections of this chapter is that polydisks are B-spaces. The results of Chapter 5 are thus applicable to polydisks, and it follows relatively easily that Theorem B holds for arbitrary Stein spaces. We prove at the same time that, for a complex space to be a Stein space, it is sufficient that it be holomorphically spreadable and weakly holomorphically convex.

Returning to the old problem of identifying domains of holomorphy in $\mathbb{C}^n$ (Chapter 1), we present a detailed characterization in the form of sixteen equivalent conditions (67.3), bringing together such varied concepts and results as Stein domains, B-spaces, pseudoconvexity, exhaustion by Runge pairs, solvability of the additive Cousin problem, Dolbeault's Lemma, the Character Theorem, and the algebraic representability of 1 in $\mathcal{O}(X)$.

§ 61 Dolbeault's Lemma

In this section we prove Theorem B for the structure sheaf of a polynomially convex region $X \Subset \mathbb{C}^n$. In order to prove that ${}_X\mathcal{O}$ is acyclic, we construct a resolution of ${}_n\mathcal{O}$ on $\mathbb{C}^n$, using the sheaves of $\mathscr{C}^\infty$-differential forms (see § 58.A):

61.1 Poincaré's Lemma ($\bar{\partial}$-version). *The sequence of sheaves on* $\mathbb{C}^n$

$$0 \to {}_n\mathcal{O} \to \mathscr{C}^\infty \xrightarrow{\bar{\partial}} \mathscr{E}^{0,1} \to \ldots \xrightarrow{\bar{\partial}} \mathscr{E}^{0,n} \to 0$$

is exact.

The proof is an immediate consequence of Lemma 61.7. ∎

The sheaf $\mathscr{E}^{0,0} = \mathscr{C}^\infty$, and thus the $\mathscr{C}^\infty$-modules $\mathscr{E}^{p,q}$, are fine sheaves (see E. 50f); hence, by 61.1, the "$\bar{\partial}$-complex" $(\mathscr{E}^{0,*}, \bar{\partial})$ induces an acyclic resolution of ${}_n\mathcal{O}$ on each subset $D \subset \mathbb{C}^n$. Thus 50.4 implies the following:

61.2 de Rham's Theorem ($\bar{\partial}$-version). *For each $D \subset \mathbb{C}^n$ and $q \geqq 0$, there is a natural isomorphism of vector spaces*

$$H^q(D, \mathcal{O}) \cong \operatorname{Ker}[\bar{\partial} : \mathscr{E}^{0,q}(D) \to \mathscr{E}^{0,q+1}(D)] \,/\, \bar{\partial}\mathscr{E}^{0,q-1}(D). \quad ∎$$

The main relut of this section is a global version of the punctual $\bar{\partial}$-Poincaré Lemma:

61.3 Dolbeault's Lemma. *If $X \Subset \mathbb{C}^n$ is polynomially convex, then*

$$0 \to {}_n\mathcal{O}(X) \to \mathscr{C}^\infty(X) \xrightarrow{\bar{\partial}} \mathscr{E}^{0,1}(X) \to \ldots \xrightarrow{\bar{\partial}} \mathscr{E}^{0,n}(X) \to 0$$

is an exact sequence of vector spaces. In particular, $H^q(X, \mathcal{O}) = 0$ holds for each $q \geqq 1$.

Let us recall from § 12 that a region $X \Subset \mathbb{C}^n$ is called *polynomially convex* if, for each compact set $K \subset X$, the polynomially convex hull

$$\hat{K}_{\mathbb{C}[z]} = \{z \in X; |P(z)| \leqq \|P\|_K, \forall P \in \mathbb{C}[z_1, \ldots, z_n]\}$$

of K in X is compact.

E. 61a. Prove these statements for a region $X \Subset \mathbb{C}^n$:

i) X is polynomially convex iff, for each compact set $K \subset X$, the set $\partial X \cap \{z \in \mathbb{C}^n; |P(z)| \leqq \|P\|_K, \forall P \in \mathbb{C}[z]\}$ is empty.

ii) If X is polynomially convex, then X is holomorphically convex.

iii) The converse of ii) does not hold (hint: $S^1 \subset \mathbb{C}^*$).

61.4 Example. For the closure $\square := \overline{\boldsymbol{P}(a; \varrho)}$ of a polydisk $\boldsymbol{P}(a; \varrho) \subset \mathbb{C}^n$ and polynomials $P_1, \ldots, P_m \in \mathbb{C}[z_1, \ldots, z_n]$, the closed "*polynomial polyhedron*"

$$\square_{P_1, \ldots, P_m} := \{z \in \square; |P_j(z)| \leqq 1, j = 1, \ldots, m\}$$

coincides with its polynomially convex hull. That implies immediately that its interior $\mathring{\square}_{P_1,\dots,P_m}$ is polynomially convex. ∎

For simplicity, let us call a subset L of a region $X \subset\subset \mathbb{C}^n$ a *β-set* in X if, for each $q \geqq 1$, all forms $\omega \in \mathscr{E}^{0,q}(X)$ that are closed near L are exact near L. Here, we use the following terminology:

"*ω is ($\bar\partial$-) closed near L*", if $\bar\partial\omega|_U = 0$ on an appropriate open neighborhood U of L;

"*ω is ($\bar\partial$-) exact near L*", if there exists an $\eta \in \mathscr{E}^{0,q-1}(X)$ such that $\bar\partial\eta|_V = \omega|_V$ for an appropriate open neighborhood V of L.

Dolbeault's Lemma states that each polynomially convex region X is a β-set in itself. The following result justifies the manner of speaking "K is a β-set" if K is *compact*:

61.5 Lemma. *If a compact set K is included in two open subsets U and V of $\mathbb{C}^n$, then K is a β-set in U iff K is a β-set in V.*

Proof. We may assume that $U \subset\subset V$. Then it suffices to show that, for every form ω on U, there exists a form η on $\mathbb{C}^n$ such that $\operatorname{supp}\eta \subset U$ and $\eta = \omega$ near K. Fix a relatively compact neighborhood W of K in U; to the open cover $(\mathbb{C}^n \setminus \bar W, U)$ of $\mathbb{C}^n$ there corresponds a smooth partition of unity $1 = f + g$ such that $f|_W = 1$ and $f|_{\mathbb{C}^n \setminus U} = 0$. Then $\eta := f\omega$ has the desired property. ∎

As a first step in the proof of 61.1, we show the following one-dimensional result:

61.6 Lemma. *If a function $f \in \mathscr{C}^\infty(\mathbb{C})$ has compact support, then the assignment*

$$z \mapsto \frac{1}{2\pi i} \int_{\mathbb{C}} f(z+w)\, \frac{dw \wedge d\bar w}{w}$$

determines a function $g \in \mathscr{C}^\infty(\mathbb{C})$ *such that* $\dfrac{\partial g}{\partial \bar z} = f$. *Moreover, if f is $\mathscr{C}^\infty$ or holomorphic in additional parameters, then so is g.*

Proof. The existence of the integral is easy to see: using polar coordinates $w = re^{i\vartheta}$, we obtain that $dw \wedge d\bar w = 2ir\,d\vartheta \wedge dr$ and thus that

$$g(z) = \frac{1}{\pi} \int_0^\infty \int_0^{2\pi} f(z + re^{i\vartheta}) e^{-i\vartheta} d\vartheta \wedge dr\,,$$

where $f(z + re^{i\vartheta})e^{-i\vartheta}$ is $\mathscr{C}^\infty$ in the parameters r and ϑ and has compact support for fixed z. By [La$_2$ V § 8], we may interchange integration and differentiation; hence, g has the same differentiability properties as f. In particular, we obtain that

$$\frac{\partial g}{\partial \bar z}(z) = \frac{1}{\pi}\int_0^\infty \int_0^{2\pi} \frac{\partial f(z+re^{i\vartheta})}{\partial \bar z} e^{-i\vartheta} d\vartheta \wedge dr$$

$$= \lim_{\varepsilon\to 0} \frac{1}{\pi}\int_\varepsilon^\infty \int_0^{2\pi} \frac{f(z+re^{i\vartheta})}{\partial \bar z} e^{-i\vartheta} d\vartheta \wedge dr$$

$$= \lim_{\varepsilon\to 0} \frac{1}{2\pi i} \int_{\mathbb{C}\setminus \boldsymbol{B}(\varepsilon)} \frac{f(z+w)}{\partial \bar z} \frac{dw\wedge d\bar w}{w}.$$

The symmetry of $f(z+w)$ with respect to z and w implies that

$$\frac{\partial f(z+w)}{\partial \bar z}\frac{dw\wedge d\bar w}{w} = \frac{\partial f(z+w)}{\partial \bar w}\frac{dw\wedge d\bar w}{w} = d\Big(-\frac{f(z+w)}{w}dw\Big).$$

Fix z and a ball $\boldsymbol{B}=\boldsymbol{B}(r)$ such that $\operatorname{supp} f(z+*)\subset \boldsymbol{B}$; then, for $\varepsilon<r$, Stoke's Theorem and $f(z+*)|_{\partial \boldsymbol{B}}=0$ imply that

$$\int_{\mathbb{C}\setminus \boldsymbol{B}(\varepsilon)} \frac{\partial f(z+w)}{\partial \bar z}\frac{dw\wedge d\bar w}{w} = \int_{\boldsymbol{B}\setminus \boldsymbol{B}(\varepsilon)} d\Big(-\frac{f(z+w)}{w}dw\Big) =$$

$$= 0 - \int_{\partial \boldsymbol{B}(\varepsilon)} -\frac{f(z+w)}{w}dw = \int_0^{2\pi} if(z+\varepsilon e^{i\vartheta})d\vartheta;$$

hence, $\dfrac{\partial g}{\partial \bar z}(z) = \dfrac{1}{2\pi i}\lim\limits_{\varepsilon\to 0}\displaystyle\int_0^{2\pi} if(z+\varepsilon e^{i\vartheta})d\vartheta = f(z)$. ■

61.7 Lemma. *Every closed polydisk $\square\subset\mathbb{C}^n$ is a β-set.*

Proof. By induction on k, we show for each $q\geqq 1$ that

> "If a form $\omega\in\mathscr{E}^{0,q}(\mathbb{C}^n)$ that depends only on $d\bar z_1,\dots,d\bar z_k$ is closed near $\square$, then it is exact near $\square$."

For $k=0$, we have that $\omega=0$, since $q\geqq 1$.

"$k-1\Rightarrow k$" There exists a unique decomposition $\omega=d\bar z_k\wedge\zeta+\xi$, where $\zeta\in\mathscr{E}^{0,q-1}(\mathbb{C}^n)$ and $\xi\in\mathscr{E}^{0,q}(\mathbb{C}^n)$ depend only on $d\bar z_1,\dots,d\bar z_{k-1}$. If we can construct a form $\tau\in\mathscr{E}^{0,q}(\mathbb{C}^n)$ that is independent of $d\bar z_k,\dots,d\bar z_n$ and such that $d\bar z_k\wedge\zeta+\tau=\bar\partial\chi$ near $\square$ for some $\chi\in\mathscr{E}^{0,q-1}(\mathbb{C}^n)$, then $\omega-(d\bar z_k\wedge\zeta+\tau)=\xi-\tau$ is closed near $\square$ and independent of $d\bar z_k,\dots,d\bar z_n$. By the induction hypothesis, there exists a form $\sigma\in\mathscr{E}^{0,q-1}(\mathbb{C}^n)$ such that $\bar\partial\sigma=\xi-\tau$; hence, we obtain $\omega=\bar\partial(\sigma+\chi)$.

Thus we have to construct τ. If ζ is of the form $\zeta=\sum f_J d\bar z_J$, where $J\in\mathbb{N}\binom{q-1}{k-1}$ and $f_J\in\mathscr{C}^\infty(\mathbb{C}^n)$, then each f_J depends holomorphically on $z_{k+1},\dots,z_n$ near $\square$: we have that $\partial f_J/\partial\bar z_i=0$ for $i>k$, since

$$0=\bar\partial\omega=\bar\partial\Big(\sum_J f_J d\bar z_k\wedge d\bar z_J\Big)+\bar\partial\xi=\sum_{i=1}^n\sum_J\frac{\partial f_J}{\partial\bar z_i}d\bar z_i\wedge d\bar z_k\wedge d\bar z_J+\bar\partial\xi$$

near $\Box$ and since $\bar\partial\xi$ contains no $d\bar z_i \wedge d\bar z_k \wedge d\bar z_J$ for $i > k$. Denote the polyradius of $\Box$ by ϱ, fix an $\varepsilon > 0$ such that $\left.\frac{\partial f_J}{\partial \bar z_i}\right|_{P(\varrho+\varepsilon)} = 0$ for each $i > k$, and choose an $h \in \mathscr{C}^\infty(\mathbb{C})$ such that $h|_{\overline{P^1(\varrho_k)}} = 1$ and $\operatorname{supp} h \subset P^1(\varrho_k + \varepsilon)$. By 61.6, there exist functions $g_J \in \mathscr{C}^\infty(\mathbb{C}^n)$ that, near $\Box$, depend holomorphically on $z_{k+1}, \dots, z_n$, such that $\frac{\partial g_J}{\partial \bar z_k} = h(z_k) f_J(z)$. For $\tau := \sum_{j=1}^{k-1} \sum_J \frac{\partial g_J}{\partial \bar z_j} d\bar z_j \wedge d\bar z_J \in \mathscr{E}^{0,q}(\mathbb{C}^n)$, we have near $\Box$ that

$$\bar\partial\left(\sum_J g_J\, d\bar z_J\right) = \tau + d\bar z_k \wedge \zeta + 0. \quad \blacksquare$$

As a particular consequence, we obtain this special case of Theorem B:

61.8 Corollary. *The structure sheaf of a polydisk is acyclic.*

Proof. A polydisk $P = P^n(a; \varrho)$ is the nested union of the closed polydisks $\Box_j := \overline{P(a; \varrho - 1/j)}$. By 61.7 and 61.2, we obtain that $H^q(\Box_j, \mathcal{O}) = 0$ for each $q \geqq 1$; hence, each homomorphism $H^q(\Box_{j+1}, \mathcal{O}) \to H^q(\Box_j, \mathcal{O})$ is the zero-mapping. Thus $H^q(P, \mathcal{O}) = 0$ for $q \geqq 2$ is an easy consequence of 56.3; for $q = 1$, it follows from 56.4 and 6.7. $\blacksquare$

The method of exhaustion of § 56 will enable us to reduce the proof of Dolbeault's Lemma for polynomially convex regions to that for polynomial polyhedra. To that end, we prove the following:

61.9 Lemma. *i) For $K \subset U \Subset \mathbb{C}^n$ such that $K = \hat K_{\mathbb{C}[z]}$ is compact, there exists a polynomial polyhedron D such that $K \subset \mathring{D} \subset D \subset U$.*

ii) If $X \Subset \mathbb{C}^n$ is polynomially convex, then there exists an exhaustion $X = \bigcup_{j=1}^{\infty} D_j$ by polynomial polyhedra D_j such that $D_j \subset \mathring{D}_{j+1}$.

Proof. i) We may assume that U is included in a closed polydisk $\Box$. For each $a \in \Box \setminus U$, there exists (as in E. 12d) a polynomial P such that $\|P\|_K < 1 < |P(a)|$; since $\Box \setminus U$ is compact, we may choose $P_1, \dots, P_m$ such that $\|P_j\|_K < 1$ for each j and $(\sup_{j=1,\dots,m} |P_j|)(a) > 1$ for each $a \in \Box \setminus U$. Then $D := \Box_{P_1,\dots,P_m}$ has the desired properties.

ii) Any compact exhaustion $X = \bigcup_{j=1}^{\infty} K_j$ according to 51 A. 2 yields a compact exhaustion $X = \bigcup_{j=1}^{\infty} L_j$ such that $L_j = \widehat{(K_j)}_{\mathbb{C}[z]}$; we may even assume that $L_j \subset \mathring{L}_{j+1}$. Now it suffices to choose polynomial polyhedra D_j according to i) such that $L_j \subset \mathring{D}_j \subset \mathring{L}_{j+1}$. $\blacksquare$

As a second step we generalize 61.7 to polynomial polyhedra $\Box_{P_1,\dots,P_m}$. The proof goes by induction on m; it uses the description

$$\square_{P_1,\ldots,P_m,P} = \varphi^{-1}((\square \times \overline{\boldsymbol{P}(1)})_{P_1,\ldots,P_m}),$$

where φ is the holomorphic mapping $\mathbb{C}^n \to \mathbb{C}^{n+1}$, $z \mapsto (z, P(z))$, and the following technical result:

61.10 Lemma. *Let K' be a compact set in $\mathbb{C}^n$ and $P \in \mathbb{C}[z_1, \ldots, z_n]$, a polynomial. If $K := K' \times \overline{\boldsymbol{P}^1(1)} \subset \mathbb{C}^{n+1}$ is a β-set, then so is $K'_P := \{z \in K'; |P(z)| \leqq 1\} \subset \mathbb{C}^n$; precisely,*

(61.10.1) $\begin{cases} \textit{If } \omega \in \mathscr{E}^{0,q}(\mathbb{C}^n) \textit{ is closed near } K'_P, \textit{ then there exists an } \eta \in \mathscr{E}^{0,q}(\mathbb{C}^{n+1}) \textit{ that} \\ \textit{is closed near } K \textit{ and such that } \varphi^* \eta = \omega \textit{ near } K'_P. \end{cases}$

Proof. By § 58, φ^* and $\bar\partial$ commute. Thus (61.10.1) implies that $K'_P = \varphi^{-1}(K)$ is a β-set if K is. Thus we only have to prove (61.10.1). The graph $\Gamma(P)$ satisfies

$$N(\mathbb{C}^{n+1}; Q) = \Gamma(P) = \varphi(\mathbb{C}^n) \quad \text{for} \quad Q(z, w) := w - P(z).$$

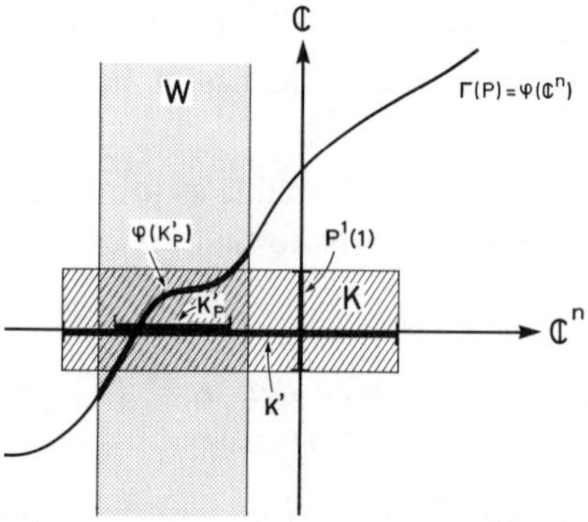

Let π denote the projection $\mathbb{C}^{n+1} \to \mathbb{C}^n, (z, w) \mapsto z$; choose an $f \in \mathscr{C}^\infty(\mathbb{C}^{n+1})$ such that f equals 1 near $\varphi(K'_P)$ and has a compact support included in $W := \{(z, w) \in \mathbb{C}^{n+1}; (\pi^* \bar\partial \omega)(z, w) = 0\}$. For $\tau \in \mathscr{E}^{0,q}(\mathbb{C}^{n+1})$, set $\eta := f \cdot \pi^* \omega - Q \cdot \tau$; then $\varphi^* \eta = \omega$ near K'_P, since $Q \circ \varphi = 0$ and $\pi \circ \varphi = \mathrm{id}_{\mathbb{C}^n}$ imply there that

$$\begin{aligned} \varphi^* \eta &= (\varphi^0 f) \cdot (\varphi^* \pi^* \omega) - (\varphi^0 Q) \cdot (\varphi^* \tau) \\ &= (f \circ \varphi) \cdot (\pi \circ \varphi)^* (\omega) - (Q \circ \varphi) \cdot (\varphi^* \tau) = 1 \cdot \omega - 0 = \omega. \end{aligned}$$

Thus it remains to find τ such that the associated η is $\bar\partial$-closed near K. Every solution τ of the differential equation

$$\bar\partial f \wedge \pi^* \omega = Q \cdot \bar\partial \tau$$

satisfies that condition: since Q is holomorphic and since $(\bar\partial \pi^* \omega)(z, w) = (\pi^* \bar\partial \omega)(z, w) \neq 0$ implies that $f(z, w) = 0$, we have that

$$\bar\partial \eta = \bar\partial (f \cdot \pi^* \omega - \mathrm{Q} \cdot \tau) = \bar\partial f \wedge \pi^* \omega + 0 - 0 - Q \bar\partial \tau = 0.$$

Finally, the differential equation has a solution: since we may divide $\bar{\partial} f \wedge \pi^* \omega$ by Q (note that $\bar{\partial} f = 0$ near $\varphi(K_P') = N(K; Q)$), we see that

$$\bar{\partial}\left(\frac{1}{Q}\,\bar{\partial} f \wedge \pi^* \omega\right) = \frac{1}{Q}\,\bar{\partial}\bar{\partial} f \wedge \pi^* \omega + 0 = 0;$$

hence, the solvability of the differential equation follows from the assumption that K is a β-set. ■

61.11 Corollary. *For a polynomial polyhedron $D := \Box_{P_1, \ldots, P_m} \subset \mathbb{C}^n$ and the holomorphic mapping $\psi: \mathbb{C}^n \to \mathbb{C}^{n+m}$, $z \mapsto (z, P_1(z), \ldots, P_m(z))$, the following statements hold:*

i) D is a β-set.

ii) The region $\mathring{D}$ is Runge.

iii) If $\omega \in \mathscr{E}^{0,q}(\mathbb{C}^n)$ is closed near D, then there exists an $\eta \in \mathscr{E}^{0,q}(\mathbb{C}^{n+m})$ that is closed near $\Box \times \overline{\boldsymbol{P}^m(1)}$ such that $\psi^* \eta = \omega$ near D.

Proof. iii) ⇒ ii) By 61.4, $\mathring{D}$ is polynomially convex; hence, by 61.9 ii), it suffices to find, for each $f \in \mathscr{C}^\infty(\mathbb{C}^n)$ holomorphic near D, and each $\varepsilon > 0$, a polynomial $P \in \mathbb{C}[z]$ such that $\|f - P\|_D < \varepsilon$. Fix an $F \in \mathscr{E}^{0,0}(\mathbb{C}^{m+n})$ according to iii) such that $\bar{\partial} F = 0$ near $\Box \times \overline{\boldsymbol{P}^m(1)}$ and $\psi^0(F) = f$ near D. Further, choose a $Q \in \mathbb{C}[z_1, \ldots, z_{n+m}]$ such that $\|F - Q\|_{\Box \times \overline{\boldsymbol{P}^m(1)}} < \varepsilon$ (see 6.7). Then we have

$$P(z) := Q(z, P_1(z), \ldots, P_m(z)) = \psi^0 Q(z) \in \mathbb{C}[z_1, \ldots, z_n],$$

and the fact that $\psi(D) \subset \Box \times \overline{\boldsymbol{P}^m(1)}$ implies that $\|f - P\|_D = \|\psi^0(F - Q)\|_D < \varepsilon$.

i) and ii) can be proved by induction on m: The case $m = 0$ has been treated in 61.7.

"$m - 1 \Rightarrow m$" Apply 61.10 to $K' := \Box_{P_2, \ldots, P_m}$ and $P := P_1$: by the induction hypothesis, $K = K' \times \overline{\boldsymbol{P}^1(1)} = (\Box \times \overline{\boldsymbol{P}^1(1)})_{P_2, \ldots, P_m} \subset \mathbb{C}^{n+1}$ is a β-set; hence, $K_P' = D$ is a β-set, and i) is proved. For iii), choose, according to (61.10.1), an $\eta_1 \in \mathscr{E}^{0,q}(\mathbb{C}^{n+1})$ that is closed near K, such that $\varphi^* \eta_1 = \omega$ near $K_P' = D$. Set $\tilde{\psi}: \mathbb{C}^{n+1} \to \mathbb{C}^{n+m}$, $(z, w) \mapsto (z, w, P_2(z), \ldots, P_m(z))$; then, by the induction hypothesis, there exists an $\eta \in \mathscr{E}^{0,q}(\mathbb{C}^{n+1+m-1})$ that is closed near $(\Box \times \overline{\boldsymbol{P}^1(1)}) \times \overline{\boldsymbol{P}^{m-1}(1)} = \Box \times \overline{\boldsymbol{P}^m(1)}$ such that $\tilde{\psi}^* \eta = \eta_1$ near K. Then $\omega = \varphi^* \eta_1 = \varphi^* \tilde{\psi}^* \eta = \psi^* \eta$ near D. ■

Proof of Dolbeault's Lemma. By 61.9, we may assume that $X = \bigcup_{j=1}^{\infty} D_j$ is an exhaustion of the polynomially convex X by polynomial polyhedra D_j such that $D_j \subset \mathring{D}_{j+1}$. We have to show that $H^q(X, \mathcal{O}) = 0$ for $q \geqq 1$. Now every restriction-homomorphism $\varrho_j^k: H^k(\mathring{D}_{j+2}, \mathcal{O}) \to H^k(\mathring{D}_j, \mathcal{O})$ can be factored through $H^k(D_{j+1}, \mathcal{O}) \underset{61.11}{=} 0$; hence, it is the zero map for $k \geqq 1$. That implies the assertion for $q \geqq 2$ by 56.3. Finally, 56.4 implies that $H^1(X, \mathcal{O}) = 0$, as each $\mathcal{O}(\mathring{D}_{j+1}) \to \mathcal{O}(\mathring{D}_j)$ has a dense image by 61.11 ii). ■■

E. 61b. A holomorphically convex region $X \subset\subset \mathbb{C}^n$ is polynomially convex iff X is Runge.

E. 61c. A Runge domain need not be holomorphically convex (not to mention polynomially convex; hint: $\mathbb{C}^{2*}$).

By 61.8, the structure sheaf of a polydisk $\boldsymbol{P}$ is acyclic. Let $\mathscr{G}$ be a coherent ${}_{\boldsymbol{P}}\mathcal{O}$-module and fix $m \in \mathbb{N}$. Then, by E. 43o, on every sufficiently small polydisk $\boldsymbol{P}'$ in $\boldsymbol{P}$, there exists an exact sequence

$$(61.12.1)\quad \mathcal{O}^{p_m} \to \ldots \to \mathcal{O}^{p_1} \to \mathcal{O}^{p_0} \to \mathscr{G} \to 0.$$

Hence, the following result implies that $\mathscr{G}|_{\boldsymbol{P}'}$ is acyclic:

61.12 Proposition. *Let $(X, \mathcal{O})$ be a complex space with a countable topology and $\mathscr{G}$, a coherent $\mathcal{O}$-module. Suppose that*

i) $\mathcal{O}$ is acyclic,

ii) $\operatorname{emb}_x(\operatorname{red} X) \leqq m$ for every $x \in X$,

iii) on X, there exists an exact sequence

$$\mathcal{O}^{p_{2m-1}} \xrightarrow{\varphi_{2m-1}} \mathcal{O}^{p_{2m-2}} \to \ldots \to \mathcal{O}^{p_0} \xrightarrow{\varphi_0} \mathscr{G} \xrightarrow{\varphi_{-1}} 0.$$

Then $\mathscr{G}$ is acyclic.

Proof. Fix $q \geqq 1$. For $0 \leqq j \leqq 2n-1$, the exact sequences

$$0 \to \mathscr{K}er\,\varphi_j \to \mathcal{O}^{p_j} \to \mathscr{K}er\,\varphi_{j-1} \to 0$$

induce exact sequences

$$H^{q+j}(X, \mathcal{O}^{p_j}) \to H^{q+j}(X, \mathscr{K}er\,\varphi_{j-1}) \to H^{q+j+1}(X, \mathscr{K}er\,\varphi_j) \to H^{q+j+1}(X, \mathcal{O}^{p_j}).$$

Since $H^k(X, \mathcal{O}^{p_j}) = 0$ for $k \geqq 1$ by E. 50d, we have that

$$H^q(X, \mathscr{G}) = H^q(X, \mathscr{K}er\,\varphi_{-1}) \cong H^{q+1}(X, \mathscr{K}er\,\varphi_0) \cong \ldots \cong$$
$$\cong H^{q+2m}(X, \mathscr{K}er\,\varphi_{2m-1}) \underset{\text{E. 50c}}{=} 0. \quad ∎$$

For the proof of 55.5 and 55.9 we have used this consequence of 61.12:

61.13 Corollary. *Let $\mathscr{F}$ be a coherent analytic sheaf on $U \subset\subset \mathbb{C}^n$ and $\mathcal{O}^p \xrightarrow{\varphi} \mathscr{F}$, a surjective homomorphism of ${}_U\mathcal{O}$-modules. Then, for every sufficiently small polydisk $\boldsymbol{P} \subset U$, the homomorphism*

$$\varphi(\boldsymbol{P}): \mathcal{O}^p(\boldsymbol{P}) \to \mathscr{F}(\boldsymbol{P})$$

is surjective.

Proof. By 61.12, we may assume that the coherent ${}_U\mathcal{O}$-module $\mathscr{K}er\,\varphi$ is acyclic on $\boldsymbol{P}$; thus $H^1(\boldsymbol{P}, \mathscr{K}er\,\varphi) = 0$. ∎

E. 61d. Let $\mathscr{G}$ be a sheaf on a complex space X, and fix $q \in \mathbb{N}$; then the presheaf $U \mapsto H^q(U, \mathscr{G})$ determines a sheaf $\mathscr{H}^q(\mathscr{G})$ on X. If $\mathscr{G}$ is a coherent ${}_X\mathcal{O}$-module, then $\mathscr{H}^0(\mathscr{G}) = \mathscr{G}$ and $\mathscr{H}^q(\mathscr{G}) = 0$ for every $q \geqq 1$ (for 74.10).

E. 61e. If $\mathscr{G}$ is a locally free ${}_M\mathcal{O}$-module on a manifold M, then $H^q(M, \mathscr{G}) = 0$ for $q > \dim M$. (Hint: use the acyclic resolution $(\mathscr{E}^{0,*} \otimes_{\mathcal{O}} \mathscr{G}, \bar{\partial} \otimes 1)$ of $\mathscr{G}$; in [Rf]$_2$, that Vanishing Theorem is proved for arbitrary analytic sheaves on complex spaces.)

§ 62 Theorem B for Strictly Pseudoconvex Domains

The goal of this section is to show that polydisks are B-spaces. The method of proof yields at the same time the Finiteness Theorem of Cartan-Serre and the solution of Levi's problem for strictly pseudoconvex domains.

By Dolbeault's Lemma, the structure sheaf of a polynomially convex region is acyclic. Since finite intersections of such regions are again polynomially convex, 61.12 and Leray's Theorem justify the following conclusion:

(62.1.1) *Let $X \Subset \mathbb{C}^n$ denote a region and $\mathscr{F}$, a coherent ${}_X\mathcal{O}$-module. If $\mathfrak{U} = (U_j)_{j \in J}$ is an open cover of X by polynomially convex subsets U_j on which there exists an exact sequence*

$$\mathcal{O}^{m_{2n-1}} \to \ldots \to \mathcal{O}^{m_0} \to \mathscr{F} \to 0,$$

then $H^*(X, \mathscr{F}) = H^*(\mathfrak{U}, \mathscr{F})$.

The restriction of a coherent ${}_X\mathcal{O}$-module $\mathscr{F}$ to a sufficiently small polynomially convex subset is acyclic. The problem, of course, is to find *large* open sets on which $\mathscr{F}$ is acyclic. One standard solution to that problem consists in synthesizing the resolutions of $\mathscr{F}$ on each of a collection of small subsets to form a single resolution on their union ("Cartan's Attaching Lemma"). We describe another approach that follows ideas of Grauert and Rossi:

62.1 Theorem. *Let G be a bounded domain in $\mathbb{C}^n$. Then $H^q(G, \mathscr{F}) = 0$ for each $q \geqq 1$ and every analytic sheaf $\mathscr{F}$ that is defined and coherent near $\bar{G}$ if G satisfies the following condition: There exist an open neighborhood U of $\bar{G}$ and a function $\varphi \in \mathscr{C}^3(U, \mathbb{R})$ such that*

i) $G = \{z \in U; \varphi(z) < 0\} \Subset U$, and

ii) the Levi form $L_{\varphi, a}$ is positive definite for each $a \in \partial G$.

For the proof of Theorem B in § 63 we only need the following consequence of 62.1:

62.2 Corollary. *Every polydisk is a B-space.*

Proof. We may assume that $\boldsymbol{P} = \boldsymbol{P}^n(1)$. Then the assignment

$$\varphi_j(z) := \frac{1}{n} \sum_{i=1}^{n} |z_i|^2 + \sum_{i=1}^{n} |z_i|^{2j} - 1$$

determines by E. 13c a sequence of strictly plurisubharmonic functions $\varphi_j \in \mathscr{C}^\infty(\mathbb{C}^n, \mathbb{R})$ such that

i) $\varphi_j|_{\partial P} > 0$, i.e., $G_j := \{\varphi_j < 0\} \subset\subset P$;

ii) $\varphi_j > \varphi_{j+1}$ on $P \backslash 0$, i.e., $G_j \subset\subset G_{j+1}$;

iii) for each $z \in P$, there exists a j such that $\varphi_j(z) < 0$, i.e., $P = \bigcup_{j=1}^{\infty} G_j$.

By 11.4, each (G_{j+1}, G_j) is a Runge pair; by 62.1 and the footnote to 56.2, P is a B-space. ∎

E. 62a. Every $B^n(r) \times \mathbb{C}^m$ and every $P^n(r) \times \mathbb{C}^m$ is a B-space.

The proof of 62.1 runs as follows (remember that $q \geqq 1$): By "swelling out" the boundary ∂G a finite number of times, we construct a relatively compact neighborhood G' of $\bar{G}$ in U, such that the restriction-homomorphisms $H^q(G', \mathscr{F}) \to$ $\to H^q(G, \mathscr{F})$ are surjective. From that we deduce that $\dim H^q(G, \mathscr{F}) < \infty$ with a lemma known to functional analysts. The final argument for the vanishing of $H^q(G, \mathscr{F})$ is provided by the structure of $H^*(G, \mathscr{F})$ as an $\mathcal{O}(G)$-module.

The "swelling out" is based on an approximation of functions $\psi \in \mathscr{C}^3(U, \mathbb{R})$ near $a \in U$ by quadratic polynomials (see E. 13d):

$$(62.3.1) \quad \psi(z) = \psi(a) + \operatorname{re} P_{\psi,a}(z) + L_{\psi,a}(z-a) + R_{\psi,a}(z), \text{ where } \lim_{z \to a} \frac{R_{\psi,a}(z)}{|z-a|^2} = 0.$$

We start with a preliminary statement:

62.3 Lemma. *If $V \subset\subset U \subset \mathbb{C}^n$ are neighborhoods of 0 and $\varphi \in \mathscr{C}^3(U, \mathbb{R})$ is such that $\varphi(0) = 0$ and $L_{\varphi,0} > 0$, then there exist a polydisk $\mathbf{P}(\delta) \subset V$ and an $\varepsilon > 0$ such that, for each $\psi \in \mathscr{C}^3(U, \mathbb{R})$ satisfying the inequality*

$$\|\psi - \varphi\|_3 := \sum_{\substack{|\nu+\mu| \leqq 3 \\ \nu, \mu \in \mathbb{N}^n}} \frac{1}{\nu! \mu!} \|D^\nu \bar{D}^\mu (\psi - \varphi)\|_V < \varepsilon,$$

the following statements hold:

i) $\mathbf{P}(\varrho) \cap \{\psi < 0\}$ is polynomially convex for each $\varrho \leqq \delta$.

ii) $L_{\psi,z} > 0$ for each $z \in \overline{\mathbf{P}(\delta)}$.

iii) $\operatorname{re} P_{\psi,a}(z) < 0$ for $z, a \in \overline{\mathbf{P}(\delta)}$ such that $z \neq a$ and $\psi(z) \leqq \psi(a)$.

Proof. ii) Obviously, the constant $c := \min\{L_{\varphi,0}(h); |h| = 1\}$ is positive. As $L_{\psi,a}(h)$ depends continuously on the parameters ψ, a, and h, there exist $\varepsilon, \delta > 0$ such that

$$\min\{L_{\psi,a}(h); |h| = 1, a \in \overline{\mathbf{P}(\delta)} \subset V, \|\varphi - \psi\|_3 < \varepsilon\} > \frac{c}{2};$$

in particular, that implies ii).

iii) By (62.3.1), it suffices to prove, for sufficiently small $\varepsilon, \delta > 0$, that

(62.3.2) $L_{\psi,a}(z-a) + R_{\psi,a}(z) > 0$, if $z, a \in \overline{\boldsymbol{P}(\delta)}$ are such that $z \neq a$.

For that inequality, we only have to verify that

(62.3.3) $\dfrac{|R_{\psi,a}(z)|}{|z-a|^2} < \dfrac{c}{2}$ if $z, a \in \overline{\boldsymbol{P}(\delta)}$ are such that $z \neq a$, since then (set $h := z - a$)

$$L_{\psi,a}(h) + R_{\psi,a}(z) \geqq |h|^2\left(L_{\psi,a}\left(\frac{h}{|h|}\right) - \frac{|R_{\psi,a}(z)|}{|h|^2}\right) > |h|^2\left(\frac{c}{2} - \frac{c}{2}\right) = 0.$$

We now prove (62.3.3) for $\varepsilon \leqq \|\varphi\|_3$ and δ so small that $\|\varphi\|_3 \cdot |z-a| < c/4$ for $z, a \in \overline{\boldsymbol{P}(\delta)}$: set $h := z - a \neq 0$; for $\chi \in \mathscr{C}^3(U, \mathbb{R})$, there exists a y between a and z [Ns$_3$ 1.1.9] such that

$$R_{\chi,a}(z) = \sum_{|\nu+\mu|=3} \frac{1}{\nu!\mu!}(D^\nu \bar{D}^\mu \chi)(y) h^\nu \bar{h}^\mu.$$

Since "$|h^\nu \bar{h}^\mu| \leqq |h|^3$ for $|\nu + \mu| = 3$" implies that $|R_{\chi,a}(z)| \leqq \|\chi\|_3 \cdot |h|^3$, we obtain the inequality

$$|R_{\chi,a}(z)| \,/\, |h|^2 \leqq \|\varphi\|_3 \cdot |h| < c/4$$

if $\|\chi\|_3 \leqq \|\varphi\|_3$. Finally, (62.3.3) follows from the equality $R_{\psi,a} = R_{\psi-\varphi,a} + R_{\varphi,a}$.

Now we derive i) for δ, ε as in (63.3.2). To that end, we prove this: if K is a compact subset of $\boldsymbol{P}(\varrho) \cap \{\psi < 0\}$, then so is the polynomially convex hull $\tilde{K}$. First of all, we have that

$$\tilde{K} \cap \overline{\boldsymbol{P}(\varrho)} \cap N(\psi) = \emptyset.$$

For if $a \in \overline{\boldsymbol{P}(\varrho)} \cap N(\psi)$, then iii) implies that $\operatorname{re} P_{\psi,a}|_{\overline{\boldsymbol{P}(\varrho)} \cap \{\psi < 0\}} < 0$ and thus that

$$\|e^P\|_K = \|e^{\operatorname{re} P}\|_K < e^0 = |(e^P)(a)|$$

for $P := P_{\psi,a}$. If $Q \in \mathbb{C}[z]$ denotes a sufficiently large section of the Taylor series of e^P, then $\|Q\|_K < |Q(a)|$, i.e., $a \notin \tilde{K}$.

Consequently, there exists a decomposition

$$\tilde{K} = \tilde{K}_+ \cup \tilde{K}_-, \quad \text{with} \quad \tilde{K}_+ = \tilde{K} \cap \{\psi > 0\} \quad \text{and} \quad \tilde{K}_- = \tilde{K} \cap \{\psi < 0\},$$

into compact subsets included in the convex set $\boldsymbol{P}(\varrho)$. Now choose disjoint neighborhoods U_+ of $\tilde{K}_+$ and U_- of $\tilde{K}_-$ in $\boldsymbol{P}(\varrho) \setminus N(\psi)$ and, according to 61.9, a polynomial polyhedron W such that $\tilde{K} \subset W \Subset U_+ \cup U_-$. Since the characteristic function χ of U_+ is holomorphic on W, 61.11 yields the existence of a polynomial $P \in \mathbb{C}[z]$ such that $\|\chi - P\|_{\tilde{K}} < 1/2$. Then the inequalities $\|P\|_K \leqq \|P\|_{\tilde{K}_-} < 1/2$ and $|P| > 1/2$ on $\tilde{K}_+$ imply that $\tilde{K}_+ = \emptyset$ and $\tilde{K} = \tilde{K}_- \subset \boldsymbol{P}(\varrho) \cap \{\psi < 0\}$. ■

We now are ready for the **proof of 62.1:**

1st step. *For a fixed $\mathscr{F}$ on U, there exists a function $\psi \in \mathscr{C}^3(U, \mathbb{R})$ such that*
i) $G \Subset G' := U \cap \{\psi < 0\} \Subset U$;
ii) $\varrho^q : H^q(G', \mathscr{F}) \to H^q(G, \mathscr{F})$ is surjective (for $q \geqq 1$).

Proof. Fix a neighborhood $V \Subset U$ of $\bar{G}$; then, according to 62.3, there exist points $a_j \in \partial G$, real numbers $\varepsilon_j, \delta_j > 0$, and functions $\psi_j \in \mathscr{C}^3(U, \mathbb{R})$ such that for $j = 1,\ldots,m$ the followings holds:

i) $\partial G \subset \bigcup_{k=1}^{m} \boldsymbol{P}(a_k; \delta_k/4) \subset \bigcup_{k=1}^{m} \boldsymbol{P}(a_k; \delta_k) \subset V$.

ii) On $\boldsymbol{P}(a_j; \delta_j)$, there exists an exact sequence as in (62.1.1).

iii) 62.3 holds for $a_j, \varepsilon_j, \delta_j$ instead of $0, \varepsilon, \delta$.

iv) $\psi_j \begin{cases} < 0 & \text{on } \boldsymbol{P}(a_j; \delta_j/4), \\ \leqq 0 & \text{on } \boldsymbol{P}(a_j; \delta_j/2), \\ = 0 & \text{elsewhere}. \end{cases}$

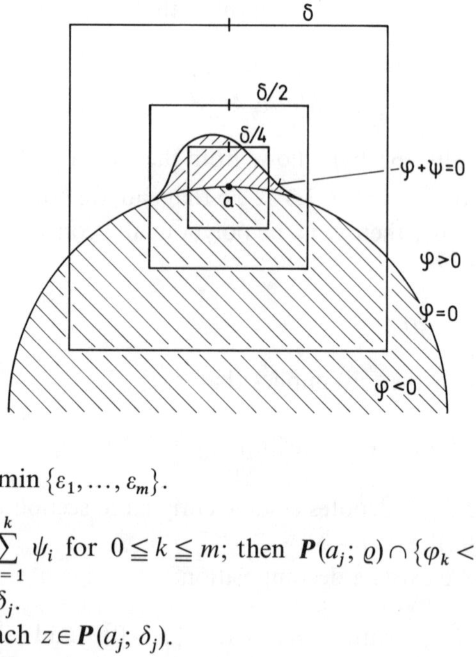

v) $\|\psi_j\|_3 < \varepsilon := \dfrac{1}{m} \min\{\varepsilon_1, \ldots, \varepsilon_m\}$.

vi) Set $\varphi_k := \varphi + \sum_{i=1}^{k} \psi_i$ for $0 \leqq k \leqq m$; then $\boldsymbol{P}(a_j; \varrho) \cap \{\varphi_k < 0\}$ is polynomially convex for each $\varrho \leqq \delta_j$.

vii) $L_{\varphi_k, z} > 0$ for each $z \in \boldsymbol{P}(a_j; \delta_j)$.

Set $G_k := U \cap \{\varphi_k < 0\}$ and $\psi := \varphi_m$; then

$$G = G_0 \Subset G_1 \Subset \ldots \Subset G_m = U \cap \{\psi < 0\} =: G',$$

and the inequalities $(\varphi + \Sigma\psi_i)|_{\bar{G}} < 0$ and $\operatorname{supp} \Sigma\psi_i \subset V$ imply that $\bar{G} \subset G' \subset V \Subset U$.
If all restriction-homomorphisms $H^q(G_k, \mathscr{F}) \to H^q(G_{k-1}, \mathscr{F})$ are surjective, then so is ϱ^q. Hence, we may assume that $m = 1$, $a = a_1$, etc. It suffices to construct an

open cover $\mathfrak{U} = (U_k)_{1 \leqq k \leqq s}$ of G' such that

α) each U_k is polynomially convex,

β) on each U_k, there exists an exact sequence as in (62.1.1),

γ) $U_1 = \boldsymbol{P}(a; \delta) \cap G'$ and $U_k \subset G$ for $k \neq 1$.

By (62.1.1), then, $\mathfrak{U}$ is a Leray cover of G' and $\mathfrak{U} \cap G$, a Leray cover of G, since $U_1 \cap G = \boldsymbol{P}(a; \delta) \cap \{\varphi < 0\}$ is polynomially convex (see vi)). The restriction-homomorphism of Čech complexes $C_a^*(\mathfrak{U}, \mathscr{F}) \to C_a^*(\mathfrak{U} \cap G, \mathscr{F})$ is bijective for $q \geqq 1$ (since $U_j \cap U_k = (U_j \cap G) \cap (U_k \cap G)$ for $j \neq k$); consequently, $H^q(G', \mathscr{F}) \to H^q(G, \mathscr{F})$ is surjective.

In the following construction of $\mathfrak{U}$, all polydisks are supposed to be so small that there exist exact sequences as in (62.1.1). The set $U_1 := \boldsymbol{P}(a; \delta) \cap \{\psi < 0\}$ is polynomially convex by vi). Now cover $\partial G \setminus \boldsymbol{P}(a; 3\delta/4)$ with polydisks $\boldsymbol{P}_2, \ldots, \boldsymbol{P}_t$ that do not intersect $\overline{\boldsymbol{P}(a; \delta/2)}$ and such that $U_j := \boldsymbol{P}_j \cap \{\varphi < 0\}$ is polynomially convex. Then it suffices to cover $G' \setminus \bigcup_{j=1}^{t} U_j = G \setminus \bigcup_{j=1}^{t} U_j$ with polydisks $U_{t+1}, \ldots, U_s \subset G \setminus \overline{\boldsymbol{P}(a; \delta/2)}$. ■

E. 62b. The existence of the function φ in 62.1 is essential: if G' is a sufficiently small domain in $\mathbb{C}^2$ that includes the domain G of E.12g and contains the point 0, then G' is not a domain of holomorphy. Thus 63.7 implies that $H^1(G', \mathcal{O}) \neq 0$.

2nd step. $\dim H^q(G, \mathscr{F}) < \infty$.

We apply the following result:

62.4 Lemma. *Let $W \subset\subset X$ denote complex spaces with a countable topology, and $\mathfrak{U} = (U_j)_{1 \leqq j \leqq s}$ and $\mathfrak{V} = (V_j)_{1 \leqq j \leqq s}$, open covers of X and W, respectively, such that $V_j \subset\subset U_j$ and $V_j \subset W$. If $\mathscr{F}$ is a coherent ${}_X\mathcal{O}$-module and $H^q(\mathfrak{U}, \mathscr{F}) \to H^q(\mathfrak{V}, \mathscr{F})$, a surjection, then $H^q(\mathfrak{V}, \mathscr{F})$ is a finite dimensional vector space.*

For $j = 1, \ldots, m$, using the notation of the first step, set

$$U_j := \boldsymbol{P}(a_j; \delta_j) \cap \{\psi < 0\} \quad \text{and} \quad V_j := \boldsymbol{P}(a_j; \delta_j/2) \cap \{\varphi < 0\}.$$

Then cover $G \setminus \bigcup_{j=1}^{m} \boldsymbol{P}(a_j; \delta_j/4)$ with polydisks $U_j = \boldsymbol{P}(a_j; \delta_j)$ in G for $j = m+1, \ldots, s$, such that the polydisks $V_j := \boldsymbol{P}(a_j; \delta_j/2)$ still cover. Since

$$H^q(\mathfrak{U}, \mathscr{F}) = H^q(G', \mathscr{F}) \to H^q(G, \mathscr{F}) = H^q(\mathfrak{V}, \mathscr{F})$$

is surjective by the first step, Lemma 62.4 yields that $\dim H^q(G, \mathscr{F}) < \infty$. ■

For the proof of 62.4 we use two lemmas:

62.5 Lemma (L. Schwartz). *If $E \underset{v}{\overset{u}{\rightrightarrows}} F$ are continuous linear mappings of Fréchet spaces such that u is compact and v, surjective, then $(u+v)(E)$ is a closed subspace of F, and $F/(u+v)(E)$ is a Hausdorff topological vector space of finite dimension.*

A proof is given for instance in [GuRo App. B. 12]. ■

62.6 Lemma. *If $\mathcal{F}$ is a coherent ${}_X\mathcal{O}$-module on a complex space X with a countable topology, then, for every $W \Subset X$, the restriction-homomorphism $\mathcal{F}(X) \to \mathcal{F}(W)$ is a compact mapping.*

Proof of 62.4. As in E. 55h v), define linear mappings

$$Z^q(\mathfrak{U}, \mathcal{F}) \times C^{q-1}(\mathfrak{V}, \mathcal{F}) \underset{v}{\overset{u}{\rightrightarrows}} Z^q(\mathfrak{V}, \mathcal{F}), \;\; u(f,g) := -f|_{\mathfrak{V}},\; v(f,g) := f|_{\mathfrak{V}} + dg\,.$$

Then u is *compact*, since each restriction $\mathcal{F}(U_{j_0 \dots j_q}) \to \mathcal{F}(V_{j_0 \dots j_q})$ is compact by 62.6; hence, $C^q(\mathfrak{U}, \mathcal{F}) \to C^q(\mathfrak{V}, \mathcal{F})$, as a finite product of such mappings, and thus its restriction $u: Z^q(\mathfrak{U}, \mathcal{F}) \to Z^q(\mathfrak{V}, \mathcal{F})$ are compact mappings (see E. 55h ii)).

Moreover, v is *surjective*: as $H^q(\mathfrak{U}, \mathcal{F}) \to H^q(\mathcal{J}, \mathcal{F})$ is, there exists for every $h \in Z^q(\mathcal{J}, \mathcal{F})$ an $f \in Z^q(\mathfrak{U}, \mathcal{F})$ such that $f|_{\mathfrak{V}} - h \in B^q(\mathcal{J}, \mathcal{F})$; thus, there is a $g \in C^{q-1}(\mathfrak{V}, \mathcal{F})$ such that $f|_{\mathfrak{V}} - h = dg$, i.e., $h = v(f, -g)$.

Thus 62.5 implies that $H^q(\mathfrak{V}, \mathcal{F}) = Z^q(\mathfrak{V}, \mathcal{F}) \,/\, B^q(\mathfrak{V}, \mathcal{F})$ is of finite dimension. ■

Proof of 62.6. Let $X \Subset \mathbb{C}^n$ be polynomially convex. For ${}_X\mathcal{O}^p$, the assertion follows from 5.4. Let us assume that there exists an exact sequence on X as in (62.1.1). Then we obtain a commutative diagram

$$\begin{array}{ccc}
{}_n\mathcal{O}^p(X) & \xrightarrow{\;\;\psi\;\;} & \mathcal{F}(X) \\
\Big\downarrow \varrho_{\mathcal{O}^p} & & \Big\downarrow \varrho_{\mathcal{F}} \\
{}_n\mathcal{O}^p(W) & \xrightarrow{\;\;\varphi\;\;} & \mathcal{F}(W),
\end{array}$$

where ψ is surjective, as $\mathcal{K}er\,\psi$ is acyclic on X by 61.12 and 61.3. As ψ is open, φ continuous, and $\varrho_{\mathcal{O}^p}$ compact, the mapping $\varrho_{\mathcal{F}}$ is compact.

For a complex space with a countable topology, choose a finite open cover $(W_j)_{1 \le j \le m}$ of $\overline{W}$ such that the following statements hold:

α) Each W_j is biholomorphic to an analytic subset of a polynomially convex set $V_j \Subset \mathbb{C}^{n_j}$.

β) The trivial extension to V_j of the image sheaf of $\mathcal{F}$ admits an exact sequence as in (62.1.1).

Now, if $(U_j)_{1 \le j \le m}$ is another open cover of $\overline{W}$ such that $U_j \Subset W_j$, then all restrictions $\varrho_j: \mathcal{F}(W_j) \to \mathcal{F}(W \cap U_j)$ are compact. There is a commutative diagram

$$\begin{array}{ccccc}
\mathcal{F}(X) & \to & \mathcal{F}\Big(\bigcup\limits_{j=1}^{m} W_j\Big) & \hookrightarrow & \prod\limits_{j=1}^{m} \mathcal{F}(W_j) \\
\Big\downarrow \varrho_{\mathcal{F}} & & & & \Big\downarrow \Pi\varrho_j \\
\mathcal{F}(W) & & \xrightarrow{\;\;\;\;\varphi\;\;\;\;} & & \prod\limits_{j=1}^{m} \mathcal{F}(U_j \cap W),
\end{array}$$

in which $\Pi\,\varrho_j$ is compact and φ, as a closed injective mapping (see E. 55fi)), is proper. Consequently, $\varrho_{\mathscr{F}}$ is compact. ■

3rd step. $H^q(G,\mathscr{F})=0$ *for analytic sheaves* $\mathscr{F}$ *that are defined and coherent near* $\bar{G}$. We proceed by downwards induction on $q \geqq 1$. For $q>2n$, the assertion follows from E. 50 c. For "$q+1 \Rightarrow q$" we show that

$$(62.1.2)\quad \left\{\begin{array}{l}\text{Near } \bar{G}\text{, there exists a homomorphism } \varphi:\mathscr{F}^m\to\mathscr{F} \text{ of } \mathscr{O}\text{-modules such}\\ \text{that } \operatorname{supp}\mathscr{F}/\varphi\mathscr{F}^m \text{ is discrete and } \varphi^q: H^q(G,\mathscr{F}^m)\to H^q(G,\mathscr{F}) \text{ is the}\\ \text{zero map.}\end{array}\right.$$

That implies $H^q(G,\mathscr{F})=0$: The exact sequences

$$0\to\mathscr{K}er\,\varphi\to\mathscr{F}^m\to\varphi(\mathscr{F}^m)\to 0 \quad\text{and}\quad 0\to\varphi(\mathscr{F}^m)\to\mathscr{F}\to\mathscr{F}/\varphi(\mathscr{F}^m)\to 0$$

induce a commutative diagram

$$\begin{array}{ccccc}
H^q(G,\mathscr{F}^m) & \searrow^{\varphi^q} & & & \\
\alpha\downarrow & & & & \\
H^q(G,\varphi(\mathscr{F}^m)) & \xrightarrow{\beta} & H^q(G,\mathscr{F}) & \xrightarrow{\gamma} & H^q(G,\mathscr{F}/\varphi(\mathscr{F}^m)) \\
\downarrow & & & & \\
H^{q+1}(G,\mathscr{K}er\,\varphi), & & & &
\end{array}$$

in which the row and the column are exact. By the induction hypothesis,

$$H^{q+1}(G,\mathscr{K}er\,\varphi)=0,\quad\text{and}\quad H^q(G,\ \mathscr{F}/\varphi(\mathscr{F}^m))=0,$$

as $\operatorname{supp}\mathscr{F}/\varphi(\mathscr{F}^m)$ is discrete near $\bar{G}$ (see 50.12). Thus α and β are surjective, and the fact that $\varphi^q=0$ implies $H^q(G,\mathscr{F})=0$.

For a proof of (62.1.2), we may assume that $\mathscr{F}$ is coherent on an open neighborhood U of $\bar{G}$. By the second step, the vector space $F:=H^q(G,\mathscr{F})$ is of finite dimension. It admits the structure of an $\mathscr{O}(U)$-module (see for instance the construction of the Čech cohomology). There is a natural homomorphism of algebras

$$\Phi:\mathscr{O}(U)\to\operatorname{End}_{\mathbb{C}}(F):=\operatorname{Hom}_{\mathbb{C}}(F,F),\ \Phi(f)(\omega):=f\omega;$$

its kernel $\mathfrak{i}$ is an ideal in $\mathscr{O}(U)$. Fix a W such that $\bar{G}\subset W\subset\subset U$; by 55.12, there are functions $f_1,\dots,f_m\in\mathfrak{i}$ such that

$$V(W;\mathscr{O}\cdot\mathfrak{i})=V(W;f_1,\dots,f_m)=:A.$$

Set

$$\varphi:\mathscr{F}^m\to\mathscr{F},\ (g_1,\dots,g_m)\mapsto\sum_{j=1}^m f_j g_j;$$

then $\varphi^q(\omega)=\sum_{j=1}^m f_j\omega_j=0$ for $\omega=(\omega_1,\dots,\omega_m)\in F^m \underset{\text{E. 50d}}{=} H^q(G,\mathscr{F}^m)$, since all f_j

belong to $\mathfrak{i}$. Hence, $\varphi^q = 0$. We have that $\operatorname{supp}\mathscr{F}/\varphi(\mathscr{F}^m) \subset A$, since, for every $x \in W \setminus A$, $f_j(x) \neq 0$ for at least one j, and thus $\mathscr{F}_x = f_{jx} \cdot \mathscr{F}_x \subset \varphi(\mathscr{F}_x^m) \subset \mathscr{F}_x$. Hence, it remains to show that A is a finite set. If $x_1, \ldots, x_k$ are k different points in A, then the evaluation-mappings $f \mapsto f(x_j)$ induce homomorphisms

$$\mu : \mathcal{O}(\mathbb{C}^n) \to \mathbb{C}^k \quad \text{and} \quad \lambda : \mathcal{O}(U)/\mathfrak{i} \to \mathbb{C}^k .$$

It is easy to see that μ is surjective; hence, so is λ. Consequently, we have that

$$k \leqq \dim_{\mathbb{C}}(\mathcal{O}(U)/\mathfrak{i}) \leqq \dim \operatorname{End}_{\mathbb{C}}(F) \leqq (\dim F)^2 < \infty . \quad ■■■$$

There is another important consequence of 62.4 (for simplicity, we use Theorem B in the proof):

62.7 Finiteness Theorem of Cartan-Serre. *If X is a compact complex space and $\mathscr{F}$, a coherent ${}_X\mathcal{O}$-module, then, for each $q \geqq 0$, the vector space $H^q(X, \mathscr{F})$ is of finite dimension.*

Proof. Denote by $\mathfrak{U} = (U_j)_{1 \leqq j \leqq m}$ an open cover of X by local models $U_j \cong A_j \hookrightarrow \boldsymbol{P}^k(1)$ such that $\mathfrak{V} := (V_j)_{1 \leqq j \leqq m}$ for $V_j := A_j \cap \boldsymbol{P}^k(1/2)$ still covers X. Then Theorem B implies that $\mathfrak{U}$ and $\mathfrak{V}$ are Leray covers of $W := X \subset\subset X$. Since we have that $H^*(\mathfrak{U}, \mathscr{F}) = H^*(X, \mathscr{F}) = H^*(\mathfrak{V}, \mathscr{F})$, the assertion follows from 62.4. ■

That theorem is the particular case $Y = \bullet$ of the following stronger version of a result that we mentioned already in § 45:

Grauert's Coherence Theorem. *If $f : X \to Y$ is a proper holomorphic mapping, then all image sheaves $f_q\mathscr{F}$ of coherent ${}_X\mathcal{O}$-modules are coherent ${}_Y\mathcal{O}$-modules (see* [FoKn]*).*

Remember that the *q-th image sheaf* $f_q\mathscr{F}$ has been defined by the presheaf

$$U \mapsto H^q(f^{-1}(U), \mathscr{F}), U \subset Y.$$

E. 62c. Prove that $\dim H^1(\mathbb{C}^{2*}, \mathcal{O}) = \infty$. (Hint: by 7.10, the exact sequence $0 \to {}_2\mathcal{O} \xrightarrow{\cdot z_1} {}_2\mathcal{O} \to {}_2\mathcal{O}/{}_2\mathcal{O} \cdot z_1 \to 0$ provides an exact sequence ${}_2\mathcal{O}(\mathbb{C}^2) \to {}_1\mathcal{O}(\mathbb{C}^*) \xrightarrow{\delta} H^1(\mathbb{C}^{2*}, {}_2\mathcal{O})$; thus $\{\delta z_2^{-j}; j \in \mathbb{N}\}$ is a linearly independent set of cohomology classes; see also E.74e.)

The method of proof that we used for 62.1 also provides the solution of *Levi's problem* (see § 14) for strictly pseudoconvex domains in $\mathbb{C}^n$ (for pseudoconvex domains, see § 63 A):

62.8 Theorem. *If G is a bounded domain in $\mathbb{C}^n$ such that*

i) G admits a description $G = \{z \in U; \varphi(z) < 0\} \subset\subset U \subset \mathbb{C}^n$, where $\varphi \in \mathscr{C}^3(U, \mathbb{R})$ and

ii) the Leviform $L_{\varphi, a}$ is positive definite for each $a \in \partial G$,

then G is holomorphically convex and thus a domain of holomorphy.

Proof. By 12.8, it suffices to construct for each $a \in \partial G$ a polydisk $\boldsymbol{P} = \boldsymbol{P}(a; \delta) \subset\subset U$ and an $f \in \mathscr{M}(G \cup \boldsymbol{P})$ that is holomorphic in G and has a pole at the point a. We may assume that $a = 0$; then the polynomial $P := P_{\varphi,0}$ satisfies $P(0) = 0$, and, for sufficiently small $\boldsymbol{P} \subset U$, 62.3 iii) ensures that $\operatorname{re} P|_{\boldsymbol{P} \cap \{\varphi < 0\}} < 0$. Thus P has no zeros in $\boldsymbol{P} \cap G$. Hence, the meromorphic function $1/P$ is holomorphic on $\boldsymbol{P} \cap G$

and has a pole at the point 0. In order to construct the global function f, we solve an additive Cousin Problem on a "swelling out" G_1 of G at the point 0: set $V := G \cup \boldsymbol{P}$ and determine ε and δ according to 62.3; moreover, there exists a $\chi \in \mathscr{C}^3(U, \mathbb{R})$ such that $\chi \leqq 0, \chi(0) < 0$, $\operatorname{supp} \chi \subset \boldsymbol{P}(\delta)$ and $\|\chi\|_3 < \varepsilon$. For $\psi := \varphi + \chi$, then, $G_1 := U \cap \{\psi < 0\} \subset\subset U$ is strictly pseudoconvex by 62.3 ii). Now we consider the additive Cousin distribution $s_1 := (0, 1/P)$ with respect to the cover $(G, \boldsymbol{P} \cap G_1)$ of G_1; by 62.1 and 53.5, s_1 admits a solution $f \in \mathscr{M}(G_1)$. It is obvious that $f|_G$ is holomorphic and that f has a pole at the point 0. ■

E. 62d. A relatively compact open subset W of a complex space X is called *strictly pseudoconvex* if, for each $a \in \partial W$, there exist near a a biholomorphic mapping f_a onto a closed subspace of some polydisk $\boldsymbol{P}_a \subset\subset \mathbb{C}^{n_a}$ and a strictly plurisubharmonic function $\varphi_a \in \mathscr{C}^3(\boldsymbol{P}_a, \mathbb{R})$ such that $W = \{\varphi_a \circ f_a < 0\}$ holds near a. Use Theorem B in order to show: *If $W \subset\subset X$ is strictly pseudoconvex and if $\mathscr{F}$ is a coherent analytic sheaf near $\bar{W}$, then* $\dim H^q(W, \mathscr{F})$ *is finite for every $q > 0$ (for* E. 62g*).*

E. 62e. Show for the mapping $\pi : X \to \mathbb{C}^2$ that corresponds to the blowing up of the point 0 in $\mathbb{C}^2$:

i) $\pi^{-1}(B^2(1))$ is strictly pseudoconvex in X.

ii) If $\mathscr{I} \subsetneq {}_X\mathscr{O}$ is a coherent ideal such that $N(\mathscr{I}) \subsetneq \pi^{-1}(0)$, then $0 < \dim H^1(X, \mathscr{I}) < \infty$.

E. 62f. Show for the line bundles $E(-k)$ on $\mathbb{P}_1$ (see E. 54 Bc) for $k > 0$:

i) The assignment $\varphi(z, \lambda) := \sum_{j=0}^{k} |z^j \lambda|^2$ determines a real-valued "polynomial function" on $E(-k)$ that is strictly plurisubharmonic outside the zero section S_0.

ii) S_0 has a fundamental system of strictly pseudoconvex neighborhoods.

E. 62g. *Every strictly pseudoconvex domain W in a complex space X is holomorphically convex.* (Hint: consider the linearly independent subset $\{s_j = (0, P^{-j}); j \in \mathbb{N}_{>0}\}$ of $(\mathscr{M}/\mathscr{O})(G_1)$ in the proof of 62.8; by E. 62d, the subset $\{\delta s_j; j \in \mathbb{N}_{>0}\}$ of $H^1(G_1, \mathscr{O})$ is linearly dependent.)

§ 63 Characterization of Stein Spaces

While we derive Theorem B from 62.2, we show that the axioms for Stein spaces can be weakened considerably:

63.1 Definition. *A complex space X is called* weakly holomorphically convex, *if every compact set K in X has an open neighborhood U such that $\hat{K}_{\mathscr{O}(X)} \cap U$ is compact.*

It is easy to see then, that U always can be chosen in such a way that $U \subset\subset X$ and $\hat{K} \cap \partial U = \emptyset$.

63.2 Theorem. *Let X be a complex space. Then the following conditions are equivalent:*

i) X is a B-space; i.e., every coherent ${}_X\mathscr{O}$-module is acyclic.

ii) X is holomorphically convex and holomorphically separable. [1]

iii) X is weakly holomorphically convex and holomorphically spreadable.

iv) X is weakly holomorphically convex, and every compact analytic subset of X is finite.

Proof. "i) $\Rightarrow$ ii)" has been demonstrated in 52.6, while "ii) $\Rightarrow$ iii) $\Rightarrow$ iv)" is a consequence of the following easy exercise:

E. 63a. Let X be a complex space. Then, X is holomorphically separable $\underset{\neq}{\Rightarrow}$ X is holomorphically spreadable $\underset{\neq}{\Rightarrow}$ Every compact analytic subset of X is finite (hint: E.51c ii); E.52h; E.51c iii), Maximum Principle, E.51Aa; E.51e iii)).

Let us prove two preliminary lemmas for the demonstration of "iv) $\Rightarrow$ i)":

63.3 Lemma. *For every complex space X, the following hold:*

i) If $K \subset U \Subset X$ is such that K is compact and satisfies $K = \hat{K}_{\mathcal{O}(X)} \cap U$ and $\hat{K} \cap \partial U = \emptyset$, then there exists a holomorphic mapping $\varphi: X \to \mathbb{C}^m$ such that

$$\varphi(\partial U) \cap \mathbf{P} = \emptyset \quad \text{and} \quad \varphi(K) \subset \mathbf{P}$$

for $\mathbf{P} := \mathbf{P}^m(1)$. Moreover, $W := U \cap \varphi^{-1}(\mathbf{P})$ is an analytic polyhedron, and the restriction $\varphi|_W: W \to \mathbf{P}$ is proper.

ii) If X is weakly holomorphically convex and has a countable topology, then there exists an exhaustion $X = \bigcup_{j=1}^{\infty} W_j$ by analytic polyhedra $W_j \Subset W_{j+1}$ given by holomorphic mappings $\varphi_j \in \operatorname{Hol}(X, \mathbb{C}^{m_j})$ such that $\varphi_j: W_j \to \mathbf{P}^{m_j}(1)$ is proper.

iii) If every compact analytic subset of X in ii) is finite, then every W_j is a B-space.

Proof. i) and ii) can be shown in the same way as 61.9, if one replaces $\square \backslash U$ with ∂U (use E. 51f).

iii) By hypothesis, $\varphi_j: W_j \to \mathbf{P}^{m_j}(1)$ is finite. Since $\mathbf{P}^{m_j}$ is a B-space by 62.2, so is W_j by 52.18. ■

63.4 Lemma. *Let $U \Subset X$ and $V \Subset Y$ be complex spaces and $\varphi: X \to Y$, a holomorphic mapping such that $\varphi(U) \subset V$. If (Y, V) is a Runge pair, and if the image of $\varphi^0: \mathcal{O}(V) \to \mathcal{O}(U)$ is dense, then (X, U) is a Runge pair.*

[1] By 51 A. 3, those conditions are equivalent to X being a Stein space.

Proof. There is a commutative diagram

$$\begin{array}{ccc} \mathcal{O}(Y) & \longrightarrow & \mathcal{O}(X) \\ {\scriptstyle\alpha}\downarrow & & \downarrow{\scriptstyle\beta} \\ \mathcal{O}(V) & \xrightarrow{\varphi^0} & \mathcal{O}(U), \end{array}$$

in which $\operatorname{Im}\beta$ is dense, since $\operatorname{Im}\alpha$ and $\operatorname{Im}\varphi_0$ are. ■

End of the proof of 63.2. iv) ⇒ i) We may assume that X is connected and thus has a *countable topology*: For $a \in X$, the set

$$\widehat{\{a\}} = \bigcap_{\substack{f \in \mathcal{O}(X) \\ f(a)=0}} f^{-1}(0) = \bigcap_{f \in \mathcal{O}(X)} f^{-1}(f(a))$$

is analytic; on the other hand, it has an open neighborhood U such that $U \cap \widehat{\{a\}}$ is compact and thus finite by assumption. Hence, X is holomorphically spreadable, and we may apply 51 A. 3.

Fix a Stein exhaustion $X = \bigcup_{j=1}^{\infty} W_j$ according to 63.3. If we can show that each (W_{j+1}, W_j) is a Runge pair, then, by 56.2, X is a B-space (at the same time we have proved the missing implication in the proof of 56.2). To that end we may assume that $j = 0$. By 52.8, there exists a $\vartheta \in \operatorname{Hol}(W_1, \mathbb{C}^n)$ such that $\vartheta|_{W_0}$ determines an embedding; we may assume that $\|\vartheta\|_{\overline{W}_0} < 1$. Using φ_0 from the construction of the exhaustion $X = \bigcup W_j$ in 63.3 ii), we obtain a commutative diagram

$$\begin{array}{ccc} W_1 & \xrightarrow{(\varphi_0,\, \vartheta)} & \mathbb{C}^m \times \mathbb{C}^n \\ \cup\!\!\!\!\cup & & \cup\!\!\!\!\cup \\ W_0 & \xrightarrow{\psi = (\varphi_0,\, \vartheta)|_{W_0}} & \boldsymbol{P}^m \times \boldsymbol{P}^n, \end{array}$$

in which ψ is an embedding: ψ is proper, since $\varphi_0 : W_0 \to \mathbb{C}^m$ is (see 63.3 i)), ψ is injective, and all homomorphisms ψ_z^0 are surjective, as ϑ has those properties. By 6.7, $(\mathbb{C}^m \times \mathbb{C}^n, \boldsymbol{P}^m \times \boldsymbol{P}^n)$ is a Runge pair; by 62.2, $\boldsymbol{P}^m \times \boldsymbol{P}^n$ is a B-space, and ψ^0 is an epimorphism by 52.5. Thus 63.4 ensures that (W_1, W_0) is a Runge pair. ■■

The idea of the proof of 63.2 has further applications:

E. 63 b. For $K \subset U \subset\subset X$, where X is a Stein space, U is $\mathcal{O}(X)$-convex, and K, compact, show that
i) there exists an analytic polyhedron $W = \varphi^{-1}(\boldsymbol{P}^m(1)) \cap U$ for an appropriate $\varphi \in \operatorname{Hol}(X, \mathbb{C}^m)$ such that $K \subset W \subset\subset U$;
ii) (X, W) is a Runge pair (hint: end of proof of 63.2 with X instead of W_1; for 63.5).

63.5 Proposition. *The following statements about an open subspace W of a Stein space X are equivalent:*
i) W is a Stein space and (X, W), a Runge pair.
ii) W is $\mathcal{O}(X)$-convex.

Proof. i) $\Rightarrow$ ii) If K is a compact subset of W, then so is $\hat{K}_{\mathcal{O}(W)}$; hence, it suffices to show that $\hat{K}_{\mathcal{O}(X)} \cap W \subset \hat{K}_{\mathcal{O}(W)}$. For $a \in W \setminus \hat{K}_{\mathcal{O}(W)}$, there exists an $f \in \mathcal{O}(W)$ such that $|f(a)| > \|f\|_K$; since $\|\cdot\|_{K \cup \{a\}}$ is a continuous seminorm on $\mathcal{O}(W)$ (see E. 55i), there exists a function g near f in the dense subset $\mathcal{O}(X)|_W$ of $\mathcal{O}(W)$ such that $|g(a)| > \|g\|_K$, i.e., $a \notin \hat{K}_{\mathcal{O}(X)}$.

ii) $\Rightarrow$ i) Obviously, W is a Stein space. We may assume that X is connected; then, by 51 A.3, X has a countable topology. From E.63b and 51 A.2iii), we derive the existence of an exhaustion $W = \bigcup W_j$ by analytic polyhedra W_j in X such that $W_j \subset\subset W_{j+1}$ and that (X, W_j) is a Runge pair for each j. It remains to show that the image of the restriction-homomorphism $\mathcal{O}(X) \to \mathcal{O}(W) \underset{\text{E.55f}}{=} \varprojlim \mathcal{O}(W_j)$ is dense. Assume the topology in $\mathcal{O}(W_j)$ to be given by a metric δ_j; then we may assume that $\delta_j \leqq \delta_{j+1}$ (see proof of 56.4). Thus each $f \in \mathcal{O}(W)$ has a fundamental system of neighborhoods of the form

$$U_j = \left\{ g \in \mathcal{O}(W);\ \delta_j(f|_{W_j}, g|_{W_j}) < \frac{1}{j} \right\}.$$

For each j, choose $h \in \mathcal{O}(X)$ such that $\delta_j(f|_{W_j}, h|_{W_j}) < 1/j$; then $h|_W$ belongs to U_j. ■

63.6 Corollary. *If W is an analytic polyhedron in a Stein space X, then (X, W) is a Runge pair.*

Proof. Apply 63.5 and E.51f.

E. 63c. Let X be a Stein space, $U \subset\subset X$, and $K = \hat{K}_{\mathcal{O}(X)} \cap U$ a compactum. Then, for every $f \in \mathcal{O}(U)$ and every $\varepsilon > 0$, there exists a $g \in \mathcal{O}(X)$ such that $\|f - g\|_K < \varepsilon$ (hint: 63.3 i), E.55i; for 63 A.2).

E. 63d. If (X_1, U_1) and (X_2, U_2) are Runge pairs of Stein spaces, then so is $(X_1 \times X_2, U_1 \times U_2)$; if, moreover, X_1, X_2 are open subspaces of a complex space X, then $(X_1 \cap X_2, U_1 \cap U_2)$ is a Runge pair.

E. 63e. Every analytic subset A of a connected Stein space X of dimension n has a fundamental system of open $\mathcal{O}(X)$-convex neighborhoods; in particular, A is a B-set in X (hint: there exists a proper holomorphic injection $f: X \to \mathbb{C}^m \times \mathbb{C}^n$ such that $A = f^{-1}(\mathbb{C}^m \times 0)$, see 57.4 and E.56e; apply E.11c and 63.5).

The problem of characterizing domains of holomorphy in $\mathbb{C}^n$ was the starting point for our investigation of Stein spaces. We finally come back to that initial problem and give a list of 16 equivalent characterizations, where we give prominence to the points 6)–8):

63.7 Theorem (Characterization of domains of holomorphy). *For a domain G in $\mathbb{C}^n$, the following statements are equivalent:*

1) G is a Stein space.

2) G is holomorphically convex.

3) G is weakly holomorphically convex.

4) G is a domain of holomorphy.

5) $H^1(G, \mathcal{I}) = 0$ for every coherent ideal $\mathcal{I}$ in ${}_G\mathcal{O}$ such that the zero set $N(\mathcal{I})$ is discrete.

6) G is a B-space.

7) G is pseudoconvex.
8) Every $a \in \partial G$ admits a neighborhood U in $\mathbb{C}^n$ such that $G \cap U$ is Stein.
9) There exists an exhaustion $G = \bigcup_{j=1}^{\infty} G_j$ by open subsets $G_j \subset\subset G_{j+1} \subset\subset G$ that are Stein.
10) Every hyperplane section H of G in $\mathbb{C}^n$ is Stein, and the induced restriction-homomorphism ${}_G\mathcal{O}(G) \to {}_H\mathcal{O}(H)$ is surjective.
11) Every hyperplane section H of G in $\mathbb{C}^n$ is Stein, and every additive Cousin problem on G has a solution.
12) For every complex line E in $\mathbb{C}^n$, the restriction-homomorphism ${}_G\mathcal{O}(G) \to$ $\to {}_{G\cap E}\mathcal{O}(G \cap E)$ is surjective.
13) The sequence

$$0 \to \mathcal{O}(G) \to \mathcal{E}^{0,0}(G) \xrightarrow{\bar\partial} \dots \xrightarrow{\bar\partial} \mathcal{E}^{0,n}(G) \to 0$$

is exact.
14) $H^q(G, \mathcal{O}) = 0$ for $q = 1, \dots, n-1$.
15) For $f_1, \dots, f_m \in \mathcal{O}(G)$ without common zeros, there exist $g_1, \dots, g_m \in \mathcal{O}(G)$ such that $1 = \sum_{j=1}^{m} f_j g_j$.
16) The mapping $e: G \to Sp(G)$, $z \mapsto \varepsilon_z$, is surjective.

We develop the proof, following this diagram:

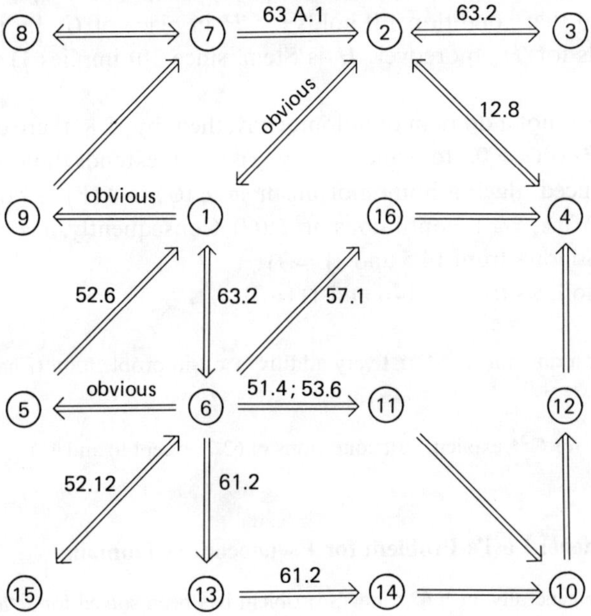

11) ⇒ 10) We have to show that ${}_G\mathcal{O}(G) \to {}_H\mathcal{O}(H)$ is surjective; we may assume that $H = N(G; z_1)$. Then every $g \in \mathcal{O}(H)$ can be lifted to the open neighborhood

$U := (\mathbb{C} \times H) \cap G$ of H in G. By assumption, the additive Cousin distribution $\{g/z_1 \in \mathcal{M}(U), 0 \in \mathcal{M}(G\backslash H)\}$ admits a solution $f \in \mathcal{M}(G)$, i.e., $f|_{G\backslash H} \in \mathcal{O}(G\backslash H)$ and $\varphi := f - g/z_1 \in \mathcal{O}(U)$. Hence, we obtain $z_1 f \in \mathcal{O}(G)$ and $z_1 f - g = z_1 \varphi$; thus $(z_1 f)|_H = g$.

10) $\Rightarrow$ 12) If H is a hyperplane section of G that includes $E \cap G$, then ${}_G\mathcal{O}(G) \to {}_H\mathcal{O}(H)$ is surjective by assumption and ${}_H\mathcal{O}(H) \to {}_{E\cap G}\mathcal{O}(E \cap G)$ is surjective by 52.5, as $E \cap G$ is a closed subspace of the Stein space H.

12) $\Rightarrow$ 4) If G is not a domain of holomorphy, then, by 12.8, there exists a polydisk $\boldsymbol{P} = \boldsymbol{P}(a; r)$ such that $\partial G \cap \boldsymbol{P} \neq \emptyset$, to which all $f \in \mathcal{O}(G)$ extend holomorphically (see 12.1). We may assume that $a = 0$ and that $b := (1, 0, \ldots, 0) \in \boldsymbol{P} \cap \partial G$ (if necessary for the latter, replace $\boldsymbol{P}$ in appropriate linear coordinates with a polydisk $\{|z_1| < 1 + \varepsilon, |z_j| < \varepsilon \text{ for } j \geqq 2\} \subset \boldsymbol{P}$). Then, for $E := \mathbb{C} \cdot b$, the assignment $z_1 \mapsto 1/(z_1 - 1)$ determines a holomorphic function on $G \cap E$ that does not extend to a function $\check{f} \in \mathcal{O}(G)$: otherwise, $\sum_{\nu \in \mathbb{N}^n} \frac{D^\nu \check{f}}{\nu!}(0) z^\nu$ would determine a function $g \in \mathcal{O}(\boldsymbol{P})$ such that $g_0 = \check{f}_0$ and thus $(g|_{E\cap \boldsymbol{P}})_0 = f_0 = 1/(z_1 - 1)_0$ for the germs at the point 0. Since $E \cap \boldsymbol{P}$ is connected, b would be a singular point for $\check{f}$! ↯

14) $\Rightarrow$ 10) We induct on n. The case $n = 1$ is obvious.

"$n \Rightarrow n + 1$" We may assume that $H = N(G; z_1)$. Then there is an exact sequence

$$\mathcal{O} \to {}_G\mathcal{O} \xrightarrow{\cdot z_1} {}_G\mathcal{O} \to {}_G\mathcal{O}/\mathcal{I} \to 0,$$

where $\mathcal{I} := {}_G\mathcal{O} \cdot z_1$. Thus $H^1(G, {}_G\mathcal{O}) = 0$ implies that ${}_G\mathcal{O}(G) \to ({}_G\mathcal{O}/\mathcal{I})(G) \cong {}_H\mathcal{O}(H)$ is surjective and that condition 14) holds for H in place of G. By induction hypothesis, 10) holds for H; moreover, H is Stein, since 10) implies 1). Thus 10) holds for G.

16) $\Rightarrow$ 4) If G is not a domain of holomorphy, then, by 12.8, there exists a polydisk $\boldsymbol{P}$ such that $\boldsymbol{P} \cap \partial G \neq \emptyset$, to which every $f \in \mathcal{O}(G)$ extends holomorphically. By E. 55g, the induced algebra-homomorphism $\varphi: \mathcal{O}(G) \to \mathcal{O}(\boldsymbol{P})$ is continuous; thus, for every $x \in \boldsymbol{P}\backslash G$, $\varepsilon_x \circ \varphi$ is continuous on $\mathcal{O}(G)$. Consequently, e is not surjective.

8) $\Leftrightarrow$ 7) This follows from 14.5 and 1) $\Leftrightarrow$ 7).

9) $\Rightarrow$ 7) This follows from E. 14a and 1) $\Rightarrow$ 7). ∎

E. 63 f. Prove for a domain $G \Subset \mathbb{C}^2$: Every additive Cousin problem on G has a solution iff G is Stein, iff $H^1(G, \mathcal{O}) = 0$.

E. 63 g. Disprove for $\mathbb{C}^{2*}$ explicitly all conditions of 63.7, except 9) and 13).

§ 63 A Supplement: Levi's Problem for Pseudoconvex Domains

In 62.8 and, more generally, in E.62g, Levi's problem has been solved for *strictly pseudoconvex domains* in complex spaces.

63 A.1 Theorem. *Every pseudoconvex domain in $\mathbb{C}^n$ is holomorphically convex.*

Proof. Fix a strictly pseudoconvex exhaustion $G = \bigcup G_j$ according to 14.7. Then $G_{j-1} = \{x \in G_j;\ \varphi_{j-1}(x) < 0\} \subset\subset G_j$ holds; hence, G_{j-1} is Stein, by 62.8, and satisfies the assumption of 63 A.2. Thus 56.2 implies the assertion. ■

63 A.2 Proposition. *If a strictly pseudoconvex region in a Stein space X is of the form $U = \{x \in X;\ \varphi(x) < 0\}$ for an appropriate $\varphi \in \mathscr{C}(X, \mathbb{R})$ that is strictly subharmonic on X, then (X, U) is a Runge pair of Stein spaces.*

There, "φ is *strictly subharmonic*" means that near each $a \in X$, φ is of the form $\varphi = \chi_a \circ f_a$, where f_a is a biholomorphic mapping of a neighborhood of a onto a closed subspace of a polydisk $\boldsymbol{P}^{n_a}$, and $\chi_a \in \mathscr{C}^3(\boldsymbol{P}^{n_a}, \mathbb{R})$, a strictly plurisubharmonic function.

Proof. By 63.5, we only have to show that U is $\mathcal{O}(X)$-convex. To that end, let K denote a compact subset of U; then $c := \max\{\varphi(z);\ z \in K\} < 0$, and we may assume that $K = \{\varphi \leqq c\}$. By assumption, $\hat{K} := \hat{K}_{\mathcal{O}(X)} \subset X$ is compact. It suffices to show that $K = \hat{K}$, or, equivalently, that $d := \max\{\varphi(z);\ z \in \hat{K}\} = c$. If there exists a $u \in \hat{K} \setminus K$, then we may even assume that $\varphi(u) = d$. Using a local model of X at the point u, we obtain in the notation of 62.3 iii) that $P_{\varphi,u}(u) = 0$ and $\operatorname{re} P_{\varphi,u}(z) < 0$ near u for $z \neq u$ and $\varphi(z) \leqq d$. Hence, $f := \exp P_{\varphi,u}$ and $\hat{K}$ satisfy the conditions of 63 A.3, and there exists a $g \in \mathcal{O}(\hat{K})$ such that $\|g\|_K < |g(u)|$. By E.63c, we may even assume that $g \in \mathcal{O}(X)$; consequently, u is not in $\hat{K}$. ↯ ■

E. 63 A a. The annulus $\boldsymbol{R} := \{z \in \mathbb{C};\ 1 < |z| < 2\}$ is a strictly pseudoconvex domain in $\mathbb{C}$, but $(\mathbb{C}, \boldsymbol{R})$ is not a Runge pair (hint: E.14b).

63 A.3 Lemma. *Let X denote a Stein space, $K = \hat{K}_{\mathcal{O}(X)}$, a compact subset, and V, an open neighborhood of a point $a \in K$. If $f \in \mathcal{O}(V)$ is such that*

$$|f|\,|_{V \cap K^*} < |f(a)|$$

for $K^* := K \setminus \{a\}$, then there exists a $g \in \mathcal{O}(K)$ such that

$$|g|\,|_{K^*} < |g(a)|\,.$$

Proof. We may assume that $f(a) = 1$; hence, after shrinking V if necessary, 54.3 implies that there exists an $F \in \mathcal{O}(V)$ such that

$$F(a) = 0,\ e^F = f, \quad \text{and} \quad V(V, F) = V(V; f-1) =: A\,.$$

Thus $A \cap K = \{a\}$; by 63.3, there exists an analytic polyhedron W in X such that $K \subset W \subset\subset X \setminus (A \cap V)$. Hence, $A \cap W$ is a closed subspace of W defined by the ideal sheaf

$$\mathscr{I} = \begin{cases} \mathcal{O} \cdot F & \text{on } V \cap W \\ \mathcal{O} & \text{on } W \setminus A\,. \end{cases}$$

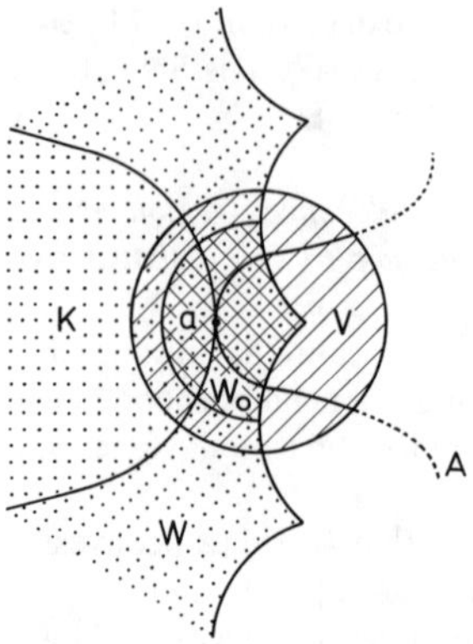

By E. 51f, W is Stein. Since, by Theorem A, the principal ideal $\mathscr{I}_a$ is generated by a function $h \in \mathscr{I}(W)$, there exists an $\eta \in \mathcal{O}^*(W_0)$ for an appropriate neighborhood $W_0 \subset W \cap V$ of a such that $h = \eta F$. If W has been chosen sufficiently small, then there exists an open neighborhood W_1 of $K \setminus W_0$ in $W \setminus A$ such that

$$W = W_0 \cup W_1 \quad \text{and} \quad \operatorname{re} F|_{W_{01}} = \log |f||_{W_{01}} < 0\,.$$

Thus $\mathfrak{U} := (W_0, W_1)$ is a cover of W, and we can define log F and

$$\frac{\log F}{h} \in \mathcal{O}(W_{01}) = C_a^1(\mathfrak{U}, \mathcal{O}) = Z^1(\mathfrak{U}, \mathcal{O})\,.$$

Then $H^1(\mathfrak{U}, \mathcal{O}) \underset{50.10}{\subset} H^1(W, \mathcal{O}) = 0$ implies that $\dfrac{\log F}{h} = g_1 - g_0$ for appropriate $g_j \in \mathcal{O}(W_j)$. Thus the assignment

$$H|_{W_0} := F \cdot \exp(g_0 h),\; H|_{W_1} := \exp(g_1 h)$$

determines a function $H \in \mathcal{O}(W)$, which satisfies the equality

$$F = H \cdot \exp(-g_0 \eta F) = H \cdot \exp(-g_0 \eta H \cdot \exp(-g_0 \eta F)) =: H \cdot \exp(H \cdot \psi)$$

on W_0, for an appropriate $\psi \in \mathcal{O}(W_0)$. On a sufficiently small neighborhood $U \subset\subset W_0$ of a, there exists an expansion

$$F = H(1 + H\psi + \ldots) =: H + H^2 \varphi, \quad \text{where} \quad \varphi \in \mathcal{O}(U)\,.$$

Set $c := \|\varphi\|_U$; then the fact that $\operatorname{re} F|_{W_0 \cap K^*} < 0$ implies that

$$\operatorname{re} H - c|H|^2 \leqq \operatorname{re}(H + H^2 \varphi) = \operatorname{re} F < 0 \quad \text{on} \quad U \cap K^*.$$

For every $\lambda \in \mathbb{C}$, these equivalences are easy to verify:

$$\operatorname{re}\lambda - c|\lambda|^2 \underset{>}{\overset{\leq}{=}} 0 \Leftrightarrow |\lambda - \frac{1}{2c}| \underset{<}{\overset{\geq}{=}} \frac{1}{2c}\,.$$

They imply that $H(U \cap K) \cap \overline{\boldsymbol{P}^1(1/2c;\, 1/2c)} = \{0\}$, and $N(K^*; H) = \emptyset$ ensures that $\delta := \inf\{|H(z)|;\, z \in K \setminus U\} > 0$. Thus (see the figure) we obtain that $H(K^*) \subset \{\lambda \in \mathbb{C};\, |\lambda - \varepsilon| > \varepsilon\}$ for an appropriate $\varepsilon > 0$, and $g := \dfrac{\varepsilon}{\varepsilon - H}$ satisfies $g \in \mathcal{O}(K), g(a) = 1$ and $|g\|_{K^*} < 1$. ■

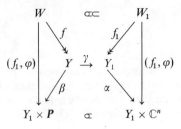

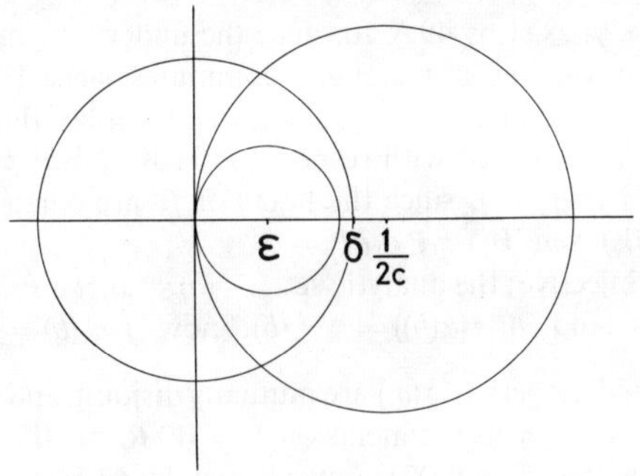

§ 63 B Supplement: Weakly Holomorphically Convex Spaces are Holomorphically Convex

We have seen in 63.2 that a *holomorphically spreadable* complex space is holomorphically convex iff it is weakly holomorphically convex. Using Grauert's Coherence Theorem, we essentially can drop that restriction:

63 B. 1 Theorem. *A complex space is holomorphically convex iff it is weakly holomorphically convex and each connected component has a countable topology.*

Proof. The implication "*only if*" follows from E. 51 Ad. *If.* We may assume that the weakly holomorphically convex space X has a countable topology. Let us fix an exhaustion $X = \bigcup_{j=0}^{\infty} W_j$ according to 63.3, and mappings $\varphi_j \in \mathrm{Hol}(X, \mathbb{C}^{n_j})$ such that $\varphi_j : W_j \to \boldsymbol{P}_j$ is proper. We show the following for the holomorphic hulls $f_j : W_j \to Y_j$ of W_j (see 57.11 and E. 51 f):

(63 B. 1.1) There exist natural commutative diagrams with Runge pairs (Y_{j+1}, Y_j) of Stein spaces

$$\begin{array}{ccc} W_j & \Subset & W_{j+1} \\ \downarrow f_j & & \downarrow f_{j+1} \\ Y_j & \Subset & Y_{j+1} \end{array} .$$

Then the projective limit Y_0 is a Stein space by 56.2, and $f := \varprojlim f_j : X \to \varprojlim Y_j$ is obviously a proper holomorphic mapping; hence, X is holomorphically convex (see E. 51 e ii)).

Proof of (63 B. 1.1). By construction, each Y_j is a Stein space such that $\mathcal{O}(Y_j) \cong \mathcal{O}(W_j)$. We may assume that $j = 0$; put $W := W_0$, $f := f_0$, etc. Thus there exists a commutative diagram of holomorphic mappings

$$\begin{array}{ccccc} W & & \Subset & & W_1 \\ & \searrow f & & f_1 \swarrow & \\ \downarrow (f_1, \varphi) & & Y \xrightarrow{\gamma} Y_1 & & \downarrow (f_1, \varphi) \\ & \swarrow \beta & & \alpha \searrow & \\ Y_1 \times \boldsymbol{P} & & \subset & & Y_1 \times \mathbb{C}^n \end{array}$$

in which β (and similarly α, γ) exist by 49 A. 10, since the underlying mapping of sets $|(f_1, \varphi)|$ is f-invariant, by construction; the lower "rectangle" commutes, since Y is a quotient space of W, and thus the equality $\beta \circ f = (f_1, \varphi)|_W = \alpha \circ f_1|_W = \alpha \circ \gamma \circ f$ implies that $\beta = \alpha \circ \gamma$.

The subspace W of W_1 is saturated with respect to f_1: If $f_1^{-1}(b) \not\subset W$ for some $b \in \gamma(Y)$, we may assume that $b \in f_1(\partial W) \cap f_1(W)$, since the fibers of f_1 are connected (see 57.11). Thus we obtain that $\mathrm{pr}_2 \circ \alpha(b) \in \varphi(W) \cap \varphi(\partial W) \subset \boldsymbol{P} \cap \partial \boldsymbol{P} = \emptyset$. ↯

Next we show that γ is injective: the analytic set $\gamma^{-1}(b) =: \{a_i; i \in I\}$ is finite, since (f_1, φ) and thus β are proper maps, and $\beta^{-1}(\alpha(b)) = \gamma^{-1}(b)$; now $f_1^{-1}(b) = f^{-1}(\gamma^{-1}(b)) = \bigcup_I f^{-1}(a_i)$ implies $|I| \leqq 1$, since the analytic sets $f^{-1}(a_i)$ are mutually disjoint, and since $f_1^{-1}(b)$ is connected. Consequently, we may interprete γ as the inclusion $Y = W/R_f \subset\subset W_1/R_{f_1} = Y_1$.

Finally, we have to show that Y is $\mathcal{O}(Y_1)$-convex; then, by 63.5, (Y_1, Y) is a Runge pair. For a compact set $K \subset Y$, it is easy to verify that

$$\hat{K}_{\mathcal{O}(Y_1)} \cap Y \subset f_1(\widehat{(f_1^{-1}(K)})_{\mathcal{O}(W_1)} \cap W);$$

now f_1 is proper and W is $\mathcal{O}(W_1)$-convex by E. 51f; hence, $\hat{K}_{\mathcal{O}(Y_1)} \cap Y$ is compact. ■

Supplement Chapter 7: Normal Complex Spaces

On manifolds, the Riemann Removable Singularity Theorems ensure that weakly holomorphic functions are holomorphic. In this supplement, we investigate a "strong" and a "weak" Riemann Removable Singularity Property for reduced complex spaces X, and consider the question, how X should be modified to yield the corresponding continuation theorem. Spaces with the strong property, called "normal complex spaces", behave, in various important aspects, in a manner similar to that of manifolds; the "normalization" (§ 71) is therefore an important step toward the "regularization" of a complex space into a manifold. (For *curves*, that step completes the regularization (74.4), whereas *surfaces* require a sequence of quadratic transformations and normalizations; in higher dimensions, it is necessary to "blow up" closed subspaces [Hr].)

The weak Riemann Removable Singularity Property provides a characterization of reduced spaces with a *maximal complex structure* (§ 72); the *maximalization* $\tilde{X} \to X$ enriches the function theory on X as much as is possible without changing the underlying topological structure.

A reduced space is a Stein space iff its normalization is Stein; we use that fact in § 73 to show that finite images of Stein spaces are Stein.

In § 74, we exhibit a criterion for normality that is used to obtain the most familiar examples of normal spaces – hypersurfaces in manifolds whose singular points form a subset of codimension at least two. For the proof, we introduce local cohomology theory, since that technique plays an important role in the solution of other problems in complex analysis as well (e. g., the extension of coherent analytic sheaves [SiTr], and Serre duality [RaRu]).

§ 71 Normalization

In this section, $(X, \mathcal{O})$ always denotes a *reduced* complex space.

In view of the importance of the Riemann Removable Singularity Theorems for the construction of holomorphic functions with specified properties, we now investigate extension theorems for holomorphic functions on complex spaces, using the sheaf $\tilde{\mathcal{O}}$ of weakly holomorphic functions on X as defined in 46 B.3. Let us call a weakly holomorphic function f on U *continuous* (resp., *holomorphic*) if it admits a continuous (resp., holomorphic) extension to U. Since such an extension is unique, we denote it with f as well.

We intend to characterize the complex spaces X for which one of the following conditions holds:

α) $\tilde{\mathcal{O}} \subset {}_X\mathcal{C}$ *(weakly holomorphic functions are continuous; see 71.9).*

β) $\tilde{\mathcal{O}} \cap {}_X\mathcal{C} = \mathcal{O}$ *(continuous weakly holomorphic functions are holomorphic; see § 72).*

Those properties are independent of each other (see E. 71a). We are particularly interested in the case in which both properties hold:

γ) $\tilde{\mathcal{O}} = \mathcal{O}$ *(weakly holomorphic functions are holomorphic).*

We say that the *weak* (resp., *strong*) *Riemann Removable Singularities Theorem* holds, if β) (resp., γ)) is satisfied.

E. 71a. Verify the following table:

	α)	β)	γ)
$\mathbb{C}^n$	+	+	+
$V(\mathbb{C}^2; z_1 z_2)$	−	+	−
$V(\mathbb{C}^2; z_1^2 - z_2^3)$	+	−	−
$V(\mathbb{C}^2; z_1 z_2 (z_1 - z_2))$	−	−	−

(hint: E.53 Ab, E.321, E.44i).

E. 71b. Test $V(\mathbb{C}^2; z_1(z_1 - z_2^3))$ and $\mathbb{C}^2/(z \sim -z)$ for properties α), β), and γ).

71.1 Definition. *The space X is called normal at the point $a \in X$ (and the germ X_a is called a normal germ) if $\tilde{\mathcal{O}}_a = \mathcal{O}_a$, that is, if every weakly holomorphic function near a is actually holomorphic. The space X is called* normal *if it is normal at every point.*

Thus, the complex structure of a normal space X is determined completely by the underlying topological structure $|X|$ and the complex structure on the regular part $X \setminus S(X)$.

If X_a is a prime germ, then $\tilde{\mathcal{O}}_a$ is included in the integral closure $\hat{\mathcal{O}}_a$ of $\mathcal{O}_a$ in $\mathcal{M}_a$ (see 46 B.4 and 46 A.3). We shall derive the following fact from 71.6 (and thus motivate the name "normal" space):

71.2 Proposition. *A germ X_a is normal iff $\mathcal{O}_a$ is a normal ring (that is, iff $\mathcal{O}_a = \hat{\mathcal{O}}_a$).*

The integral closure $\hat{R}$ of the reduced analytic algebra $R = {}_X\mathcal{O}_a$ is a finite ring extension of R (see the proof of 71.4), and hence a finite ring-direct product of normal analytic algebras: $\hat{R} = R_1 \oplus \ldots \oplus R_m$, where the R_j are the normalizations of the residue class rings $R/\mathfrak{p}_j$ that correspond to the prime components of X_a (see E.71h). Geometrically, there corresponds to the inclusion $R \hookrightarrow \hat{R}$ a finite holomorphic surjection $\dot{\bigcup}_{j=1}^{m} \hat{X}_j \twoheadrightarrow X_a$, with normal germs $\hat{X}_j$. Since X is normal outside the thin analytic subset $S(X)$ by 23.8, that motivates the following notion:

71.3 Definition. *A finite holomorphic mapping $\pi: \hat{X} \to X$ is called a* normalization *of X, if it satisfies the following conditions:*

i) $\hat{X}$ is normal.

ii) There exists a thin analytic subset A of X such that $\pi^{-1}(A)$ is thin in $\hat{X}$ and $\pi: \hat{X} \setminus \pi^{-1}(A) \to X \setminus A$ is biholomorphic.

In particular, the normalization-mapping π is closed and thus surjective.

E. 71c. Determine normalizations for the examples of E.71a and E.71b.

For a normal space Y and a holomorphic mapping $\pi: Y \to X$ such that $\pi^{-1}(S(X))$ is a thin subset of Y, the homomorphism ${}_X\mathcal{O} \to \pi({}_Y\mathcal{O})$ of 45 A.2 induces a homomorphism of ${}_X\mathcal{O}$-modules

$$(71.4.1) \quad \vartheta: {}_X\widetilde{\mathcal{O}} \to \pi({}_Y\mathcal{O}),$$

since Y is normal; it follows that $\pi^{-1}({}_X\widetilde{\mathcal{O}}) \subset {}_Y\widetilde{\mathcal{O}} = {}_Y\mathcal{O}$. Obviously, ϑ is injective if π is surjective. We intend to show that, for a normalization $\pi: \hat{X} \to X$, the induced homomorphism ϑ is an isomorphism; then we shall deduce the following result:

71.4 Normalization Theorem. *Every reduced complex space X admits (up to isomorphism) exactly one normalization: the $\mathcal{O}$-algebra $\widetilde{\mathcal{O}}$ is a coherent $\mathcal{O}$-module, and* Specan $\widetilde{\mathcal{O}} \to X$ *is the normalization.*

For the proof of 71.4, it is again advisable to replace the geometric version with an algebraic formulation. Hence, we replace the $\mathcal{O}$-algebra $\widetilde{\mathcal{O}}$, which was constructed in a geometric-analytic manner, with the $\mathcal{O}$-algebra $\hat{\mathcal{O}}$, which we construct in the following algebraic way: for a Stein open subset $U \Subset X$, let $\widehat{\mathcal{O}(U)}$ denote the integral closure of $\mathcal{O}(U)$ in $Q(\mathcal{O}(U)) \underset{53.1}{=} \mathcal{M}(U)$. By 51.5 and § 30, that determines a subsheaf $\hat{\mathcal{O}}$ of $\mathcal{M}$. We have that

$$\hat{\mathcal{O}}_a = \lim_{U \to a} \widehat{\mathcal{O}(U)} \overset{!}{=} \hat{\mathcal{O}}_a \subset \mathcal{M}_a,$$

since, in a suitable neighborhood W of a, each $\varphi \in \mathcal{M}_a$ admits a representation $\varphi = f/g$, where f, g lie in $\mathcal{O}(W)$, and g induces a non-zero-divisor in each stalk $\mathcal{O}_w$. If φ satisfies an equation of integral dependence at the point a, then it does so near a.

In oder to show that $\hat{\mathcal{O}} \cong \widetilde{\mathcal{O}}$, we prove the following result:

71.5 Proposition (Existence of universal denominators). *For each $a \in X$, there exist an open neighborhood W of a and a $u \in \mathcal{O}(W)$ that induces in each stalk $\mathcal{O}_w$ a non-zero-divisor such that*

$$u \cdot \widetilde{\mathcal{O}}|_W \subset \mathcal{O}|_W.$$

Proof. For a fixed a, we may assume that $X \hookrightarrow \boldsymbol{P} \Subset \mathbb{C}^n$ has a decomposition $X = \bigcup_{j=1}^{m} X_j$ into irreducible components such that each germ X_{ja} is prime. Then, by 52.4, every holomorphic function on X (resp., X_j) is the restriction of a holomorphic function on $\boldsymbol{P}$. By 46 B.4, we may assume that there exist functions $u_j \in \mathcal{O}(\boldsymbol{P})$ such that each $u_j|_{X_j}$ is a universal denominator for X_j. According to 52.5, there are functions $f_j \in \mathcal{O}(\boldsymbol{P})$ such that $f_j|_{X_j} \neq 0 \in \mathcal{O}(X_j)$ and $f_j|_{\bigcup_{k \neq j} X_k} = 0$. Then $u := \sum_{j=1}^{m} f_j u_j$ is a universal denominator for X: at no point does it induce a zero-divisor, as no X_j is contained in $N(X; u)$ (see E.53a). For each $h \in \widetilde{\mathcal{O}}(X)$, we have that $uh \in \mathcal{O}(X)$: obviously, it suffices to show that $f_j u_j h \in \mathcal{O}(X)$; for an extension H of $u_j h \in \mathcal{O}(X_j)$ to $\boldsymbol{P}$, we see that $f_j H \in \mathcal{O}(\boldsymbol{P})$, and therefore $f_j u_j h = f_j H|_X$. ∎

71.6 Corollary. *There exists exactly one isomorphism $\hat{\mathcal{O}} \cong \widetilde{\mathcal{O}}$ of $\mathcal{O}$-algebras.*

Proof. By 71.5, we may assume that u is a universal denominator for X. Then 46 A.3 ii) implies that $\widetilde{\mathcal{O}}$ is uniquely an $\mathcal{O}$-subalgebra of $\hat{\mathcal{O}}$. We still have to show that a meromorphic function $f/g \in \mathcal{O}(U)$ is weakly holomorphic if it satisfies an equation of integral dependence

$$(f/g)^b = \sum_{j=0}^{b-1} a_j \cdot (f/g)^j$$

(where, without loss of generality, every $a_j \in \mathcal{O}(U)$ is bounded). By E.53a, $N(U;g)$ is a thin subset of U, and it follows from E.33a that

$$\|f/g\|_{U \setminus N(g)} \leq \max\left(1, \sum_{j=0}^{b-1} \|a_j\|_U\right);$$

hence, f/g is bounded at $N(g)$. ■

E. 71 d. Show that $z_1 + z_2$ induces a universal denominator for $V(\mathbb{C}^2; z_1 z_2)$.

Proof of 71.2. We know that X is normal iff $\mathcal{O} = \tilde{\mathcal{O}}$; by 71.6, that happens iff $\mathcal{O} = \hat{\mathcal{O}}$. ■■

Thus we can use the homomorphism (71.4.1)

$$\hat{\mathcal{O}} = \tilde{\mathcal{O}} \to \pi({}_{\hat{X}}\mathcal{O})$$

to give an algebraic characterization of a normalization:

71.7 Definition. *A finite holomorphic mapping $\pi: \hat{X} \to X$ is called an* algebraic normalization *if the following conditions hold:*

i) $\hat{X}$ is a normal space.

ii) $\hat{\mathcal{O}} \cong \pi({}_{\hat{X}}\mathcal{O})$.

By 45.16, there exists at most one algebraic normalization π of X. By ii), $\mathcal{O}_b = \hat{\mathcal{O}}_b = \pi({}_{\hat{X}}\mathcal{O})_b$ for every $b \in X \setminus S(X)$, so 45 A.1 implies that $\pi^{-1}(b)$ contains only one point; π is even biholomorphic near b, since $\pi^0: {}_X\mathcal{O}_b \to \pi({}_{\hat{X}}\mathcal{O})_b$ is an isomorphism (see 44.4). Thus $\pi: \hat{X} \setminus \pi^{-1}(S(X)) \to X \setminus S(X)$ is biholomorphic, and the closed mapping π is surjective.

71.8 Proposition. *Normalization and algebraic normalization of X coincide.*

Proof. In order to show that an algebraic normalization $\pi: \hat{X} \to X$ is a normalization, we still have to show that $\pi^{-1}(S(X))$ is a thin subset of $\hat{X}$. We may assume that X is Stein, and that u is a universal denominator for X such that $u|_{S(X)} = 0$ (if need be, after multiplication with an f according to E.53h). Then, for every $a \in X$, u_a is a non-zero-divisor in $\tilde{\mathcal{O}}_a \cong (\pi({}_{\hat{X}}\mathcal{O}))_a = \bigoplus_{b \in \pi^{-1}(a)} {}_{\hat{X}}\mathcal{O}_b$ and $u \circ \pi$ is a non-zero-divisor in ${}_{\hat{X}}\mathcal{O}_b$; hence, $\pi^{-1}S(X) \subset N(u \circ \pi)$ is a thin subset of $\hat{X}$, by E.53a.

If $\pi: \hat{X} \to X$ is a normalization, then the induced homomorphism (71.4.1) $\vartheta: {}_X\hat{\mathcal{O}} \to \pi({}_{\hat{X}}\mathcal{O})$ is injective. For $U \Subset X$, $V := \pi^{-1}(U)$, and $g \in \pi({}_{\hat{X}}\mathcal{O})(U) = {}_{\hat{X}}\mathcal{O}(V)$, we have that

$$g \circ (\pi|_{V \setminus \pi^{-1}S(X)})^{-1} \in {}_X\tilde{\mathcal{O}}(U) = {}_X\hat{\mathcal{O}}(U);$$

thus ϑ is an isomorphism, and π must be the algebraic normalization. ■

Proof of 71.4. Since the algebraic normalization is unique, it suffices to show its existence locally. Then it coincides automatically with $(\hat{X}, \pi) = \operatorname{Specan} \tilde{\mathcal{O}}$, as $\tilde{\mathcal{O}} = \hat{\mathcal{O}} = \pi({}_{\hat{X}}\mathcal{O})$ is coherent, by 45.1.

For each point $a \in X$, the integral closure $\hat{R}$ of $R = {}_X\mathcal{O}_a$ is a finite R-module (by 71.5 and 46 A.3 i)), and hence a ring-direct product $\hat{R} = R_1 \oplus \ldots \oplus R_m$ of analytic algebras (see 24.4). The rings R_j are integrally closed, by E.71f, and thus normal, by 53 A.6; that is, the corresponding germs $\hat{X}_{a_j}$ are normal. Corresponding to the finite homomorphisms of rings $R \hookrightarrow \hat{R} \twoheadrightarrow R_j$, there exist finite holomorphic mappings

$$\hat{X}_{\hat{a}_j} \hookrightarrow \dot{\bigcup_{j=1,\ldots,m}} \hat{X}_{\hat{a}_j} \xrightarrow{\pi} X_a.$$

After choosing a sufficiently small representative $\pi: \hat{X} := \cup \hat{X}_j \to X$, we may assume that $\hat{X}$ is normal, since $\pi^{-1}(a)$ contains only normal points of $\hat{X}$ (see 71.10). In order to show that

$\hat{\mathcal{O}} = \pi({}_{\hat{X}}\mathcal{O})$ near a, we verify the hypotheses of 46 A.3. After shrinking X and $\hat{X}$, we may assume that

i) π is finite (by 45.4), and the inclusion

$$ {}_X\mathcal{O}_a = R \hookrightarrow \hat{R} = R_1 \oplus \ldots \oplus R_m = \bigoplus_{j=1}^{m} {}_{\hat{X}}\mathcal{O}_{\hat{a}_j} $$

is induced by π^0 (see 44.3);

ii) The canonical homomorphism $\mathcal{O} \to \pi({}_{\hat{X}}\mathcal{O})$ is injective (by i) and 41.17);

iii) There exists a universal denominator for X (by 71.5);

iv) Every u_x induces a non-zero-divisor in $\pi({}_{\hat{X}}\mathcal{O})_x$ (that holds at a, and hence near a; see the proof of 46.1 iii));

v) $u \cdot \pi({}_{\hat{X}}\mathcal{O}) \subset \mathcal{O} \subset \pi({}_{\hat{X}}\mathcal{O})$ (by 41.17, since it holds for $x = a$).

Now 46 A.3 tells us that, for each $x \in X$, the inclusion $\mathcal{O}_x \hookrightarrow \pi({}_{\hat{X}}\mathcal{O})_x$ "is" the integral closure of $\mathcal{O}_x$; consequently, $\hat{\mathcal{O}} = \pi({}_{\hat{X}}\mathcal{O})$. ∎

With similar methods, we can investigate property α):

71.9 Proposition. *For each point a in X, the following statements are equivalent:*

i) Every weakly holomorphic function $f \in \tilde{\mathcal{O}}_a$ *is continuous at a.*[1]

ii) The germ X_a *is irreducible.*

iii) $\mathcal{O}_a$ *is an integral domain.*

Proof. The implication "i) ⇒ ii)" follows from E.53 Ab, and "ii) ⇒ iii)" is a consequence of 44.14. iii) ⇒ i) Let $\pi : \hat{X} \to X$ be the normalization. Then $\hat{\mathcal{O}}_a = \pi({}_{\hat{X}}\mathcal{O})_a$ is an integral domain as well (by E.71f ii)); consequently, its ring-direct decomposition has only one factor, and the fiber $\pi^{-1}(a)$ therefore contains just one point, say b. That determines an isomorphism

$$ \tilde{\mathcal{O}}_a = \pi({}_{\hat{X}}\mathcal{O})_a \cong {}_{\hat{X}}\mathcal{O}_b, f \mapsto \hat{f} := f \circ \pi . $$

Since π is finite and $\hat{f}$ is continuous at b, we obtain a continuous extension of f at the point a by setting $f(a) := \hat{f}(b)$. ∎

E. 71 e. A germ X_a has a representative X without reducible points iff $\tilde{\mathcal{O}}_a \subset \mathscr{C}_a$. (Hint: if $\hat{X}$ is Stein, then, for each $x \in X$, there exists an $f \in \mathcal{O}(\hat{X})$ that is not constant along $\pi^{-1}(x)$ (see 52.5). Construct such an f explicitly for $X = \mathbb{C}^2/((w,0) \sim (0,w))$ and $a = \dot{0}$. In E. 72e, an embedding $X \hookrightarrow \mathbb{C}^3$ is given).

E. 71 f. Prove the following statements for a ring-direct product $R = R_1 \oplus \ldots \oplus R_m$:

i) $r = r_1 + \ldots + r_m$ is a zero-divisor in R iff at least one r_j is a zero-divisor in R_j.

ii) The total quotient rings satisfy the equality $Q(R) = Q(R_1) \oplus \ldots \oplus Q(R_m)$. In particular, $m = 1$ if R is an integral domain.

iii) R is integrally closed iff each R_j is integrally closed (for 71.2).

It remains to prove the following:

71.10 Proposition. *The set of nonnormal points of a (not necessarily reduced) complex space is analytic.*

Proof. By E.47c, the set of nonreduced points is analytic; hence, by E.43c, it suffices to find for each reduced space X a coherent $\mathcal{O}$-module whose support is the set of nonnormal points of X.

---

[1] i.e., $\lim_{U \setminus S(U) \ni x \to a} f(x)$ exists.

We may assume that there exists a universal denominator u for X; by 47.1 and 47.14, then, the $\mathcal{O}$-modules $\mathcal{I} := \sqrt{\mathcal{O}\cdot u}$ and $\mathcal{S} := \mathcal{H}om_{\mathcal{O}}(\mathcal{I}, \mathcal{I})$ are coherent. Finally, 71.11 provides an inclusion of $\mathcal{O}$-modules $\mathcal{O} \subset \mathcal{S} \subset \hat{\mathcal{O}}$ such that

$$X_x \text{ is normal} \Leftrightarrow \mathcal{O}_x = \hat{\mathcal{O}}_x \underset{71.11\text{ iii)}}{\Leftrightarrow} \mathcal{O}_x = \mathcal{S}_x \Leftrightarrow (\mathcal{S}/\mathcal{O})_x = 0. \quad \blacksquare$$

71.11 Lemma. *Let R be a noetherian ring, and consider an ideal $\mathfrak{a}$ in R that contains a non-zero-divisor u of R. Then the following statements hold for $S := \mathrm{Hom}_R(\mathfrak{a}, \mathfrak{a})$:*

i) For every $\varphi \in S$, there exists a unique $q \in Q(R)$ such that $\varphi = \mu_q : a \mapsto qa$; moreover, $q = \varphi(u)/u$. That determines a canonical identification of S with the subring $\{q \in Q(R);\, q\mathfrak{a} \subset \mathfrak{a}\}$ of $Q(R)$, so that $R \subset S \subset \hat{R}$.

ii) If $\mathfrak{a} = \sqrt{\mathfrak{a}}$, then $S = \{q \in \hat{R};\, q\mathfrak{a} \subset R\}$.

iii) If u is a universal denominator for R and $\mathfrak{a} = \sqrt{R\cdot u}$, then $R = S$ iff $R = \hat{R}$.

Proof. i) For every $\varphi \in S$ and $a \in \mathfrak{a}$, we have that $u\varphi(a) = \varphi(ua) = a\varphi(u)$; hence, $\varphi = \mu_q$ for $q := \varphi(u)/u \in Q(R)$. On the other hand, if $\varphi = \mu_p$ for some $p \in Q(R)$, then $pu = \mu_p(u) = \mu_q(u) = qu$, and $p = q$, as u is not a zero-divisor. Finally, S is finitely-generated over the noetherian ring R; hence, by 23 A.6, S is integral over R.

ii) As we know already that $S = \{q \in \hat{R};\, q\mathfrak{a} \subset \mathfrak{a}\}$, it remains to prove this implication:

$$\text{if } q \in \hat{R} \text{ and } q\mathfrak{a} \subset R, \text{ then } q\mathfrak{a} \subset \mathfrak{a}.$$

Now q satisfies an equation of integral dependence over R,

$$q^b = \sum_{j=0}^{b-1} r_j q^j;$$

so the fact that $a \in \mathfrak{a}$, $qa \in R$, and $r_j a^{b-j} \in \mathfrak{a}$ implies that

$$(qa)^b = \sum_{j=0}^{b-1} (r_j a^{b-j})\,(qa)^j \in \mathfrak{a};$$

hence, $qa \in \sqrt{\mathfrak{a}} = \mathfrak{a}$.

iii) "$\Rightarrow$" As R is a noetherian ring, we have that $\mathfrak{a}^m = (\sqrt{R\cdot u})^m \subset R\cdot u$ for every sufficiently large m. Since the fact that $h \in \hat{R}$ implies that $uh \in R$, there exists a minimal number $d \in \mathbb{N}$ such that $h\mathfrak{a}^d \subset R$. We have to show that $d = 1$. If $d \geqq 2$, fix an $a \in \mathfrak{a}^{d-1}$ such that $ha \notin R$; then, since $ha \in \hat{R}$ and $(ha)\mathfrak{a} \subset R$, we find that $ha \in S = R$ (see ii)). ↯ ■■

E. 71 g. Why is it not possible to show the coherence of $\tilde{\mathcal{O}}$ directly, following the proof of 46 A.3 i)?

We have called a reduced space X *normal*, if its open subsets satisfy the *first* (strong) *Riemann Removable Singularities Theorem.* For such a space, we can generalize 7.7:

71.12 Second Riemann Removable Singularities Theorem. *For a normal space X with an analytic subset A of codimension at least 2, every holomorphic function on $X \setminus A$ has a unique holomorphic extension to X. Thus the inclusion $X \setminus A \hookrightarrow X$ induces an isomorphism of topological algebras $\mathcal{O}(X) \cong \mathcal{O}(X \setminus A)$.*

Proof. We have to show that every $\varphi \in \mathcal{O}(X \setminus A)$ is locally bounded at A. By 46.1, we may assume that there exists a commutative diagram

$$\begin{array}{ccccc} X' & \subset\subset & X\setminus A & \subset\subset & X \\ \downarrow f & & \downarrow f & & \downarrow f \\ G' & \subset\subset & G\setminus f(A) & \subset\subset & G \subset\subset \mathbb{C}^n, \end{array}$$

where $f: X \to G$ is a finite surjective holomorphic mapping, $f: X' \to G'$ is a covering of degree b, and $A \underset{7.7}{\subset} S(X) \subset X \setminus X'$. Then the elementary symmetric functions

$$\sigma_1(z) = \sum \varphi(x_j),\ \sigma_2(z) = \sum_{i \neq j} \varphi(x_i)\varphi(x_j), \ldots, \sigma_b(z) = \prod \varphi(x_j)$$

for $\{x_1, \ldots, x_b\} = f^{-1}(z)$ are holomorphic on G' and locally bounded on $G\setminus f(A)$, as f is a finite mapping and φ is a continuous function on $X\setminus A$. Hence, by 7.3, every σ_j admits a holomorphic extension to $G\setminus f(A)$; by 7.7, it even has an extension to G, since $\operatorname{codim}_G f(A) \geqq 2$, according to E.49p. In $\mathcal{O}(G')[T]$, we have that

$$T^b - \sigma_1(z)T^{b-1} + \ldots \pm \sigma_b(z) = \prod_{j=1}^{b} (T - \varphi(x_j))\,;$$

hence, it follows that, in $\mathcal{O}(G')$, and hence also in $\mathcal{O}(X\setminus A)$, that

$$\varphi^b - (\sigma_1 \circ f)\varphi^{b-1} + \ldots \pm \sigma_b \circ f = 0\,.$$

Every $\sigma_j \circ f$ is locally bounded at A; by E.33a, the same holds for φ. If the topology of X is countable, then $\mathcal{O}(X) \to \mathcal{O}(X\setminus A)$ is a homeomorphism, by 55.6 and 55.8 ii). For general X, one obtains that result by using a projective limit. ∎

71.13 Corollary. *Let A be a thin analytic subset of a normal space X, and suppose that $f: X\setminus A \to Y$ is a holomorphic mapping. If f admits a continuous extension $\bar{f}: X \to Y$, then $\bar{f}$ is holomorphic.*

Proof. We may assume that $Y \hookrightarrow G \subset\subset \mathbb{C}^n$. Then f is determined by holomorphic functions on $X\setminus A$ that are locally bounded at A. As X is normal, they admit a unique holomorphic extension to X. ∎

71.14 Corollary. *For the normalization $\pi: \hat{X} \to X$, the following statements hold:*

i) If $\hat{X} = \bigcup_{j\in J} \hat{X}_j$ is the decomposition of $\hat{X}$ into connected (and hence into irreducible) components, then $X = \bigcup_{j\in J} \pi(\hat{X}_j)$ is the decomposition of X into irreducible components, and each $\pi: \hat{X}_j \to \pi(\hat{X}_j)$ is the normalization of $\pi(\hat{X}_j)$.

ii) For each $a \in X$, there exists a canonical bijection between $\pi^{-1}(a)$ and the set of prime components of X_a.

Proof. i) Let $X = \bigcup_{i\in I} X_i$ be the decomposition into irreducible components, and set $X_i' := X_i \setminus S(X)$ and $\hat{X}_j' := \hat{X}_j \setminus \pi^{-1}(S(X))$. Then $\pi: \bigcup_{j\in J} \hat{X}_j' \to \bigcup_{i\in I} X_i'$ is clearly biholomorphic. Since $\hat{X}\setminus \pi^{-1}(S(X)) = \bigcup_J \hat{X}_j'$ and $X\setminus S(X) = \bigcup_I X_i'$ are decompositions into connected components, we may assume that $I = J$ and that every $\pi_j: \hat{X}_j' \to X_j'$ is biholomorphic. Now $\pi: \hat{X}_j \to X$ is a closed mapping; hence, $\pi(\hat{X}_j) = \pi(\overline{\hat{X}_j'}) = \overline{X_j'} = X_j$, and $\pi: \hat{X}_j \to X_j$ is the normalization of X_j.

ii) For a prime germ X_a, we saw in the proof of 71.9 that $\pi^{-1}(a)$ contains only one point. Let $U = \bigcup_{j=1}^{m} U_j$ be a decomposition of an open neighborhood of a into irreducible components such that each U_{ja} is a nonempty prime germ. According to i), denote by $\pi^{-1}(U) = \hat{U} = \bigcup_{j=1}^{m} \hat{U}_j$ the decomposition of $\hat{X}$ according to i) into connected components such that $\pi : \hat{U}_j \to U_j$ is the normalization of U_j. Then $\pi^{-1}(a) = \bigcup_{j=1}^{m} (\pi|_{X_j})^{-1}(a)$ contains m points. ∎

E. 71h. If $\mathfrak{p}_1, \ldots, \mathfrak{p}_m$ denote the minimal prime ideals in a reduced analytic algebra R, then $\hat{R} \cong \widehat{R/\mathfrak{p}_1} \oplus \ldots \oplus \widehat{R/\mathfrak{p}_m}$ (integral closures).

The normalization cannot be extended so as to form a functor on the category of reduced complex spaces:

E. 71i. The equivalence relation $(z, 0) \sim (-z, 0)$ determines a complex space $\mathbb{C}^2/\sim$ with $\pi: \mathbb{C}^2 \to \mathbb{C}^2/\sim$ as its normalization. Then $f: \mathbb{C} \to \mathbb{C}^2/\sim, z \mapsto \pi(\sqrt{z}, 0)$ is a holomorphic mapping that cannot be lifted to the normalizations (hint: $\mathbb{C}/z \sim -z \cong \mathbb{C}$).

The situation is different for open mappings:

71.15 Proposition. *If $f: X \to Y$ is a holomorphic mapping between reduced spaces, and $f^{-1}(\{y \in Y; {}_Y\mathcal{O}_y$ is not normal$\})$ is a thin subset of X, then there exists exactly one holomorphic mapping $\hat{f}$ between the normalizations such that the diagram*

$$\begin{array}{ccc} \hat{X} & \xrightarrow{\hat{f}} & \hat{Y} \\ \downarrow & & \downarrow \pi \\ X & \xrightarrow{f} & Y \end{array}$$

commutes.

Proof. For the set A of nonnormal points of Y, the set $X \setminus f^{-1}(A)$ is dense in X; hence, there exists at most one $\hat{f}$. Now it suffices to prove the existence for a local model near the point 0. We may assume that X is normal and consequently that Y is irreducible at the point 0 (see E. 49j); we may suppose further that $\hat{Y} \hookrightarrow \boldsymbol{P} \subset\subset \mathbb{C}^n$, and that $f(0) = 0$. Then the mapping

$$f: X \setminus f^{-1}(A) \to Y \setminus A \cong \hat{Y} \setminus \pi^{-1}(A) \subset\subset \mathbb{C}^n$$

admits a holomorphic extension $\hat{f}: X \to \mathbb{C}^n$. Now the fact that $\pi^{-1}(0) = 0$ implies that $\hat{f}(0) = 0$; by virtue of continuity, $\hat{f}$ maps all points near 0 into $\boldsymbol{P}$, and hence into Y. Obviously, the associated diagram commutes. ∎

E. 71j. A holomorphic mapping $f: Y \to X$ of reduced spaces is called a (continuous) *modification of* X if there exists a thin analytic subset A of X such that $f^{-1}(A)$ is a thin subset of Y and $f: Y \setminus f^{-1}(A) \to X \setminus A$ is biholomorphic. Prove the following statements:

i) Each finite modification of a normal space is biholomorphic.

ii) A surjective modification need not be proper.

iii) Proper modifications need not be finite. (Hint for ii) and iii): E. 32 Bh.)

E. 71k. *Construction of germs with a given normalization.* Let $R \subset S$ be analytic algebras such that $\mathfrak{m}_S^k \subset \mathfrak{m}_R$ for some k (*Strukturausdünnung*, see E. 24d).

i) If S is normal, then $S = \hat{R}$.

ii) Determine for small k (up to isomorphism) all subalgebras $R \subset {}_1\mathcal{O}_0$ such that ${}_1\mathfrak{m}_0^k \subset \mathfrak{m}_R$, and compute their embedding-dimensions.

E. 71 l. Examples of normalizations: Prove the following statements for $X_1 = V(\mathbb{C}^3; z_1^2 - z_2^2 z_3)$ (Whitney's umbrella) and $X_2 = V(\mathbb{C}^3; z_1 z_2 z_3 - z_1^2 - z_2^3)$:

i) $z_1/z_2 \in \widetilde{\mathcal{O}}(X_j)$. ii) ${}_{X_i}\mathcal{O}_0[z_1/z_2] \cong {}_2\mathcal{O}_0$, ${}_{X_i}\widetilde{\mathcal{O}}_0 = {}_{X_i}\mathcal{O}_0[z_1/z_2]$.

(Hint: the mapping

$$\mathbb{C}\{z_1, z_2, z_3\} \to \mathbb{C}\{w, z_2, z_3\}; \qquad z_1 \mapsto wz_2, z_2 \mapsto z_2, z_3 \mapsto z_3,$$

induces finite morphisms

$$ {}_{X_1}\mathcal{O}_0 \to \mathbb{C}\{w, z_2, z_3\}/(w^2 - z_3) \cong {}_{X_1}\mathcal{O}_0[\bar{w}] \quad \text{and}$$
$$ {}_{X_2}\mathcal{O}_0 \to \mathbb{C}\{w, z_2, z_3\}/(z_3 w - w^2 - z_2) \cong {}_{X_2}\mathcal{O}_0[\bar{w}]).$$

iii) The normalizations are

$$\mathbb{C}^2 \to X_1, (u, v) \mapsto (uv, u, v^2) \quad \text{and} \quad \mathbb{C}^2 \to X_2, (u, v) \mapsto (u^2 v, uv, u + v).$$

§ 72 Maximal Complex Structures

We now come back to the investigation of the property $\mathcal{O} = \widetilde{\mathcal{O}} \cap {}_X\mathcal{C}$. In this section, we always assume that $X = (X, \mathcal{O})$, Y, and Z are *reduced* complex spaces.

72.1 Definition. *We say that the structure of a complex space X is maximal (or that X is maximal) if* $\mathcal{O} = \widetilde{\mathcal{O}} \cap {}_X\mathcal{C}$.

Thus X has a maximal structure iff the weak Riemann Removable Singularities Theorem holds; in particular, normal spaces are maximal.

E. 72 a. Prove these statements for the reduced space $(X, \mathcal{O})$:

i) $(X, \mathcal{O})$ is maximal iff $\mathcal{O}$ is the only reduced structure sheaf on $|X|$ that includes $\mathcal{O}$.

ii) There exists a nonreduced structure sheaf $\mathcal{H}$ on $|X|$ that includes $\mathcal{O}$ (hint: E. 51h).

72.2 Theorem. *If Y is maximal, and if* $f: X \to Y$ *is a holomorphic mapping, then the following statements hold:*

i) If f is a homeomorphism, then f is biholomorphic.

ii) If Y is normal, and f is an injective mapping such that $\dim X_x = \dim Y_{f(x)}$ *for each* $x \in X$, *then f is an open mapping, and* $f: X \to f(X)$ *is biholomorphic.*

iii) A continuous mapping $g: Y \to Z$ *is holomorphic iff the graph* $\Gamma(g)$ *is an analytic subset of* $Y \times Z$.

Proof. i) We may assume that $|f| = \mathrm{id}: |X| = |Y|$; then ${}_Y\mathcal{O} \subset {}_X\mathcal{O}$. As the restriction of f to $X \setminus S(Y)$ is biholomorphic (see 46 A.1), we obtain that ${}_Y\widetilde{\mathcal{O}} = {}_X\widetilde{\mathcal{O}}$, and consequently that ${}_X\widetilde{\mathcal{O}} \cap {}_X\mathcal{C} = {}_Y\widetilde{\mathcal{O}} \cap {}_Y\mathcal{C} = {}_Y\mathcal{O} \subset {}_X\mathcal{O} \subset {}_X\widetilde{\mathcal{O}} \cap {}_X\mathcal{C}$.

ii) By 48.10, f is open, since Y has no reducible points (see 71.9). Hence, we may apply i).

iii) *Only if.* This is a consequence of 49 A.6 iv). *If.* By i), the holomorphic homeomorphism $\mathrm{pr}_Y: \Gamma(g) \to Y$ is biholomorphic; so $f = \mathrm{pr}_Z \circ (\mathrm{pr}_Y)^{-1}$ is a holomorphic mapping. ■

In contrast with the situation of normalization, there exists a functor "maximalization" on the category of reduced complex spaces:

72.3 Maximalization Theorem. *i) If* $X = (X, \mathcal{O})$ *is a (reduced) complex space, then* $\widetilde{X} := (X, \widetilde{\mathcal{O}} \cap {}_X\mathcal{C})$ *is a complex space with maximal structure.*

ii) If $f: X \to Y$ is a holomorphic mapping, then so is $\tilde{f} = f: \tilde{X} \to \tilde{Y}$.
We call $\tilde{X}$ the "maximalization" (or "weak normalization") of X.

Proof. i) Let us first show that $\tilde{X} \cong \hat{X}/R_\pi$, where $\pi: \hat{X} \to X$ denotes the normalization: as R_π is a finite equivalence relation, there is a natural identification $|\hat{X}/R_\pi| = |X|$. We have that $\mathcal{O} \subset \tilde{\mathcal{O}} \cap {}_X\mathcal{C} \subset \tilde{\mathcal{O}} \underset{71.8}{=} \pi({}_{\hat{X}}\mathcal{O})$; moreover, for $U \subset\subset X$ and $f \in \tilde{\mathcal{O}}(U) = {}_{\hat{X}}\mathcal{O}(\pi^{-1}(U))$,

$$f \in (\tilde{\mathcal{O}} \cap {}_X\mathcal{C})(U) \Leftrightarrow f \quad \text{is } R_\pi\text{-invariant}.$$

Thus the ringed spaces $\tilde{X}$ and $\hat{X}/R_\pi$ are isomorphic. As X is locally holomorphically separable, so is $\tilde{X}$; by 49 A.16, $\tilde{X}$ is a complex space, that obviously is maximal.

ii) By 72.2 iii), we have to show that $\Gamma(f) = \Gamma(\tilde{f})$ is an analytic subset of $\tilde{X} \times \tilde{Y}$. That is obvious, since $\tilde{X} \times \tilde{Y} \to X \times Y$, as a product of holomorphic mappings, is holomorphic, and since $\Gamma(f)$ is an analytic subset of $X \times Y$. ■

E. 72b. For $X := V(\mathbb{C}^2; z_1 z_2)$, show that $(X, \tilde{\mathcal{O}})$ is not a ringed space.

E. 72c. For a holomorphic homeomorphism $f: X \to Y$, show that X is maximal iff $f({}_X\mathcal{O}) = {}_Y\tilde{\mathcal{O}} \cap {}_Y\mathcal{C}$.

E. 72d. i) ${}_X\tilde{\mathcal{O}} \cap {}_X\mathcal{C}$ is a coherent ${}_X\mathcal{O}$-module.

ii) Specan $({}_X\tilde{\mathcal{O}} \cap {}_X\mathcal{C})$ is the maximalization of X.

iii) The set $\{x \in X; \mathcal{O}_x \neq \tilde{\mathcal{O}}_x \cap \mathcal{C}_x\}$ of points at which X is not maximal is a thin analytic subset of X.

E. 72e. *Examples of maximal structures.* i) Whitney's umbrella X is maximal, and

$$\mathbb{C}^2/((0, v) \sim (0, -v)) \to X, \quad (u, v) \mapsto (uv, u, v^2),$$

is a biholomorphic mapping (hint: E.71 l), $\left(\frac{z_1}{z_2}\right)^2 = z_3 \in {}_X\mathcal{O}_0$, E.49 Am ii), proof of 72.3).

ii) The structure of $X = V(\mathbb{C}^3; z_1 z_2 z_3 - z_1^2 - z_2^3)$ is maximal, and

$$\mathbb{C}^2/((u,0) \sim (0,u)) \to X, \quad (u, v) \mapsto (u^2 v, uv, u+v),$$

is a biholomorphic mapping.

72.4 Proposition. *Let R be an equivalence relation on a complex space X such that X/R is a complex space; then the following statements hold:*

i) If X is maximal, then so is X/R.

ii) If R is open and X is normal, then X/R is normal.

Proof. Let $\pi: X \to \bar{X} := X/R$ denote the natural projection.

i) By 72.2 i), we only have to show that id : $\bar{X} \to \tilde{\bar{X}}$ is a holomorphic mapping. That is a consequence of the fact that $R_\pi = R_{\tilde{\pi}}$: by 49 A.10, the holomorphic mapping $\tilde{\pi}: \tilde{X} = X \to \tilde{\bar{X}}$ admits a holomorphic factorization over π.

ii) It suffices to prove that the normalization $\tau: \hat{\bar{X}} \to \bar{X}$ is injective, for then τ, being a finite mapping, is in fact biholomorphic, since $\bar{X}$ is maximal, by i) (see 72.2). By 71.14 ii), we need only show that $\bar{X}$ is irreducible at every point. To that end, we use the diagram

$$\begin{array}{ccc} X & \xrightarrow{\hat{\pi}} & \hat{X} \\ & \pi \searrow \quad \swarrow \tau & \\ & \bar{X} & \end{array}$$

($\hat{\pi}$ exists, according to 71.15, as R, and thus π, is open). Suppose that there exists a fiber $\tau^{-1}(a)$ that contains more than one point, say $b_1, \ldots, b_m$. We may assume that $\bar{X}$ is connected and that $\hat{X}$ consists of pairwise disjoint open neighborhoods $\hat{V}_j$ of the b_j's. There exists a b_j in $\operatorname{Im}\hat{\pi}$, and $U := \hat{\pi}^{-1}(\hat{V}_j)$ is an open subset of X, but $\pi(U) = \tau\hat{\pi}(U) = \tau(\hat{V}_j)$ is not open in $\bar{X}$. ↯ ∎

72.5 Corollary. *If G is a transformation group acting on a normal space X in such a manner that X/G is a complex space, then X/G is normal.*

Proof. By E.33 Be, the equivalence relation R_G is open. ∎

72.6 Proposition. *Let R denote an equivalence relation on a normal space X such that X/R is a complex space. If every equivalence class $R(x)$ is connected, then X/R is normal.*

Proof. By 72.4, $\bar{X} := X/R$ is maximal; hence, we only have to show that $\bar{X}$ is irreducible at every point. We may assume that $\bar{X}$ and X are connected (as the open sets X_j in a disjoint decomposition $X = X_1 \cup X_2$ are saturated with respect to R (every $R(x)$ is connected), they induce a decomposition of $\bar{X}$ of the same kind). If there exists a (without loss of generality global) decomposition $\bar{X} = \bar{A} \cup \bar{B}$ into proper analytic subsets, then it induces a similar decomposition of X, although the connected normal space X is irreducible. ↯ ∎

E. 72f. Give an equivalence relation R on a manifold X such that X/R is a nonnormal complex space.

72.7 Corollary. *The holomorphic hull of a normal holomorphically convex space X is a normal Stein space.*

Proof. In 57.11, the holomorphic hull of X has been constructed as a quotient X/R_φ in which every class $R_\varphi(x)$ is connected. ∎

72.8 Examples. By 72.5 and 72.7, the following spaces are normal surfaces for $k \geqq 1$:

$$Y_k := \mathbb{C}^2/\{\zeta I_2;\ \zeta^k = 1\}, \quad X(-\mathrm{k}) := E(-k)/S_0 \ \text{(see E. 57h).}$$

E. 72g. In the notations of 72.8, construct a biholomorphic mapping $\varphi: Y_k \to X(-k)$. Then $E(-k)$ determines for $k \geqq 2$ a resolution of the singularity of Y_k. (Hint: extend

$$\varphi: \mathbb{C}^2 \setminus \{0\} \to E(-k) \setminus S_0 \xrightarrow{\cong} X(-k) \setminus 0,$$

$$(x, y) \mapsto \begin{cases} ([1, x/y], y^k) \in U_0 \times \mathbb{C}, & y \neq 0 \\ ([y/x, 1], x^k) \in U_1 \times \mathbb{C}, & x \neq 0 \end{cases} \quad \text{(see E.54 Bc),}$$

holomorphically by setting $\varphi(0) := 0$, and apply 72.2 to the induced mapping $\bar{\varphi}$).

§ 73 Finite Mappings on Stein Spaces

Although finite surjective holomorphic mappings preserve neither holomorphic convexity nor holomorphic spreadability (see E. 51e), they do preserve the property of being a Stein space.

73.1 Proposition. *If $f: X \to Y$ is a finite surjective holomorphic mapping, then X is Stein iff Y is Stein.*

Proof. By 52.19, we may assume that X and Y are reduced. *If.* This is a consequence of 52.18, or of 63.2 iv), even if f is not surjective. *Only if.* Each compact analytic subset of Y is finite; hence, it suffices to show that Y is holomorphically convex by verifying the hypotheses of the criterion (51.2.1). To that end, we proceed in several steps:

a) If X and Y are *normal* and connected, then $|Y|$ has a countable basis of open sets (see 51 A. 3 and E. 33 Bg). For a discrete sequence $(y_j)_{j \in \mathbb{N}}$ in Y, every sequence of inverse images x_j in X is discrete, as well; hence, there exists a $\varphi \in \mathcal{O}(X)$ such that $\sup_j |\varphi(x_j)| = \infty$. By E. 73c, φ satisfies an equation of then form

$$\varphi^m = \sum_{k=0}^{m-1} a_k \varphi^k, \; a_k \in \mathcal{O}(Y);$$

consequently, $\sup_j |a_k(y_j)| = \infty$ for at least one k (see E. 33a).

b) We now prove 73.1 for spaces Y of *finite (global) dimension* n by induction: Since every discrete space is Stein, we may assume that $n \geqq 1$ for the step "$n-1 \Rightarrow n$". Let Y_j, $j \in J$, denote the irreducible components of Y; for each j, then, there exists an irreducible component X_j of X such that $f(X_j) = Y_j$ (since f is a finite surjection, see E. 49j). As the subspace $\bigcup_{j \in J} X_j \hookrightarrow X$ is Stein, we may assume that $X = \bigcup_J X_j$. Then, by 71.15, the normalization $\hat{f}: \hat{X} \to \hat{Y}$ of f exists, and $\hat{X}$ is Stein; hence, by a) and 71.14, $\hat{Y}$ is a Stein space as well. To deduce that Y is Stein, we use the following result:

73.2 Lemma. *Let $\pi: \hat{Y} \to Y$ denote the normalization of a reduced complex space Y. Then $A := V(Y; \mathcal{O} : \hat{\mathcal{O}})$ is a complex subspace of Y such that the functions $\hat{\varphi} \in \mathcal{O}(\hat{Y})$ induced by a holomorphic function on Y are precisely those for which $\hat{\varphi}|_{\pi^{-1}(A)}$ is induced by a holomorphic function on A. Moreover, $|A| = \{y \in Y; {}_Y\mathcal{O}_y \text{ nonnormal}\}$.*

For a discrete sequence $(y_i)_{i \in \mathbb{N}}$ in Y, it suffices to consider the following two cases:

α) Every point y_i lies in A. As $\hat{A} := \pi^{-1}(A) \hookrightarrow \hat{Y}$ is Stein and $\dim \hat{A} < \dim \hat{Y} = \dim Y$, we may assume by the induction hypothesis that A is Stein. Hence, there exists a $\psi \in \mathcal{O}(A)$ such that $\sup_i |\psi(y_i)| = \infty$. By 52.4, there is a holomorphic extension $\hat{\varphi} \in \mathcal{O}(\hat{Y})$ of $\hat{\psi} := \pi^0(\psi) \in \mathcal{O}(\hat{A})$, and 73.2 implies the existence of a $\varphi \in \mathcal{O}(Y)$ such that $\pi^0(\varphi) = \hat{\varphi}$; it follows that $\sup |\varphi(y_i)| = \infty$.

β) No y_i lies in A. By 52.5, then, there exists a $\hat{\varphi} \in \mathcal{O}(\hat{Y})$ such that $\hat{\varphi}|_{\hat{A}} = 0$ and $\hat{\varphi}|_{\pi^{-1}(y_i)} \geqq i$. Applying 73.2 again, we see that $\hat{\varphi} = \pi^0(\varphi)$ for an appropriate $\varphi \in \mathcal{O}(Y)$ with $\sup |\varphi(x_i)| = \infty$.

c) Finally, we assume that Y is connected and has infinitely many irreducible components $Y_j, j \in J$, of arbitrary dimension. As in the first step of the proof of 51 A. 3, we may assume that $J = \mathbb{N}$ (note that each Y_j is Stein). Let us denote the Stein spaces $\bigcup_{j=0}^{k} Y_j$ with Z_k, so that $Y = \bigcup_{k=0}^{\infty} Z_k$. Now, let (y_i) be a discrete sequence of pairwise distinct points in Y. According to 52.5, we define recursively a sequence of functions $f_k \in \mathcal{O}(Z_k)$ such that

$$f_k|_{Z_{k-1}} = f_{k-1} \quad \text{and} \quad f_k(y_i) = i \quad \text{for} \quad y_i \in Z_k.$$

They determine an $f \in \mathcal{O}(Y)$ such that $f|_{Z_k} = f_k$ and $f(y_i) = i$. ■

Proof of 73.2. As $\hat{\mathcal{O}} = \pi({}_{\hat{Y}}\mathcal{O})$ is a coherent $\mathcal{O}$-module, the $\mathcal{O}$-ideal $\mathcal{I} := \mathcal{O} : \hat{\mathcal{O}}$ is coherent (see E. 41 g); it contains all of the universal denominators for $\mathcal{O}$. Then we see that $A = V(Y; \mathcal{I}) \hookrightarrow Y$, and that

$${}_Y\mathcal{O}_y \text{ is nonnormal} \Leftrightarrow {}_Y\mathcal{O}_y \neq {}_Y\hat{\mathcal{O}}_y \Leftrightarrow {}_Y\mathcal{O}_y \neq \mathcal{I}_y \Leftrightarrow y \in N(\mathcal{I}).$$

Let $\tau : \hat{A} := \pi^{-1}(A) \to A$ denote the restriction of π; we have to show for each $f \in \mathcal{O}(\hat{Y})$ and the commutative diagram

$$\begin{array}{ccc} \mathcal{O}(Y) & \xrightarrow{\pi^0} & \mathcal{O}(\hat{Y}) \\ \downarrow & & \downarrow \\ \mathcal{O}(A) & \xrightarrow{\tau^0} & \mathcal{O}(\hat{A}) \end{array}$$

that $f \in \operatorname{Im} \pi^0$ iff $\hat{f}|_{\hat{A}} \in \operatorname{Im} \tau^0$.

We prove the nontrivial implication "*if*" for a fixed $y \in Y$. With the notation $\mathfrak{a} := \mathcal{I}_y$, there is a natural commutative diagram

$$\begin{array}{ccccc} R = {}_Y\mathcal{O}_y & \hookrightarrow & \pi({}_{\hat{Y}}\mathcal{O})_y = \hat{R} & & \\ \downarrow p & & \downarrow \hat{p} & & \\ R/\mathfrak{a} = {}_A\mathcal{O}_y & \xrightarrow{\alpha} & \tau({}_{\hat{A}}\mathcal{O})_y & \underset{31.8\text{ iv)}}{=} & \hat{R}/\mathfrak{a}, \end{array}$$

for which we verify that $\hat{p}^{-1}(\operatorname{Im} \alpha) \subset R$. Given $\hat{\varphi} \in \hat{p}^{-1}(\alpha(R/\mathfrak{a}))$, we fix a $\psi \in R$ such that $\alpha p(\psi) = \hat{p}(\hat{\varphi})$. Then $\hat{\varphi} - \psi \in \mathfrak{a} \cdot \hat{R} \subset R$, since $\mathfrak{a}$ contains the universal denominators; now the fact that $\psi \in R$ implies that $\hat{\varphi}$ lies in R as well. ■■

73.3 Corollary. *If a finite transformation group G acts on a Stein space X, then X/G is Stein.*
That is an immediate consequence of E. 49 Ao ii) and 73.1. ■

E. 73a. For each reduced Stein space X, there exists a universal denominator (hint: apply 53.1 to $\mathcal{O} : \tilde{\mathcal{O}}$).

E. 73b. For a proper holomorphic surjection $f : X \to Y$, prove that, if X is holomorphically convex and every compact analytic subset of Y is finite, then Y is Stein (hint: 57.11, 49 A.9).

E. 73c. For an irreducible and reduced space X and a normal space Y, let $\varphi : X \to Y$ be a finite surjective holomorphic mapping. Then $\mathcal{O}(X)$ is integral over $\mathcal{O}(Y)$ (hint: for $f \in \mathcal{O}(X)$, apply E. 49o and 33.7 to $\mathrm{pr}_Y : (\varphi, f)(X) \to Y$, and construct a polynomial $Q \in \mathcal{O}(Y)[T]_{\mathrm{mon}}$ such that $Q(f) = 0$, as in the proof of 33.6; for 73.1).

§ 74 A Criterion for Normality

If X_0 is the germ of a hypersurface in a manifold, then X_0 is normal iff $\operatorname{codim}_{X_0} S(X_0)$ is at least two. In order to prove that fact in a more general setting, we investigate germs defined by a system of equations $f_1 = \ldots = f_p = 0$ of the following type:

74.1 Definition. *A finite sequence $(f_j)_{1 \leq j \leq p}$ of nonunits in a ring R is called an R-sequence, if the equivalence class $\bar{f}_j$ in $R/(f_1, \dots, f_{j-1})$ is a non-zero-divisor for each j.*

74.2 Remark. *If $(f_j)_{1 \leq j \leq p}$ is an R-sequence in an analytic algebra R, then*

$$\dim R \,/\, R \cdot (f_1, \dots, f_p) = \dim R - p.$$

Proof. We obviously may assume that $p = 1$. By E. 48a, f_1 is an active element, so the Active Lemma 48.5_{alg} yields the assertion. ■

In particular, an ${}_m\mathcal{O}_0$-sequence determines a complete intersection $V(\mathbb{C}_0^m; f_1, \dots f_p)$ in $\mathbb{C}_0^m$. Also the converse is true; see [GrRe Satz III. 1.7 and III. 1.9].

74.3 Theorem. *For every germ $X_a = V(\mathbb{C}_0^m; f_1, \dots, f_p)$, the following statements hold:*

i) If X_a is normal, then $\operatorname{codim}_{X_a} S(X_a)$ is at least 2.

ii) If $(f_j)_{1 \leq j \leq p}$ is an ${}_m\mathcal{O}_0$-sequence, then X_a is normal iff $\operatorname{codim}_{X_a} S(X_a) \geqq 2$.

We first give some applications of 74.3.

74.4 Corollary. *Every connected normal space X of pure dimension one is a Riemann surface.* That obviously follows, as the fact that $\dim S(X) \leqq -1$ means that $S(X) = \emptyset$. ■

74.5 Corollary. *If a reduced subspace X of a manifold is of codimension 1 at each point, then X is normal iff $\operatorname{codim}_X S(X) \geqq 2$.*

Proof. By E. 48l, X admits at every point a description by a single equation in an algebra ${}_n\mathcal{O}_0$. ■

The example $V(\mathbb{C}^3; z_1 z_2, z_3, z_2 z_3^2)$ shows that the assumption "reduced" in 74.5 is essential.

Consider the particular case of a *two-dimensional* germ X_a. If X_a can be defined by means of an ${}_n\mathcal{O}_0$-sequence, then 74.3 justifies the following criterion: *X_a is normal iff X_a has at most an isolated singularity.* If X_a admits a decomposition into two-dimensional primary germs, then, by E. 48l, that criterion applies for $X_a \hookrightarrow \mathbb{C}_0^3$, but not for $X_a \hookrightarrow \mathbb{C}_0^4$:

74.6 Examples. For the ideals $\mathfrak{a}$, $\mathfrak{b}$, and $\mathfrak{p}$ as defined in E. 45e, we obtain the following:

i) The vertex of the cone $V(\mathbb{C}_0^4; \mathfrak{p})$ is an *isolated reduced irreducible two-dimensional singularity* (see E. 48g and E. 49i) *that is not normal;* its normalization is given by the restriction to $V(\mathbb{C}_0^5; \mathfrak{b}) \to$ $\to V(\mathbb{C}_0^4; \mathfrak{p})$ of the projection $\mathbb{C}^5 \to \mathbb{C}^4$, $z \mapsto (z_1, z_2, z_4, z_5)$ (see E. 49 As and 72.8).

ii) The vertex of the cone $V(\mathbb{C}_0^4; \mathfrak{a})$ is an isolated *normal two-dimensional singularity* (see E. 49 As and 72.8) *that is not a complete intersection* (see E. 48g).

74.7 Proposition. *For a complex space X, the following statements are equivalent:*

i) X is normal.

ii) For each $U \Subset X$, the restriction-mapping $\mathcal{O}(U) \to \mathcal{O}(U \setminus S(X))$ is bijective.

Proof. i) $\Rightarrow$ ii) This is a consequence of 74.3 i) and 71.12.

ii) $\Rightarrow$ i) Obviously, $S(X) \neq X$. Moreover, X is reduced: we may assume that X is Stein; if the sheaf $\mathcal{N}$ of nilpotent elements had a nonzero stalk, then Theorem A would imply the existence of a $0 \neq \varphi \in \mathcal{N}(X)$ such that $\operatorname{supp} \varphi \subset \operatorname{supp} \mathcal{N} \subset S(X)$, and it would follow that $\varphi|_{X \setminus S(X)} = 0$. ↯ We see that $\mathcal{O} = \tilde{\mathcal{O}}$, since $\tilde{\mathcal{O}}(U) \subset \mathcal{O}(U \setminus S(X)) = \mathcal{O}(U)$ for each $U \Subset X$. ■

E. 74a. For each $n \geqq 1$, construct a nonnormal reduced space X of dimension n without redicuble points that contains only one singularity (hint: E. 24d).

Proof of 74.3 i). We may assume that $S = S(X)$ has the nullstellen ideal $\mathcal{J} = \mathcal{O} \cdot (f_1, \dots, f_m)$ for appropriate $f_j \in \mathcal{O}(X)$; for a proof by contradiction, we may suppose further that S is a submanifold of X with codimension one. For each $a \in S$, (0) is the only prime ideal in $R := {}_X\mathcal{O}_a$ that is properly included in $\mathfrak{p} := \mathcal{I}_a$ (see 48.11); by 74.8, the ideal $R_\mathfrak{p} \cdot \mathfrak{p}$ in the *localization*

$$R_\mathfrak{p} := (R \setminus \mathfrak{p})^{-1} \cdot R = \left\{ \frac{f}{g} \in Q(R); \, g \notin \mathfrak{p} \right\}$$

of R by $\mathfrak{p}$ is a principal ideal $R_\mathfrak{p} \cdot \mathfrak{p} = R_\mathfrak{p} \cdot f$ for an appropriate $f \in \mathfrak{p}$. For each j, after shrinking X, we can find a representation

$$f_j = \frac{\varphi_j}{\psi_j} \cdot f \quad \text{such that} \quad f, \varphi_j, \psi_j \in \mathcal{O}(X) \quad \text{and} \quad \psi_j \notin \mathfrak{p}.$$

Thus we obtain that $\psi_j f_j = \varphi_j f$, and we may assume that $\mathcal{I} = \mathcal{O} \cdot f$ on $X \setminus \bigcup_{j=1}^{m} N(\psi_j)$. Since $\Pi \psi_j \notin \mathfrak{p}$, we have that $S \not\subset \bigcup N(\psi_j)$; consequently, there exists a point $b \in S$ such that $\mathcal{I}_b = \mathcal{O}_b \cdot f_b$. Finally, E. 48k implies that $\mathcal{O}_b$ is regular, since $\mathcal{O}_b / \mathcal{I}_b$ is regular. ↯ ■

74.8 Lemma. *Let $\mathfrak{p}$ be a prime ideal in a normal noetherian ring. If there is no prime ideal $\mathfrak{q}$ such that $(0) \subsetneq \mathfrak{q} \subsetneq \mathfrak{p}$, then $R_\mathfrak{p} \cdot \mathfrak{p}$ is a principal ideal: $R_\mathfrak{p} \cdot \mathfrak{p} = R_\mathfrak{p} \cdot f$ for an appropriate $f \in \mathfrak{p}$.*

Proof. Certainly, $\mathfrak{p}$ is a minimal prime ideal over (g) for each $0 \neq g \in \mathfrak{p}$. It follows from 44.15 that there exists an $s \in R$ such that $\mathfrak{p} = (g):(s)$. We now prove that there exist an $f \in \mathfrak{p}$ and an $r \in R \setminus \mathfrak{p}$ such that $\mathfrak{p} = (f):(r)$. Then r is a unit in $R_\mathfrak{p}$, and 74.8 follows from the fact that

$$R_\mathfrak{p} \cdot f \subset R_\mathfrak{p} \cdot \mathfrak{p} = R_\mathfrak{p} \cdot r \cdot \mathfrak{p} = R_\mathfrak{p} \cdot r \cdot [(f):(r)] \subset R_\mathfrak{p} \cdot f.$$

Now $q := s/g$ is not in R, since $1 \notin (g):(s)$. But there exists a $j \geqq 1$ such that $gq^j \in R$ and $gq^{j+1} \notin R$, for, if not, we would obtain that $q\mathfrak{a} \subset \mathfrak{a}$ for $\mathfrak{a} := R \cdot (gq^j; j \geqq 1)$, and thus that $q \in S = \hat{R} = R$, according to 71.11 i). Set

$$f := gq^{j-1} \quad \text{and} \quad r := gq^j;$$

then $f/r = g/s$, $\mathfrak{p} = (f):(r)$, $f \in \mathfrak{p}$, and finally $r \notin \mathfrak{p}$, as $r^2/f = gq^{j+1} \notin R$. ■

The proof of 74.3 ii) is based on the following results, which we prove later, using 74.12 and 74.13:

74.9 Proposition. *For a region $G \subset\subset \mathbb{C}^n$, an ${}_n\mathcal{O}_0$-sequence $(f_j)_{j=1,\ldots,p}$, and $X := V(G; f_1, \ldots, f_p)$, the trivial extension ${}_X\mathcal{O}^G$ admits a free resolution of length $\leqq p$ near $0 \in G$; that is, there exists an exact sequence of the following form near 0:*

$$0 \to {}_G\mathcal{O}^{m_p} \to {}_G\mathcal{O}^{m_{p-1}} \to \ldots \to {}_G\mathcal{O}^{m_0} \to {}_X\mathcal{O}^G \to 0.$$

74.10 Proposition (Riemann Continuation Theorem for Cohomology Classes). *Let $\mathcal{G}$ be a coherent analytic sheaf on a manifold M. If free resolutions of $\mathcal{G}$ of length $\leqq p$ exist locally, then, for every analytic subset $A \hookrightarrow M$, the restriction-homomorphisms*

$$\varrho^j : H^j(M, \mathcal{G}) \to H^j(M \setminus A, \mathcal{G})$$

are bijective for $j \leqq \operatorname{codim}_M A - (p+2)$, and injective for $j = \operatorname{codim}_M A - (p+1)$.

74.11 Corollary. *If $X \hookrightarrow G \subset\subset \mathbb{C}^n$ can be defined at each point $x \in X$ by means of an ${}_n\mathcal{O}_x$-sequence, and if $\operatorname{codim}_X S(X) \geqq 2$, then the restriction-homomorphism*

$$\varrho : \mathcal{O}(X) \to \mathcal{O}(X \setminus S(X))$$

is bijective.

Proof of 74.11. We may assume that the hypotheses of 74.9 are satisfied. Then 74.10 implies that the mapping ${}_X\mathcal{O}(X) = H^0(G, {}_X\mathcal{O}^G) \to H^0(G \backslash S(X), {}_X\mathcal{O}^G) = {}_X\mathcal{O}(X \backslash S(X))$ is bijective, since, locally,

$$\operatorname{codim}_G S(X) - (p+2) = \operatorname{codim}_G X + \operatorname{codim}_X S(X) - (p+2) \geqq p+2-(p+2) = 0$$

(see 74.2 and 49.13). ■

Proof of 74.3 ii). In order to apply 74.7, we have to show that every $\varrho_U : \mathcal{O}(U) \to \mathcal{O}(U \backslash A)$ is bijective; but that follows immediately from 74.11. ■

By 41.17 and E. 41c, it suffices to *prove* 74.9 for the ${}_n\mathcal{O}_0$-module ${}_n\mathcal{O}_0/{}_n\mathcal{O}_0 \cdot (f_1, \ldots, f_p)$; it is therefore a consequence of 74.12 below. In that lemma, we use the "Koszul complex": for elements $f_1, \ldots, f_p$ of a ring R, we define (see § 58A)

$$K_j := A^{p-j}(R^p); \quad d_j : K_j \to K_{j-1}, \quad \omega \mapsto \sum_{i=1}^{p} f_i e_i \wedge \omega,$$

where $\{e_1, \ldots, e_p\}$ is the canonical basis of $A^1(R^p)$. By a simple computation, one verifies that $d_j \circ d_{j+1} = 0$; therefore, (K_*, d_*) is a (chain) complex that permits treatment analogous to that used for the (cochain) complexes in § 50; it is called the *Koszul complex* $K_*(R; f_1, \ldots, f_p)$. Its homology

$$H_j(K_*) := \operatorname{Ker} d_j / \operatorname{Im} d_{j+1}$$

satisfies this condition (where $A^p(R^p) = R$):

$$H_0(K_*(R; f_1, \ldots, f_p)) = R/R \cdot (f_1, \ldots, f_p).$$

74.12 Lemma. *If $(f_j)_{j=1,\ldots,p}$ is an R-sequence, then the Koszul complex $K_*(R; f_1, \ldots, f_p)$ provides a free resolution of length p of $R/R \cdot (f_1, \ldots, f_p)$; that is, the sequence*

$$0 \to K_p \xrightarrow{d_p} K_{p-1} \to \ldots \to K_0 \to R/R \cdot (f_1, \ldots, f_p) \to 0$$

is exact.

Proof by induction. For $p = 1$, $d_1 : R \to R$ is multiplication by f_1. Since f_1 is not a zero-divisor, the sequence

$$0 \to R \xrightarrow{f_1} R \to R/R \cdot f_1 \to 0$$

is exact.

"$p-1 \Rightarrow p$" We have to show that $H_j(K_*) = 0$ for $j \geqq 1$. For that purpose, we construct an exact sequence of complexes

$$(74.12.1) \quad 0 \to K'_* \xrightarrow{\alpha_*} K_* \xrightarrow{\beta_*} K'_{*-1} \to 0,$$

where $K'_* = K_*(R; f_1, \ldots, f_{p-1})$; the latter induces an exact homology sequence

$$\begin{array}{c} H_{j+1}(K'_*) \to H_{j+1}(K_*) \to H_{j+1}(K'_{*-1}) \xrightarrow{\partial_j} H_j(K'_*) \to \ldots \\ \qquad\qquad\qquad\qquad \| \\ \qquad\qquad\qquad\qquad H_j(K'_*) \end{array}$$

Further we show that the connecting homomorphism ∂_0 is injective. Then the induction hypothesis, "$H_j(K'_*) = 0$ for every $j \geqq 1$", implies that $H_j(K_*) = 0$ for $j \geqq 2$, and the exactness of the sequence

$$0 = H_1(K'_*) \to H_1(K_*) \to H_0(K'_*) \xrightarrow{\partial_0} H_0(K'_*)$$

implies that $H_1(K_*) = \operatorname{Ker} \partial_0 = 0$. For the construction of (74.12.1) we use the fact that each $\omega \in A^{p-j}(R^p)$ has a unique decomposition

$$\omega = \omega_0 \wedge e_p + \omega_1, \ \omega_0 \in A^{p-j-1}(R^{p-1}), \ \omega_1 \in A^{p-j}(R^{p-1}).$$

Consider the mappings

$$\alpha_j : K_j' \to K_j, \ \omega_0 \mapsto \omega_0 \wedge e_p, \quad \text{and} \quad \beta_j : K_j \to K_{j-1}', \ \omega \mapsto \omega_1 ;$$

then it is easy to check that the diagram

$$\begin{array}{ccccccccc}
0 & \to & K_j' & \xrightarrow{\alpha_j} & K_j & \xrightarrow{\beta_j} & K_{j-1}' & \to & 0 \\
 & & \downarrow{\scriptstyle d_j'} & & \downarrow{\scriptstyle d_j} & & \downarrow{\scriptstyle d_{j-1}'} & & \\
0 & \to & K_{j-1}' & \xrightarrow{\alpha_{j-1}} & K_{j-1} & \xrightarrow{\beta_{j-1}} & K_{j-2}' & \to & 0
\end{array}$$

has exact rows and commutes. In order to prove that ∂_0 is injective, we recall the definition of ∂_0 (similar to that of δ^q in §50): every $\omega_1 \in K_0'$ may be interpreted as an element of K_1. By definition,

$$d_1 \omega_1 = (-1)^{p-1} f_p \omega_1 \wedge e_p = \alpha_0((-1)^{p-1} f_p \omega_1),$$

so $\partial_0 \bar{\omega}_1 = [(-1)^{p-1} \bar{f}_p] \bar{\omega}_1$. Then ∂_0, being multiplication with the non-zero-divisor $\pm \bar{f}_p$ in $R/R\cdot(f_1, \ldots, f_{p-1}) = H_0(K_*')$, is injective. ■

The rest of this section contains the *proof of 74.10*. We begin by introducing a new tool, namely, the local cohomology that we apply to locally closed analytic subsets.

Let $\mathscr{G}$ be sheaf of R-modules on a topological space T, and S, a locally closed subset. Then $S = U \cap A$ for appropriate subsets $U \subset\!\!\!\!\circ\, T$ and $A = \bar{A}$ in T. The R-module

$$\Gamma_S(T, \mathscr{G}) := \{g \in \mathscr{G}(U);\ g|_{U \setminus A} = 0\}$$

of *sections of $\mathscr{G}$ with supports in S* is independent of the particular choices of U and A (proof?). Application of the section-functor Γ_S to the canonical resolution $\mathscr{C}^*(\mathscr{G})$ of 50.3 yields a complex of R-modules

$$\Gamma_S(T, \mathscr{C}^*(\mathscr{G})) \colon \ \Gamma_S(T, \mathscr{C}^0(\mathscr{G})) \xrightarrow{d^0} \Gamma_S(T, \mathscr{C}^1(\mathscr{G})) \xrightarrow{d^1} \ldots,$$

with a cohomology module for each $q \geqq 0$

$$H^q(\Gamma_S(T, \mathscr{C}^*(\mathscr{G}))) =: H^q_S(T, \mathscr{G}) =: H^q(T, T \setminus S; \mathscr{G}),$$

called the qth *module of local cohomology of T with supports in S and coefficients in $\mathscr{G}$* (or the qth relative cohomology of T mod $T \setminus S$). Obviously, $H^q_T(T, \mathscr{G}) = H^q(T, \mathscr{G})$; it is not difficult to generalize 50.1–50.7:

E. 74b. *Properties of local cohomology.* i) If $0 \to {}'\mathscr{G} \to \mathscr{G} \to {}''\mathscr{G} \to 0$ is an exact sequence, then so is $0 \to \Gamma_S(T, {}'\mathscr{G}) \to \Gamma_S(T, \mathscr{G}) \xrightarrow{p} \Gamma_S(T, {}''\mathscr{G})$; if, moreover, ${}'\mathscr{G}$ is flabby, then p is surjective.

ii) $\mathscr{G}$ is a flabby sheaf iff $\mathscr{G}$ is Γ_S-acyclic for each locally closed subset S of T (that is, iff $H^q_S(T, \mathscr{G}) = 0$ for each $q \geqq 1$).

iii) For an exact sequence $0 \to {}'\mathscr{G} \to \mathscr{G} \to {}''\mathscr{G} \to 0$, construct a connecting homomorphism $\delta^q : H^q_S(T, {}''\mathscr{G}) \to H^{q+1}_S(T, {}'\mathscr{G})$ and show that, *mutatis mutandis*, $H^*_S = \{H^q_S, \delta^q;\ q \in \mathbb{N}\}$ satisfies the axioms 50.1 for a cohomology theory.

Observe also that H^*_S depends only on the germ of T along S:

E. 74c. *Excision Theorem.* Show that $H_S^*(T,\mathscr{G}) = H_S^*(V,\mathscr{G})$ for $S \subset V \Subset T$; in particular, $H_V^*(T,\mathscr{G}) = H^*(V,\mathscr{G})$ (for 74.10).

We have to compare the local cohomologies with respect to different subsets of supports:

E. 74d. If S' is a closed subset of S, then S' and $S \backslash S'$ are locally closed in T. Moreover, there exists a natural exact sequence

$$0 \to \Gamma_{S'}(T,\mathscr{G}) \to \Gamma_S(T,\mathscr{G}) \xrightarrow{\varrho} \Gamma_{S\backslash S'}(T,\mathscr{G})$$

in which ϱ is surjective if $\mathscr{G}$ is flabby.

Thus we obtain an exact sequence

$$0 \to \Gamma_{S'}(T,\mathscr{C}^*(\mathscr{G})) \to \Gamma_S(T,\mathscr{C}^*(\mathscr{G})) \to \Gamma_{S\backslash S'}(T,\mathscr{C}^*(\mathscr{G})) \to 0$$

of complexes of R-modules, and thus an associated exact cohomology sequence

$$(74.10.1)\quad 0 \to \Gamma_{S'}(T,\mathscr{G}) \to \Gamma_S(T,\mathscr{G}) \to \Gamma_{S\backslash S'}(T,\mathscr{G}) \xrightarrow{\delta^0} H_{S'}^1(T,\mathscr{G}) \to H_S^1(T,\mathscr{G}) \to \dots .$$

As A is *closed* in T, the modules $H_A^q(T,\mathscr{G})$ contain the obstruction to the extension of a cohomology class from $T\backslash A$ to T, as the Excision Theorem transforms (74.10.1) into the exact sequence

$$(74.10.2)\quad \dots \to H_A^q(T,\mathscr{G}) \to H^q(T,\mathscr{G}) \xrightarrow{\varrho^q} H^q(T\backslash A,\mathscr{G}) \to H_A^{q+1}(T,\mathscr{G}) \to \dots .$$

We have thus reduced 74.10 to the following assertion:

$$(74.10.3)\quad H_A^q(M,\mathscr{G}) = 0 \quad \text{for} \quad q \leqq c-(p+1) \quad \text{where} \quad c := \operatorname{codim}_M A .$$

We apply once again a local-global principle: the presheaf

$$V \mapsto H_{S\cap V}^q(V,\mathscr{G})$$

(with natural restrictions $H_{S\cap V}^q(V,\mathscr{G}) \to H_{S\cap W}^q(W,\mathscr{G})$ for $W \Subset V \Subset T$) determines a sheaf

$$\mathscr{H}_S^q(\mathscr{G}) = {}_T\mathscr{H}_S^q(\mathscr{G}),$$

called the qth <u>local cohomology sheaf</u> *of T with supports in S and coefficients in $\mathscr{G}$.* In [SiTr 0.6], there is a proof of the following result:

74.13 Lemma. *If the sheaves ${}_T\mathscr{H}_S^j(\mathscr{G})$ are acyclic for $j = 0, \dots, p$, then the natural homomorphisms*

$$H_S^q(T,\mathscr{G}) \to H^0(T,\mathscr{H}_S^q(\mathscr{G}))$$

are bijective for every $q \leqq p+1$. ∎

Thus (74.10.3) is a consequence of the following:

$$(74.10.4)\quad {}_M\mathscr{H}_A^q(\mathscr{G}) = 0 \quad \text{for} \quad q \leqq c-(p+1).$$

Proof of 74.10. We may assume that M is a (sufficiently small) polydisk $\boldsymbol{P} \Subset \mathbb{C}^n$ on which $\mathscr{G}$ admits a free resolution of length $\leqq p$; we then verify (74.10.4) by inducting on p.

"$p = 0$" We have that $\mathscr{G} \cong {}_n\mathscr{O}^m$; by E. 50d, we may assume that $\mathscr{G} = \mathscr{O}$. Then (74.10.2) provides an exact sequence on $\boldsymbol{P}$,

$$(74.10.5)\quad \dots \to \mathscr{H}_A^q(\mathscr{O}) \to \mathscr{H}^q(\mathscr{O}) \to \mathscr{H}_{\boldsymbol{P}\backslash A}^q(\mathscr{O}) \to \mathscr{H}_A^{q+1}(\mathscr{O}) \to \dots .$$

Since $\mathscr{H}^0(\mathscr{O}) = \mathscr{O}$ and $\mathscr{H}^q(\mathscr{O}) = 0$ for each $q \geqq 1$ by E. 61d, we have the exact sequence

$$0 \to \mathscr{H}_A^0(\mathscr{O}) \to \mathscr{O} \xrightarrow{\varrho} \mathscr{H}_{\boldsymbol{P}\backslash A}^0(\mathscr{O}) \to \mathscr{H}_A^1(\mathscr{O}) \to 0 .$$

As ϱ is injective for $c = 1$ (see 6.1) and bijective for $c \geqq 2$ (see 7.7), we have to verify (74.10.4) only for $2 \leqq q \leqq c-1$ and a point $a \in \operatorname{supp} \mathscr{H}_A^q(\mathscr{O}) \subset A$.

α) If a is a regular point of A, we may assume that $A = N(\boldsymbol{P}; z_1, \dots, z_c)$. By (74.10.5), it suffices to show that ${}_{\boldsymbol{P}}\mathscr{H}^{q-1}_{\boldsymbol{P}\setminus A}(\mathcal{O}) = 0$ for $2 \leqq q \leqq c-1$. Set $U_j := \boldsymbol{P}\setminus N(z_j)$; then $\mathfrak{U} := (U_j)_{1\leqq j \leqq c}$ is a Leray cover of $\boldsymbol{P}\setminus A$ (see 51.8). By 74.14, we have that

$$H^i(\boldsymbol{P}\setminus A, \mathcal{O}) = H^i(\mathfrak{U}, \mathcal{O}) = 0 \quad \text{for} \quad 1 \leqq i \leqq c-2,$$

since we may use the Laurent series expansion with respect to z_j according to E. 7e in order to define homomorphisms

$$\varphi_I^{jI}: \mathcal{O}(U_{jj_0\dots j_q}) \to \mathcal{O}(U_{j_0\dots j_q}), \quad \sum_{m=-\infty}^{\infty} g_m z_j^m \mapsto \sum_{m=0}^{\infty} g_m z_j^m .$$

We finally obtain the assertion by passing to the direct limit.

β) For $a \in S(A)$, we use the exact sequence (see 74.10.1)

$$(74.10.6) \quad \dots \to \mathscr{H}^q_{S(A)}(\mathcal{O}) \to \mathscr{H}^q_A(\mathcal{O}) \to \mathscr{H}^q_{A\setminus S(A)}(\mathcal{O}) \to \mathscr{H}^{q+1}_A(\mathcal{O}) \to \dots$$

on A to induct on $\dim A$: for $q \leqq c-1$ and $V \Subset \boldsymbol{P}$, we have that

$$0 = H^0(V\setminus S(A), \mathscr{H}^q_{A\setminus S(A)}(\mathcal{O})) \underset{74.10;\,\alpha)}{=} H^q_{A\setminus S(A)}(V\setminus S(A), \mathcal{O}) \underset{\text{E. 74c}}{=} H^q_{A\setminus S(A)}(V, \mathcal{O});$$

as a result, $\mathscr{H}^q_{A\setminus S(A)}(\boldsymbol{P}, \mathcal{O}) = 0$. By the induction hypothesis, we may assume that $\mathscr{H}^q_{S(A)}(\mathcal{O}) = 0$ for $q \leqq n - \dim S(A) - 1$. Then (74.10.6) implies that $\mathscr{H}^q_A(\mathcal{O}) = 0$ for $q \leqq n - \dim A - 1 = c - 1$.

"$p-1 \Rightarrow p$" A free resolution of length p of $\mathscr{G}$ provides an exact sequence

$$0 \to \mathscr{K} \to \mathcal{O}^m \to \mathscr{G} \to 0,$$

and thus a free resolution of length $p-1$ of the coherent $\mathcal{O}$-module $\mathscr{K}$. There is an exact sequence (see E. 74b iii))

$$\dots \to \mathscr{H}^q_A(\mathscr{K}) \to \mathscr{H}^q_A(\mathcal{O}^m) \to \mathscr{H}^q_A(\mathscr{G}) \to \mathscr{H}^{q+1}_A(\mathscr{K}) \to \dots$$

in which $\mathscr{H}^j_A(\mathcal{O}^m) = 0$ for $j \leqq c-1$; by the induction hypothesis, $\mathscr{H}^j_A(\mathscr{K}) = 0$ for $j \leqq c-p$; we conclude that $\mathscr{H}^q_A(\mathscr{G}) = 0$ for $q \leqq c-(p+1)$. ■

We have used the Vanishing Lemma 74.14 for the Čech cohomology. There, $\mathscr{G}$ is a sheaf of R-modules on a topological space T, and $\mathfrak{U} := (U_j)_{j\in J}$, $J = \{0, \dots, m\}$ is an open cover of T. For $I := (j_0, \dots, j_q) \in J^{q+1}$, set

$$\mathscr{G}(I) := \mathscr{G}(U_I), \quad I_{\hat{k}} := (j_0, \dots, \hat{j}_k, \dots, j_q) \in J^q, \; jI := (j, j_0, \dots, j_q) \in J^{q+2}$$

and

$$\varrho_I^{I'}: \mathscr{G}(I') \to \mathscr{G}(I) \quad \text{if} \quad I' \subset I$$

for the restriction-homomorphisms. Then we obtain the following:

74.14 Lemma. *If, for each $I \in J^{q+1}$ and each $j \in J\setminus I$, there exist homomorphisms $\varphi_I^{jI}: \mathscr{G}(jI) \to \mathscr{G}(I)$ such that*

i) $\varphi_I^{jI} \circ \varrho_{jI}^I = \mathrm{id}_{\mathscr{G}(I)}$, and

ii) $\varrho_I^{I'} \circ \varphi_{I'}^{jI'} = \varphi_I^{jI} \circ \varrho_{jI}^{jI'}$ if $I' \subset I$ and $j \notin I$,

then $H^q(\mathfrak{U}, \mathscr{G}) = 0$ for $1 \leqq q \leqq m-1$.

Proof. For each $j \in J$, and each $q \geqq 1$, we construct an endomorphism π_j of $C^q_a := C^q_a(\mathfrak{U}, \mathscr{G})$ such that every cocycle $f \in Z^q_a$ is cohomologous to $\pi_j(f)$, and such that $\pi_j(f)_I = 0$ if $j \notin I$ or if $f_I = 0$. Then $\pi := \pi_0 \circ \dots \circ \pi_m = 0$ on Z^q_a for $1 \leqq q \leqq m-1$, and the assertion follows, since each $f \in Z^q_a$ is cohomologous to $\pi(f)$. Set

$$\pi_j := \mathrm{id} - \delta^{q-1} \circ \psi_j, \quad \text{with} \quad \psi_j: C^q_a \to C^{q-1}_a; \; \psi_j(f)_I = \begin{cases} \varphi_I^{jI}(f_{jI}) & j \notin I \\ 0 & j \in I. \end{cases}$$

Then it remains to show for $I \in J^{q+1}$ and $f \in Z_a^q$ that

$$(\delta^{q-1} \circ \psi_j(f))_I = f_I \quad \text{if} \quad j \notin I \quad \text{or} \quad f_I = 0.$$

For $j \notin I$, we find that

$$(\delta^{q-1} \circ \psi_j(f))_I = \sum_{k=0}^{q} (-1)^k \varrho_I^{I_k}(\psi_j(f)_{I_k}) = \sum_{k=0}^{q} (-1)^k \varrho_I^{I_k} \varphi_{I_k}^{jI_k}(f_{jI_k})$$

$$\underset{\text{ii)}}{=} \sum_{k=0}^{q} (-1)^k \varphi_I^{jI} \varrho_{jI}^{jI_k}(f_{jI_k}) = \varphi_I^{jI}[\varrho_{jI}^{I}(f_I) - (\varrho_{jI}^{I}(f_I) + \sum_{k=0}^{q} (-1)^{k+1} \varrho_{jI}^{jI_k}(f_{jI_k}))]$$

$$\underset{\text{i)}}{=} f_I - \varphi_I^{jI}(\delta^q(f)_{jI}) = f_I, \quad \text{since} \quad \delta^q(f) = 0.$$

For $j \in I$, and $f_I = 0$, there exists precisely one k such that $j \notin I_k$; therefore,

$$(\delta^{q-1} \circ \psi_j(f))_I = (-1)^k \varrho_I^{I_k}(\varphi_{I_k}^{jI_k}(f_{jI_k})) = 0. \quad ■■$$

E. 74e. Prove that $H^q((\mathbb{C}^{n+1})^*, \mathcal{O}) = 0$ for $q \geqq 1$, $q \neq n$, and that

$$H^n((\mathbb{C}^{n+1})^*, \mathcal{O}) = \{f \in \mathcal{O}(\mathbb{C}^{*n+1});\ f(z) = \sum_{\nu < 0} a_\nu z^\nu\}$$

(hint: for $q \leqq n-1$ apply 74.10 and Theorem B; for $q \geqq n$ use the Leray cover $\mathfrak{U} := (U_i)_{1 \leqq i \leqq n}$, $U_i := \{z \in \mathbb{C}^n; z_i \neq 0\}$, and the alternating Čech complex $C_a^q(\mathfrak{U}, \mathcal{O})$. Decompose $h = \sum_\nu a_\nu z^\nu \in \mathcal{O}(\mathbb{C}^{*n+1}) = C_a^n(\mathfrak{U}, \mathcal{O})$ as $h = h_- + \sum_{j=0}^{n} g_j$, with $h_- = \sum_{\nu<0} a_\nu z^\nu$ and $g_j = \sum_{\nu \in (\mathbb{Z}_{<0}^{j}) \times \mathbb{N} \times \mathbb{Z}^{n-j}} a_\nu z^\nu \in \mathcal{O}(U_{0 \ldots \hat{j} \ldots n})$; then the cochain $g = ((-1)^j g_j) \in C_a^{n-1}(\mathfrak{U}, \mathcal{O})$ satisfies the equality $h = h_- + \delta g$; for E. 74f).

E. 74f. Prove that $H^q(\mathbb{P}_n, \mathcal{O}) = 0$ for each $q \geqq 1$ (hint: E.74e; with the notation of 32.4 iv), the homomorphism

$$H^q(\tau): H^q(\mathbb{P}_n, \mathcal{O}) \cong H_a^q(\mathfrak{U}, \mathcal{O}) \to H_a^q(\tau^{-1}\mathfrak{U}, \mathcal{O}) \cong H^q((\mathbb{C}^{n+1})^*, \mathcal{O})$$

is injective, since it admits a left inverse that is induced by the homomorphism

$$\psi^q : C_a^q(\tau^{-1}\mathfrak{U}, \mathcal{O}) \to C_a^q(\mathfrak{U}, \mathcal{O}), (f_I = \sum_\nu a_\nu^I z^\nu) \mapsto (g_I = \sum_{|\nu|=0} a_\nu^I z^\nu),$$

that is compatible with the coboundary operator δ. For $q = n$, every cochain $C^n(\tau)(f)$ is cohomologous to 0; see E.74e).

E. 74g. If A is an analytic subset of a manifold M of codimension at least 3 then *each additive Cousin distribution on $M \backslash A$ extends to M* (hint: use the Five Lemma for the cohomology of M and $M \backslash A$ associated with the exact sequence $0 \to \mathcal{O} \to \mathcal{M} \to \mathcal{M}/\mathcal{O} \to 0$, and apply 53 A.9 and E.50g; for $\operatorname{codim}_M A = 2$, see 53.4).

E. 74h. *Riemann Continuation Theorem for Cohomology Classes.* For a closed subset S in a manifold M and an $a \in S$, define

$$\dim_{M_a} S_a := \min\{\dim A_a;\ S_a \subset A_a \hookrightarrow M_a\}.$$

That determines an *analytic codimension* $\operatorname{codim}_M S$ of S in M. Generalize 74.10 for S in place of A (hint: consider $H^*(M, \mathcal{G}) \to H^*(M \backslash S, \mathcal{G}) \to H^*(M \backslash A, \mathcal{G})$).

List of Examples

A *pointwise convergent* sequence in $\mathcal{O}(X)$ that does *not converge* E. 2f
Simply connected domains in $\mathbb{C}^2$ that are *topologically*, but *not biholomorphically equivalent* 3.11
A biholomorphic image of a *Runge domain* that is not a Runge domain 6.8
A real plane H in $\mathbb{C}^2$ such that $\mathbb{C}^2 \setminus H$ is not holomorphically convex 12 A. 5
Multiple points 31.4iv)
A *nonreduced fiber* of a holomorphic mapping between manifolds E. 31h
A nonreduced ringed space that is *algebraically reduced* 31.12
A holomorphic map between manifolds that is a *mono*morphism and an *epi*morphism but *not* an *isomorphism* E. 31n
A holomorphic mapping to which the *Riemann Removable Singularity Theorems* do not carry over E. 32g
Achsenkreuz $N(\mathbb{C}^2; z_1 z_2)$ E. 32i
Neil's parabola $N(\mathbb{C}^2; z_1^2 - z_2^3)$ E. 32i
Whitney's umbrella $N(\mathbb{C}^3; z_1^2 z_2 - z_3^2)$ E. 32i
An *irreducible polynomial* that induces the germ of a *reducible holomorphic function* E. 32j
A normal *non factorial* analytic algebra without zero-divisors E. 32 la iii)
A *nonnormal* analytic algebra E. 32 l b)
A two-dimensional compact manifold that is not *projective-algebraic* 32 B. 9i)
A holomorphic *group action* on a manifold such that the orbit space is not Hausdorff 32 Bf
A connected curve X such that the real part $X_{\mathbb{R}}$ is not connected § 33
Two submanifolds of a manifold such that the *intersection* is not reduced E. 43b
A one-dimensional analytic algebra ${}_X\mathcal{O}_a$ such that all nonunits are zero-divisors (*line with a double point*) E. 43l
A *reducible* ideal such that the primary decomposition is not unique, though its zero set is irreducible E. 44g
A continuous function that is not holomorphic, though its restriction to every irreducible component is E. 44i
A mapping that is *open* at a point, but not near the point E. 45h
An exact sequence of coherent analytic sheaves such that the *image sheaves* do not form an exact sequence § 45A
A subset that is topologically *thin*, but not analytically E. 46Ba
An *active* zero-divisor E. 48b
An injective algebra-homomorphism ${}_3\mathcal{O}_0 \to {}_2\mathcal{O}_0$ § 48
A subgerm in a normal complex space of *codimension* 1 that is *not* the germ of a *hypersurface* E. 48l
A reducible not reduced complex space with an *irreducible* reduction E. 49b
A holomorphic mapping of manifolds with nonconstant fiber dimension § 49
A pair of morphisms of complex spaces for which the ringed *cokernel* differs from the complex cokernel 49 A. 8ii)
A pair of holomorphic mappings that does not admit a complex *cokernel* 49 A. 8iii)
Two complex *structure sheaves* such that the *intersection* is not a complex structure sheaf E. 49 Ai iii)

Bibliography

[Ab] Abhayankar, S.S.: Local Analytic Geometry. New York: Academic Press 1964.

[AnFr] Andreotti, A., Frankel, Th.: The Lefschetz Theorem on Hyperplane Sections. Ann. Math. **69**, 713–717 (1959).

[AnSt] Andreotti, A., Stoll, W.: Analytic and Algebraic Dependence of Meromorphic Functions. Lecture Notes in Math., 234. Heidelberg: Springer-Verlag 1971.

[AtMac] Atiyah, M.F., MacDonald, I.G.: Introduction to Commutative Algebra. Reading, Mass.: Addison-Wesley 1969.

[BeTh] Behnke, H., Thullen, P.: Theorie der Funktionen mehrerer komplexer Veränderlicher. 2. Aufl. Erg. Math. 51. Heidelberg: Springer-Verlag 1970.

[BaSt] Bănică, C., Stănăşilă, O.: Algebraic Methods in the Global Theory of Complex Spaces. London: J. Wiley 1976.

[BoHu] Bombieri, E., Husemöller, D.: Classification and Embeddings of Surfaces. Proc. Symp. Pure Math., **29**, 329–420 (1975).

[Bou EVT] Bourbaki, N.: Espaces vectoriels topologiques, chap. I/II. Paris: Hermann 1953.

[Bou GT] Bourbaki, N.: General Topology, Parts I/II and X. Paris: Hermann, Reading Mass.: Addison-Wesley 1966.

[Br] Bredon, G.E.: Sheaf Theory. New York: MacGraw-Hill 1967.

[BrJä] Bröcker, Th., Jänich, K.: Einführung in die Differentialtopologie. Heidelberger Taschenbücher 143. Heidelberg: Springer-Verlag 1973.

[Ca] Cartan, H.: Séminaire 13e année: 1960/61 I/II 2e éd. Paris 1962.

[Ch] Chow, W.L.: On the Geometry of Algebraic Homogeneous Spaces. Ann. Math., **50**, 32–67 (1949).

[Co] Conway, J.B.: Functions of one Complex Variable. Graduate Texts in Math. 11. New York: Springer-Verlag 1975.

[CoHe] Coleff, N.R., Herrera, M.E.: Les courants résiduel associés à une forme méromorphe. Lecture Notes in Math. 633. Heidelberg: Springer-Verlag 1978.

[Cz] Cazacu, C.A.: Theorie der Funktionen mehrerer komplexer Veränderlicher. Berlin: VEB Deutscher Verlag der Wissenschaften 1975.

[Dr] Demailly, J.P.: Un example de fibré holomorphe non de Stein à fibre $\mathbb{C}^2$ ayant pour base le disque ou le plan. Invent. math. **48**, 293–302 (1978).

[Di] Dieudonné, J.: Foundations of Modern Analysis. New York: Academic Press 1967.

[Do] Douady, A.: Le problème des modules pour les sous-espaces analytiques compactes d'un espace analytique donné. Ann. Inst. Four. **16**, 1–95 (1966).

[Fh] Frisch, J.: Introduction à la géométrie analytique complexe. Corso estivo di matematica. Scuola normale superiore. Pisa: 1971.

[Fi] Fischer, G.: Complex Analytic Geometry. Lecture Notes in Math. 538. Heidelberg: Springer-Verlag 1976.

[Fo] Forster, O.: Lectures on Riemann Surfaces. Graduate Texts in Math. 81. New York: Springer-Verlag 1981.

[Fo]$_2$ Forster, O.: Plongement des variétés de Stein. Comm. Math. Helv. **45**, 170–184 (1970).

[Fo]$_3$ Forster, O.: Zur Theorie der steinschen Algebren und Moduln. Math. Zeitschr. **97**, 376–405 (1967).

[FoKn] Forster, O., Knorr, K.: Ein Beweis des Grauertschen Bildgarbensatzes nach Ideen von B. Malgrange. Man. math. **5**, 19–44 (1971).

[Gb] Greub, W.: Multilinear Algebra. New York: Springer-Verlag 1978.

[Go] Godement, R.: Topologie algébrique et théorie des faisceaux. Paris: Hermann 1964

[Gr] Grauert, H.: Über Modifikationen und exzeptionelle analytische Mengen. Math. Ann. **146**, 331–368 (1962).

[Gr]$_2$ Grauert, H.: Ein Theorem der analytischen Garbentheorie und die Modulräume komplexer Strukturen. Pub. Math. IHES **5**, 233–292 (1960).

[GrFr] Grauert, H., Fritzsche, K.: Several Complex Variables. Graduate Texts in Math. 38. New York: Springer-Verlag 1976.

[GrKe] Grauert, H., Kerner, H.: Deformationen von Singularitäten komplexer Räume. Math. Ann. **153**, 236–260 (1964).

[GrRe] Grauert, H., Remmert, R.: Analytische Stellenalgebren. Unter Mitarbeit von O. Riemenschneider. Grundlehren math. Wiss. 176. Heidelberg: Springer-Verlag 1971.

[GrRe]$_2$ Grauert, H., Remmert, R.: Theory of Stein Spaces. Grundlehren math. Wiss. 236. Heidelberg: Springer-Verlag 1979. Translated from: Theorie der Steinschen Räume. Grundlehren math. Wiss. 227. Heidelberg: Springer-Verlag 1977.

[GrRe]$_3$ Grauert, H., Remmert, R.: Komplexe Räume. Math. Ann. **136**, 245–318 (1958).

[GsHa] Griffiths, Ph., Harris, J.: Principles of Algebraic Geometry. New York: John Wiley & Sons 1978.

[Gu] Gunning, R.C.: Lectures on Riemann Surfaces. Princeton, N. J.: Princeton University Press 1966.

[GuRo] Gunning, R.C., Rossi, H.: Analytic Functions of Several Complex Variables. Englewood Cliffs, N.J.: Prentice-Hall 1965.

[Hs] Horst, C.: Konstruktion komplexer Räume mit vorgegebenen globalen Eigenschaften. Bayer. Akad. Wiss., math. nat. Klasse, 43–48 (1979).

[Ha] Hartogs, F.: Zur Theorie der analytischen Funktionen mehrerer unabhängiger Veränderlicher, insbesondere über die Darstellung derselben durch Reihen, welche nach Potenzen einer Veränderlichen fortschreiten. Math. Ann. **62**, 1–88 (1906).

[Hi] Hirzebruch, F.: Über eine Klasse von einfach zusammenhängenden komplexen Mannigfaltigkeiten. Math. Ann. **124**, 77–86 (1951).

[HiMa] Hirzebruch, F., Mayer, K.H.: O(n)-Mannigfaltigkeiten, exotische Sphären und Singularitäten. Lecture Notes in Math. 57. Heidelberg: Springer-Verlag 1968.

[Hm] Hamm, H.: Zum Homotopietyp Steinscher Räume. J. f. d. reine u. angew. Math. **338**, 121–135 (1983).

[Ho] Holmann, H.: Komplexe Räume mit komplexen Transformationsgruppen. Math. Ann. **150**, 327–360 (1963).

[Hö] Hörmander, L.: An Introduction to Complex Analysis in Several Variables. Princeton, N.J.: van Nostrand 1966.

[Hr] Hironka, H.: Desingularization of Complex-Analytic Varieties. Actes Congrès intern. Math., tome 2, 627–631, Paris (1970).

[Ht] Hartshorne, R.: Algebraic Geometry. Graduate Texts in Math. 52. New York: Springer-Verlag 1977.

[Is] Iss'sa, H.: On the Meromorphic Function Field of a Stein Variety. Ann. Math. **83**, 34–46 (1966).

[Ka]$_B$ Kaup, B.: Über offene analytische Äquivalenzrelationen und komplexe Basen. Habilitationsschrift. Univ. Freiburg/Üchtland 1973.

[Ka]$_L$ Kaup, L.: Eine topologische Eigenschaft Steinscher Räume. Nachr. Göttinger Akad. Wiss. 213–226 (1966).

[KaUp] Kaup, W., Upmeier, H.: An Infinitesimal Version of Cartan's Uniqueness Theorem. Manuscripta math. **22**, 381–401 (1977).

[Kn] Kuhlmann, N.: Über holomorphe Abbildungen komplexer Räume. Arch. Math. **15**, 81–90 (1964).

[Ko] Kodaira, K.: Complex Structures on $S^1 \times S^3$. Proc. nat. Acad. Sci. USA, **55**, 240–243 (1966).

$[Ko]_2$ Kodaira, K.: On Compact Complex Analytic Surfaces I, II, III. Ann. Math. **71**, 111–152 (1960); **77**, 563–626 (1963); **78**, 1–40 (1963).

$[Ko]_3$ Kodaira, K.: On the Structure of Compact Complex Analytic Surfaces I–IV. Am. Journ. Math. **86**, 751–798 (1964); **88**, 682–721 (1966); **90**, 55–83; 1048–1066 (1968).

[Ku] Kultze, R.: Garbentheorie. Stuttgart: Teubner 1970.

[La] Lang, S.: Algebra. Reading, Mass.: Addison-Wesley 1971.

$[La]_2$ Lang, S.: Analysis II. Reading, Mass.: Addison-Wesley 1969.

$[La]_3$ Lang, S.: Introduction to Differentiable Manifolds. New York: John Wiley and Sons 1962.

[Mu] Mumford, D.: Algebraic Geometry I. Complex Projective Varieties. Grundlehren math. Wiss. 221. Heidelberg: Springer-Verlag 1976.

[Na] Nachbin, L.: Holomorphic Functions, Domains of Holomorphy and Local Properties. Amsterdam: North Holland Publ. Comp. 1970.

[Ns] Narasimhan, R.: Several Complex Variables. Chicago: University of Chicago Press 1971.

$[Ns]_2$ Narasimhan, R.: Introduction to the Theory of Analytic Spaces. Lecture Notes in Math. 25, Heidelberg: Springer-Verlag 1966.

$[Ns]_3$ Narasimhan, R.: Analysis on Real and Complex Manifolds. Amsterdam: North-Holland 1968.

[Os] Osgood, W.F.: Lehrbuch der Funktionentheorie Band II, New York: Chelsea Publ. Comp. 1965.

[OkScSp] Okonek, Ch., Schneider, M., Spindler, H.: Vector Bundles on Complex Projective Spaces. Progress in Math. 3, Boston: Birkhäuser 1980.

[Pf] Pflug, R.P.: Holomorphiegebiete, pseudokonvexe Gebiete und das Levi-Problem. Lecture Notes in Math. 432. Heidelberg: Springer-Verlag 1975.

[Pr] Prill, D.: The Divisor Class Group of Some Rings of Holomorphic Functions. Math. Z. **121**, 58–80 (1971).

[RaRu] Ramis, J.P., Ruget, G.: Résidus et dualité. Inv. math. **26**, 89–113 (1974).

[ReScVe] Reiffen, H.J., Scheja, G., Vetter, U.: Algebra Band I. BI-Hochschultaschenbuch Bd. 110. Mannheim: Bibl. Inst. 1969.

[Rf] Reiffen, H.J.: Das Lemma von Poincaré für holomorphe Differentialformen auf komplexen Räumen. Math. Zeitschr. **101**, 269–284 (1967).

$[Rf]_2$ Reiffen, H.J.: Riemannsche Hebbarkeitssätze für Kohomologieklassen mit kompaktem Träger. Math. Ann. **164**, 272–279 (1966).

[Rn] Rothstein, W.: Vorlesungen über Einführung in die Funktionentheorie mehrerer komplexer Veränderlicher I und II. Münster: Aschendorffsche Verlagsbuchhandlung 1965.

[Ro] Rolewicz, S.: On Spaces of Holomorphic Functions. Studia math. **21**, 135–160 (1962).

[Ru] Rudin, W.: Lectures on the Edge-of-the-Wedge-Theorem. Am. Math. Soc.: Regional conference series in Math. **6** (1971).

[Ry] Royden, H.L.: Real Analysis. New York: Macmillan Comp. 1963.

[Se] Serre, J.P.: Lie Algebras and Lie Groups. New York: Benjamin 1965.

$[Se]_2$ Serre, J.P.: Faisceaux algébriques cohérents. Ann. Math. **61**, 197–278 (1955).

[Se]$_3$ Serre, J.P.: Un théorème de dualité. Comm. Math. Helv. **29**, 9–26 (1955).

[Sh] Shafarevich, I.R.: Basic Algebraic Geometry. Grundlehren math. Wiss. 213. Heidelberg: Springer Verlag 1974.

[Sh]$_2$ Shafarevich, I.R. et al.: Algebraic Surfaces. Proc. Steklov Inst. Math. 75. Providence: Am. Math. Soc. 1967.

[SiTr] Siu, Y.-T., Trautmann, G.: Gap-Sheaves and Extension of Coherent Analytic Subsheaves. Lecture Notes in Math. 172. Heidelberg: Springer-Verlag 1971.

[Sp] Spanier, E.H.: Algebraic Topology. New York: McGraw-Hill 1966.

[StWi] Streater, R.F., Wightman, A.S.: PCT, Spin and Statistics and all that. New York: Benjamin Inc. 1964.

[Te] Tennison, B.R.: Sheaf Theory. London Math. Soc. Lecture Notes Series 20. Cambridge: Univ. Press 1975.

[Ue] Ueno, K.: Classification Theory of Algebraic Varieties and Compact Complex Spaces. Lecture Notes in Math. 439. Heidelberg: Springer-Verlag 1975.

[Vl] Vladimirov, V.S.: Methods of the Theory of Functions of Many Complex Variables. Cambridge Mass.: M.I.T. Press 1966.

[We] Wells, R.O.Jr.: Differential Analysis on Complex Manifolds. Englewood Cliffs, N.J.: Prentice Hall 1973.

[Wh] Whitney, H.: Complex Analytic Varieties. Reading, Mass.: Addison-Wesley 1972.

[Wh]$_2$ Whitney, H.: Local Properties of Analytic Varieties. Differential and Combinatorial Topology, a symposium in honor of Marston Morse. Princeton, N.J.: Princeton Univ. Press. 1965.

[Wi] Wiegmann, K.W.: Einbettungen komplexer Räume in Zahlenräume. Invent. math. **1**, 229–242 (1966).

[Wl]$_1$ Weyl, H.: Die Idee der Riemannschen Fläche. Stuttgart: Teubner 1955.

[Wl]$_2$ Weyl, H.: The Concept of a Riemann Surface. Reading, Mass.: Addison-Wesley 1955.

[WhZy] Wheeden, R.L., Zygmund, A.: Measure and Integral. An Introduction to Real Analysis. New York – Basel: Marcel Dekker Inc. 1977.

Glossary of Notations

$\mathbb{N}, \mathbb{Z}, \mathbb{Q}, \mathbb{R}, \mathbb{C}$	natural numbers (including 0); integers; rational, real, complex numbers
$A := B$, $B =: A$	A is defined to be equal to B
$P :\Leftrightarrow Q$, $Q \Leftrightarrow: P$	P is to be understood to mean Q
■	end of proof or absence of proof
■■	end of final proof in a chain of results
$\\|z\\|$	Euclidean norm 2
$\|z\|$	maximum norm 2
$\frac{\partial f}{\partial z}, \frac{\partial f}{\partial \bar{z}}$	matrix of partial derivatives with respect to z_j, resp. $\bar{z}_j$ 2, 26
$\mathcal{O}(X), \mathcal{O}(\bar{X})$	algebra of holomorphic functions on X, resp. near $\bar{X}$ 2, 48
$A \subset\!\!\!\!\circ\; B$	open subset or subspace 2, 98, 148
$\mathbb{C}[z], \mathbb{C}[z_1,\ldots,z_n]$	polynomial algebra in $z_1,\ldots,z_n$ 3
R^*	set of units in the ring R 3
$d_a f$	derivative of f at the point a 4
$\mathscr{C}(T) = \mathscr{C}(T, \mathbb{C})$	algebra of ($\mathbb{C}$-valued) continuous functions 4
$\\|f\\|_A$	A-norm of f 4
$A \subset\subset T$	relatively compact subset 5
$\bar{A}$	closure of A 5
$W(v, \varepsilon; s_j)$	s_j-ball around v with radius ε 5
$\mathring{A}$	interior of A 5
im, re	imaginary part, real part 6,
$\varprojlim_{j \in J}$	projective limit 6
$\mathrm{Hol}(X,Y)$	set of holomorphic mappings 7, 106, 148
f^0, f_z^0	comorphism to f 7, 101
$f\|_A$	restriction of f to A 7, 93, 102, 161
pr_i	i-th projection 7, 152
$\boldsymbol{P}^n(a;\varrho), \boldsymbol{P}(a;\varrho), \boldsymbol{P}(\varrho), \boldsymbol{P}$	(open) polydisks 8
$\boldsymbol{B}^n(a;r), \boldsymbol{B}(a;r), \boldsymbol{B}(r), \boldsymbol{B}$	(open) balls 9
$\mathbb{C}^{n*}$	$\mathbb{C}^n \setminus \{0\}$ 8
$\\|f\\|, \|f\|$	norms of $f = (f_1,\ldots,f_m)$ 9
$\mathrm{Aut}(X)$	group of automorphisms of X 9, 119
$\mathrm{Hom}_{\mathrm{alg}}(E, F)$	vector space of continuous algebra-homomorphisms 10
ε_x	evaluation mapping 10, 97
∂D	topological boundary of D 10
$\boldsymbol{T}^n(a;\varrho)$	distinguished boundary of the polydisk $\boldsymbol{P}^n(a;\varrho)$ 10
$\|v\|$	$\sum_{j=1}^{n} v_j$ 12

$v!$	$v_1! \cdot \ldots \cdot v_n!$ 12
z^v	$z_1^{v_1} \cdot \ldots \cdot z_n^{v_n}$ 12
$v+1$	$(v_1+1, \ldots, v_n+1)$ 12
$D^v f$	partial derivative 12
$M \cdot w$	for $M \subset \mathbb{C}$: $\{mw; m \in M\}$ 15
I_n	identity matrix 16
↯	contradiction 21
$N(U; f_1, \ldots) = N(f_1, \ldots)$	zero-set of the family $(f_1, \ldots)$ 21
$\operatorname{codim} A$, $\operatorname{codim}_a A$	codimension 22, 190
$\Gamma(f)$	graph of f 25, 201
$J_f(a)$	Jacobian determinant 26
$\dim X$, $\dim_a X$	dimension 30, 106, 183, 189, 321
$\check{X}$	logarithmically convex complete hull 34
$\hat{K}_{\mathfrak{F}}$, $\hat{K}$	$\mathfrak{F}$-convex hull 37, 223
$\operatorname{dist}(A, B)$	distance between A and B 39
δ_X	boundary distance function on X 39
$\operatorname{bd} D$	boundary of the complex disk D 43
h_D	harmonic extension of $h \in \mathscr{C}(\operatorname{bd} D, \mathbb{R})$ to D 43
$\mathscr{C}(G, \mathbb{R})$	vector space of continuous real-valued functions 50
$\mathscr{C}^p(G, \mathbb{R})$	vector space of p-fold continuously partially differentiable functions 49
Δ	Laplace operator 52
g_z	$\frac{\partial g}{\partial z}$ 53
$L_{f,a}$	Levi form of f at the point a 55
$L_{f,a} > 0$	positive definite Levi form 55
$\tilde{B}$	pseudoconvex hull of B 57
$\mathbb{C}[\![X]\!] = \mathbb{C}[\![X_1, \ldots, X_n]\!] = {}_n\mathscr{F}$	algebra of formal power series 65
$\mathfrak{m}_{[\![X]\!]}$, $\mathfrak{m}_{\{X\}}$, ${}_n\mathfrak{m}$, $\mathfrak{m}$, etc.	maximal ideals 66, 67
$o(P)$	order of a power series 66
$\Vert \cdot \Vert_r$	r-pseudonorm of a power series 67
$\mathbb{C}\{X\} = \mathbb{C}\{X_1, \ldots, X_n\} = {}_n\mathcal{O}_0$	algebra of convergent power series 67
${}_n\mathcal{O}_a$, ${}_X\mathcal{O}_a$	local algebra at the point a 67
$\varinjlim$	inductive limit 67
f_a	germ of f at the point a 67, 95, 154
$\mathfrak{m}^j$	$\mathfrak{m} \cdot \ldots \cdot \mathfrak{m}$ or $\mathfrak{m} \oplus \ldots \oplus \mathfrak{m}$ 69
$R[X]_g$	polynomials of (multi-)degree $< g$ 71
$R \cdot (m_1, \ldots, m_k)$, $\mathscr{R} \cdot (m_1, \ldots, m_k)$	R-submodule (resp., $\mathscr{R}$-submodule) generated by $m_1, \ldots, m_k$ 79, 137
$R[X]_{\text{mon}}$	set of monic polynomials 80
$a \mid b$	a divides b 80
$R[s]$	subring generated by s 87
$\hat{R}$	integral closure of R 88

$Q(R)$	total ring of quotients of R 88
$a_{\not j}$	a_j is to be omitted 89
${}_n\mathcal{O}$	structure sheaf of $\mathbb{C}^n$ 93
$\mathcal{G}(U) = \Gamma(U, \mathcal{G})$	space of sections in $\mathcal{G}$ 93, 95
ϱ_V^U	restriction homomorphism 93
$U_{i_0 \ldots i_p}$	$U_{i_0} \cap \ldots \cap U_{i_p}$ 93
$\mathcal{G}\vert_U$	restriction of the sheaf $\mathcal{G}$ to U 94, 97
$\mathrm{Hom}(\mathcal{F}, \mathcal{G})$	homomorphisms from $\mathcal{F}$ to $\mathcal{G}$ 94
$\mathcal{H}om(\mathcal{F}, \mathcal{G})$	sheaf of homomorphisms from $\mathcal{F}$ to $\mathcal{G}$ 94
$\mathcal{G}_t$	stalk of the sheaf $\mathcal{G}$ at the point t 94
$\dot\cup$	disjoint union 95
$\vert\mathcal{G}\vert$	topological space associated to the sheaf $\mathcal{G}$ 95
Γ	section-functor 95
$f(a)$	value of the section f at the point a 95
supp	support 96, 218
$\varphi^{-1}\mathcal{G}$	topological inverse image sheaf 97
${}_X\mathcal{C}$	sheaf of continuous functions 97
$(T, \mathcal{A})$	ringed space 97
Red	reduction 97, 98, 103
$\vert T\vert$	underlying topological space for a ringed space 97
$\bullet$	simple point 98
$\vec{\bullet}$	double point 98
$V(T; \mathcal{I}), V(X; f_1, \ldots, f_m)$	subvariety determined by $\mathcal{I}$ or by $f_1, \ldots, f_m$, respectively 99, 100
$A \hookrightarrow T$	closed subspace of T 99, 102
${}_A\mathcal{I}$	nullstellen ideal of A 99, 100
$\vert\varphi\vert$	continuous mapping underlying a morphism of ringed spaces 101
φ^*	analytic (or ringed) inverse image 102
$\varphi^{-1}(B)$	inverse image of a subspace 102, 149
$\varphi^{-1}(t)$	fiber of a morphism 102, 149
$\sqrt{\mathfrak{a}}$	radical of the ideal $\mathfrak{a}$ 105
$\mathfrak{n}_R, {}_T\mathcal{N}$	nilradical 105
$(X, \mathcal{O})$	complex space 106, 148
${}_X\mathcal{O}, \mathcal{O}$	structure sheaf of a complex space 106, 148
$\mathbb{P}_n$	projective space 107
S^j	sphere of real dimension j 107
$[z_0, \ldots, z_n]$	homogeneous coordinates of a point in a projective space 108
$\mathbb{P}GL(\mathbb{C}^{n+1})$	projective general linear group 108
$S(X)$	subset of singular points of X 109, 191
M'	dual vector space $\mathrm{Hom}(M, \mathbb{C})$ 111
$TR,\ T_aX,\ TX_a$	tangent spaces 111, 112, 166
T_af	tangent mapping 112

Der R	space of derivations of R 113
$\mathbb{C}^{n\times m}$	vector space of $n\times m$ matrices 117
$SL(n,\mathbb{C})$	special linear group 117
$GL(k,n-k)$	117
G/H	set of left cosets 118
$G_k(n)$	Grassmann manifold 119
$U(n)$	unitary group 119
F_n	flag manifold 119
$(X\setminus A)\cup_\varphi U'$	surgery on X 123
Σ_m	Hirzebruch surface 124
$\mathrm{Res}(P,Q)$	resultant 132
$\mathrm{Dis}(P)$	discriminant 133
$x_j\to\partial X$	convergence to the "boundary" of X 133
R_f, R_G	equivalence relation, defined by a mapping f or a transformation group G 134
$G(x)$	orbit of x under the action of G 134
$X/R_G, X/G$	orbit spaces 134, 207
$R(A)$	R-saturated hull of A 134
$\subset\subset$	open relatively compact subset 135
$\mathscr{K}er\ (s_1,\ldots,s_p)$	sheaf of relations between $s_1,\ldots,s_p$ 138
$\mathscr{G}^T$	trivial extension of $\mathscr{G}$ to T 144
$\mathscr{A}nn(\mathscr{F})$	annihilatior of $\mathscr{F}$ 144
$\mathscr{F}:\mathscr{G},\ (t):(s)$	submodule quotient, ideal quotient 145, 161
$\mathscr{T}or^{\mathscr{R}}(\mathscr{F})$	torsion-sheaf of $\mathscr{F}$ 145
$N(T;\mathscr{I}),\ N(\mathscr{I})$	zero set of the ideal $\mathscr{I}$ 145
$X^{(j)}$	j-th infinitesimal neighborhood 148
$X_1\cap X_2, X_1\cup X_2$	intersection and union of complex subspaces 148, 157
$\mathscr{H}ol\ (X,\mathbb{C}^m)$	sheaf of $\mathbb{C}^m$-valued holomorphic mappings on X 150
$X_1\times X_2$	product of X_1 and X_2 152
X_a	germ of X at the point a 154
$\mathrm{Hol}(X_a, Y_b)$	holomorphic mappings of germs 154
$\mathfrak{B}$	lattice of germs of analytic subspaces of X_a 157
$\mathfrak{N}$	lattice of germs of analytic subsets of X_a 157
$\mathfrak{I}$	lattice of ideals of ${}_X\mathcal{O}_a$ 157
emb X_a	embedding-dimension of X_a 166
$\mathrm{Hol}_X(Y,Z)$	holomorphic mappings between spaces over X 168
$f\mathscr{G}$	image sheaf of $\mathscr{G}$ 170
Specan $\mathscr{A}$	analytic spectrum of $\mathscr{A}$ 172
$\tilde{\mathcal{O}}, {}_X\tilde{\mathcal{O}}$	sheaf of weakly holomorphic functions 179
dim R	dimension of the analytic algebra R 183
X_{reg}	$X\setminus S(X)$ 191
$\underset{g}{\overset{f}{\rightrightarrows}}$	pair of morphisms 198

$\operatorname{Ker} \underset{g}{\overset{f}{\rightrightarrows}}$	kernel of $\underset{g}{\overset{f}{\rightrightarrows}}$ 199
$\operatorname{Coker} \underset{g}{\overset{f}{\rightrightarrows}}$	cokernel of $\underset{g}{\overset{f}{\rightrightarrows}}$ 199
$\operatorname{Fix} f$	fixed point space of f 200
$X_1 \times_Y X_2$	fiber product of X_1 and X_2 over Y 200
Δ_X	diagonal of $X \times X$ 201
$(T, \mathscr{A})/R,\ T/R$	quotient space of $(T, \mathscr{A})$ 204
R^X	trivial extension of the equivalence relation R to X 205
G_x	isotropy subgroup of G at the point x 206
$\operatorname{Hol}(X, Y)^G$	set of G-invariant holomorphic mappings 207
$\mathbb{P}_{\mathfrak{q}}$	weighted projective space 208
N_f	equivalence relation of level sets of f 210
$H^q(T, \mathscr{F})$	q^{th} cohomology of T with coefficients in $\mathscr{F}$ 215
δ^q	connecting homomorphism in the cohomology sequence 215
$\mathscr{C}(T, \mathscr{F}) = \mathscr{C}(\mathscr{F})$	sheaf of all sections in $\mathscr{F}$ 215
$\mathscr{C}^*(\mathscr{F})$	canonical acyclic resolution of $\mathscr{F}$ 216
$H^*(E^*)$	cohomology module of the complex E^* 216
$Z^q(E^*),\ B^q(E^*)$	cochain module, coboundary module of E^* 216
$C^q(\mathfrak{U}, \mathscr{F}),\ Z^q(\mathfrak{U}, \mathscr{F}),\ B^q(\mathfrak{U}, \mathscr{F})$ $H^q(\mathfrak{U}, \mathscr{F}),\ \check{H}^q(T, \mathscr{F})$	Čech cochains, etc. 218
$\tau^{\mathfrak{U}}_{\mathfrak{B}}$	induced homomorphism in cohomology for a refinement $\mathfrak{B}$ of $\mathfrak{U}$ 218
$C^q_a,\ Z^q_a,\ B^q_a,\ H^q_a,\ \check{H}^q_a$	alternating Čech-theory 219
$A(a)$	set of a-automorphic functions 221
$\\|f\\|_K$	$\\|\operatorname{Red} f\\|_K$ for $f \in \mathcal{O}(X)$ 223
$\mathscr{M}$	sheaf of meromorphic functions 238
$S^{-1}R$	$\{r/s \in Q(R);\ s \in S\}$ 239
$\mathscr{D}_m$	sheaf of denominators of $m \in \mathscr{M}(X)$ 239
$\mathcal{O}^*,\ \mathscr{M}^*$	sheaves of multiplicatively invertible elements 244
$\mathscr{D}$	sheaf of divisors 245
(m)	divisor of $m \in \mathscr{M}(X)$ 245
$c(D)$	Chern class of the divisor D 246
$\mathcal{O}(\zeta)$	sheaf of holomorphic sections in the line bundle ζ 247
$\mathcal{O}(-D)$	sheaf of ideals associated to the holomorphic divisor D 249
$\mathscr{GL}(r, \mathcal{O})$	sheaf of invertible holomorphic $r \times r$ -matrices 251
TX	tangent bundle of X 252
$T'X$	contangent bundle of X 252
${}^E_X\mathcal{O}$	sheaf of holomorphic sections in the vector bundle E 252
${}_X\Theta$	tangent sheaf of X 252
${}_X\Omega^p$	sheaf of holomorphic p-forms 252, 273, 274
$E(m)$	line bundle on $\mathbb{P}_1$ with Chern class m 254
$\mathcal{O}(m)$	${}^{E(m)}_{\mathbb{P}_1}\mathcal{O}$ 254
$\mathcal{O} \cdot M$	analytic subsheaf of $\mathscr{G}$ generated by $M \subset \mathscr{G}(X)$ 260

$Sp(X), Sp_c(X)$	spectrum and continuous spectrum of X 267
$\mathbb{N}\binom{n}{p}$	$\{(j_1,\dots,j_p)\in\mathbb{N}^p; 1\leqq j_1<\dots<j_p\leqq n\}$ 272
$dz_I, d\bar{z}_I$	$dz_{i_1}\wedge\dots\wedge dz_{i_p}, d\bar{z}_{i_1}\wedge\dots\wedge d\bar{z}_{i_p}$ 272
$\mathscr{E}^p, \mathscr{E}^{p,q}$	sheaves of $\mathscr{C}^\infty$-forms 272
$\partial, \bar{\partial}$	∂-operator and $\bar{\partial}$-operator 272
$\omega\wedge\chi$	Grassmann product 275
$A^*(M)$	Grassmann algebra of M 275
φ_J	$\varphi_{s_1}\wedge\dots\wedge\varphi_{s_r}$ 275
$d: \mathscr{E}^r\to\mathscr{E}^{r+1}$	exterior derivative 276
φ^*	morphism between complexes of forms 277
$\square$	$\overline{\boldsymbol{P}(a;\varrho)}$ 278
$\square_{P_1,\dots,P_r}$	polynomial polyhedron 278
$\lVert\chi\rVert_3$	$\mathscr{C}^3$-norm of χ 286
$f_q\mathscr{F} = \mathscr{R}^q f\mathscr{F}$	q-th image sheaf of $\mathscr{F}$ 292
$\hat{\mathscr{O}}$	integral closure of $\mathscr{O}$ 304
$R_\mathfrak{p}$	localization of R by $\mathfrak{p}$ 316
$K_*(R; f_1,\dots,f_p)$	Koszul complex 317
$\Gamma_S(T,\mathscr{G})$	module of sections in $\mathscr{G}$ with supports in S 318
$H^q_S(T,\mathscr{G}) = H^q(T, T\setminus S; \mathscr{G})$	local cohomology with supports in S 318
${}_T\mathscr{H}^q_S(\mathscr{G})$	local cohomology sheaf with supports in S 319

For variables, we use letters from the following alphabets:

Fraktur		Latin		Script		Greek		
$\mathfrak{A}$	$\mathfrak{a}$	A	a	$\mathscr{A}$	𝒶	A	α	alpha
$\mathfrak{B}$	$\mathfrak{b}$	B	b	$\mathscr{B}$	𝒷	B	b	beta
$\mathfrak{C}$	$\mathfrak{c}$	C	c	$\mathscr{C}$	𝒸	Γ	γ	gamma
$\mathfrak{D}$	$\mathfrak{d}$	D	d	$\mathscr{D}$	𝒹	Δ	δ	delta
$\mathfrak{E}$	$\mathfrak{e}$	E	e	$\mathscr{E}$	ℯ	E	ε	epsilon
$\mathfrak{F}$	$\mathfrak{f}$	F	f	$\mathscr{F}$	𝒻	Z	ζ	zeta
$\mathfrak{G}$	$\mathfrak{g}$	G	g	$\mathscr{G}$	ℊ	H	η	eta
$\mathfrak{H}$	$\mathfrak{h}$	H	h	$\mathscr{H}$	𝒽	Θ	θ, ϑ	theta
$\mathfrak{I}$	$\mathfrak{i}$	I	i	$\mathscr{I}$	𝒾	I	ι	iota
$\mathfrak{J}$	$\mathfrak{j}$	J	j	$\mathscr{J}$	𝒿	K	κ	kappa
$\mathfrak{K}$	$\mathfrak{k}$	K	k	$\mathscr{K}$	𝓀	Λ	λ	lambda
$\mathfrak{L}$	$\mathfrak{l}$	L	l	$\mathscr{L}$	𝓁	M	μ	mü
$\mathfrak{M}$	$\mathfrak{m}$	M	m	$\mathscr{M}$	𝓂	N	ν	nü
$\mathfrak{N}$	$\mathfrak{n}$	N	n	$\mathscr{N}$	𝓃	Ξ	ξ	xi
$\mathfrak{O}$	$\mathfrak{o}$	O	o	$\mathscr{O}$	ℴ	O	o	omicron
$\mathfrak{P}$	$\mathfrak{p}$	P	p	$\mathscr{P}$	𝓅	Π	π	pi
$\mathfrak{Q}$	$\mathfrak{q}$	Q	q	$\mathscr{Q}$	𝓆	P	ϱ	rho
$\mathfrak{R}$	$\mathfrak{r}$	R	r	$\mathscr{R}$	𝓇	Σ	σ	sigma
$\mathfrak{S}$	$\mathfrak{s}$	S	s	$\mathscr{S}$	𝓈	T	τ	tau
$\mathfrak{T}$	$\mathfrak{t}$	T	t	$\mathscr{T}$	𝓉	Y	υ	ypsilon
$\mathfrak{U}$	$\mathfrak{u}$	U	u	$\mathscr{U}$	𝓊	Φ	ϕ, φ	phi
$\mathfrak{V}$	$\mathfrak{v}$	V	v	$\mathscr{V}$	𝓋	X	χ	chi
$\mathfrak{W}$	$\mathfrak{w}$	W	w	$\mathscr{W}$	𝓌	Ψ	ψ	psi
$\mathfrak{X}$	$\mathfrak{x}$	X	x	$\mathscr{X}$	𝓍	Ω	ω	omega
$\mathfrak{Y}$	$\mathfrak{y}$	Y	y	$\mathscr{Y}$	𝓎			
$\mathfrak{Z}$	$\mathfrak{z}$	Z	z	$\mathscr{Z}$	𝓏			

Subject Index